Fault Detection and Diagnosis in Engineering Systems

Fault Detection and Diagnosis in Engineering Systems

Janos J. Gertler

CRC Press is an imprint of the
Taylor & Francis Group, an **informa** business

CRC Press
Taylor & Francis Group
6000 Broken Sound Parkway NW, Suite 300
Boca Raton, FL 33487-2742

First issued in paperback 2019

ISBN-13: 978-0-8247-9427-9 (hbk)
ISBN-13: 978-0-367-40043-9 (pbk)

Library of Congress Cataloging-in-Publication Data

Gertler, Janos.
Fault detection and diagnosis in engineering systems / Janos J. Gertler.
p. cm.
Includes bibliographical references and index.
ISBN: 0-8247-9427-3 (alk. paper)
1. Fault location (Engineering). 2. System analysis. I. Title.
TA169.6.G47 1998
620'.004—dc21 98-16716
CIP

Visit the Taylor & Francis Web site at
http://www.taylorandfrancis.com

and the CRC Press Web site at
http://www.crcpress.com

Preface

The detection and isolation (diagnosis) of faults in engineering systems is of great practical significance. The systems concerned encompass a broad spectrum of human-made machinery, including industrial production facilities (power plants, chemical plants, oil refineries, steel mills, paper mills, etc.), transportation vehicles (ships, airplanes, automobiles) and household appliances (heating/air conditioning equipment, refrigerators, washing machines, etc.). The early detection of the occurrence of faults is critical in avoiding product deterioration, performance degradation, major damage to the machinery itself and damage to human health or even loss of lives. The quick and correct diagnosis of the faulty component then facilitates proper and optimal decisions on emergency and corrective actions and on repairs.

The traditional approaches to fault detection and diagnosis involve the limit checking of some variables or the application of redundant sensors (physical redundancy). More advanced methods rely on the spectral analysis of signals emanating from the machinery or on the comparison of the actual plant behavior to that expected on the basis of a mathematical model (analytical redundancy). The latter approach includes methods which are more deterministically framed (such as parity relations and observers) and those formulated more on a statistical basis (Kalman filtering and parameter estimation). The boundaries between the various approaches are rather blurred and, most recently, several methods have been shown to be closely related to one another and even to produce identical results under broad conditions.

This book is devoted to the model based approach to fault detection and isolation. Its focus is on dynamic consistency (parity) relations, with a side focus on parameter estimation. These methods are conceptually and computationally straightforward, and are posed in a framework familiar to most practicing engineers. Due to the equivalences referred to above, the same fault detection and isolation task may be implemented by a variety of techniques, resulting in different computational algorithms inside but identical outside behavior and performance. With this in mind, the family of methods presented in this book provides a complete answer to a wide class of fault detection and isolation problems - within the limits of our present knowledge, of course.

Relatively few books have been written so far on the subject of model-based engineering fault detection and diagnosis. The earliest ones are by D.M. Himmelblau (1978) and L.F. Pau (1981), the former concentrating on chemical engineering systems. The books edited by M. Basseville and A. Benveniste (1986) and by R.J. Patton et al. (1989) are collections of chapters written by various authors. The work of Viswanadham and coauthors (1987b) focuses mostly on reliability issues while the book by M. Basseville and I. Nikiforov (1993) has a strong statistical orientation. A.D. Pouliezos and G.S. Stavrakakis (1994) wrote a broad survey of the existing fault detection literature. This book is unique in the sense that it provides a deep and unified treatment of the subject, based on the author's own research and extensive interaction with the other workers of the field.

The book may be used as the text in an intermediate level graduate course on fault detection and diagnosis. The potential audience includes students in electrical engineering, mechanical engineering, chemical engineering, and in more specialized disciplines of engineering such as control, aerospace, vehicular, marine, nuclear, etc. The book may also serve as a useful self-study and reference source for practicing engineers working in the listed areas.

The book is self-contained in that the background material it requires, in linear systems, probability and statistics, and parameter estimation, is presented in the introductory chapters.

The first chapter defines the fault detection and isolation task. It then provides an overview of the methods which do not rely on an explicit model of the plant, and a brief introduction to those which do, the latter constituting the subject of the book. This is followed by two motivational examples, concerned with diagnosis in automotive engines and in a chemical reaction process.

The next three chapters contain the theoretical background material for the rest of the book. In Chapter 2, the description techniques of discrete linear dynamic systems are reviewed. The input-output description relies on the shift operator idea, introduced by Karl Johan Åström (see Åström and Wittenmark, 1997). The state-space sections of the chapter have been adopted mostly from the seminal book of Thomas Kailath (1980) and the texts of William L. Brogan

(1985) and Chi-Tsong Chen (1984). Chapter 3 reviews some fundamental concepts of probability and statistics, drawing from the classical work of Maurice G. Kendall and Alan Stuart (1969) and from some modern texts such as Louis L. Scharf's (1990). Chapter 4 is an introduction to parameter estimation, building on Lennart Ljung's fundamental work (1987) and the author's own teaching and research experience.

Chapters 5 through 9 constitute the kernel of the book. The fundamental concepts of analytical redundancy are introduced in Chapter 5. The equivalence between parity relations and observers is demonstrated here. Chapter 6 is devoted to the parity relation implementation of residual generators, including direct implementation from transfer functions, state-space implementation by the Chow-Willsky scheme, and systematic input-output design using the fault system matrix. In Chapter 7, the concepts and design of structured residual sets is discussed, together with their parity relation implementation. Chapter 8 deals with the design and implementation of directional residual sets. Finally, the above ideas and design techniques are extended and applied to parametric faults in Chapter 9. These chapters feature a chain of numerical examples to illustrate the various ideas and methods.

Chapters 10 through 13 contain a number of important extensions. Chapter 10 deals with the subject of robustness in residual generation. Several approximate design techniques, utilizing singular value decomposition or constrained optimization, are described here. The statistical evaluation of parity equation residuals is discussed in Chapter 11, including the propagation of random noise in the residual generator, testing the residuals for detection, and isolation decisions based on statistical analysis. The two final chapters are devoted to parameter estimation issues. Model building for the detection of additive faults is treated in Chapter 12, followed in Chapter 13 by the estimation of parametric faults and a link to parity relations. Again, concepts and techniques are illustrated by numerical examples throughout.

This book has been the result of four years of writing - preceded by more than a decade of learning and research. Especially in the latter respect, I am indebted to a succession of students and coworkers who have helped me tremendously in understanding fundamental concepts, formulating methods and seeing how they work (or do not work). It would be hard to rank their contribution so I am just listing them here in alphabetical order: Zerihun Abate, Tamer Barkana, Gus DiPierro, Xiaowen Fang, Jim Heron, Ron Hira, Zdzislaw Kowalczuk, Moid Kunwer, Weihua Li, Qiang Luo, Luis de Miguel, Ramin Monajemy, Ziyang Qiu, John Shutty, Adam Strassel, Amartur Sundar and Kewen Yin.

I am indebted to my long-time friend Tibor Vámos, who started me on the path of process control and was my mentor for decades, and who provided fatherly advice and support to this endeavor as well. My thanks also go to

Gedeon Almásy, another long-time friend, who first infected me with the ideas of balance equations and fault detection. I am grateful for the valuable help I received, in the course of writing the book, from my friends and colleagues József Bokor, Yariv Ephraim, Andre Manitius and Peter Paris. I have also benefited substantially from discussions, some brief, some more extensive, with those who share with me an interest in fault detection, including Michèle Basseville, Albert Benveniste, Jie Chen, Paul Frank, David Himmelblau, Rolf Isermann, Michel Kinnaert, Thomas McAvoy, Igor Nikiforov, Ron Patton, Yaubin Peng, Giorgio Rizzoni, Marcel Staroswiecki and N. Viswanadham.

Special thanks are due to Sami Fadali and Giorgio Rizzoni, who taught from a part of the manuscript, and to Angelo Alessandri, who read a couple of chapters; they called my attention to several errors and provided a number of valuable suggestions. Hassene Ould Sidatt helped with the proofreading of the material and caught many inaccuracies. The more complex figures were artfully produced by Qiang Luo.

My special thanks go also to Rod Learmonth and Rita Lazazzaro at Marcel Dekker for their support and patience.

Any project of this kind is driven by the author's desire of self-realization. It requires significant sacrifices, not only of the writer but also of those close to him. I am indebted to my wife Judit and son Nick (Balazs), for their understanding and support in this endeavor.

Janos J. Gertler

Table of Contents

Notation and Symbols

GENERAL NOTATION

M	matrix	M'	transpose of M
m	(column) vector	m'	row vector; transpose of m
$m_{i\cdot}$	i-th row of M	$m_{\cdot j}$	j-th column of M
m_{ij}	ij-th element of M	m_i	i-th element of m
$AdjM$	adjoint of M	$RankM$	rank of M

$DetM = |M|$ determinant of M

t	discrete time
τ	continuous time; discrete time-shift
s	Laplace transform variable (see symbol list for additional meanings)
z	z-transform variable (see symbol list for additional meanings)
ϕ	shift operator

$m(\phi)$	polynomial or rational function in ϕ^{-1}
$m^{+}(\phi)$	polynomial (or rational function) in ϕ
m^k	the coefficient of ϕ^{-k} in $m(\phi)$
$Deg\ m(\phi)$	the polynomial degree of $m(\phi)$

$x(t)$	scalar time function	$\boldsymbol{x}(t)$	vector time function
$X(t)$, $X(t,K)$	finite sequence of $x(t)$	$\boldsymbol{X}(t)$, $\boldsymbol{X}(t,K)$	finite sequence of $\boldsymbol{x}(t)$
$\mathrm{E}\{x(t)\}$	expectation of $x(t)$	$Var\{x(t)\}$	variance of $x(t)$
$\bar{x}(t)$	window average of $x(t)$	$f[x(t)]$	function of $x(t)$

SPECIFIC SYMBOLS

a, b, c	constant
d	delay
$f_x(\xi)$	probability density function of $x(t)$ at ξ
$g(\phi)$	numerator of transfer function $m(\phi)$
$\tilde{g}(\phi)$	expanded numerator of $m(\phi)$, see (2.64)
$h(\phi)$	denominator of transfer function $m(\phi)$
$\bar{h}(\phi)$	common denominator in multiple-output system, see (2.64)
$k(\phi)$	residual filter (Chapter 11)
m	window length (Chapter 3); rank of fault transfer matrix, see (10.14)
m^*	rank of rank-reduced approximation
$m(\phi)$	input to output transfer function
$m_{ij}(\phi)$	element of $\boldsymbol{M}(\phi)$
n	number of residuals in the residual vector
$n(\tau)$	continuous-time step response
$\boldsymbol{n}_{\cdot j}(t)$	j-th underlying parameter gain in the primary residuals, see (5.19)
$n_{ij}(\phi)$	numerator of $s_{ij}(\phi)$
$\boldsymbol{o}(t)$	generic (ARMA) primary residual vector, see (6.8)
$\boldsymbol{o}^*(t)$	MA primary residual vector, see (6.13)
p_j	system pole
$\boldsymbol{p}(t)$	fault input vector
$\boldsymbol{p}(t,K)$	data vector in recursive/sliding parameter estimation, see (4.59)
$\boldsymbol{q}(t)$	disturbance input vector
$\boldsymbol{r}(t)$	residual vector, see (5.26)
$s(\phi)$	denominator of noise-to-residual transfer function, see (11.18)
$s_{ij}(\phi)$	fault/disturbance to output transfer function
$\boldsymbol{u}(t)$	(observed) input vector; $\boldsymbol{u}^{\circ}(t)$: true value of $\boldsymbol{u}(t)$
$\blacktriangle\boldsymbol{u}(t)$	fault associated with $\boldsymbol{u}(t)$; $\delta\boldsymbol{u}(t)$: noise associated with $\boldsymbol{u}(t)$
$\boldsymbol{v}(t)$	noise input vector

$v'_i(\phi)$	input transformation in the i-th residual, see (6.29)
w	test statistic
w'_i	i-th residual transformation
$x(t)$	random variable (Chapter 4)
$\boldsymbol{x}(t)$	state vector; random variable vector (Chapter 4)
$y(t)$	random variable (Chapter 4)
$\boldsymbol{y}(t)$	(observed) output vector; $\boldsymbol{y}^\circ(t)$: true value of $\boldsymbol{y}(t)$
$\blacktriangle\boldsymbol{y}(t)$	fault associated with $\boldsymbol{y}(t)$; $\delta\boldsymbol{y}(t)$: noise associated with $\boldsymbol{y}(t)$
$z_{Fij}(\phi)$	response specification (i-th residual to j-th fault), see (5.39)
$\boldsymbol{z}_{Fi\cdot}(\phi)$	response specification (i-th residual to all faults), see (5.40)
$\boldsymbol{z}_{F\cdot j}(\phi)$	response specification (full residual vector to j-th fault), see (5.42)
$\boldsymbol{A}$	state transition matrix
$\boldsymbol{B}$	input-to-state gain matrix
$\boldsymbol{C}$	state-to-output gain matrix
D	decision rule
$\boldsymbol{D}$	input-to-output direct gain matrix
D	*in subscript*: of disturbances
$\boldsymbol{E}$	disturbance (or fault)-to-state gain matrix
$\boldsymbol{F}$	disturbance (or fault)-to-output direct gain matrix
F	*in subscript*: of faults
$F_x(\xi)$	probability distribution function of $x(t)$ at ξ
$\boldsymbol{G}(\phi)$	numerator of the input-output transfer function $\boldsymbol{M}(\phi)$
$\bar{\boldsymbol{G}}(\phi)$	expanded numerator matrix, see (2.64)
H	hypothesis
$\boldsymbol{H}(\phi)$	diagonal matrix of denominator polynomials, see (2.62)
$\boldsymbol{I}$	unit matrix
J	quadratic performance index
$\boldsymbol{J}$	Jordan matrix, see (2.149); repeated unit matrix, see (11.100)
$\boldsymbol{J}$	matrix in the Chow-Willsky scheme, see (6.20)
K	length of the identification dataset
K	number of canonical row-structures (Chapter 7)
$\boldsymbol{K}$	observer gain matrix, see (5.47)
$\boldsymbol{K}$	noise-to-state gain matrix, see (11.9)
$\boldsymbol{K}$	matrix in the Chow-Willsky scheme, see (6.20)

$L_x(\xi, \theta)$	likelihood function, see (3.186)
$L_j(t)$	likelihood function, see (11.75)
$\boldsymbol{L}$	noise-to-output direct gain matrix, see (11.9)
$\boldsymbol{L}$	matrix in the Chow-Willsky scheme, see (6.20)
$\boldsymbol{M}(\phi)$	input-output transfer function
$\boldsymbol{M}^{\circ}(\phi)$	true input-output transfer function
$\Delta\boldsymbol{M}(\phi)$	transfer function discrepancy, see (5.13)
$N(t)$	underlying parameter to primary residual gain matrix, see (5.19)
$N(\phi)$	numerator of transfer function $S(\phi)$, see (6.10)
N	*in subscript*: of noise
P	probability
$\boldsymbol{Q}(t,K)$	data matrix in recursive/sliding parameter estimation, see (4.58)
$\boldsymbol{Q}(\phi)$	numerator of noise-to-residual transfer function, see (11.18)
$\boldsymbol{R}$	parameter Jacobian, see (13.49)
$\boldsymbol{R}(t)$	finite sequence of $\boldsymbol{r}(t)$, see (11.20)
$\boldsymbol{S}$	matrix of singular values, see (10.35)
$\boldsymbol{S}(\phi)$	disturbance (or fault, noise) to output transfer function, see (5.9)
$\boldsymbol{S}_i(\phi)$	fault transfer matrix for the i-th structured residual, see (7.16)
$\boldsymbol{T}_j$	template matrix, see (11.80)
$\boldsymbol{V}(\phi)$	input-to-residual transformation, see (5.23)
$\boldsymbol{W}_c$	controllability matrix, see (2.203)
$\boldsymbol{W}_o$	observability matrix, see (2.192)
$\boldsymbol{W}(t)$	residual transformation matrix for parametric faults, see (9.41)
$\boldsymbol{W}(\phi)$	generic residual transformation matrix for additive faults, see (6.7)
$\boldsymbol{W}^*(\phi)$	MA residual transformation matrix, see (6.15)
$\boldsymbol{Z}(\phi)$	complete response specification, see (5.44)
α	false alarm rate (test size), see (3.200)
α, $\alpha(\phi)$	coefficient
ß	detection power, see (3.204)
β, $\beta(\phi)$	coefficient
$\boldsymbol{\beta}_j$	direction of the j-th residual response, see (5.37)
$\gamma_j(\phi)$	dynamic of the j-th residual response, see (5.37)
$\boldsymbol{\Upsilon}_j$	right singular vector, see (10.34)
δ	polynomial degree, see (6.113)

$\delta(t)$	unit impulse function
$\epsilon(t)$, $\epsilon(\tau)$	unit-step function
$\epsilon(t)$	equation error, see (4.1); binary test result, see (5.36)
ζ	invariant zero, see 2.6.3.; sensitivity ratio, see (5.32)
η	value of the random variable $y(t)$ (Chapter 3)
η	triggering limit, see (5.31)
θ	parameter (Chapter 3); $\hat{\theta}$: estimate of θ
$\boldsymbol{\theta}$	vector of underlying (physical) parameters
$\boldsymbol{\theta}^{\circ}$	true value of $\boldsymbol{\theta}$ (Chapter 9)
$\blacktriangle\boldsymbol{\theta}$	uncertainty (fault) of $\boldsymbol{\theta}$
$\boldsymbol{\theta}^{\#}$	estimated parameter values (Chapter 5, 9)
$\blacktriangle\theta_j^{\circ}$	nominal size of discrepancy, see (10.75)
κ	number of observed inputs; test threshold
λ_i	eigenvalue
λ	forgetting factor, see (4.63); core size, see (7.19)
μ	number of observed outputs
μ_x	mean of $x(t)$, see (3.12)
ν	(true) system order
$\tilde{\nu}$	apparent system order, see (2.64)
ξ	value of the random variable $x(t)$ (Chapter 3)
ξ	sensitivity condition (Chapter 5, see (5.33))
$\boldsymbol{\xi}_i$	right-side eigenvector, see (2.134)
$\boldsymbol{\xi}_{.j}(t)$	j-th underlying parameter gain in MA primary residuals, see (9.30)
$\boldsymbol{\pi}$	vector of model parameters; $\hat{\boldsymbol{\pi}}$: parameter estimates (Chapter 4)
$\boldsymbol{\pi}^{\circ}$	true parameter value; $\blacktriangle\boldsymbol{\pi}$: parameter discrepancy
π^{i}	partial fraction coefficient, see (2.221)
$\pi(\phi)$	determinant of the fault system matrix, see (6.82), (6.114)
$\pi_S(\phi)$	stable factor of $\pi(\phi)$; $\pi_U(\phi)$: unstable factor of $\pi(\phi)$, see (6.119)
ρ	order of persistent excitation (Chapter 3)
ρ	number of faults and disturbances
ρ_I	number of strictly input faults and disturbances, see 6.2.
σ_x^2	variance of $x(t)$, see (3.14)
σ	window width (residual degree), see 6.1.3. and (10.19)
σ_j	singular value, see (10.35)

τ	number of zeroes per column in the structure matrix, see (7.5)
$\upsilon_{ij}(\phi)$	element of the response modifier matrix $\mathbf{\Upsilon}(\phi)$, see (6.105)
$\psi(\tau)$	covariance function, see (3.20), (3.56)
$\boldsymbol{\psi}_i'$	left-side eigenvector, see (2.137); left singular vector, see (10.34)
$\boldsymbol{\psi}_i'(t)$	regression vector, see (4.49)
$\omega_{ij}(\phi)$	element of $\mathbf{\Omega}(\phi)$
$\vartheta_{ij}(t)$	limit model error, see (5.35)
$\vartheta^{+}(\phi)$	repeated pole-factors in multiple-output realization, see (2.175)
$\varphi(\tau)$	correlation function, see (3.19), (3.55)
$\boldsymbol{\varphi}'(t)$	regression vector, see (4.24), (4.31), (4.39)
$\boldsymbol{\varphi}_{\cdot j}$	ideal fault code for the j-th fault, see (7.3)
$\mathbf{B}$	fault direction matrix, see (8.6)
$\mathbf{\Gamma}$	matrix of right singular vectors, see (10.34)
$\mathbf{\Gamma}(\phi)$	system matrix, see (2.181); fault system matrix, see (6.80)
$\mathbf{\Gamma}(\phi)$	dynamic response matrix, see (8.6)
$\Lambda_x(\xi, \ldots)$	likelihood ratio, see (3.214)
$\mathbf{\Lambda}$	matrix of eigenvalues, see (2.145)
$\mathbf{\Lambda}(t,K)$	regression matrix, see (11.122)
Ξ	value-set for $X(t)$
$\mathbf{\Xi}$	matrix of right-side eigenvectors, see (2.140)
$\mathbf{\Xi}(t)$	underlying parameter gains in MA residual, see (9.30)
$\mathbf{\Pi}$	matrix subject to singular value decomposition (Chapter 10)
$\mathbf{\Pi}$	gain matrix from the residual to the mean estimate, see 11.3.2.
$\mathbf{\Upsilon}(\phi)$	response modifier matrix, see (6.104)
$\mathbf{\Upsilon}_P(\phi)$	modifier for stable/polynomial generator
$\mathbf{\Upsilon}_R(\phi)$	modifier for causal (realizable) generator
$\mathbf{\Phi}$	incidence (structure) matrix, see 7.1.1.
$\mathbf{\Phi}$, $\mathbf{\Phi}(\tau)$	correlation matrix, see (3.83), (3.91), (3.98), (3.104), (3.114)
$\mathbf{\Phi}$, $\mathbf{\Phi}(t,K)$	regression matrix, see (4.11), (4.57)
$\mathbf{\Psi}$	matrix of left-side eigenvectors, see (2.141)
$\mathbf{\Psi}$	matrix of left singular vectors, see (10.34)
$\mathbf{\Psi}$, $\mathbf{\Psi}(\tau)$	covariance matrix, see (3.88)
$\mathbf{\Psi}'(t)$	regression matrix, see (4.52)
$\mathbf{\Omega}(\phi)$	adjoint of the fault system matrix, see (6.82), (6.114)

Fault Detection and Diagnosis in Engineering Systems

1
Introduction to Fault Detection and Diagnosis

1.0. INTRODUCTION

Ever since humans have been building machines, they have been naturally concerned about their condition. For centuries, the only way to learn about malfunctions and their location was by biological senses (an approach still widely practiced); looking for changes in shape or color, listening to sounds unusual in strength or pitch, touching to feel heat or vibration, and smelling for fumes from leaks or overheating. Later, measuring devices were introduced, which provided more exact information about important physical variables. However, these devices (sensors) also proved prone to malfunction, raising the dilemma of false alarms. The potential for faults in the sensors became even more critical when they were applied in the automatic control of the machines, where the effects of such malfunctions may be more direct and devastating, and where the human operator is frequently removed from the process.

A dramatic development took place with the arrival of the computer and the proliferation of its real-time application. The computer could be entrusted with the automatic supervision of machinery. Its intelligence enabled it to integrate information originating from diverse sources, opening the possibility of locating the faulty component even in a complex system, including the sensors responsible for false alarms. The speed of computers also made it realistic to capture faults while they are developing, before they lead to significant disruptions in the machine's operation.

Initially, computers were complex and expensive and needed a special operating environment. They could only be applied to machinery which itself was complex and expensive enough to justify them. Large industrial production facilities, such as power plants, oil refineries, chemical plants, steel mills and paper making machines were the first to be equipped with computerized condition monitoring. This was followed by similar applications in major transportation equipment, such as large ships and airplanes. In such systems, a central computer would read hundreds or even thousands of measurements and supervise tens or hundreds of control loops.

With the advent of the microprocessor, the situation changed substantially. Processors became mere components, which are inexpensive enough to be built into various devices and which can operate under broad ambient conditions. This development, coupled with the arrival of advanced communication, resulted in the decentralization of the large industrial monitoring and control systems. More importantly, microcomputers appeared in mass-produced consumer products, such as automobiles, heating and air conditioning devices and even household appliances. Thus computerized condition monitoring, though invisible to many, has become part of everyday life in the developed consumer society.

In this initial chapter, some fundamental concepts and approaches of fault detection and diagnosis will be introduced first. This will be followed by two motivational examples of real-life systems. In the first example, the on-board detection and isolation of component faults in automotive engines will be discussed. The second example will outline the fault-detection problems in a complex chemical plant, consisting of a reaction process and auxiliary equipment, and suggest some solutions.

1.1. THE SCOPE OF FAULT DETECTION AND DIAGNOSIS

We are concerned with the detection and diagnosis of faults in engineering systems, such as production facilities, machines, vehicles, appliances, etc., whether they occur in the plant (the technical equipment proper) or in its measurement and control instruments. In this section, we will describe what is meant by faults and will specify the tasks of detection and diagnosis.

Types of faults. In general, faults are deviations from the normal behavior in the plant or its instrumentation. The faults of interest belong to one of the following categories:

- *Additive process faults.* These are unknown inputs acting on the plant, which are normally zero and which, when present, cause a change in the plant outputs independent of the known inputs. Such faults best describe plant leaks, loads, etc.

- *Multiplicative process faults.* These are changes (abrupt or gradual) in some plant parameters. They cause changes in the plant outputs which depend also on the magnitude of the known inputs. Such faults best describe the deterioration of plant equipment, such as surface contamination, clogging, or the partial or total loss of power.
- *Sensor faults.* These are discrepancies between the measured and actual values of individual plant variables. These faults are usually considered additive (independent of the measured magnitude), though some sensor faults (such as sticking or complete failure) may be better characterized as multiplicative.
- *Actuator faults.* These are discrepancies between the input command of an actuator and its actual output. Actuator faults are usually handled as additive though, again, some kinds (sticking or complete failure) may be better described as multiplicative.

Fault detection and diagnosis systems implement the following tasks (Fig. 1.1):

1. *Fault detection*, that is, the indication that something is going wrong in the monitored system;

2. *Fault isolation*, that is, the determination of the exact location of the fault (the component which is faulty);

3. *Fault identification*, that is, the determination of the magnitude of the fault.

The isolation and identification tasks together are referred to as *fault diagnosis*. While detection is an absolute must in any practical system and isolation is almost equally important, fault identification (though useful) may not justify the extra effort it requires. Therefore, most practical systems contain only the *fault detection and isolation* stages (and are referred to as FDI systems). Also, in many cases "diagnosis" is used simply as a synonym to "isolation."

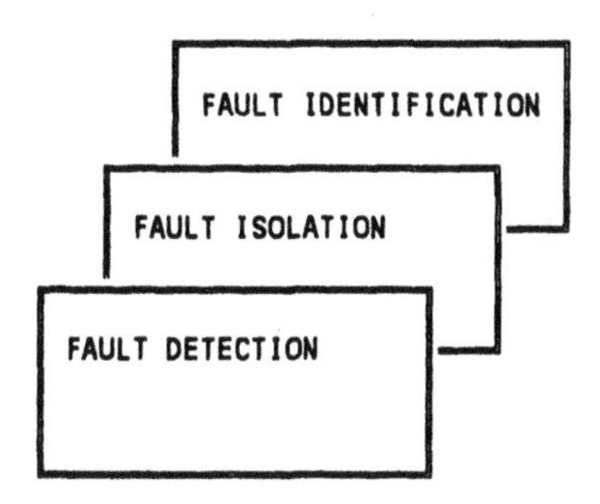

FIGURE 1.1. Tasks of fault detection and diagnosis.

Usually, the fault detection and diagnosis activity takes place on-line, in real time. The two tasks, detection and diagnosis, may be performed in parallel or sequentially. In some diagnostic systems, a single decision conveys not only the fact that a fault is present but also its location. In other systems, the detection task is running permanently while the diagnostic task is triggered only upon the detection of the presence of a fault.

Particularly in model-based fault detection and diagnosis (see below), the following conventions are usually adopted:

1. It is assumed that *faults* are not present initially in the system but *arrive* at some later time. The faults are generally described by deterministic time-functions which are unknown. Important special cases are the jump-fault (step function) and the drift-fault (ramp function), with both their magnitude and *arrival time* being unknown.

2. One may speak of *additive disturbances* as well, which are also deterministic and unknown inputs to the system. The distinction between additive faults and disturbances is subjective: faults are those unknown inputs which we wish to detect and isolate while disturbances are nuisances which we wish to ignore.

3. Any *noise*, originating from the plant or from sensors and actuators, is considered random with zero mean (any nonzero mean is handled as a fault or disturbance).

4. *Modeling errors* are discrepancies between the model (model parameters) and the true system. They may be present ever since the origins of the system or may arise due to operating-point changes. Model errors are nuisances the effect of which we want to suppress. They may be considered as *multiplicative disturbances*, in contrast to multiplicative faults which are also discrepancies between the model and the true system, but which we wish to detect.

The *detection performance* of the diagnostic technique is characterized by a number of important and quantifiable benchmarks, namely:

- *Fault sensitivity*, that is, the ability of the technique to detect faults of reasonably small size;
- *Reaction speed*, that is, the ability of the technique to detect faults with reasonably small delay after their arrival;
- *Robustness*, that is, the ability of the technique to operate in the presence of noise, disturbances and modeling errors, with few false alarms.

Fault sensitivity, reaction speed and robustness arise from an interplay between faults on the one hand and noise, disturbances and model errors on the other hand, and are affected by the design of the detection algorithm. In most cases, there are design trade-offs between the various properties.

Isolation performance, that is, the ability of the diagnostic system to distinguish faults depends on the physical properties of the plant, on the size of faults, noise, disturbances and model errors, and on the design of the algorithm. Multiple simultaneous faults are, in general, more difficult to isolate than single faults. Also, the interplay between faults and disturbances, noise and model errors may lead to uncertain or incorrect isolation decisions. Further, some faults may be *non-isolable* from one another because they act on the physical plant in an undistinguishable way.

1.2. APPROACHES TO FAULT DETECTION AND DIAGNOSIS

The methods of fault detection and diagnosis may be classified into two major groups: those which do not utilize the mathematical model of the plant and those which do. This book is devoted to the model-based methods. Before introducing these, the model-free techniques will be briefly reviewed below.

1.2.1. Model-Free Methods

The fault detection and isolation methods which do not use the mathematical model of the plant range from physical redundancy and special sensors through limit-checking and spectrum analysis to logical reasoning.

Physical redundancy. In this approach, multiple sensors are installed to measure the same physical quantity. Any serious discrepancy between the measurements indicates a sensor fault. With only two parallel sensors, fault isolation is not possible. With three sensors, a voting scheme can be formed which isolates the faulty sensor. Physical redundancy involves extra hardware cost and extra weight, the latter representing a serious concern in aerospace applications.

Special sensors may be installed explicitly for detection and diagnosis. These may be limit sensors (measuring e.g., temperature or pressure), which perform limit checking (see below) in hardware. Other special sensors may measure some fault-indicating physical quantity, such as sound, vibration, elongation, etc.

Limit checking. In this approach, which is widely used in practice, plant measurements are compared by computer to preset limits. Exceeding the threshold indicates a fault situation. In many systems, there are two levels of limits, the first serving for pre-warning while the second triggering an emergency reaction. Limit checking may be extended to monitoring the *time-trend* of selected variables. While simple and straightforward, the limit checking approach suffers from two serious drawbacks:

- Since the plant variables may vary widely due to normal input variations, the test thresholds need to be set quite conservatively.
- The effect of a single component fault may propagate to many plant varia-

bles, setting off a confusing multitude of alarms and making isolation extremely difficult.

Spectrum analysis of plant measurements may also be used for detection and isolation. Most plant variables exhibit a typical frequency spectrum under normal operating conditions; any deviation from this is an indication of abnormality. Certain types of faults may even have their characteristic signature in the spectrum, facilitating fault isolation.

Logic reasoning techniques form a broad class which is complementary to the methods outlined above, in that they are aimed at *evaluating* the symptoms obtained by the detection hardware or software. The simplest techniques consist of trees of logical rules of the "IF-symptom-AND-symptom-THEN-conclusion" type. Each conclusion can, in turn, serve as a symptom in the next rule, until the final conclusion is reached. The system may process the information presented by the detection hardware/software or may interact with a human operator, inquiring from him/her about particular symptoms and guiding him/her through the entire logical process.

1.2.2. Model-Based Methods

Model-based fault detection and diagnosis methods utilize an explicit mathematical model of the monitored plant. The engineering systems (production facilities, machines, etc.) dealt with in this book are dynamic systems characterized by continuous-time operation. Their natural mathematical description is in the form of differential equations, or equivalent transformed representations. However, the monitoring computers operate in a sampled fashion, using sampled data. Therefore it is customary and practical to describe the monitored plants in discrete time, in the form of difference equations or their transformed equivalents. This is the main approach followed in this book as well, though we will refer to the continuous-time model at some instances. Also, though most physical systems are nonlinear, their mathematical description usually relies on linear approximations. Again, we will mostly follow this approach, with occasional references to nonlinear models.

Most of the model-based fault detection and diagnosis methods rely on the concept of *analytical redundancy*. In contrast to physical redundancy, when measurements from parallel sensors are compared to each other, now sensory measurements are compared to analytically computed values of the respective variable. Such computations use present and/or previous measurements of other variables, and the mathematical plant model describing their nominal relationship to the measured variable. The idea can be extended to the comparison of two analytically generated quantities, obtained from different sets of variables. In either case, the resulting differences, called *residuals*, are indicative of the presence of faults in the system. Another class of model-based

FIGURE 1.2. Stages of model-based fault detection and diagnosis.

methods relies directly on *parameter estimation.*

The generation of residuals needs to be followed by *residual evaluation*, in order to arrive at detection and isolation decisions (Fig. 1.2). Because of the presence of noise and model errors, the residuals are never zero, even if there is no fault. Therefore the detection decision requires testing the residuals against thresholds, obtained empirically or by theoretical considerations.

To facilitate fault isolation, the residual generators are usually designed for *isolation enhanced* residuals, exhibiting structural or directional properties. The isolation decisions then can be obtained in a structural (Boolean) or directional (geometric) framework, with or without the inclusion of statistical elements.

Robustness issues. As it was mentioned above, the residuals generated to indicate faults may also react to the presence of noise, disturbances and model errors. Desensitizing the residuals to these sources is a most important aspect in the design of the detection and diagnosis algorithm. In particular:

- To deal with the effects of *noise*, the residuals may be filtered and statistical techniques may be applied to their evaluation. The latter may, though, be hampered by the lack of sufficient information concerning the statistical properties of the noise and the noise-transfer dynamics of the plant.
- *Disturbance decoupling* may be built into the design of the residual generator, but it competes with isolation enhancement for the available design freedom.
- *Robustness* in the face of *modeling errors* is the most fundamental problem in model-based fault detection and isolation. Several methods are available which usually rely on some sort of optimization. Unfortunately, this problem does not lend itself to easy solution and the known techniques are effective only under limited circumstances.

Residual generation methods. There are four, somewhat overlapping approaches to residual generation in model based fault detection and isolation:

1. Kalman filter. The innovation (prediction error) of the Kalman filter can be used as a fault detection residual; its mean is zero if there is no fault (and disturbance) and becomes nonzero in the presence of faults. Since the innovation sequence is white, statistical tests are relatively easy to construct. However, fault isolation is somewhat awkward with the Kalman filter; one needs to run a bank of "matched filters," one for each suspected fault and for each possible arrival time, and check which filter output can be matched with the actual observations.

2. Diagnostic observers. Observer innovations also qualify as fault detection residuals. "Unknown input" design techniques may be used to decouple the residuals from (a limited number of) disturbances. The residual sequence is colored which makes statistical testing somewhat complicated. The freedom in the design of the observer can be utilized to enhance the residuals for isolation. The dynamics of the fault response can be controlled, within certain limits, by placing the poles of the observer.

3. Parity (consistency) relations. Parity relations are rearranged direct input-output model equations, subjected to a linear dynamic transformation. The transformed residuals serve for detection and isolation. The residual sequence is colored, just like in the case of observers. The design freedom provided by the transformation can be used for disturbance decoupling and fault isolation enhancement. Also, the dynamics of the response can be assigned, within the limits posed by the requirements of causality and stability.

4. Parameter estimation. Parameter estimation is a natural approach to the detection and isolation of parametric (multiplicative) faults. A reference model is obtained by first identifying the plant in a fault-free situation. Then the parameters are repeatedly re-identified on-line. Deviations from the reference model serve as a basis for detection and isolation. Parameter estimation may be more reliable than the analytical redundancy methods, but it is also more demanding in terms of on-line computation and input excitation requirements.

As it has been realized recently, there is a fundamental equivalence between parity relation and observer based designs, in that the two techniques produce identical residuals if the generators have been designed for the same specification. A relationship, though weaker, has been found between parity relations and parameter estimation as well, in that parity relations designed for the isolation of parametric faults are the minimum data-length least-squares estimators for the same.

1.2.3. Historical Notes and References

The model based approach takes one of its origins from chemical process control, where the traditional material and energy balance calculations evolved into systematic data reconciliation and the detection of gross errors. Some of the pioneers of this effort were David Himmelblau (1978) of the University of Texas, Richard Mah of Northwestern University (Mah et al, 1976), George Stephanopoulos, then with the University of Minnesota (Romagnoli and Stephanopoulos, 1981), and Vladimir Vaclavek (1974) of the Institute of Chemical Technology in Prague.

Another root can be traced to aerospace related research, primarily sponsored by NASA in the seventies. This was spearheaded by Alan Willsky and his coworkers at MIT (Willsky and Jones, 1976; Deckert et al, 1977) and by J.E. Potter and M.C. Suman (1977) of the Northrop Corporation. This effort lead to the fundamental formulation of parity (consistency) relation concepts (Chow and Willsky, 1984; Lou et al, 1986).

An independent activity, with surprisingly similar results, is due to L.A. Mironovskii of Leningrad (1979). Other contributors to the field of parity relations include Yacov Ben-Haim (1980) of the Technion, Israel, Asok Ray of Penn State (Desai and Ray, 1984), Mohammad-Ali Massoumnia and Wallace Vander Velde of MIT (1988), N. Viswanadham of Bangalore, India, in collaboration with engineers at GE Corporate R&D (Viswanadham et al, 1987a), Ron Patton, then in York, England (Patton and Chen, 1991) and Marcel Staroswiecki of Lille, France (Staroswiecki et al, 1993). The author and his coworkers have also been active in this area of research (Gertler and Singer, 1985 and 1990; Gertler and Luo, 1989; Gertler and Kunwer, 1995; Gertler and Monajemy, 1995).

In parallel, and partially overlapping with the above efforts, several researchers were looking into the possibility of applying Kalman filters and observers to fault detection and isolation problems. The Kalman filter idea can be traced to R.K. Mehra and J. Peschon (1971) of Scientific Systems, Inc. Alan Willsky (1976 and 1986) made important contributions here as well. Other early contributors include Bernard Friedland (1979), then with the Singer Company, and Garry Leininger (1981), then at Purdue, followed more recently by Michelle Basseville (1986) of IRISA, France, and by Ramine Nikoukhah (1994) of INRIA, France, among others.

The original ideas of diagnostic observers probably came from R.V. Beard (1971) and Harold Jones (1973) of MIT and Robert Clark of the University of Washington (Clark et al, 1975). They were followed by a long line of researchers, including Paul Frank of Duisburg, Germany (Frank and Keller, 1980; Frank and Wünnenberg, 1989), M.A. Massoumnia (Massoumnia, 1986; Massoumnia et al, 1989), N. Viswanadham (Viswanadham and Srichander, 1987), Ron Patton (Patton and Kangethe, 1989), Jason Speyer of UCLA (White and Speyer, 1987), and Georgio Rizzoni of Ohio State (Park and Rizzoni, 1994).

In the area of fault detection and isolation by parameter estimation, substantial work has been done by Rolf Isermann and his colleagues in Darmstadt, Germany (Isermann, 1984 and 1993; Isermann and Freyermuth, 1991). Other contributors include A. Rault and coworkers (1984) of Adersa-Gerbios, France and B. Ninness and G.C. Goodwin of Newcastle, Australia (1991). An important related activity is due to M. Basseville, A. Benveniste and coworkers, concerning the detection of small parametric faults by the statistical analysis of residuals obtained over extended sets of observations (Basseville et al, 1987; Benveniste et al, 1987; Zhang et al, 1994).

The fundamental equivalence between parity relation and observer based designs was demonstrated by the author in a survey paper (Gertler, 1991). Similar results, some partial, others supplementary, were obtained by N. Viswanadham (Viswanadham et al, 1987a), Paul Frank (1990), J.F. Magni and P. Mouyon (1994) and others. The clear link between parity relations and parameter estimation was pointed out in (Gertler, 1995b), again confirmed by other researchers, including T. Höfling and R. Isermann (1995) and, in a somewhat limited form, G. Delmaire and coauthors (1994).

1.3. APPLICATION EXAMPLE: CAR ENGINE DIAGNOSIS

The following is a simplified account of a real-life example concerned with automobile engines. It has been drawn from an application project undertaken by the author and a team of coworkers, and funded by the General Motors Corporation (Gertler and Costin, 1994; Gertler et al, 1995). The objective of the project was to develop an on-board system for the detection and isolation of faults in the sensors and actuators participating in the engine's emission control system. Its inclusion here will hopefully motivate the reader to study the rest of the book.

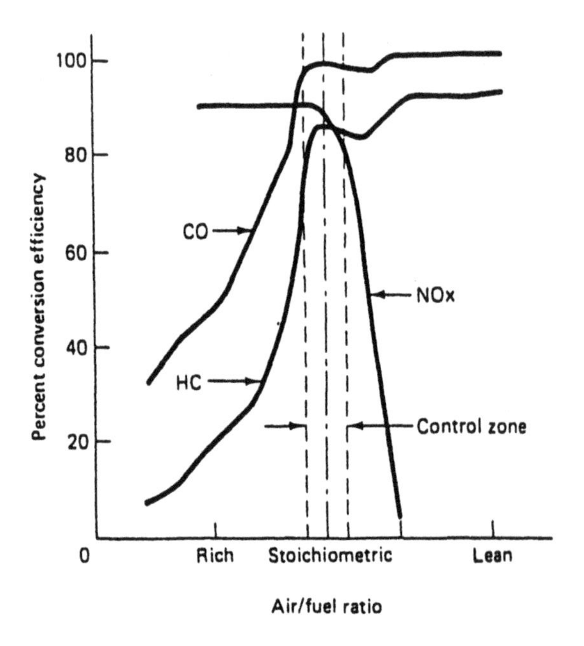

FIGURE 1.3. Efficiency of single-bed catalyst (General Motors Corporation).

1.3.1. Emission Control System

Today's cars are equipped with an extensive emission control system, meant to substantially reduce the pollution emitted by the vehicle. Of concern are three major pollutants, namely carbon monoxide and unburnt hydrocarbons, both originating from the imperfect burning of the fuel, and nitrous oxides, generated in the combustion process under high pressure and temperature. The two center-pieces of the emission control system are the catalytic converter and the exhaust gas recirculation valve.

The function of the *catalytic converter* is to remove the three pollutants from the exhaust gas. On most cars, a single-bed catalyst is employed. While the removal of carbon monoxide and hydrocarbons requires an oxidizing atmosphere (excess oxygen), nitrous oxides can be removed in a reducing atmosphere (no excess oxygen). Reasonable performance for all three pollutants can be achieved if the air-to-fuel ratio is kept close to the stoichiometric value of 14.7 (Fig. 1.3). To maintain this stoichiometric ratio is the main objective of the engine's fuel control system, which is implemented as a function of the on-board computer.

The air-to-fuel ratio is measured by an *excess oxygen sensor* inserted into the exhaust pipe. The commercially used oxygen sensor has a highly nonlinear (and temperature dependent) static characteristic (Fig. 1.4) and a time-varying time delay. The basic fuel-control scheme combines the relay-like sensor with an integrating controller (Fig. 1.5), resulting in a permanently limit-cycling operation. The driver sets the air flow entering the engine, by operating the

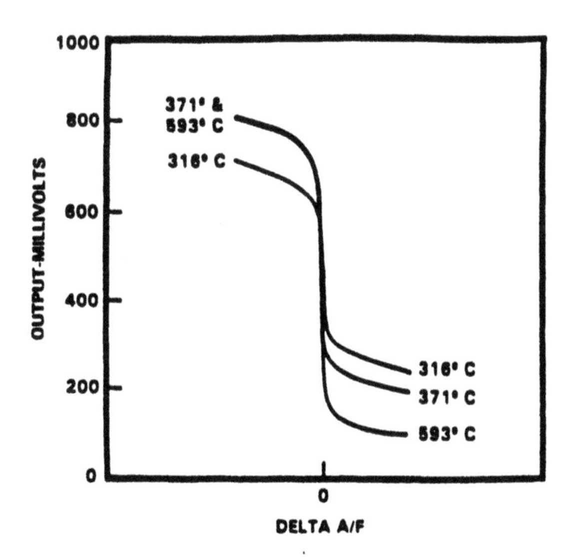

FIGURE 1.4. Oxygen sensor characteristic (General Motors Corporation).

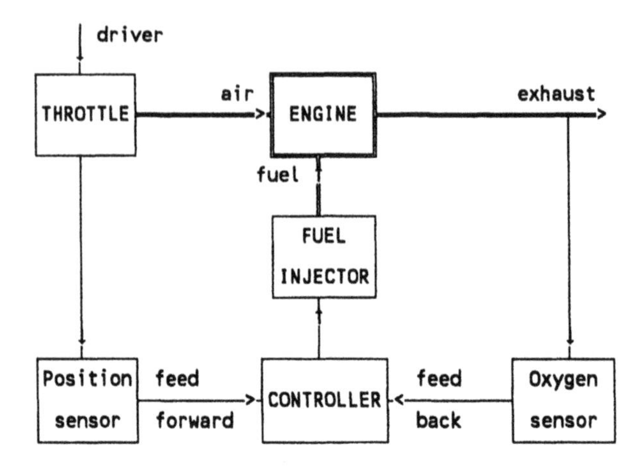

FIGURE 1.5. Fuel control.

throttle, and the controller adjusts the fuel flow, by affecting the injection duration of the *fuel injectors*, to restore the desired ratio. Since this basic feedback operation involves a significant physical time delay in the engine, exhaust pipe and oxygen sensor, it is combined with a less accurate but fast feed-forward action utilizing a *throttle position sensor* (Fig. 1.5).

The function of the *exhaust gas recirculation* (EGR) *valve* is to dilute with inert exhaust gas the air/fuel mix entering the engine, in order to reduce the generation of nitrous oxides by lowering the temperature of combustion. The EGR valve is also computer controlled, based upon the measurements of the manifold absolute pressure (MAP) and engine speed (RPM) (Fig. 1.6).

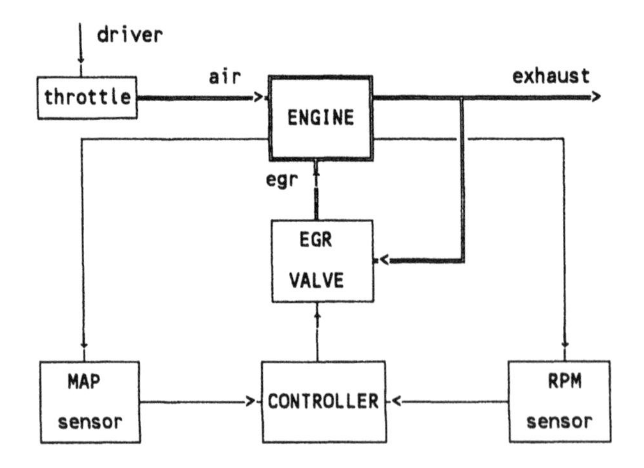

FIGURE 1.6. Exhaust gas recirculation.

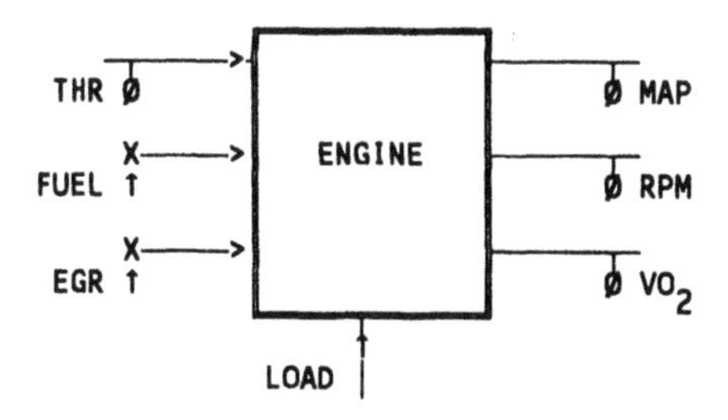

FIGURE 1.7. Engine variables.

The scope of the diagnostic task. Clearly, a fault in any of the sensors and actuators which participate in the emission control system causes the system to move away from its desired operating point and thus may result in increased emissions. Realizing this, the newest environmental regulations enacted in the United States (OBD-2) require that any emission control component fault which results in an emission increase of 50% or more must be detected and isolated on-board, during the vehicle's normal operation. Detection should turn on a warning signal for the driver, advising him to have the car serviced, while the isolation decision should be stored as a fault code for the servicing technician. OBD-2 allows three consecutive "drive cycles" for the diagnosis of such faults, where a drive cycle is a 24 minute standard sequence of varying driving conditions.

In the reported project, the scope of diagnosis was limited to two groups of actuators and four sensors, namely (Fig. 1.7)

- the fuel injectors (FUEL)
- the exhaust gas recirculation valve (EGR)
- the throttle position sensor (THR)
- the intake manifold pressure sensor (MAP)
- the engine speed sensor (RPM)
- the exhaust oxygen sensor (VO_2).

The load-torque acting on the engine (LOAD) is a major unmeasurable disturbance; it is desirable for the diagnostic algorithm to be insensitive to this disturbance.

1.3.2. The Fault Detection and Isolation Scheme

Structured parity relations. Diagnostic residuals have been generated using structured parity relations.

The functional model of the engine can be presented in the form of three interacting single-output subsystems (Fig. 1.8).

- *The Manifold Subsystem* contains the gas mechanics of the intake manifold, including the engine as a pump; its output is the manifold absolute pressure.

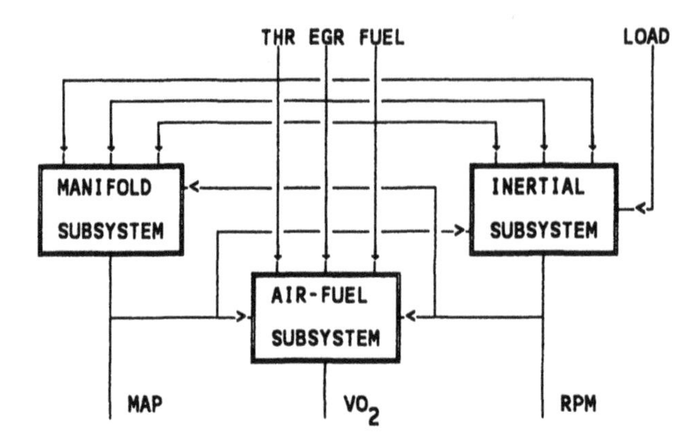

FIGURE 1.8. Engine subsystems (© 1995 IEEE).

- *The Inertial Subsystem* contains the dynamics of movement of the powertrain and the vehicle. These dynamics depend on the vehicle mass, air drag, transmission gear, etc. The subsystem's inputs include the unmeasurable load torque; its output is the engine speed (RPM).
- *The Air-Fuel Subsystem* contains the reaction chemistry of the engine; its output is the oxygen sensor voltage.

Since only the inertial subsystem is affected by the load torque (and the time varying parameters such as the vehicle mass), insensitivity with respect to those can easily be achieved by omitting this subsystem from the algorithm. This leaves two subsystems from which parity relations can be derived. For these, the engine speed (RPM) is simply handled as a measured input variable.

Now observe that the Manifold Subsystem does not contain the VO_2 variable. Thus a residual computed from the input-output model of this subsystem is insensitive to any fault of the oxygen sensor. Similarly, the output of the combined Manifold and Air-Fuel Subsystems can be computed without taking the MAP measurement into account. Thus the residual obtained from the model of the combined subsystems is insensitive to any fault of the MAP sensor. These model equations can thus be interpreted as structured parity relations and the residuals they provide are structured residuals.

Further structured residuals can be obtained by the algebraic manipulation of the two "primary residuals" described above. Express for example the RPM input from one of the model equations and substitute it into the other. The resulting equation does not contain the RPM variable and so the residual it returns is insensitive to the faults of the RPM sensor. Similarly, express the EGR (or FUEL) input from one of the model equations and substitute it into the other; the residual the resulting equation returns is insensitive to the faults of the EGR (or FUEL) actuator.

The structure of the equation set is shown in Table 1.1. Each row corre-

TABLE 1.1. Residual structures for the car engine (© 1995 IEEE).

	THR	EGR	Fuel	MAP	RPM	VO_2
#76	1	1	1	1	1	0
#73	1	1	1	0	1	1
#37	0	1	1	1	1	1
#57	1	0	1	1	1	1
#67	1	1	0	1	1	1
#75	1	1	1	1	0	1

sponds to a parity equation (residual) and each column to a component. A 1 in the table means that a fault of the component affects the residual, a 0 means it does not. (The row identifiers are octal representations of the row-struture.) Clearly, each equation (residual) is insensitive to a different single fault.

Residual evaluation. To obtain a detection and diagnosis decision, the six residuals are computed on-line and compared to appropriate thresholds. The residuals are first low-pass filtered to reduce the effect of noise. Thresholds are determined empirically, so that fault-free data does not cause false alarms. The results of the tests are represented as 1 (fired) or 0 (not fired). As a second stage of filtering, the test firings, carrying +/- signs for direction, are counted (or the excesses over the thresholds added up). The sums are periodically checked and the counters reset. If any of the counters reaches a "counter threshold" then a fault may be suspected.

For isolation, the six bits from the counter-tests are put together and compared to the "fault codes" of the various faults (which are the columns of the structure matrix). If there is a match, the respective fault is declared. Note that in the above structure, each column of the structure matrix is different, thus each fault has a distinct fault code. Further, each valid code contains exactly five 1's, so if a set of tests returns fewer then the isolation decision is uncertain but never incorrect. (Such a structure will be called "strongly isolating.") If there are multiple faults, then all the six tests fire. Thus a multiple fault can be declared but the faulty components cannot be further isolated.

Model identification. Since a theoretical model of sufficient accuracy is not available for the engine, the models are obtained from empirical data by systems identification. One possible approach is to identify the models of the two physical subsystems (Manifold and Air-Fuel Subsystems) and then derive the additional four equations algebraically. An alternative approach is to identify all the six model equations directly from the experimental data. For those equations which do not describe a physical subsystem, the Boolean structure provides guidance; one of the variables (reasonably VO_2) is selected as the output and the others handled as (pseudo) inputs. The great advantage of direct identification is that any nonlinear model-structure may be selected (subject only to the limitations of the identification technique), while nonlinearities would make the algebraic transformations, which arise with the indirect approach, very difficult.

Some experimental results. Figure 1.9 shows two residuals (#57 and #76), and their counters, in a fault-free situation. The residuals vary, due to disturbances and model errors, but the counters do not exceed their thresholds. In Figure 1.10, there is a small EGR fault. As expected, #57 does not respond (it was designed to be insensitive to this fault) but #76 does (occasionally, because the fault is small). Of course, one should see all the six residuals (counters) to form an isolation decision.

1.4. EXAMPLE: COMPLEX CHEMICAL PLANT

As a second motivational example, we have chosen a celebrated test problem, the Tennessee Eastman process. This is a real-life chemical process of intermediate complexity which has been published, in somewhat modified form to protect proprietary information, by the process control group of the Eastman Chemical Company (Downs and Vogel, 1993). The main intention of the authors was to offer the problem to the academic community as a testbed for control, optimization and diagnostic strategies. The process will be briefly described here, based on the original publication, then some potential diagnostic tasks and solutions will be outlined.

1.4.1. Process Description

The Tennessee Eastman process (Fig. 1.11) produces two products (G and H) from four reactants (A, C, D and E). Also present is an inert (B) and a byproduct (F), making a total of eight components. The major technological units of the process are the reactor, condenser, stripper, vapor/liquid separator and compressor. There are 12 manipulated variables (actuators) in the system and 41 variables are measured. Various disturbances act on the process and all measurements are subject to random measurement noise.

Instead of considering the entire process, we will concentrate on the main technological unit, the reactor. The reactor is equipped with an internal heat exchanger for removing the heat of the reaction; the coolant flow is controlled. The reactor is also provided with a variable-speed agitator for mixing its contents. The flow-rate and composition of the input flow are measured. Also measured are the temperature, pressure and level of the reactor, as well as the outlet temperature of the coolant and the agitator speed.

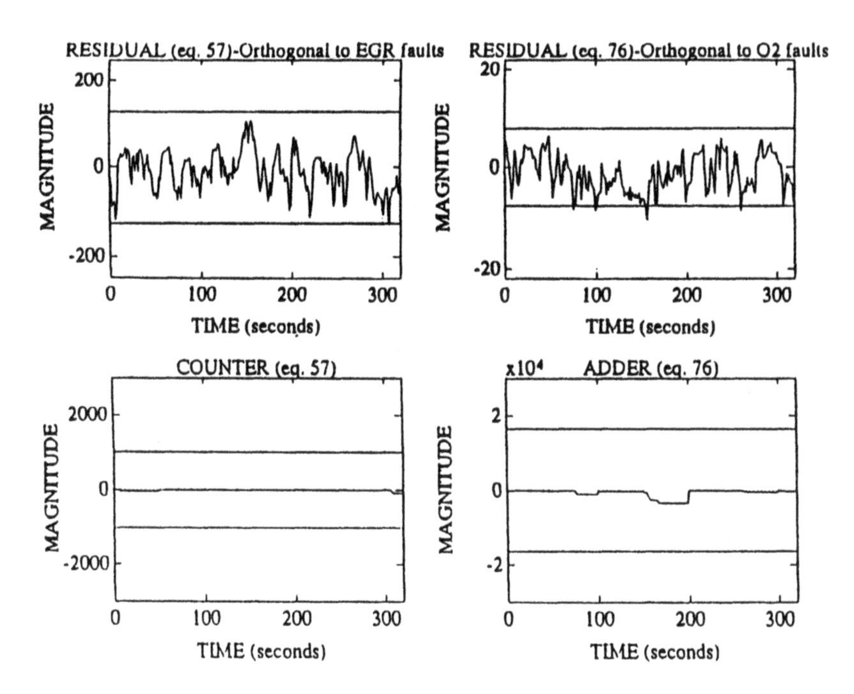

Figure 1.9. Residuals and counters with no fault (© 1995 IEEE).

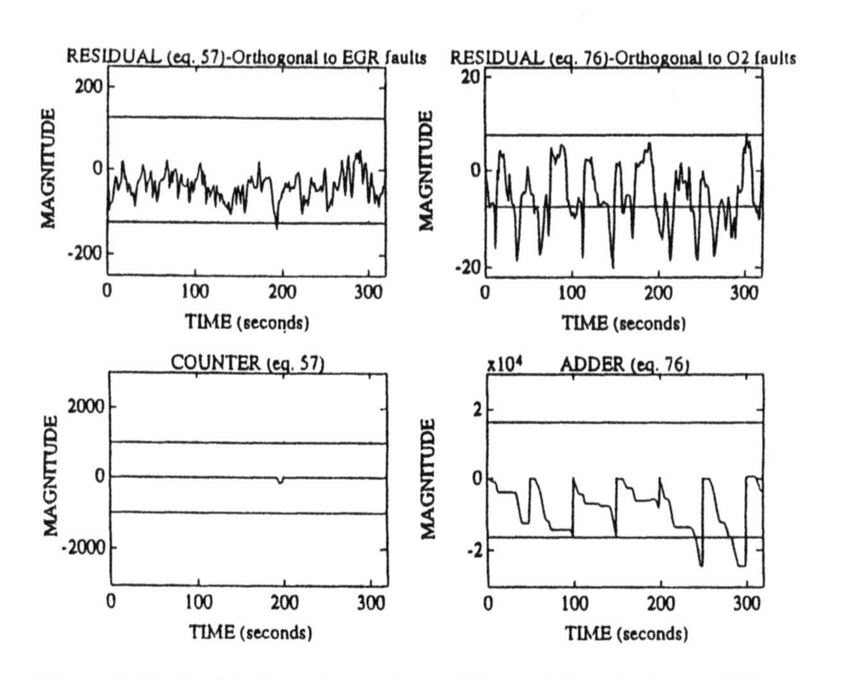

Figure 1.10. Residuals and counters with small EGR fault (© 1995 IEEE).

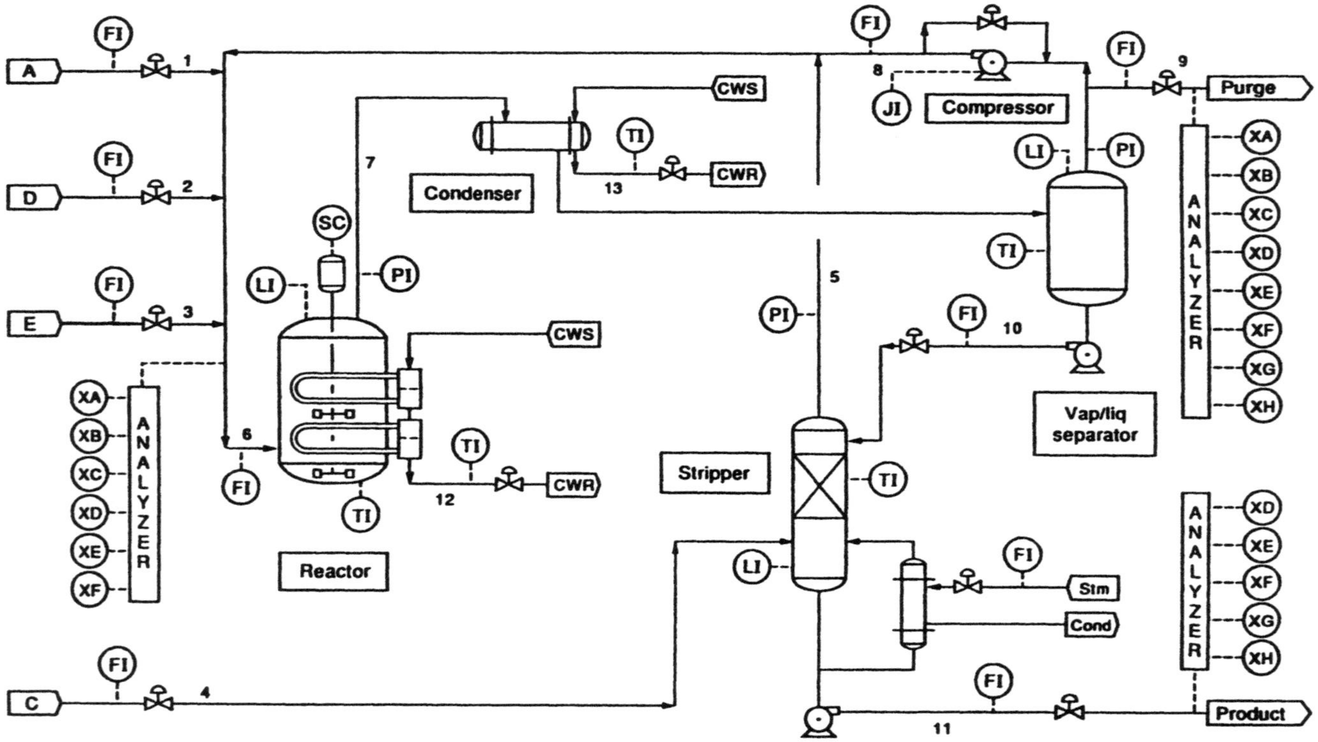

FIGURE 1.11. The Tennessee Eastman test problem (© 1993 Elsevier Science).

We make the following observations here:

- If we restrict ourselves to the reactor (and condenser, see below), then the upstream control and disturbance actions are all reflected in the measurements of the input flow and thus need not be considered.
- The reactor pressure, and via this the other outputs, are affected by the condenser temperature downstream. Therefore it is reasonable to include the condenser in the subsystem of interest, together with the reactor, with the outlet temperature of the condenser coolant as an additional input.
- The variation of the coolant inlet temperatures is a potential disturbance in the subsystem. Both the removal of reaction heat and the condenser pressure are related to some average value of the respective coolant temperature (between the inlet and outlet temperatures). Yet the outlet temperatures are measured and used in any model calculation. Thus any variation of the inlet temperature changes the relationship between the measured and average values, acting as a disturbance.

1.4.2. Fault Detection and Diagnosis

The following are some tentative thoughts about the fault detection and diagnosis problems and solutions in the reactor-condenser subsystem. The actual design of the diagnostic algorithm would require more detailed studies of the plant. We will divide our discussion into two parts: the individual control loops and the "core" of the subsystem (the reactor and condenser proper).

Individual control loops. First let us consider the three controlled variables in the reactor-condenser subsystem: the reactor coolant temperature, the condenser coolant temperature and the agitator speed. These are controlled by manipulating the respective coolant flows or the motor voltage, in individual control loops not shown in the process diagram. The reference values for these control loops originate from some outside controllers in cascade schemes (McAvoy and Ye, 1994). The effect of the rest of the plant (for example, the temperature inside the reactor) on the individual control loops may be considered as disturbances for the local controllers. The schematic of a simple control loop is shown in Fig. 1.12.

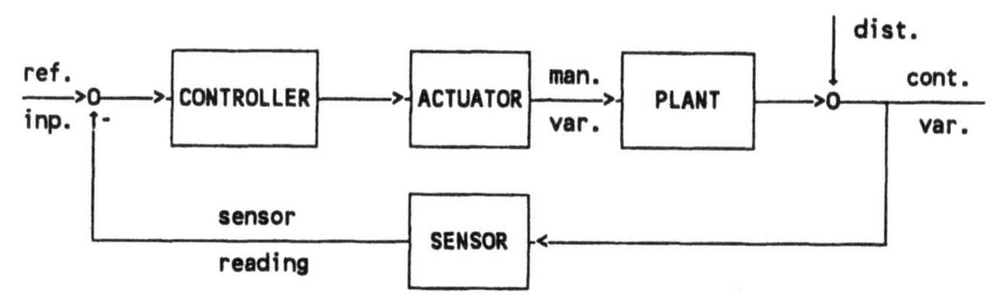

FIGURE 1.12. Single control loop.

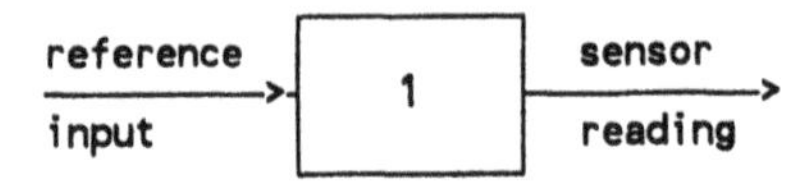

a. External model

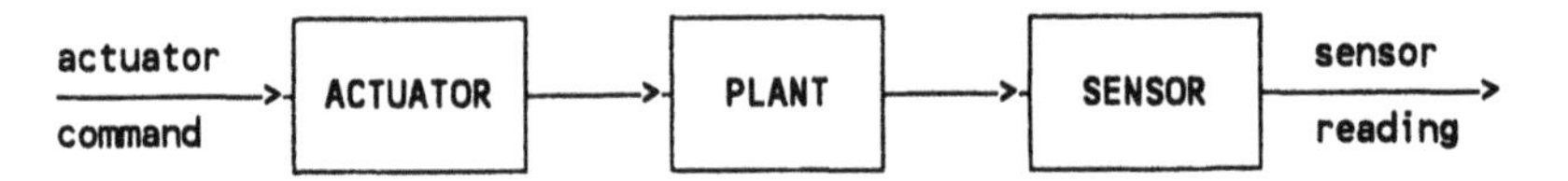

b. Internal model

FIGURE 1.13. Diagnostic models for a single control loop.

If the actuator command is not known (as is usually the case with analog control), and the actuator position is not measured, then no information is available from inside the single loop. The only diagnostic check such loops allow is seeing if the measured value of the controlled variable agrees with the reference (Fig. 1.13.a). Thus actuator faults cannot be detected, as long as the controller can compensate for them. Further, sensor faults cannot be detected either (at least not from observations of the single loop) because the check is based on the same sensor reading that the controller uses. The only conclusion one may draw from observing such single loops is whether the loop is functioning or not.

However, if the actuator command is known (which is the case with digital control), or the actuator position is measured, then a mathematical relationship describing the inside operation of the control loop may be available (Fig. 1.13.b). This can be utilized to detect faults of the actuator and of the output sensor, but it still does not support their isolation. With additional information (from the subsystem core), isolation of these faults may also become possible.

Subsystem core. Let us turn now to the main reactor-condenser equipment. The subsystem core has three controlled inputs, six measured inputs and three outputs, namely (Fig. 1.14):

Controlled inputs: reactor coolant temperature (T_{RC})
condenser coolant temperature (T_{CC})
agitator speed (V_A)

Measured inputs: intake flowrate (F_I)
intake concentrations ($X_{IA} \dots X_{IF}$)

Measured outputs: reactor temperature (T_R)
reactor pressure (P_R)
reactor level (L_R).

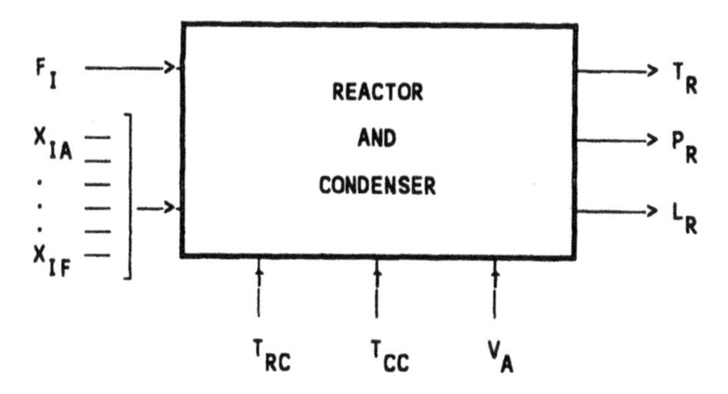

FIGURE 1.14. Reactor-Condenser subsystem core.

Note that the analyzer returns six composition values of which five are independent; we handle these as five separate sensors.

A diagnostic algorithm may be designed to detect and isolate faults in any of the twelve sensors (including those which measure controlled variables). Internal models of the control loops may be included, if they are available, extending the diagnosis to actuators. The design, however, may be restricted by properties of the plant (such as linear dependencies among the ways the variables act) which would require deeper analysis. Some indication of the nature of such restrictions is given below.

In addition to the sensor faults, one may consider two process faults, namely

- Reactor leak. This is an additive fault. Given the instrumentation of the plant, this fault is difficult to detect if the diagnosis is limited to the reactor-condenser subsystem. However, the inclusion of readings from the flow meters installed downstream would allow writing a simple material balance relation which would indicate such a fault.

- Heat exchanger fouling. This is a typical multiplicative fault and it is completely within the monitored subsystem. It is possible to design a diagnostic algorithm (parity relations enhanced for this parametric fault) which allows for its detection and isolation. It should be observed, however, that the only effect of an agitator speed change (fault) is a change in the heat transfer coefficient (Downs and Vogel, 1993). Thus it may not be possible to isolate coil fouling from a fault of the agitator speed sensor. (We refer to this situation as the two faults being "non-isolable".)

Recall that variations in the coolant inlet temperatures act as disturbances on the system. Unfortunately, their effect on the input-output relations of the subsystem core is the same as that of the incorrect measurement of the coolant outlet temperatures. Therefore temperature variations may be mis-diagnosed as sensor faults.

TABLE 1.2. Residual structures for the Tennessee Eastman process.

	inputs				outputs		
	F_I	T_{RC}	T_{CC}	V_A	T_R	P_R	L_R
r_1	0	0	1	1	1	1	1
r_2	1	0	0	1	1	1	1
r_3	1	1	0	0	1	1	1
r_4	1	1	1	0	0	1	1
r_5	1	1	1	1	0	0	1
r_6	1	1	1	1	1	0	0
r_7	0	1	1	1	1	1	0

Table 1.2 shows a parity (consistency) relation structure designed for seven sensors, without utilizing the internal models of the control loops. For simplicity, we assumed that the concentration readings are accurate and therefore outside the scope of the diagnostic algorithm. This system requires seven parity relations. It should be noted that it is not guaranteed that equations with all the structures in the Table can actually be designed. Seeing if this is the case would require more detailed analysis and the knowledge of the plant model. If the initially selected structure is not "attainable," it may still be possible to design other, also satisfactory structures - unless there are fundamental limitations in the plant which render some faults "non-isolable."

2
Discrete Linear Systems

2.0. INTRODUCTION

For the purpose of engineering analysis and design, physical systems are usually represented in some mathematical form; this representation is also called the *model* of the system. The properties of the model reflect the nature of the system, though in many cases the model may just be an approximation of the true system behavior. The most important properties of the model, and of the system they represent, are the following:

1. Static vs. dynamic models. A system is static if its outputs at any time depend only on its inputs acting at the same time. In contrast, the outputs of a dynamic system are affected by present and/or past inputs. The model of a static system consists of *algebraic equations* while dynamic systems are described by *differential or difference equations*. Dynamic systems in their steady state can also be characterized by static models.

2. Linear vs. nonlinear models. A linear model, static or dynamic, is one to which the *principle of superposition* applies, that is, where the response to a combination of inputs is the same as the sum of the individual responses. This implies that algebraically the model contains only terms which are linear in the variables. Nonlinear models have terms which are nonlinear in the variables, such as the square of a variable, the product of two or more variables, the sine or cosine function of a variable, etc. A system is linear if it can be exactly characterized by a linear model. While most physical systems are, in fact, nonlinear, approximate linear models can usually be derived which are valid in the vicinity of an operating point.

3. Continuous-time vs. discrete-time models. Continuous-time variables are those which exist (their values are defined) for any value of time. In contrast, *discrete-time variables* exist (are defined) only for certain time instants. Continuous-time models relate continuous variables to each other while discrete-time models relate discrete ones. Continuous models are differential equations or other model forms derived from the latter, discrete models are difference equations or their derivatives. Most physical systems are continuous by nature. However, in the computational equipment and algorithms monitoring and controlling them, their variables are represented by their sampled values which are discrete. Therefore in many cases discrete models are developed for those continuous systems, describing the relationship among their sampled variables. Other systems, mostly the ones existing on computers, are discrete by nature.

In this book, as in most recent work of engineering analysis and design, we will deal with linear discrete systems. In general, dynamic systems will be considered and static ones handled as their special case. Nonlinear models and their linear approximation will be briefly addressed. Also, we will show how discrete equivalent representations can be found for continuous systems.

We will start our discussion with static models, linear and nonlinear. Then we will proceed to the difference equation model of single-input single-output discrete dynamic systems. With the introduction of the shift operator, a discrete transfer function representation will be derived and then re-interpreted in terms of the z-transform. As an alternative model form, the discrete state-space representation will be discussed and fundamental concepts of dynamic systems, such as stability, controllability and observability, will be reviewed. The realization problem, that is, finding a minimal state-space representation for multiple-input multiple-output dynamic systems, will be briefly treated and poles, zeroes and system order defined in this context.

2.1. STATIC MODELS

In this section, we will gradually build up static models, starting with single-input single-output systems and moving first to multiple-input single-output then to multiple-input multiple-output systems. In each case, we will discuss linear models then briefly review nonlinear structures and show how their linear approximation can be obtained.

2.1.1. Single-Input Single-Output Systems

Consider a system shown in Figure 2.1, with single input $u(t)$ and single output $y(t)$. A static model has the general form

FIGURE 2.1

$$y(t) = f[u(t)] \tag{2.1}$$

where f [...] represents any functional relationship. The model is linear if it satisfies, for any value-set $x_1, x_2, \ldots$ and constants $c_1, c_2, \ldots$, the following linearity property:

$$f[c_1 x_1 + c_2 x_2 + \ldots] = c_1 f[x_1] + c_2 f[x_2] + \ldots \tag{2.2}$$

This implies that a linear model must be in the form

$$y(t) = a\, u(t) \tag{2.3}$$

Any other model is nonlinear. Note that, strictly speaking, this includes the model-form $y(t) = a_0 + a_1 u(t)$ which does not satisfy (2.2).

Examples of nonlinear models are the *polynomial* model

$$y(t) = b_0 + b_1 u(t) + b_2 u^2(t) + b_3 u^3(t) + \ldots \tag{2.4}$$

or the *exponential* model

$$y(t) = exp[b_0 + b_1 u(t)] \tag{2.5}$$

Polynomial models may be used as generic nonlinear models, even if the true nonlinear relationship is different; the degree of the polynomial expression is increased until the desired accuracy of the representation is achieved. A linear approximation (*"linearized model"*), in the form of (2.3), can be obtained for the general nonlinear system (2.1), as

$$y(t) - y^{\circ} = a[u(t) - u^{\circ}] \tag{2.6}$$

where

$$y^{\circ} = f[u^{\circ}] \qquad a = \frac{df[u(t)]}{du(t)}\bigg|_{u(t)=u^{\circ}} \tag{2.7}$$

and u° is the *operating point* where the linearization takes place.

2.1.2. Multiple-Input Single-Output Systems

Consider a static system with multiple inputs $u_1(t)$, $u_2(t)$, ... , $u_K(t)$ and single output $y(t)$, as shown in Figure 2.2.a. Introduce a vector notation for the inputs, in accordance with Figure 2.2.b:

$$\boldsymbol{u}(t) = [u_1(t) \;\; u_2(t) \;\; \dots \;\; u_K(t)]' \tag{2.8}$$

Now the general input-output relationship is

$$y(t) = f\,[\boldsymbol{u}(t)] \tag{2.9}$$

The model is linear if it satisfies, for any vector value-set $\boldsymbol{x}_1, \boldsymbol{x}_2, \dots$ and constants $c_1, c_2, \dots$, the following linearity property:

$$f\,[c_1\boldsymbol{x}_1 + c_2\boldsymbol{x}_2 + \dots] = c_1 f\,[\boldsymbol{x}_1] + c_2 f\,[\boldsymbol{x}_2] + \dots \tag{2.10}$$

This implies that the only possible linear model is

$$y(t) = a_1u_1(t) + a_2u_2(t) + \dots + a_Ku_K(t) = \boldsymbol{a}'\boldsymbol{u}(t) \tag{2.11}$$

where

$$\boldsymbol{a}' = [a_1 \;\; a_2 \;\; \dots \;\; a_K] \tag{2.12}$$

There is an infinite variety of nonlinear models. Of these, most important are the polynomial models and, in particular, the quadratic model

$$\begin{aligned} y(t) = b_o &+ b_1u_1(t) + \dots + b_Ku_K(t) \\ &+ b_{11}u_1^2(t) + b_{12}u_1(t)u_2(t) + \dots + b_{1K}u_1(t)u_K(t) \\ &\qquad + b_{22}u_2^2(t) + \dots + b_{2K}u_2(t)u_K(t) \\ &\qquad + \dots + b_{KK}u_K^2(t) \end{aligned} \tag{2.13}$$

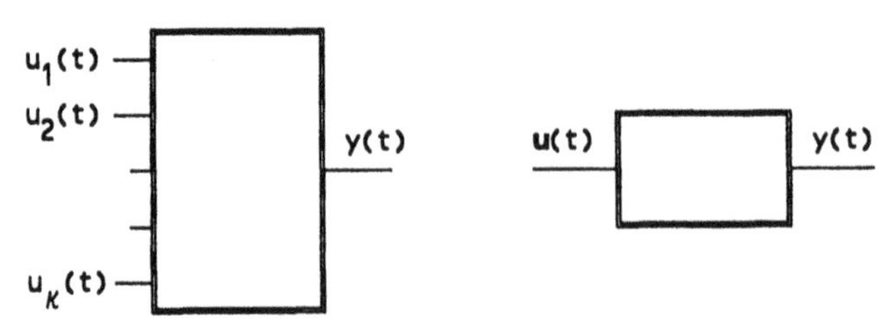

FIGURE 2.2.a

FIGURE 2.2.b

This model can also be written as

$$y(t) = \boldsymbol{u}^{*\prime}(t)\, \boldsymbol{B}\boldsymbol{u}^{*}(t) \tag{2.14}$$

where

$$\boldsymbol{u}^{*}(t) = [1 \;\; u_1(t) \;\; u_2(t) \ldots u_K(t)]' \tag{2.15}$$

and

$$\boldsymbol{B} = \begin{bmatrix} b_o & b_1 & b_2 & \cdots & b_K \\ 0 & b_{11} & b_{12} & \cdots & b_{1K} \\ 0 & 0 & b_{22} & \cdots & b_{2K} \\ & \cdots & \cdots & \cdots & \\ 0 & 0 & 0 & \cdots & b_{KK} \end{bmatrix} \tag{2.16}$$

A linear approximation to the general nonlinear system (2.9), in the operating point $\boldsymbol{u}^{\circ}$, is

$$y(t) - y^{\circ} = \boldsymbol{a}'[\boldsymbol{u}(t) - \boldsymbol{u}^{\circ}] \tag{2.17}$$

where

$$y^{\circ} = f[\boldsymbol{u}^{\circ}] \qquad a_j = \frac{\partial f[\boldsymbol{u}(t)]}{\partial u_j(t)}\Big|_{\boldsymbol{u}(t)=\boldsymbol{u}^{\circ}} \qquad j=1\ldots K \tag{2.18}$$

2.1.3. Multiple-Input Multiple-Output Systems

A multiple-input multiple-output system (Figure 2.3) has inputs (c.f. (2.8)) $\boldsymbol{u}(t)=[u_1(t) \;\; u_2(t) \ldots u_K(t)]'$ and outputs

$$\boldsymbol{y}(t) = [y_1(t) \;\; y_2(t) \;\; \ldots \;\; y_\mu(t)]' \tag{2.19}$$

The general input-output relationship is

$$\boldsymbol{y}(t) = \boldsymbol{f}[\boldsymbol{u}(t)] \tag{2.20}$$

where $\boldsymbol{f}[\ldots]$ represents a set of scalar functions $f_1[\ldots]$, $f_2[\ldots]$, ..., $f_\mu[\ldots]$, and the i-th output is obtained as

$$y_i(t) = f_i[\boldsymbol{u}(t)] \qquad i=1\ldots\mu \tag{2.21}$$

The multiple-input multiple-output system is linear if (2.10) is satisfied for all f_i, $i=1\ldots\mu$. In this case

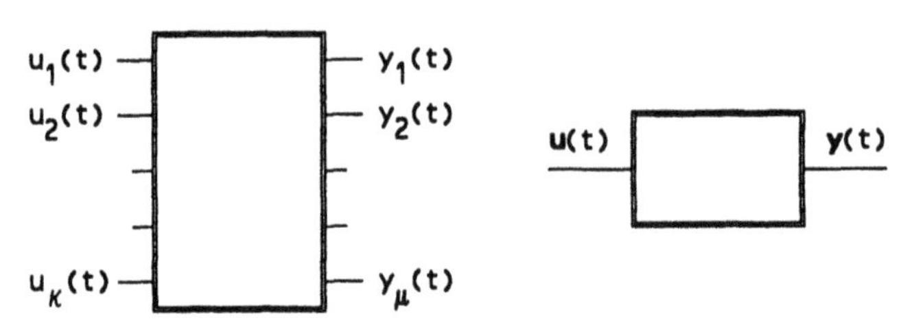

FIGURE 2.3.a FIGURE 2.3.b

$$y_i(t) = \boldsymbol{a}_{i\cdot}\, \boldsymbol{u}(t) \qquad i=1\ldots\mu \tag{2.22}$$

where

$$\boldsymbol{a}_{i\cdot} = [a_{i1} \;\; a_{i2} \;\cdots\; a_{i\kappa}] \tag{2.23}$$

and

$$\boldsymbol{y}(t) = \begin{bmatrix} \boldsymbol{a}_{1\cdot} \\ \boldsymbol{a}_{2\cdot} \\ \cdot \\ \cdot \\ \boldsymbol{a}_{\mu\cdot} \end{bmatrix} \boldsymbol{u}(t) = \boldsymbol{A}\boldsymbol{u}(t) \tag{2.24}$$

Polynomial models can be constructed separately for each row (2.21). A linear approximation of the general nonlinear model (2.20) is

$$\boldsymbol{y}(t) - \boldsymbol{y}^\circ = \boldsymbol{A}\, \boldsymbol{f}\, [\boldsymbol{u}(t) - \boldsymbol{u}^\circ] \tag{2.25}$$

where

$$\boldsymbol{y}^\circ = \boldsymbol{f}\,[\boldsymbol{u}^\circ] \qquad a_{ij} = \frac{\partial f_i\,[\boldsymbol{u}(t)]}{\partial u_j(t)} \Big|_{\boldsymbol{u}(t)=\boldsymbol{u}^\circ} \qquad i=1..\mu,\; j=1\ldots\kappa \tag{2.26}$$

2.2. DISCRETE TRANSFER FUNCTION

In this section, the input-output representation of linear discrete dynamic systems will be addressed. We will start with the difference equation model of single-input single-output systems. With the introduction of the shift operator, this will be converted into concise polynomial forms which lead then to the discrete transfer function in terms of the shift operator. The polynomial and transfer function representation will then be extended to multiple-input single-output and multiple-input multiple-output systems.

The systems discussed in this section, and in most of the book, have three important properties:

1. They are *linear*, in that they satisfy the linearity conditions (2.2) or (2.10), respectively.

2. They are *time invariant*. Time invariance means that the relationship between the input(s) and output(s) is permanent, it does not vary with time. In the mathematical representation of the system this is reflected by the model parameters being *constant*.

3. They are *causal*. By causality we mean that the system output at any time t depends only on inputs at time t and before and is not affected by future inputs.

Causality is a natural property of all physical systems. Linearity and time invariance may or may not really be exhibited by the system; in some cases, though, they may be assumed, in order to simplify the mathematical treatment.

2.2.1. Single-Input Single-Output Systems

Consider a single-input single-output causal linear time-invariant discrete dynamic system. The input $u(t)$ and output $y(t)$ are discrete-time variables, that is, their independent (time) variable t can only take integer values. The most direct way of characterizing a discrete dynamic system is by its *difference equation* which relates the present value and a number of past values of the output to a number of past values of the input (which may include its most recent value as well). The general form of the time-invariant linear difference equation is

$$y(t) + h^1 y(t-1) + h^2 y(t-2) + \ldots + h^{\upsilon} y(t-\upsilon) =$$

$$g^d u(t-d) + g^{d+1} u(t-d-1) + \ldots + g^{d+\iota} u(t-d-\iota) \qquad (2.27)$$

Here $h^1, h^2 \ldots h^{\upsilon}$ and $g^d, g^{d+1} \ldots g^{d+\iota}$ are coefficients while d is the *time delay* of the system. Solving the above equation for $y(t)$ yields the *recursive form* of the difference equation:

$$y(t) = \sum_{k=0}^{\iota} g^{d+k} u(t-d-k) - \sum_{k=1}^{\upsilon} h^k y(t-k) \qquad (2.28)$$

The difference equation can be made more concise by the use of the *shift operator*. Define the shift operator ϕ as

$$\phi^{-k} x(t) = x(t-k) \qquad (2.29)$$

where $x(t)$ denotes any discrete-time variable. With this, Eq. (2.27) can be written as

$$y(t) + h^1 \phi^{-1} y(t) + h^2 \phi^{-2} y(t) + \ldots + h^{\upsilon} \phi^{-\upsilon} y(t) =$$

$$= g^d \phi^{-d} u(t) + g^{d+1} \phi^{-d-1} u(t) + \ldots + g^{d+\ell} \phi^{-d-\ell} u(t) \qquad (2.30)$$

or

$$(1 + h^1 \phi^{-1} + h^2 \phi^{-2} + \ldots + h^{\upsilon} \phi^{-\upsilon})y(t) =$$

$$\phi^{-d} (g^d + g^{d+1} \phi^{-1} + \ldots + g^{d+\ell} \phi^{-\ell})u(t) \qquad (2.31)$$

Defining the polynomials

$$h(\phi) = 1 + h^1 \phi^{-1} + h^2 \phi^{-2} + \ldots + h^{\upsilon} \phi^{-\upsilon} \qquad (2.32a)$$

$$g(\phi) = \phi^{-d} (g^d + g^{d+1} \phi^{-1} + \ldots + g^{d+\ell} \phi^{-\ell}) \qquad (2.32b)$$

Eq. (2.27) finally becomes

$$h(\phi)y(t) = g(\phi)u(t) \qquad (2.33)$$

Eq. (2.33) is the *polynomial form* of the input-output relationship.

Note that $h(\phi)$ and $g(\phi)$ are *relative prime,* that is, they have no common factor $1-\alpha\phi^{-1}$. In case they do have such a factor, it may be canceled out without altering the relation (2.33).

Expressing $y(t)$ from (2.33) yields

$$y(t) = \frac{g(\phi)}{h(\phi)} u(t) = m(\phi)u(t) \qquad (2.34)$$

where

$$m(\phi) = g(\phi)/h(\phi) \qquad (2.35)$$

is the *discrete transfer function* of the system, given in terms of the shift operator. Eq. (2.34) is the *transfer function form* of the input-output relationship.

There are two important special cases of discrete linear dynamic systems:

If $h(\phi)=1$ then $m(\phi)=g(\phi)$; such systems are called *moving average* (*MA*);

If $g(\phi)=1$ then $m(\phi)=1/h(\phi)$; such systems are called *autoregressive* (*AR*).

In the general case, the system may be referred to as *autoregressive-moving average* (*ARMA*).

Usually it is more convenient to write the $h(\phi)$ and $g(\phi)$ polynomials as

$$h(\phi) = 1 + h^1\phi^{-1} + \ldots + h^{\nu}\phi^{-\nu} \tag{2.36a}$$

$$g(\phi) = g^o + g^1\phi^{-1} + \ldots + g^{\nu}\phi^{-\nu} \tag{2.36b}$$

where

$$\nu = max(d+\iota,\ \upsilon) \tag{2.37}$$

is the *order of the system*. Obviously, if $d \neq 0$ then the d leading coefficients in $g(\phi)$ are zero. Also, unless $d+\iota=\nu$, some of the tailing coefficients are zero in one of the polynomials. Alternatively, it is possible to separate the time delay ϕ^{-d} from the polynomial $g(\phi)$ and define the order of the remaining system as $\nu = max(\iota,\ \upsilon)$.

In some cases it may be advantageous to express the difference equation polynomials in terms of positive powers of the shift operator. These polynomials, which we will denote as $h^+(\phi)$ and $g^+(\phi)$, are obtained as

$$h^+(\phi) = \phi^{\nu}h(\phi) = \phi^{\nu} + h^1\phi^{\nu-1} + \ldots + h^{\nu} \tag{2.38a}$$

$$g^+(\phi) = \phi^{\nu}g(\phi) = g^d\phi^{\nu-d} + g^{d+1}\phi^{\nu-d-1} + \ldots + g^{\nu} \tag{2.38b}$$

Of course, Eq. (2.33) holds also in the form $h^+(\phi)y(t) = g^+(\phi)u(t)$, and the discrete transfer function is the same whether the polynomials are written in terms of ϕ^{-1} or ϕ since

$$m^+(\phi) = \frac{g^+(\phi)}{h^+(\phi)} = \frac{\phi^{\nu}g(\phi)}{\phi^{\nu}h(\phi)} = \frac{g(\phi)}{h(\phi)} = m(\phi) \tag{2.39}$$

The roots of the $h^+(\phi)$ polynomial, $p_1, p_2, \ldots, p_{\nu}$, are the *poles* of the system. The roots of the $g^+(\phi)$ polynomial, $z_1, z_2, \ldots, z_{\nu-d}$, are the (finite) *zeroes* of the system. With the poles and zeroes, the $h^+(\phi)$ and $g^+(\phi)$ polynomials are expressed as

$$h^+(\phi) = (\phi - p_1)(\phi - p_2)\ldots(\phi - p_{\nu}) \tag{2.40a}$$

$$g^+(\phi) = g^d(\phi - z_1)(\phi - z_2)\ldots(\phi - z_{\nu-d}) \tag{2.40b}$$

2.2.2. Multiple-Input Single-Output Systems

Consider a system with a single output $y(t)$ and multiple inputs $u_1(t)$, $u_2(t)$, ..., $u_{\kappa}(t)$. The input-output relationship (difference equation) for this system, in shift-operator polynomial form, is

$$h(\phi)y(t) = g_1(\phi)u_1(t) + g_2(\phi)u_2(t) + \dots + g_\kappa(\phi)u_\kappa(t) \tag{2.41}$$

Here $h(\phi)$ is as shown in Eq. (2.32a) and each $g_j(\phi)$, $j=1\dots\kappa$, is in the form of Eq. (2.32b), in general with a different delay d_j and upper limit ι_j. Define the order ν of the multiple-input system as

$$\nu = max(d_1+\iota_1, \dots, d_\kappa+\iota_\kappa, \upsilon) \tag{2.42}$$

Then the polynomials $h(\phi)$, $g_1(\phi) \dots g_\kappa(\phi)$ can be expanded to the common degree ν so that $h(\phi)$ is written as in (2.36a) and $g_j(\phi)$ becomes, in accordance with (2.36b),

$$g_j(\phi) = g_j^o + g_j^1\phi^{-1} + \dots + g_j^\nu\phi^{-\nu} \qquad j=1\dots\kappa \tag{2.43}$$

The polynomials $h(\phi)$, $g_1(\phi) \dots g_\kappa(\phi)$ are relative prime, even in their expanded form, in the sense that there is no $1\text{-}\alpha\phi^{-1}$ factor common to all of them.

The *poles of the multiple-input single-output system* are the roots of the polynomial $h^+(\phi)=\phi^\nu h(\phi)$. The polynomials $g_j^+(\phi)=\phi^\nu g_j(\phi)$ have, in general, different sets of zeroes; we will consider as *zeroes of the multiple-input single-output system* those values, if any, which are common roots of all polynomials $g_j^+(\phi)$, $j=1\dots\kappa$.

Solving Eq. (2.41) for $y(t)$ yields the transfer function form of the input-output relationship

$$y(t) = m_1(\phi)u_1(t) + m_2(\phi)u_2(t) + \dots + m_\kappa(\phi)u_\kappa(t) \tag{2.44}$$

where

$$m_j(\phi) = g_j(\phi)/h(\phi) \qquad j=1\dots\kappa \tag{2.45}$$

is the discrete transfer function from the j-th input to the output. As can be seen, $h(\phi)$ is the common denominator for all transfer functions $j=1\dots\kappa$.

Introduce the vector notation

$$u(t) = [u_1(t) \quad u_2(t) \quad \dots \quad u_\kappa(t)]' \tag{2.46}$$

$$g'(\phi) = [g_1(\phi) \quad g_2(\phi) \quad \dots \quad g_\kappa(\phi)] \tag{2.47}$$

$$m'(\phi) = [m_1(\phi) \quad m_2(\phi) \quad \dots \quad m_\kappa(\phi)] \tag{2.48}$$

Then the polynomial relationship (2.41) can be written as

$$h(\phi)y(t) = g'(\phi)u(t) \tag{2.49}$$

while the transfer function relationship (2.44) becomes

$$y(t) = m'(\phi)u(t) \tag{2.50}$$

2.2.3. Multiple-Input Multiple-Output Systems

Consider now a system with multiple inputs $u_1(t)$, $u_2(t)$, ..., $u_\kappa(t)$ and multiple outputs $y_1(t)$, $y_2(t)$, ..., $y_\mu(t)$. For each subsystem, that is each output $y_i(t)$, $i=1\ldots\mu$, there is a polynomial input-output relationship

$$h_i(\phi)y_i(t) = g_{i.}(\phi)u(t) \qquad i=1\ldots\mu \tag{2.51}$$

where

$$g_{i.}(\phi) = [g_{i1}(\phi) \quad g_{i2}(\phi) \quad \ldots \quad g_{i\kappa}(\phi)] \qquad i=1\ldots\mu \tag{2.52}$$

and

$$h_i(\phi) = 1 + h_i^1\phi^{-1} + h_i^2\phi^{-2} + \ldots + h_i^{\nu i}\phi^{-\nu i} \qquad i=1\ldots\mu \tag{2.53}$$

$$g_{ij}(\phi) = g_{ij}^o + g_{ij}^1\phi^{-1} + \ldots + g_{ij}^{\nu i}\phi^{-\nu i} \qquad i=1\ldots\mu,\ j=1\ldots\kappa \tag{2.54}$$

The subsystem orders ν_i (denoted in the equations as νi) may be different. Within each subsystem, the polynomials $h_i(\phi)$ and $g_{i1}(\phi) \ldots g_{i\kappa}(\phi)$ are relative prime. The transfer function form of the i-th equation is

$$y_i(t) = m_{i.}(\phi)u(t) \qquad i=1\ldots\mu \tag{2.55}$$

where

$$m_{i.}(\phi) = [m_{i1}(\phi) \quad m_{i2}(\phi) \quad \ldots \quad m_{i\kappa}(\phi)] \qquad i=1\ldots\mu \tag{2.56}$$

and

$$m_{ij}(\phi) = g_{ij}(\phi)/h_i(\phi) \qquad i\ldots\mu,\ j=1\ldots\kappa \tag{2.57}$$

Let us introduce the following vector-matrix notation

$$y(t) = [y_1(t) \quad y_2(t) \quad \ldots \quad y_\mu(t)]' \tag{2.58}$$

$$H(\phi) = Diag[h_1(\phi) \quad h_2(\phi) \ldots \; h_\mu(\phi)] \tag{2.59}$$

$$G(\phi) = \begin{bmatrix} g_{1.}(\phi) \\ g_{2.}(\phi) \\ \cdot \\ \cdot \\ g_{\mu .}(\phi) \end{bmatrix} \tag{2.60}$$

$$M(\phi) = \begin{bmatrix} m_{1.}(\phi) \\ m_{2.}(\phi) \\ \cdot \\ \cdot \\ m_{\mu .}(\phi) \end{bmatrix} \tag{2.61}$$

With this, the set of polynomial input-output equations (2.51) can be written in the concise form

$$H(\phi)y(t) = G(\phi)u(t) \tag{2.62}$$

while the set of transfer function equations (2.55) becomes

$$y(t) = M(\phi)u(t) \tag{2.63}$$

Sometimes it is more advantageous to write the polynomial matrix equation (2.62) as

$$\bar{h}(\phi)y(t) = \bar{G}(\phi)u(t) \tag{2.64}$$

where $\bar{h}(\phi)$ is the *least* common multiple of $h_1(\phi) \ldots h_\mu(\phi)$ and where the rows of $\bar{G}(\phi)$ are obtained as $\bar{g}_{i.}(\phi)=g_{i.}(\phi)\bar{h}(\phi)/h_i(\phi)$. Of course, the polynomials appearing in any particular row of (2.64) are not relative prime. As it will be shown later, this form is related to the input-output relationship as derived from a state-space representation. The degree $\bar{\nu}$ of the polynomial $\bar{h}(\phi)$ will be referred to as the *apparent order* of the multiple-input multiple-output system. This, as it will be seen in Section 2.6, may be lower than the *true system order* ν. The poles and zeroes of the system will also be defined in Section 2.6.

The output $y(t)$ can be expressed from Eq. (2.62), yielding

$$y(t) = H^{-1}(\phi)G(\phi)u(t) \tag{2.65}$$

Similarly, from Eq. (2.64)

$$y(t) = \frac{1}{\tilde{h}(\phi)} \tilde{G}(\phi)u(t) \tag{2.66}$$

A comparison with (2.63) suggests

$$M(\phi) = H^{-1}(\phi)\, G(\phi) = \frac{1}{\tilde{h}(\phi)} \tilde{G}(\phi) \tag{2.67}$$

This of course is easy to verify. $H^{-1}(\phi)$ is diagonal and its i-th element is $1/h_i(\phi)$; this multiplies into the i-th row of $G(\phi)$, yielding $g_{ij}(\phi)/h_i(\phi)=m_{ij}(\phi)$ for the ij-th element of the product. Also, $\tilde{g}_{ij}(\phi)/\tilde{h}(\phi)=g_{ij}(\phi)/h_i(\phi)=m_{ij}(\phi)$.

2.3. CONVENTIONAL INPUT-OUTPUT REPRESENTATIONS AND SYSTEM STABILITY

In this section, we will briefly discuss two techniques, conventionally used to represent the input-output relationship of linear time-invariant discrete dynamic systems, and will show how these are related to the shift operator approach. First the concept of the *impulse response*, and the *discrete convolution* representation will be discussed and linked to our previous results. Then z-transformation, a technique widely used in the analysis and design of discrete-time systems, will be reviewed. It will be shown that z-transformation leads to the same discrete transfer function as the one we obtained earlier with the shift operator. Using z-transformation, the *response* of linear discrete systems will be derived in closed form. The analysis of the response will then lead naturally to the definition of *system stability*. The discussion here will be limited to single-input single-output systems; generalization to multivariable situations follows naturally.

2.3.1. Discrete Impulse Response and Convolution

Define the *discrete impulse function* $\delta(t)$, as shown in Figure 2.4.a,

$$\delta(t) = \begin{cases} 1 & \text{for } t=0 \\ 0 & \text{otherwise} \end{cases} \tag{2.68}$$

The response of a system as the discrete (unit) impulse function is applied to its input is the system's *impulse response* $m(t)$ (Figure 2.4.b). The discrete impulse response uniquely characterizes the input-output behavior of the linear time-invariant system.

FIGURE 2.4.a. Discrete impulse function. FIGURE 2.4.b. Impulse response.

Any discrete (input) signal $u(t)$ can be described with the unit impulse as

$$u(t) = \sum_{t'=0}^{\infty} u(t')\, \delta(t - t') \tag{2.69}$$

where $\delta(t - t')$ is the impulse function shifted by t'. (Note that it is assumed here that $u(t)=0$ for $t<0$, which is not a real restriction since "zero time" can be chosen arbitrarily.) It follows from the time invariance of the system that its response to an input $\delta(t - t')$ is $m(t - t')$, the shifted impulse response. Further it follows from linearity that the response to a weighted sum of shifted impulses is the appropriate weighted sum of shifted impulse responses. That is, the output in response to an arbitrary input $u(t)$ is

$$y(t) = \sum_{t'=0}^{\infty} u(t')\, m(t - t') \tag{2.70}$$

Equation (2.70) is one form of *discrete convolution.*

With slight re-arrangements and re-interpretations of (2.70), several other forms can be obtained. Notice first that, due to causality, $m(t) = 0$ for $t<0$. Thus the terms $t'>t$ in (2.70) are superfluous, yielding

$$y(t) = \sum_{t'=0}^{t} u(t')\, m(t - t') \tag{2.71}$$

Further, re-naming the time variables as $t - t'=k$, $t'=t - k$ leads to

$$y(t) = \sum_{k=0}^{t} u(t - k)\, m(k) \tag{2.72}$$

Finally, re-interpreting the discrete time function $m(k)$ as a series of coefficients m_k results in

$$y(t) = \sum_{k=0}^{t} m_k u(t - k) \tag{2.73}$$

This can also be written, with the shift operator, as

$$y(t) = \sum_{k=0}^{t} m_k \phi^{-k} u(t) \tag{2.74}$$

2.3.2. z-Transformation

The (one-sided) z-transform of the discrete-time sequence $x(0)$, $x(1)$, $x(2)$, ... , $x(t)$, ..., with the implied assumption that $x(t)=0,\ t<0$, is defined as

$$X(z) = Z[x(t)] = \sum_{t=0}^{\infty} x(t)\, z^{-t} \tag{2.75}$$

We will utilize some important properties of the z-transform, all following directly from the above definition.

1. *Linearity property.*

$$Z[c_1 x_1(t) + c_2 x_2(t) + \ldots] = c_1 X_1(z) + c_2 X_2(z) + \ldots \tag{2.76}$$

2. *Shift property.*

$$Z[x(t-k)] = \begin{cases} z^{-k} X(z) & \text{for } k \geq 0 \\ z^{-k} X(z) - \sum_{t=0}^{-k-1} x(t)\, z^{-k-t} & \text{for } k<0 \end{cases} \tag{2.77}$$

3. *Transforms of a few elementary functions.*

$$Z[\delta(t)] = 1 \tag{2.78}$$

$$Z[\epsilon(t)] = 1/(1-z^{-1}) = z/(z-1) \tag{2.79}$$

$$Z[a^t \epsilon(t)] = 1/(1-az^{-1}) = z/(z-a) \tag{2.80}$$

$$Z[ta^t \epsilon(t)] = az^{-1}/(1-az^{-1})^2 = za/(z-a)^2 \tag{2.81}$$

where

$$\epsilon(t) = \begin{cases} 0 & \text{for } t<0 \\ 1 & \text{for } t \geq 0 \end{cases} \tag{2.82}$$

is the *discrete unit step function* and a is a real or complex number.

Consider now the difference equation (2.27). Apply z-transformation to both sides of the equation and make use of the linearity and shift property. This yields the following relationship between the transform $U(z)$ of the input and the transform $Y(z)$ of the output:

$$Y(z) + h^1 z^{-1} Y(z) + h^2 z^{-2} Y(z) + \dots + h^\upsilon z^{-\upsilon} Y(z)$$
$$= g^d z^{-d} U(z) + g^{d+1} z^{-d-1} U(z) + \dots + g^{d+\iota} z^{-d-\iota} U(z) \tag{2.83}$$

Note that zero initial conditions are implicitly assumed here. (2.83) can be written as

$$H(z)Y(z) = G(z)U(z) \tag{2.84}$$

where

$$H(z) = 1 + h^1 z^{-1} + h^2 z^{-2} + \dots + h^\upsilon z^{-\upsilon} \tag{2.85a}$$

$$G(z) = z^{-d} (g^d + g^{d+1} z^{-1} + \dots + g^{d+\iota} z^{-\iota}) \tag{2.85b}$$

Finally, expressing $Y(z)$ from (2.84) yields

$$Y(z) = \frac{G(z)}{H(z)} U(z) = M(z)\, U(z) \tag{2.86}$$

where

$$M(z) = G(z)/H(z) \tag{2.87}$$

is the discrete transfer function given in terms of the z-transform variable. Now comparison of Eqs. (2.85) with (2.32) reveals that $h(\phi) = H(z)_{z \leftarrow \phi}$ and $g(\phi) = G(z)_{z \leftarrow \phi}$, implying

$$m(\phi) = M(z)_{z \leftarrow \phi} \tag{2.88}$$

That is, the discrete transfer function in terms of the shift operator ϕ or of the transform variable z are formally the same and can be obtained from one another by direct substitution.

An important result can be obtained by applying the z-transform input-output relationship (2.86) to the situation when the system input is the impulse function. Change the notation for the impulse response temporarily to $m'(t)$ and keep $M(z)$ for the discrete transfer function, as derived from the difference equation. Then $Y(z) = M'(z)$ and $U(z) = \Delta(z)$, where $\Delta(z) = Z[\delta(t)]$. Thus $M'(z) = M(z)\Delta(z)$. But $\Delta(z) = 1$, yielding $M'(z) = M(z)$, that is, the discrete transfer function $M(z)$ is the z-transform of the discrete impulse response $m(t)$. (We chose $m(t)$ to denote the impulse response already with this in mind.) One important implication is that the weighting polynomial in (2.74), $\Sigma_{k=0}^{\iota} m_k \phi^{-k}$ (actually $\Sigma_{k=0}^{\infty} m_k \phi^{-k}$), can be obtained from $m(\phi) = g(\phi)/h(\phi)$ by long division.

2.3.3. System Response

The method of z-transformation offers a simple way to obtain the response of a linear discrete system in closed form. The particular approach we will use here is the *partial fraction expansion* method. The essence of this approach is the decomposition of the signal's z-transform into simple components (partial fractions) which are easily recognizable as transforms of elementary discrete time functions. This is not only a convenient technique to obtain numerical solutions for concrete systems but it also facilitates an insight into some fundamental system properties. We will first study the impulse response of the system, obtained as the inverse z-transform of the discrete transfer function, then will extend the results to systems with arbitrary inputs.

There are some technicalities that need to be kept in mind when applying the partial fractions technique to discrete systems. Namely:

1. The technique is easier to apply to z-transforms expressed in terms of positive powers of z. This simply implies that the transform, in our case the discrete transfer function, needs to be converted into z-form as $M(z)=G^+(z)/H^+(z)$.

2. The expansion can always be performed by the comparison of coefficients; a more direct technique however, which we will use below, requires that the transform be strictly proper, that is, the degree of the numerator be smaller than that of the denominator. Therefore, if $G^+(z)$ has full degree, a preliminary step is necessary in which $M(z)$ is converted as

$$M(z) = \frac{G^+(z)}{H^+(z)} = g^o + \frac{G^+(z) - g^o H^+(z)}{H^+(z)} \tag{2.89}$$

which makes the right hand side expression strictly proper.

3. Since the z-transform of most elementary functions has z in its numerator, the inverse transformation is performed more conveniently if z can be factored out of the numerator. If this is not the case, the z factor needs to be introduced as

$$M(z) = z^{-1}[zM(z)] \tag{2.90}$$

Assume that $M(z)$ is given in terms of z, it is strictly proper and z can be factored out from the numerator as $G^+(z)=z[G^+(z)/z]$. Assume also that it has only distinct poles $p_1, p_2, \ldots, p_\nu$. Then $M(z)$ can be expanded as

$$M(z) = \frac{G^+(z)}{H^+(z)} = z\frac{G^+(z)/z}{(z-p_1)(z-p_2)\ldots(z-p_\nu)}$$

$$= \frac{\pi_1 z}{(z - p_1)} + \frac{\pi_2 z}{(z - p_2)} + \ldots + \frac{\pi_\nu z}{(z - p_\nu)} \tag{2.91}$$

Here $\pi_j z/(z - p_j)$, $j = 1 \ldots \nu$, are the partial fractions of $M(z)$ and the coefficients π_j can be obtained as

$$\pi_j = (z - p_j) \frac{G^+(z)/z}{H^+(z)} \Bigg|_{z=p_j} \tag{2.92}$$

Now the comparison of (2.91) to (2.80) reveals that the impulse response is

$$m(t) = Z^{-1}[M(z)] = [\pi_1 p_1^t + \pi_2 p_2^t + \ldots + \pi_\nu p_\nu^t]\epsilon(t) \tag{2.93}$$

If the pre-manipulation (2.90) is necessary then the coefficients are computed with $G^+(z)$ in (2.92) (instead of $G^+(z)/z$) and the z^{-1} multiplier causes a one step shift in the time-response, yielding

$$m(t) = [\pi_1 p_1^{t-1} + \pi_2 p_2^{t-1} + \ldots + \pi_\nu p_\nu^{t-1}]\epsilon(t - 1) \tag{2.94}$$

Finally, if the pre-manipulation (2.89) is necessary then the constant term g^o in the transform yields the additional term $Z^{-1}[g^o] = g^o\delta(t)$ in the time response and the π coefficients are computed with $[G^+(z) - g^o H^+(z)]/z$ in (2.92).

The discrete exponential time-functions $p_j^t\,\epsilon(t)$, called the *system modes*, characterize the nature of the system response. They are completely determined by the system poles and it is only their weight in the response that depends on the system zeroes. A system mode belonging to a real pole is a real exponential function. Complex poles always appear in conjugate pairs and are accompanied by conjugate complex π coefficients; a pair of complex poles yields a pair of complex exponential modes which combine to form a real exponential-sinusoidal signal. Consider a pair of complex poles $p_{j,\,j+1} = \sigma e^{\pm i\varphi}$, with coefficients $\pi_{j,\,j+1} = \xi \pm i\eta$ (where i is the imaginary unit); it can be shown by elementary trigonometry that the combined signal is

$$\pi_j p_j^t + \pi_{j+1} p_{j+1}^t = 2c\sigma^t \cos(\varphi t + \gamma) \tag{2.95}$$

where $c = \sqrt{(\xi^2 + \eta^2)}$ and $\gamma = tn^{-1}(\eta/\xi)$.

If some pole p_j has a multiplicity of ρ then the partial fraction expansion contains the terms $\pi_{j1} z/(z - p_j)$, $\pi_{j2} z/(z - p_j)^2$, ..., $\pi_{j\rho}(z - p_j)^\rho$. (The computation of the π coefficients is more complicated now than in the case of distinct poles.) By inverse z-transformation, these then lead to the time functions (system modes) $p_j^t\,\epsilon(t)$, $tp_j^t\,\epsilon(t)$, ..., $t^{\rho-1}p_j^t\,\epsilon(t)$.

Consider now the response of the system to a general input $u(t)$. By Eq. (2.86), the z-transform of the output is $Y(z)=M(z)U(z)$. Thus the denominator of $Y(z)$ has all the poles of the system as its roots, plus the roots of the denominator of $U(z)$. Partial fraction expansion and inverse z-transformation can be applied the same way as above. The system response will contain all system modes, plus additional elementary time functions present in the input ("input modes"). The weight of the various system modes in the response will also depend on the input signal. Note that if an input "pole" is identical with a system pole, they combine to act as a multiple pole resulting in output modes present neither in the impulse response nor in the input.

2.3.4. System Stability

Stability is a very important system property. Loosely speaking, a system is stable if its output stays bounded (finite) at all times. There are several ways of defining stability in more rigorous terms; we will introduce two definitions here which follow naturally from the foregoing discussion of system response.

Stable impulse response systems. A system has a stable impulse response if all system modes are bounded. A mode $p_j^t\,\epsilon(t)$, belonging to a real pole p_j is bounded if $|p_j| \leq 1$. Similarly, for a pair of complex poles, the combined response computed in (2.95) is bounded if $\sigma = |p_j| = |p_{j+1}| \leq 1$. Note that $|p_j| = 1$ yields a system mode with constant magnitude (constant signal or periodic signal with constant magnitude). In the case of multiple poles, however, the modes $tp_j^t\epsilon(t)$, ..., $t^{\rho-1}p_j^t\epsilon(t)$ are bounded only if $|p_j| < 1$. Thus a system has a stable impulse response if and only if

$$|p_j| \leq 1 \quad \textit{for all single poles;} \quad |p_j| < 1 \quad \textit{for all multiple poles} \qquad (2.96)$$

Bounded-input bounded-output stability. A system is bounded-input bounded-output stable if its output remains bounded in response to any bounded input. The system output contains the system modes, the input modes and the occasional interaction modes (in case an input mode is identical with a system mode). The input modes are bounded by definition. For the system modes, the stable impulse response considerations apply. The interactions, however, require special attention; since the bounded input may contain constant magnitude signal components, any system mode that may combine with these has to be excluded. Thus the system is bounded-input bounded-output stable if and only if

$$|p_j| < 1 \quad \textit{for all poles} \qquad (2.97)$$

2.4. DISCRETE REPRESENTATION OF CONTINUOUS SYSTEMS

As we mentioned in the Introduction of this chapter, the physical systems we monitor and control are usually characterized by continuous-time operation. Thus their natural representation involves continuous-time models, which describe how the continuous output variables depend on the continuous input variables. However, the computing equipment which interacts with these physical systems, and on which the monitoring and control algorithms are implemented, is digital. Such digital algorithms operate on discretized representations of the variables and therefore require a discrete "equivalent" model of the continuous plant.

In this section, we will briefly review the modeling of continuous-time systems. Then we will describe a technique to obtain the "step equivalent" discrete representation for such systems. There are other equivalent representations which rely on different approximate assumptions; these are primarily used in the design of digital filters and controllers and lie outside the scope of this book.

2.4.1. Continuous-Time Systems

In this subsection, we will review the methods used to represent the input-output behavior of continuous-time linear time-invariant dynamic systems. The discussion will include the differential equation, the continuous-time transfer function obtained by Laplace transformation, the impulse and step response and the continuous-time convolution. Our treatment will be limited to single-input single-output systems.

Consider a continuous-time system with a continuous input signal $u(\tau)$ and a continuous output signal $y(\tau)$, where τ is the continuous time variable. The most straightforward model of a such a system is its differential equation

$$y(\tau) + h^{1}\frac{dy(\tau)}{d\tau} + \ldots + h^{\upsilon}\frac{d^{\upsilon}y(\tau)}{d\tau^{\upsilon}} = g^{o}u(\tau) + g^{1}\frac{du(\tau)}{d\tau} + \ldots + g^{\iota}\frac{d^{\iota}u(\tau)}{d\tau^{\iota}} \tag{2.98}$$

Here $h^{1}\ldots h^{\upsilon}$ and g^{o}, $g^{1}\ldots g^{\iota}$ are coefficients. υ is the order of the system; for any causal system, $\upsilon \geq \iota$. Usually the differential equation is not handled in its original form but is replaced by its Laplace transform.

The Laplace transform $X(s)$ of a continuous-time variable $x(\tau)$ is defined as

$$X(s) = L\,[x(\tau)] = \int_{\tau=0}^{\infty} x(\tau)e^{-s\tau}d\tau \tag{2.99}$$

where s is a complex variable. The Laplace transform is a linear transform, that is, it has the following linearity property:

$$L[c_1 x_1(\tau) + c_2 x_2(\tau) + \dots] = c_1 X_1(s) + c_2 X_2(s) + \dots \tag{2.100}$$

Further, the transform of the integral of a time-function $x(\tau)$ is obtained as:

$$L[\int_{\tau'=0}^{T} x(\tau')d\tau'] = \frac{X(s)}{s} \tag{2.101}$$

while the transform of its k-th derivative is

$$L[d^k x(\tau)/d\tau^k] = s^k X(s) - \sum_{j=0}^{k-1} s^{k-j-1}[d^j x(\tau)/d\tau^j]_{\tau=0} \tag{2.102}$$

If the initial conditions are zero, this latter equation simplifies to

$$L[d^k x(\tau)/d\tau^k] = s^k X(s) \tag{2.103}$$

Take the Laplace transform of the differential equation (2.98) and utilize the linearity property and, assuming zero initial conditions, Eq. (2.103). This yields

$$H(s)Y(s) = G(s)U(s) \tag{2.104}$$

where

$$H(s) = 1 + h^1 s + \dots + h^\upsilon s^\upsilon \tag{2.105a}$$

$$G(s) = g^o + g^1 s + \dots + g^\iota s^\iota \tag{2.105b}$$

From this

$$Y(s) = M(s)U(s) \tag{2.106}$$

where

$$M(s) = G(s)/H(s) \tag{2.107}$$

is the continuous-time transfer function.

The continuous-time step-function $\epsilon(\tau)$ (Figure 2.5.a) is usually defined as

$$\epsilon(\tau) = \begin{cases} 0 & \text{for } \tau<0 \\ 1 & \text{for } \tau\geq 0 \end{cases} \tag{2.108}$$

while the continuous-time impulse function $\delta(\tau)$ (Figure 2.5.b) is

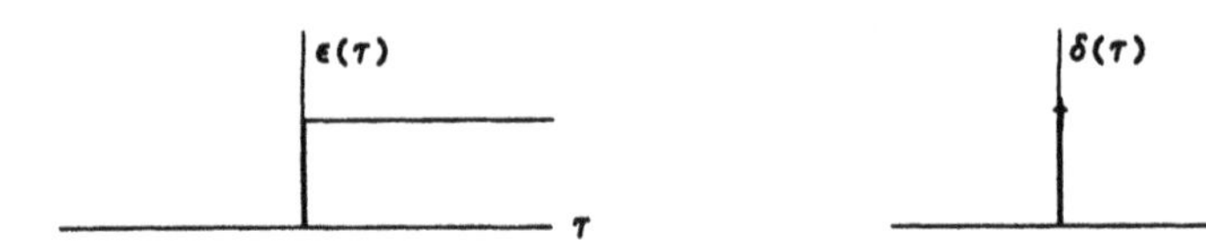

FIGURE 2.5.a. Step function. FIGURE 2.5.b. Impulse function.

$$\delta(\tau) = d\epsilon(\tau)/d\tau \tag{2.109}$$

$\delta(\tau)$, also called the ***Dirac-function***, is infinitely tall and infinitely narrow while its area is 1. It has the important property that

$$\int_{\tau=\tau_1}^{\tau_2} x(\tau)\, \delta(\tau - \tau_0)\, d\tau = \begin{cases} x(\tau_0) & \text{if } \tau_1 \leq \tau_0 < \tau_2 \\ 0 & \text{otherwise} \end{cases} \tag{2.110}$$

The response of the system to the continuous (unit) impulse function is the *continuous-time impulse response* $m(\tau)$, which completely characterizes the linear time-invariant continuous-time system. The response to an arbitrary input $u(\tau)$ can be expressed with the *continuous-time convolution*

$$y(\tau) = \int_{\tau'=0}^{\tau} m(\tau - \tau')u(\tau')d\tau' = \int_{\tau'=0}^{\tau} m(\tau')u(\tau - \tau')d\tau' \tag{2.111}$$

It follows from (2.99) and (2.110) that $L[\delta(\tau)]=1$. Thus the application of the transfer function relationship (2.106) to $u(\tau)=\delta(\tau)$, $y(\tau)=m(\tau)$ reveals that the continuous-time transfer function $M(s)$ is the Laplace transform of the continuous-time impulse response $m(\tau)$.

Alternatively, the (unit) *step response* $n(\tau)$, the response of the system to a (unit) step input, can also be used. Since the step function is the integral of the impulse function, the linearity of the system implies that

$$n(\tau) = \int_{\tau'=0}^{\tau} m(\tau')d\tau' \tag{2.112}$$

It follows from the above and (2.101) that the Laplace transform of the step response, $N(s)$, is related to the transfer function $M(s)$ as

$$N(s) = M(s)/s \tag{2.113}$$

2.4.2. Step-Equivalent Discrete Transfer Function

As mentioned before, the discrete control and monitoring algorithms operate on discrete (discretized) representations of the variables acting on or arising from the continuous plant. Strictly speaking, there are two slightly different ways how these variables are present in the system:

1. The continuous plant (unit) is controlled by a digital controller. Then the control input signal is discrete and is passed to the continuous plant by means of a holding device. The output of the plant is a continuous signal which is sampled by the computer for use in the algorithms, including feedback to the control algorithm.
2. The input signal comes from an analog controller, another technological unit or from "outside" and is continuous. Such input signals are sampled by the computer and presented to the algorithms in discrete (discretized) form, the same way as the also sampled output signal.

In either situation, the algorithms running on the computer deal with discrete input and output signals. Thus they need a discrete equivalent model representation of the continuous plant, describing how the samples taken from the continuous output are related to the discrete or sampled input signal. In case 1, such a discrete equivalent representation may be exact since the discrete input, via the holding device, completely determines the input signal acting on the plant. In case 2, however, the discrete "equivalent" can only be approximate since the samples taken from the physically continuous input signal do not provide complete information about the signal which acts on the plant input.

Technically, we will use the same approach no matter which case we are dealing with. We will assume that the input signal acting on the plant is constant (maintains the value of the latest sample) between two consecutive samples. The same assumption can be presented also as passing the continuous input signal through a sampler and a zero-order hold (Figure 2.6). Of course, this assumption describes exactly what happens in situation 1 (provided the physical holding device is of zero order, which is usually the case). In situation 2, the stepwise constant assumption amounts to the rectangular approximation of the continuous input signal. The discrete model describing the input-output relationship under the above assumption is usually referred to as the *zero-order-hold discrete equivalent* or *step-equivalent* model of the continuous system. In the sequel, this model will be derived.

FIGURE 2.6. Continuous block in discrete system.

The stepwise continuous input signal $\bar{u}(\tau)$, applied to the continuous block is

$$\bar{u}(\tau) = \sum_{t'=0}^{\infty} u(t')[\epsilon(\tau - t'\Delta) - \epsilon(\tau - t'\Delta - \Delta)] \tag{2.114}$$

Here Δ is the sampling interval. $u(t')$ is the discrete signal input to the physical holding device (case 1) or the samples taken from the continuous input signal, $u(\tau)\mid_{\tau=t'\Delta}$ (case 2). The response of the continuous block to the shifted step function $\epsilon(\tau - t'\Delta)$ is the shifted step response $n(\tau - t'\Delta)$. The complete response, due to linearity, is

$$y(\tau) = \sum_{t'=0}^{\infty} u(t')[n(\tau - t'\Delta) - n(\tau - t'\Delta - \Delta)] \tag{2.115}$$

The discrete output $y(t)$, obtained by sampling the continuous output $y(\tau)$ at $\tau=t\Delta$ is

$$y(t) = \sum_{t'=0}^{\infty} u(t')[n(t\Delta - t'\Delta) - n(t\Delta - t'\Delta - \Delta)]$$

$$= \sum_{t'=0}^{t} u(t')n(t\Delta - t'\Delta) - \sum_{t'=0}^{t-1} u(t')n(t\Delta - t'\Delta - \Delta) \tag{2.116}$$

Here we took into account also the causality of the system. Introduce $k=t - t'$ in the first sum and $k=t - t' - 1$ in the second. Then

$$y(t) = \sum_{k=0}^{t} u(t - k)n(k\Delta) - \sum_{k=0}^{t-1} u(t - k - 1)n(k\Delta) \tag{2.117}$$

The samples $n(k\Delta)$ taken from the continuous step response can be reinterpreted as coefficients n_k, leading to

$$y(t) = \sum_{k=0}^{t} n_k u(t - k) - \sum_{k=0}^{t-1} n_k u(t - k - 1) \tag{2.118}$$

Notice that the second sum is a shifted version of the first, with $u(-1)=0$. Applying the shift operator to the entire expression finally yields

$$y(t) = (1 - \phi^{-1}) \sum_{k=0}^{t} n_k \phi^{-k} u(t) \tag{2.119}$$

The comparison of this latest expression to Eq. (2.74) reveals that we are dealing here with a discrete convolution type relationship. Indeed, the difference with respect to (2.74), apart from the multiplier $(1 - \phi^{-1})$, is that now the weights are samples from a step response $n(\tau)$. Thus (2.120) can also be written as

$$y(t) = (1 - \phi^{-1})N(\phi)u(t) \tag{2.120}$$

where $N(\phi)=N(z)_{z\leftarrow\phi}$ and $n(t)=n(\tau)_{\tau=t\Delta}$. Since $n(\tau)$ is the integral of $m(\tau)$,

$$N(z) = Z\,[L^{-1}\{M(s)/s\}] \tag{2.121}$$

where L^{-1} denotes the inverse Laplace transform.

2.5. MATRIX ALGEBRA FUNDAMENTALS

In this section, some fundamental results of matrix algebra will be summarized. The features of interest concern the rank of a matrix and the linear dependence of vectors, and the eigenvalues and eigenvectors of square matrices. These results will then be utilized in the following section, which deals with the state-space representation of linear systems, and throughout the rest of the book.

2.5.1. Rank and Linear Independence

In this subsection, the basic concepts of the *rank of a matrix* and the *linear independence of vectors* will be reviewed. We will consider vectors/matrices the elements of which are constants (not polynomials) and ones which have polynomial elements. No proofs and derivations will be provided here and in the next subsection; the interested Reader is referred to texts on linear algebra.

There are several equivalent definitions (and tests) of the rank of a matrix, including

- the size of the largest square minor with non-zero determinant
- the number of linearly independent rows
- the number of linearly independent columns.

Constant matrices and vectors. Consider a constant matrix $\boldsymbol{M}$

$$\boldsymbol{M} = \begin{bmatrix} m_{11} \dots m_{1\kappa} \\ \dots \\ m_{\mu 1} \dots m_{\mu\kappa} \end{bmatrix} = \begin{bmatrix} \boldsymbol{m}_{1\cdot} \\ \dots \\ \boldsymbol{m}_{\mu\cdot} \end{bmatrix} = [\, \boldsymbol{m}_{\cdot 1} \dots \boldsymbol{m}_{\cdot\kappa} \,] \tag{2.122}$$

For such a constant matrix, all the determinants are constant. The rows $\boldsymbol{m}_{i\cdot}$, $i=1\dots\mu$, and columns $\boldsymbol{m}_{\cdot j}$, $j=1\dots\kappa$, are vectors with constant elements. A set of constant vectors (for example the columns) is linearly independent if no $\beta_1 \dots \beta_\kappa$ set of constants, at least one of them non-zero, exists that makes

$$\beta_1 \boldsymbol{m}_{\cdot 1} + \dots + \beta_\kappa \boldsymbol{m}_{\cdot\kappa} = 0 \tag{2.123}$$

The above linear independence condition is violated if
- one (or more) of the vectors is zero
- one (or more) of the vectors can be expressed as a linear combination of the others.

Polynomial matrices and vectors. Consider now a polynomial matrix $M(z)$

$$M(z) = \begin{bmatrix} m_{11}(z) \ldots m_{1\kappa}(z) \\ \ldots \\ m_{\mu 1}(z) \ldots m_{\mu\kappa}(z) \end{bmatrix} = \begin{bmatrix} m_{1\cdot}(z) \\ \ldots \\ m_{\mu\cdot}(z) \end{bmatrix} = [\, m_{\cdot 1}(z) \ldots m_{\cdot\kappa}(z) \,] \tag{2.124}$$

The determinants formed from the elements of a polynomial matrix are, in general, polynomials. Also, the rows and columns of a polynomial matrix, in general, have polynomial elements. The rank the matrix $M(z)$ has for almost all values of the variable z is called its *normal rank.* $M(z)$ has a defect in its normal rank if the determinants of its largest minors are zero for all values of z or, equivalently, if its rows/columns are linearly dependent for all values of z. Linear dependence for polynomial vectors is meant in the polynomial sense. The set of columns $m_{\cdot 1}(z) \ldots m_{\cdot\kappa}(z)$, for example, is linearly independent if no $\beta_1(z) \ldots \beta_\kappa(z)$ set of polynomials, at least one of them non-zero, exists that makes

$$\beta_1(z) m_{\cdot 1}(z) + \ldots + \beta_\kappa(z) m_{\cdot\kappa}(z) = 0 \tag{2.125}$$

Constant coefficients are included as a special case of (2.125):

$$\beta_1 m_{\cdot 1}(z) + \ldots + \beta_\kappa m_{\cdot\kappa}(z) = 0 \tag{2.126}$$

The *exceptional rank* of a polynomial matrix, at certain $z=z_j$ values, may be lower than its normal rank. These special z_j values can be computed from the determinant polynomials or from the row/column properties. With the $z=z_j$ substitution, $M(z_j)$ becomes a constant matrix.

A square polynomial matrix $M(z)$ is called *unimodular* if its determinant $|M(z)|$ is constant (does not depend on z). The inverse of a unimodular matrix, $M^{-1}(z)=AdjM(z)/|M(z)|$, is also polynomial. Further, the inverse is also unimodular, since $[M^{-1}(z)]^{-1}=AdjM^{-1}(z)/|M^{-1}(z)|=M(z)$, which is polynomial.

Further rank relations. If the matrices M and N have compatible dimensions then

$$Rank(MN) = Rank(N'M') \leq Min(RankM, RankN) \tag{2.127}$$

There are two special cases of particular interest:

a. when N is square and has full rank then

$$Rank(MN) = RankM \tag{2.128a}$$

b. for any rectangular matrix M,

$$Rank(M'M) = Rank(MM') = RankM \tag{2.128b}$$

Another useful result concerns the determinant of block matrices. Let M be

$$M = \begin{bmatrix} P & Q \\ R & S \end{bmatrix} \tag{2.129}$$

where P and S are square matrices and P has full rank. Then

$$|M| = |P|\,|S - RP^{-1}Q| \tag{2.130}$$

2.5.2. Eigenvalues and Eigenvectors

In this subsection, we will review the basic concepts of eigenvalues and eigenvectors, spectral decomposition and the Cayley-Hamilton theorem.

Eigenvalues. Consider a $\nu \cdot \nu$ (square) matrix A. The determinant

$$|\lambda I - A| = \alpha_0 + \alpha_1 \lambda + \ldots + \alpha_{\nu-1} \lambda^{\nu-1} + \lambda^{\nu} \tag{2.131}$$

is called the *characteristic polynomial* of the matrix. The solutions λ_j of the *characteristic equation*

$$|\lambda I - A| = 0 \tag{2.132}$$

are the *eigenvalues* of matrix A. The eigenvalues are real or complex scalars, the latter appearing in conjugate pairs. With all simple (distinct) eigenvalues, the characteristic polynomial can be written as

$$|\lambda I - A| = (\lambda - \lambda_1)(\lambda - \lambda_2) \ldots (\lambda - \lambda_\nu) \tag{2.133}$$

Multiple eigenvalues appear as $(\lambda - \lambda_j)^{\pi_j}$ factors, where π_j is the *algebraic multiplicity* of λ_j ; the sum of algebraic multiplicities (including $\pi_j = 1$ for simple eigenvalues) is ν. The *geometric multiplicity* (or degeneracy) of λ_j is defined as $\rho_j = \nu - Rank(\lambda_j \boldsymbol{I} - \boldsymbol{A})$, and $1 \leq \rho_j \leq \pi_j$.

Eigenvectors. A vector $\boldsymbol{\xi}_j$ which satisfies the equation

$$\lambda_j \boldsymbol{\xi}_j = \boldsymbol{A}\boldsymbol{\xi}_j \tag{2.134}$$

is a *right-side eigenvector* of $\boldsymbol{A}$, belonging to the eigenvalue λ_j. Eq. (2.134) can also be written as

$$(\lambda_j \boldsymbol{I} - \boldsymbol{A})\boldsymbol{\xi}_j = 0 \tag{2.135}$$

In fact, (2.132) expresses the requirement that, for the homogeneous equation (2.135) to yield non-trivial solutions for $\boldsymbol{\xi}_j$, the determinant of $(\lambda_j \boldsymbol{I} - \boldsymbol{A})$ has to be zero. The eigenvectors are not unique; if $\boldsymbol{\xi}_j$ is an eigenvector, so is $\beta\boldsymbol{\xi}_j$, where β is any scalar. For a multiple eigenvalue λ_j, there are ρ_j "true" eigen-vectors, satisfying (2.134), and $\pi_j - \rho_j$ "generalized" eigenvectors which satis-fy

$$-\boldsymbol{\xi}_{j,k-1} = (\lambda_j \boldsymbol{I} - \boldsymbol{A})\boldsymbol{\xi}_{jk} \qquad k = \rho_j + 1 \ldots \pi_j \tag{2.136}$$

where $\boldsymbol{\xi}_{j,k-1}$, $k - 1 = \rho_j$, is a true eigenvector. The ν eigenvectors, including the generalized eigenvectors if any, are *linearly independent*. They can also be *normalized*, so that the length of each vector $\| \boldsymbol{\xi}_j \|^2 = \boldsymbol{\xi}_j' \boldsymbol{\xi}_j = 1$.

A vector $\boldsymbol{\psi}_j$ which satisfies the equation

$$\boldsymbol{\psi}_j' \lambda_j = \boldsymbol{\psi}_j' \boldsymbol{A} \tag{2.137}$$

is a *left-side eigenvector* of $\boldsymbol{A}$, belonging to the eigenvalue λ_j. Eq. (2.137) can also be written as

$$\boldsymbol{\psi}_j'(\lambda_j \boldsymbol{I} - \boldsymbol{A}) = 0 \tag{2.138}$$

The left eigenvectors are not unique either; any scalar multiple is also an eigenvector. For a multiple eigenvalue, the set of left eigenvectors usually contains generalized eigenvectors; these are defined by the left-side analog of Eq. (2.136). The ν left-side eigenvectors are linearly independent and may be normalized.

The set of right-side eigenvectors and that of the left-side eigenvectors are bi-orthogonal in the sense that

$$\boldsymbol{\psi}_i' \boldsymbol{\xi}_j = \begin{bmatrix} 0 \;\; for\; i \neq j \\ \sigma_j \neq 0 \;\; for\; i=j \end{bmatrix} \qquad i=1 \ldots \nu,\; j=1 \ldots \nu \tag{2.139}$$

The eigenvectors can be bi-normalized so that $\sigma_j = 1,\; j=1 \ldots \nu$. Now form the matrices

$$\boldsymbol{\Xi} = [\boldsymbol{\xi}_1 \;\; \boldsymbol{\xi}_2 \; \cdots \; \boldsymbol{\xi}_\nu] \tag{2.140}$$

$$\boldsymbol{\Psi} = \begin{bmatrix} \boldsymbol{\psi}_1' \\ \boldsymbol{\psi}_2' \\ \cdot \\ \cdot \\ \boldsymbol{\psi}_\nu' \end{bmatrix} \tag{2.141}$$

Then, according to (2.139) and for bi-normalized sets, $\boldsymbol{\Psi} \boldsymbol{\Xi} = \boldsymbol{I}$, that is

$$\boldsymbol{\Psi} = \boldsymbol{\Xi}^{-1} \tag{2.142}$$

Spectral decomposition. Consider a square matrix $\boldsymbol{A}$ with ν distinct eigenvalues and with bi-normalized eigenvectors. Write Eq. (2.134) for $j=1 \ldots \nu$ as

$$[\boldsymbol{A}\boldsymbol{\xi}_1 \;\; \boldsymbol{A}\boldsymbol{\xi}_2 \; \ldots \; \boldsymbol{A}\boldsymbol{\xi}_\nu] = [\lambda_1 \boldsymbol{\xi}_1 \;\; \lambda_2 \boldsymbol{\xi}_2 \; \ldots \; \lambda_\nu \boldsymbol{\xi}_\nu] \tag{2.143}$$

That is

$$\boldsymbol{A}\,\boldsymbol{\Xi} = \boldsymbol{\Xi}\,\boldsymbol{\Lambda} \tag{2.144}$$

where

$$\boldsymbol{\Lambda} = Diag(\lambda_1, \lambda_2, \ldots, \lambda_\nu) \tag{2.145}$$

From this,

$$\boldsymbol{A} = \boldsymbol{\Xi}\boldsymbol{\Lambda}\boldsymbol{\Xi}^{-1} = \boldsymbol{\Xi}\boldsymbol{\Lambda}\boldsymbol{\Psi} \tag{2.146}$$

or

$$\boldsymbol{A} = \sum_{j=1}^{\nu} \lambda_j \boldsymbol{\xi}_j \boldsymbol{\psi}_j' \tag{2.147}$$

(2.146) and (2.147) are called the *spectral decomposition* of matrix $\boldsymbol{A}$.

If $\boldsymbol{A}$ has multiple eigenvalues with generalized eigenvectors then the diagonal $\boldsymbol{\Lambda}$ matrix is replaced by a block-diagonal *Jordan matrix* $\boldsymbol{J}$, so that (2.146) becomes

$$A = \Xi J \Psi \tag{2.148}$$

where

$$J = \begin{bmatrix} \lambda_1 & 0 & 0 & 0 & \dots \\ 0 & \lambda_2 & 0 & 0 & \dots \\ 0 & 0 & J_3 & 0 & \dots \\ 0 & 0 & 0 & J_4 & \dots \\ & & & & \dots \end{bmatrix} \tag{2.149}$$

is the Jordan matrix. In this, each simple eigenvalue λ_j appears diagonally. A multiple eigenvalue λ_j with $\rho_j = 1$ is represented by a "Jordan block" J_j while one with $\rho_j > 1$ is represented by a mix of diagonal elements and Jordan blocks. For example, for $\pi_j = 3$ and $\rho_j = 1$, the Jordan block has the form

$$J_j = \begin{bmatrix} \lambda_j & 1 & 0 \\ 0 & \lambda_j & 1 \\ 0 & 0 & \lambda_j \end{bmatrix} \tag{2.150}$$

Within each Jordan block subsystem, the first ξ_{jk} and the last ψ_{jk} are the true eigenvectors.

If $f(A)$ is a function of matrix A which preserves the size $\nu \cdot \nu$ then $f(A)$ has the same eigenvectors as A and its eigenvalues are $f(\lambda_j)$, $j = 1 \dots \nu$. Its spectral decomposition, for distinct eigenvalues, is obtained as

$$f(A) = \sum_{j=1}^{\nu} f(\lambda_j)\, \xi_j \psi_j' \tag{2.151}$$

Cayley-Hamilton theorem. A very important result, known as the *Cayley-Hamilton theorem*, states that any square matrix satisfies its own characteristic equation (2.132). That is

$$\alpha_0 I + \alpha_1 A + \dots + \alpha_{\nu-1} A^{\nu-1} + A^{\nu} = 0 \tag{2.152}$$

The Cayley-Hamilton theorem has many applications and implications in linear algebra. One of its consequences can be seen by expressing A^{ν} from (2.152)

$$A^{\nu} = -\alpha_0 I - \alpha_1 A - \dots - \alpha_{\nu-1} A^{\nu-1} \tag{2.153}$$

That is, A^{ν} (and any power of A beyond this) is a linear combination of $I, A, \ldots, A^{\nu-1}$.

2.6. REPRESENTATION IN THE STATE SPACE

An alternative to the input-output representation of linear time-invariant (discrete-time) dynamic systems is their description (or "realization") in the state space. The state-space representation hinges on a vector variable $x(t)=[x_1(t), x_2(t), \ldots, x_{\nu}(t)]'$, called the *state vector* or the vector of *state variables*, which characterizes the internal state of the system at any instant t. The size of the state vector, ν, is the *order of the system* (or, more accurately, the order of the particular state-space realization). The state vector is not unique for a given system; in fact, there is an infinite number of ways how it can be selected.

With the state vector, the behavior of a multiple-input multiple-output system is described by a first-order vector difference equation

$$x(t) = Ax(t-1) + Bu(t-1) \tag{2.154}$$

which we will call the state equation, and a static vector equation

$$y(t) = Cx(t) + Du(t) \tag{2.155}$$

which we will refer to as the output equation. The matrices A, B, C and D are the parameters of the state-space realization. With ν states, κ inputs and μ outputs, the size of matrix A is $\nu \cdot \nu$, matrix B is $\nu \cdot \kappa$, matrix C is $\mu \cdot \nu$ and matrix D is $\mu \cdot \kappa$. The parameter matrices in (2.154) and (2.155) are constant, reflecting the fact that the described system is *time invariant*.

Multiple-input single-output and single-input single-output systems are special cases of Equations (2.154) and (2.155). With a single output, $y(t)=y(t)$ becomes a scalar variable and the matrices C and D have only a single row. With a single input, $u(t)=u(t)$ is a scalar variable and the matrices B and D have only a single column.

The state equation (2.154) can also be written as

$$x(t+1) = Ax(t) + Bu(t) \tag{2.156}$$

The conversion from (2.154) to (2.156) may be interpreted as a $t'=t-1$ substitution.

The state-space representation is a powerful tool in the analysis of the internal properties of linear dynamic systems. Since polynomial manipulations are not involved, both algebraic treatments and numerical computations may be simpler than in the input-output framework. However, practical system models are usually first generated in input-output form, in many cases by numerical parameter identification. In this situation, conversion into state space requires an additional, sometimes not very simple operation. In particular, finding a minimum order state-space realization for a multiple-input multiple-output system is a non-trivial task.

The transformations of the state equation will be discussed first, together with the conversion to input-output form. This will be followed by the classical concepts of observability and controllability. The system matrix will be introduced, together with the useful concept of invariant zeroes. The issues of constructing minimum realizations will be addressed in the next section.

2.6.1. Transformation of the State Vector

As we mentioned earlier, the choice of the state vector for a given system is not unique. In particular, if $\boldsymbol{x}(t)$ is a state vector, so is

$$\grave{\boldsymbol{x}}(t) = \boldsymbol{T}\boldsymbol{x}(t) \tag{2.157}$$

for any square $(\nu \cdot \nu)$ and invertible transforming matrix $\boldsymbol{T}$. Then the state and output equations (2.154) and (2.155) apply to the new state vector as

$$\grave{\boldsymbol{x}}(t) = \grave{\boldsymbol{A}}\grave{\boldsymbol{x}}(t-1) + \grave{\boldsymbol{B}}\boldsymbol{u}(t-1) \tag{2.158}$$

$$\boldsymbol{y}(t) = \grave{\boldsymbol{C}}\grave{\boldsymbol{x}}(t) + \grave{\boldsymbol{D}}\boldsymbol{u}(t) \tag{2.159}$$

It can be shown by direct substitution that the parameter matrices in (2.158) and (2.159) are related to the original parameters as

$$\grave{\boldsymbol{A}} = \boldsymbol{T}\boldsymbol{A}\boldsymbol{T}^{-1} \quad \grave{\boldsymbol{B}} = \boldsymbol{T}\boldsymbol{B} \quad \grave{\boldsymbol{C}} = \boldsymbol{C}\boldsymbol{T}^{-1} \quad \grave{\boldsymbol{D}} = \boldsymbol{D} \tag{2.160}$$

Of course, the distinction expressed by the accent is meaningless outside the context of the transformation since any qualifying state vector may be the "original" one.

Of particular interest is the transformation $\boldsymbol{T}=\boldsymbol{\Xi}^{-1}$, where $\boldsymbol{\Xi}$ contains the right-side eigenvectors of $\boldsymbol{A}$, see Eq. (2.134) and (2.140). Then, if all the eigenvectors are distinct,

$$\grave{\boldsymbol{A}} = \boldsymbol{\Xi}^{-1}\boldsymbol{A}\boldsymbol{\Xi} = \boldsymbol{\Lambda} \tag{2.161}$$

and the state equation (2.158) becomes

$$\begin{bmatrix} \grave{x}_1(t) \\ \grave{x}_2(t) \\ \cdot \\ \cdot \\ \grave{x}_\nu(t) \end{bmatrix} = \begin{bmatrix} \lambda_1 & 0 \ldots 0 \\ 0 & \lambda_2 \ldots 0 \\ & \ldots \\ & \ldots \\ 0 & 0 \ldots \lambda_\nu \end{bmatrix} \begin{bmatrix} \grave{x}_1(t-1) \\ \grave{x}_2(t-1) \\ \ldots \\ \ldots \\ \grave{x}_\nu(t-1) \end{bmatrix} + \begin{bmatrix} \grave{b}_{1\cdot} \\ \grave{b}_{2\cdot} \\ \cdot\cdot \\ \cdot\cdot \\ \grave{b}_{\nu\cdot} \end{bmatrix} u(t-1) \tag{2.162}$$

That is, the transformed state variables are completely decoupled from each other and the vector equation (2.162) can be replaced by a set of uncoupled scalar equations

$$\grave{x}_j(t) = \lambda_j \grave{x}_j(t-1) + \grave{b}_{j\cdot} u(t-1) \qquad j=1 \ldots \nu \tag{2.163}$$

If A has multiple eigenvalues then

$$\grave{A} = \Xi^{-1} A\, \Xi = J \tag{2.164}$$

and the decoupling of the state variables is not complete. For example, with a simple eigenvalue λ_1 and a double eigenvalue λ_2 ($\pi_2=2$, $\rho_2=1$), the state equation (2.158) is

$$\begin{bmatrix} \grave{x}_1(t) \\ \grave{x}_2(t) \\ \grave{x}_3(t) \end{bmatrix} = \begin{bmatrix} \lambda_1 & 0 & 0 \\ 0 & \lambda_2 & 1 \\ 0 & 0 & \lambda_2 \end{bmatrix} \begin{bmatrix} \grave{x}_1(t-1) \\ \grave{x}_2(t-1) \\ \grave{x}_3(t-1) \end{bmatrix} + \begin{bmatrix} \grave{b}_{1\cdot} \\ \grave{b}_{2\cdot} \\ \grave{b}_{3\cdot} \end{bmatrix} u(t-1) \tag{2.165}$$

Note that with complex eigenvalues, not only a part of the $\grave{\mathbf{A}}$ matrix is complex but so are the respective rows of the $\grave{B}$ and columns of the $\grave{C}$ matrices. It is possible to find a transformation, however, which replaces a pair of complex eigenvalues in $\grave{\mathbf{A}}$ with a $2 \cdot 2$ real block and makes the concerned rows/columns in the other matrices real as well.

2.6.2. Computing the Transfer Function from the State-Space Form

The discrete transfer function of a system can be obtained from its state-space representation in a straightforward way. Consider first the state equation (2.154) and apply the shift operator:

$$\boldsymbol{x}(t) = \boldsymbol{A}\phi^{-1}\boldsymbol{x}(t) + \boldsymbol{B}\phi^{-1}\boldsymbol{u}(t) \tag{2.166}$$

Solving this for $\boldsymbol{x}(t)$ yields

$$x(t) = (I - \phi^{-1}A)^{-1}\phi^{-1}Bu(t) \tag{2.167}$$

Now substitute this into the output equation (2.155):

$$y(t) = [C(I - \phi^{-1}A)^{-1}\phi^{-1}B+D]u(t) = \frac{CAdj(I - \phi^{-1}A)\phi^{-1}B+D|I - \phi^{-1}A|}{|I - \phi^{-1}A|}u(t) \tag{2.168}$$

A comparison of (2.168) with the definition of the multiple-input multiple-output discrete transfer function as given in Eq. (2.63) suggests that

$$M(\phi) = \frac{CAdj(I - \phi^{-1}A)\phi^{-1}B+D|I - \phi^{-1}A|}{|I - \phi^{-1}A|} \tag{2.169}$$

The same result is obtained in a slightly different form if the shift operator is applied to the forward version (2.156) of the state equation. Now

$$\phi x(t) = Ax(t) + Bu(t) \tag{2.170}$$

$$x(t) = (\phi I - A)^{-1}Bu(t) \tag{2.171}$$

$$y(t) = [C(\phi I - A)^{-1}B+D]u(t) = \frac{CAdj(\phi I - A)B+D|\phi I - A|}{|\phi I - A|}u(t) \tag{2.172}$$

Obviously, (2.172) provides the positive shift form of the transfer function, that is

$$M^{+}(\phi) = M(\phi) = \frac{CAdj(\phi I - A)B+D|\phi I - A|}{|\phi I - A|} \tag{2.173}$$

As it will be seen in Section 2.7, the numerator and denominator in (2.173) are not necessarily equal to $\tilde{G}^{+}(\phi)$ and $\tilde{h}^{+}(\phi)$. The reason is that the minimum realizations of multiple-input multiple-output systems may have some of the poles of $\tilde{h}^{+}(\phi)$ with additional multiplicities. Thus we can only state here that

$$CAdj(\phi I - A)B + D|\phi I - A| = \vartheta^{+}(\phi)\tilde{G}^{+}(\phi) \tag{2.174}$$

$$|\phi I - A| = \vartheta^{+}(\phi)\tilde{h}^{+}(\phi) \tag{2.175}$$

where the polynomial $\vartheta^{+}(\phi)$ contains the extra multiplicities of the poles.

The polynomial matrix $\vartheta^{+}(\phi)\bar{G}^{+}(\phi)$ and the polynomial $\vartheta^{+}(\phi)\bar{h}^{+}(\phi)$, as obtained above, are in general not relative prime. Further, individual rows $\bar{g}_{i.}^{+}(\phi)$ of $\bar{G}^{+}(\phi)$ usually have common factors with $\bar{h}^{+}(\phi)$. Such situations are related to the observability and controllability properties of the full system and to the observability of the multiple-input single-output subsystems. These properties will be discussed in subsections 2.6.4 and 2.6.5.

Solution of the state equation. The transfer function derived above offers a simple way to finding a closed-form solution of the state equations. Recall (from Section 2.3.2) that the z-transform relationship between inputs and outputs, under zero initial conditions, is formally identical with the transfer function relationship in terms of the shift operator. Thus Eq. (2.172) may be rewritten as

$$Y(z) = [C(zI - A)^{-1}B+D]U(z) \tag{2.176}$$

From this, the zero initial condition response to any input $u(t)$ can be computed by inverse z-transformation. Nonzero initial conditions can be incorporated by applying z-transformation directly to Eq. (2.156). According to the shift property (2.77) for $k=-1$,

$$zX(z) - zx(0) = AX(z) + BU(z) \tag{2.177}$$

from which

$$Y(z) = [C(zI - A)^{-1}B+D]U(z) + Cz(zI - A)^{-1}x(0) \tag{2.178}$$

2.6.3. System Matrix and Invariant Zeroes

The *system matrix* introduced in this subsection allows a concise description of the state and output equations. It contains a full characterization of the given state-space representation and provides a powerful tool in the analysis of the internal properties of the system. One such property, the *invariant zeroes* of the realization, plays an important role in the design of diagnostic algorithms, as it will be shown in later chapters.

System matrix. Define the (backward shift) system matrix as

$$\Gamma(\phi) = \begin{bmatrix} I - \phi^{-1}A & -\phi^{-1}B \\ C & D \end{bmatrix} \tag{2.179}$$

With this system matrix, equations (2.154) and (2.155) are expressed as

$$\begin{bmatrix} I - \phi^{-1}A & -\phi^{-1}B \\ C & D \end{bmatrix} \begin{bmatrix} x(t) \\ u(t) \end{bmatrix} = \begin{bmatrix} 0 \\ y(t) \end{bmatrix} \tag{2.180}$$

Similarly, the forward shift version of the system matrix can be defined as

$$\Gamma^+(\phi) = \begin{bmatrix} \phi I - A & -B \\ C & D \end{bmatrix} \tag{2.181}$$

Now Eqs. (2.156) and (2.155) can be written as

$$\begin{bmatrix} \phi I - A & -B \\ C & D \end{bmatrix} \begin{bmatrix} x(t) \\ u(t) \end{bmatrix} = \begin{bmatrix} 0 \\ y(t) \end{bmatrix} \tag{2.182}$$

While both $\Gamma(\phi)$ and $\Gamma^+(\phi)$ bear the same system properties, it is usually easier to study the forward-shift version.

An important rank relation. Now it will be shown that

$$Rank\, \Gamma^+(\phi) = Rank\, \bar{G}^+(\phi) + \nu \qquad for\ all\ \phi \neq \lambda_j \tag{2.183}$$

Consider first the case when $\kappa \leq \mu$. Then the system matrix can be decomposed as

$$\underbrace{\begin{bmatrix} \phi I - A & -B \\ C & D \end{bmatrix}}_{\Gamma^+(\phi)} = \underbrace{\begin{bmatrix} \phi I - A & 0 \\ C & I \end{bmatrix}}_{\Gamma_I^+(\phi)} \underbrace{\begin{bmatrix} I & -(\phi I - A)^{-1}B \\ 0 & C(\phi I - A)^{-1}B + D \end{bmatrix}}_{\Gamma_{II}^+(\phi)} \tag{2.184}$$

Here $\Gamma_I^+(\phi)$ is a square matrix; according to (2.130)

$$|\Gamma_I^+(\phi)| = |\phi I - A|\,|I| \tag{2.185}$$

Thus, except at $\phi = \lambda_j$ (where λ_j is an eigenvalue of A), $\Gamma_I^+(\phi)$ has full rank. Then with (2.128a)

$$Rank\, \Gamma^+(\phi) = Rank\, \Gamma_{II}^+(\phi) \tag{2.186}$$

Now $\Gamma_{II}^+(\phi)$ is a $(\nu + \mu)\cdot(\nu + \kappa)$ matrix so its maximum size square minors

$\Gamma_{II}^{+i1,i2,..}(\phi)$ can be obtained by omitting μ-κ rows. If any of the first ν rows is omitted then a zero column results. Otherwise

$$|\Gamma_{II}^{+i1,i2,..i\kappa}(\phi)| = |I|\,|\bar{G}^{+i1,i2,..i\kappa}(\phi)/\bar{h}^{+}(\phi)| \tag{2.187}$$

where I is a $\nu\cdot\nu$ unit matrix and $\bar{G}^{+i1,i2,..i\kappa}(\phi)/\bar{h}^{+}(\phi)$ are $\kappa\cdot\kappa$ size minors of $\bar{G}^{+}(\phi)/\bar{h}^{+}(\phi)$ (c.f. (2.174) and (2.175)). That is, $\Gamma_{II}^{+}(\phi)$ and thus $\Gamma^{+}(\phi)$ has full rank $\nu+\kappa$ if and only if $\bar{G}^{+}(\phi)$ has full rank κ. In the case of rank defect, one (two, etc.) column at a time can be left out of B and D, and of $\bar{G}^{+}(\phi)$, until the first non-singular minor is found with which the reduced $\bar{G}^{+}(\phi)$ and $\Gamma^{+}(\phi)$ matrices have full rank; these will be the reduced ranks of the original matrices which obey (2.183).

In the case of $\kappa \geq \mu$, the decomposition (2.184) can be replaced with

$$\begin{bmatrix} \phi I - A & -B \\ C & D \end{bmatrix} = \begin{bmatrix} I & 0 \\ C(\phi I\text{-}A)^{-1} & C(\phi I\text{-}A)^{-1}B+D \end{bmatrix} \begin{bmatrix} \phi I\text{-}A & -B \\ 0 & I \end{bmatrix} \tag{2.188}$$

Then a similar argument as above leads again to (2.183).

Invariant zeroes. As described in Subsection 2.5.1, a polynomial matrix may drop its rank relative to its normal rank at some values of ϕ. The values ζ_j which make the system matrix $\Gamma^{+}(\phi)$ drop its rank are called the *invariant zeroes* of the realization. The rank relation (2.183) applies both to the normal ranks of the matrices $\Gamma^{+}(\phi)$ and $\bar{G}^{+}(\phi)$ and to their behavior at the invariant zeroes; this implies that $\bar{G}^{+}(\phi)$, too, drops its rank at the invariant zeroes. However, (2.183) loses its validity at $\phi=\lambda_i$; the important case when an eigenvalue of A is also an invariant zero will be discussed in the subsequent subsections.

In general, with κ inputs and μ outputs, the size of the system matrix is $(\nu+\mu)\cdot(\nu+\kappa)$. If $\kappa=\mu$ then $\Gamma^{+}(\phi)$ is square. Now the invariant zeroes are the roots ζ_1, ζ_2,... of the determinant $|\Gamma^{+}(\phi)| = \alpha(\phi - \zeta_1)(\phi - \zeta_2)$... The number of zeroes is $0 \leq Deg|\Gamma^{+}(\phi)| \leq \nu$.

If $\kappa<\mu$ then $\Gamma^{+}(\phi)$ is a "tall" matrix. Its largest minors are size $(\nu+\kappa)\cdot(\nu+\kappa)$ and there may be a large number of such minors (as many as the number of ways $\nu+\kappa$ rows can be chosen from a total of $\nu+\mu$ rows). For a value ζ_j to qualify as an invariant zero of the system, it has to be a root of all the maximum size minor determinants. Similarly, if $\kappa>\mu$ then $\Gamma^{+}(\phi)$ is a "flat" matrix. Its largest minors are $(\nu+\mu)\cdot(\nu+\mu)$ and a value ζ_j has to be a root of all the "choose $\nu+\mu$ from $\nu+\kappa$" minor determinants to qualify as an invariant system zero.

2.6.4. Observability

A system (or realization) is called *observable* if the state at a time t_o, $x(t_o)$, can be computed from a finite set of output observations $y(t_o)$, $y(t_o+1)$, ..., $y(t_o+k-1)$ and the respective inputs $u(t_o)$, $u(t_o+1)$, ..., $u(t_o+k-1)$. Observability is an internal property of the system (realization) which is determined by and can be judged from the system parameters A and C.

Write the output and state equations (2.154) and (2.155) for $t=t_o$, t_o+1, ..., t_o+k-1:

$$y(t_o) = Cx(t_o) + Du(t_o)$$

$$y(t_o+1) = Cx(t_o+1) + Du(t_o+1) = CAx(t_o) + CBu(t_o) + Du(t_o+1)$$

$$y(t_o+2) = \dots = CA^2x(t_o) + CABu(t_o) + CBu(t_o+1) + Du(t_o+2)$$

$$\dots$$

$$\dots$$

$$y(t_o+k-1) = CA^{k-1}x(t_o) + CA^{k-2}Bu(t_o) + \dots + CBu(t_o+k-2) + Du(t_o+k-1) \tag{2.189}$$

Move all the known terms to the left-hand side and re-name their sum as $\blacktriangle y(t_o)$, $\blacktriangle y(t_o+1)$, ..., $\blacktriangle y(t_o+k-1)$:

$$\blacktriangle y(t_o) = y(t_o) - Du(t_o) = Cx(t_o)$$

$$\blacktriangle y(t_o+1) = y(t_o+1) - CBu(t_o) - Du(t_o+1) = CAx(t_o)$$

$$\blacktriangle y(t_o+2) = y(t_o+2) - CABu(t_o) - CBu(t_o+1) - Du(t_o+2) = CA^2x(t_o)$$

$$\dots$$

$$\dots$$

$$\blacktriangle y(t_o+k-1) = y(t_o+k-1) - CA^{k-2}Bu(t_o) - \dots - CBu(t_o+k-2) - Du(t_o+k-1) = CA^{k-1}x(t_o) \tag{2.190}$$

That is

$$\begin{bmatrix} \blacktriangle y(t_o) \\ \blacktriangle y(t_o+1) \\ \blacktriangle y(t_o+2) \\ \cdots \\ \cdots \\ \blacktriangle y(t_o+k-1) \end{bmatrix} = \begin{bmatrix} C \\ CA \\ CA^2 \\ \cdots \\ \cdots \\ CA^{k-1} \end{bmatrix} x(t_o) \tag{2.191}$$

Rank condition. Eq. (2.191) is a set of simultaneous equations which we wish to solve for $x(t_o)$. Since $x(t_o)$ has ν elements, a unique solution exists if and only if there are ν independent equations, that is, if the matrix multiplying $x(t_o)$ has ν independent rows. One may attempt to introduce new independent rows by adding new observations, that is, by increasing k. However, according to the Cayley-Hamilton theorem (2.153), A^{ν}, $A^{\nu+1}$, ... linearly depend on I, A, ..., $A^{\nu-1}$; thus going beyond $k=\nu$ is pointless. Therefore, the solution exists, that is, the system is observable if the *observability matrix*

$$W_o = \begin{bmatrix} C \\ CA \\ CA^2 \\ \cdots \\ \cdots \\ CA^{\nu-1} \end{bmatrix} \tag{2.192}$$

has ν independent rows, that is

$$Rank W_o = \nu \tag{2.193}$$

Since W_o has ν columns, condition (2.193) means it needs to have full (column) rank. (Note that the task of computing $x(t_o)$ may, in general, be performed from fewer than ν observations but if it can not be done from ν then it can not be done at all.)

Invariance to transformation. The observability property is invariant to the transformation of the state variable. If $\grave{x}(t)=Tx(t)$ then, by (2.160), $\grave{C}=CT^{-1}$ and $\grave{A}=TAT^{-1}$. It follows that

$$\grave{C}\grave{A}^j = CA^jT^{-1} \qquad j=0 \ldots \nu-1 \tag{2.194}$$

and

$$\hat{W}_o = W_o T^{-1} \tag{2.195}$$

Since T^{-1} has full rank, it follows from (2.128a) that

$$Rank\hat{W}_o = Rank W_o \tag{2.196}$$

Pole-zero cancellation. If all other state variables are decoupled from a state variable $x_j(t)$ then the only "channel" from this state to the outputs is through the j-th column $c_{.j}$ of the output matrix C. If $c_{.j}=0$ then $x_j(t_o)$ can not be computed from the outputs and the system (realization) is unobservable.

While the above decoupling may happen in any representation, it definitely arises when a system is transformed into diagonal or block-diagonal form. Now other transformed states are completely decoupled from the (transformed) states which are accompanied by a diagonal element in the Λ or J matrices and from the *first* state variable in every Jordan group. Then the lack of observability manifests itself in the respective $\hat{c}_{.j}$ vector(s) being zero. Obviously, due to multiplication with a zero vector, the pole belonging to any of these unobservable states (modes) will be missing from the input-output representation of the system. The true order of the input-output system is reduced accordingly. When deriving the transfer functions from a non-diagonal state-space representation, this order reduction appears in the form of a pole-zero cancellation between the numerator and denominator.

Invariant zeroes. Consider a system represented in (block) diagonal form. With (2.160) and $T^{-1}=\Xi=[\,\xi_1 \;\; \xi_2 \;\; \cdots \;\; \xi_\nu]$, one can write $\hat{c}_{.j}=C\xi_j$, where ξ_j is the right eigenvector of A belonging to a diagonally appearing λ_j eigenvalue, or the true right eigenvector belonging to a Jordan block. The j-th state in the transformed system (and so the system itself) is unobservable if

$$C\xi_j = 0 \tag{2.197}$$

Combining this with the definition (2.135) of the right eigenvector yields

$$\begin{bmatrix} \lambda_j I - A \\ C \end{bmatrix} \xi_j = 0 \tag{2.198}$$

This set of homogeneous equations allows non-trivial solutions for ξ_j only if its matrix has less than full rank. Thus, if $x_j(t)$ is unobservable then λ_j is an invariant zero of the *observability system matrix*

$$\Gamma_o^+(\phi) = \begin{bmatrix} \phi I - A \\ C \end{bmatrix} \tag{2.199}$$

Note that an invariant zero of $\Gamma_o^+(\phi)$ is also an invariant zero of $\Gamma^+(\phi)$ if $\kappa \leq \mu$. This is not usually the case, however, if $\kappa > \mu$.

2.6.5. Controllability

A system (or realization) is called *controllable* if it is possible to move it from an arbitrary (known) initial state $x(t_o)$ to an arbitrary target state $x(t_o+k)$ in a finite number of steps, that is, if it is possible to find a sequence of inputs $u(t_o)$, $u(t_o+1)$, ..., $u(t_o+k-1)$ which results in the desired transition. Controllability is an internal property of the system (realization) which is determined by and can be judged from the system parameters A and B.

Write the state equation (2.156) for $t=t_o$, t_o+1, ..., t_o+k-1:

$$x(t_o+1) = Ax(t_o) + Bu(t_o)$$

$$x(t_o+2) = Ax(t_o+1) + Bu(t_o+1) = A^2x(t_o) + ABu(t_o) + Bu(t_o+1)$$

$$x(t_o+3) = \ldots = A^3x(t_o) + A^2Bu(t_o) + ABu(t_o+1) + Bu(t_o+2)$$

$$\ldots$$

$$\ldots$$

$$x(t_o+k) = A^kx(t_o) + A^{k-1}Bu(t_o) + \ldots + ABu(t_o+k-2) + Bu(t_o+k-1) \tag{2.200}$$

Define $\blacktriangle x(t_o+j)=x(t_o+j) - A^j x(t_o)$, $j=1\ldots k$; then

$$\blacktriangle x(t_o+1) = Bu(t_o)$$

$$\blacktriangle x(t_o+2) = ABu(t_o) + Bu(t_o+1)$$

$$\blacktriangle x(t_o+3) = A^2Bu(t_o) + ABu(t_o+1) + Bu(t_o+2)$$

$$\ldots$$

$$\ldots$$

$$\blacktriangle x(t_o+k) = A^{k-1}Bu(t_o) + \ldots + ABu(t_o+k-2) + Bu(t_o+k-1) \tag{2.201}$$

That is

$$\blacktriangle x(t_o+k) = [B \;\; AB \;\; A^2B \; \dots \; A^{k-1}B] \begin{bmatrix} u(t_o+k-1) \\ u(t_o+k-2) \\ u(t_o+k-3) \\ \dots \\ \dots \\ u(t_o) \end{bmatrix} \tag{2.202}$$

Rank condition. $\blacktriangle x(t_o+k)$ is a vector of dimension ν; with the initial and target states given, it is known for any selected k. Eq. (2.202) presents this vector as a linear combination of the columns of the $[B \;\; AB \;\; A^2B \; \dots \; A^{k-1}B]$ matrix, with the input values $u_1(t_o+k-1)$, $u_2(t_o+k-1)$, ..., $u_1(t_o+k-2)$, $u_2(t_o+k-2)$, etc. as their weights. An *arbitrary* ν-dimensional vector $\blacktriangle x(t_o+k)$ can be generated as such a linear combination, if and only if ν of the columns are linearly independent. In geometric terms, the columns of the matrix $[B \;\; AB \;\; A^2B \; \dots \; A^{k-1}B]$ have to span the ν-dimensional space. One may attempt to introduce new independent columns by adding new inputs, that is, by increasing k. However, according to the Cayley-Hamilton theorem (2.153), A^{ν}, $A^{\nu+1}$, ... linearly depend on I, A, ..., $A^{\nu-1}$; thus going beyond $k=\nu$ is pointless. The system therefore is controllable if the *controllability matrix*

$$W_c = [B \;\; AB \;\; A^2B \; \dots \; A^{\nu-1}B] \tag{2.203}$$

has ν independent columns, that is, if

$$Rank W_c = \nu \tag{2.204}$$

Since W_c has ν rows, condition (2.204) means it needs to have full (row) rank. (Note that the system may, in general, be brought to the desired target state in fewer than ν samples but if it can not be done in ν then it can not be done at all. Also, the input sequence providing the desired transition is, in general, not unique.)

Invariance to transformation. The controllability property is invariant to the transformation of the state variable. If $\grave{x}(t)=Tx(t)$ then, by (2.160), $\grave{B}=TB$ and $\grave{A}=TAT^{-1}$. It follows that

$$\grave{A}^j \grave{B} = TA^jB \qquad j=0\dots\nu-1 \tag{2.205}$$

$$\grave{W}_c = TW_c \tag{2.206}$$

Since T has full rank, it follows from (2.128a) that

$$Rank\grave{W}_c = RankW_c \tag{2.207}$$

Pole-zero cancellation. If a state variable $x_i(t)$ is decoupled from all other state variables then the only "channel" from the inputs to this state is through the i-th row $b_{i.}$ of the input matrix B. If $b_{i.} = 0$ then $x_i(t)$ can not be affected by the inputs and the system (realization) is uncontrollable.

While the above decoupling may happen in any representation, it definitely arises when a system is transformed into diagonal or block-diagonal form. Now the transformed states which are accompanied by a diagonal element in the Λ or J matrices and the *last* state variable in every Jordan group are completely decoupled from the other (transformed) states. Then the lack of controllability manifests itself in the respective $\grave{b}_{i.}$ vector(s) being zero. Just like in the case of unobservability, multiplication with a zero vector eliminates the pole accompanying any such state (mode) from the input-output representation. Thus the true order of the input-output system is reduced. This reduction comes about in the form of a pole-zero cancellation when the transfer function is derived from a non-diagonal state-space representation.

Invariant zeroes. Consider a system described in (block) diagonal form. With (2.160), (2.141) and $T^{-1} = \Xi$, one may write $\grave{b}_{i.} = \psi_i' B$, where ψ_i' is the left eigenvector of A belonging to a diagonally appearing λ_i eigenvalue, or the true left eigenvector belonging to a Jordan block. The i-th state in the transformed system (and so the system itself) is uncontrollable if

$$\psi_i' B = 0 \tag{2.208}$$

Changing the sign of (2.208) and combining it with the definition (2.138) of the left eigenvector yields

$$\psi_i' [\, \lambda_i I - A \quad -B \,] = 0 \tag{2.209}$$

This set of homogeneous equations allows non-trivial solutions for ψ_i' only if its matrix has less than full rank. Thus, if $x_i(t)$ is uncontrollable then λ_i is an invariant zero of the *controllability system matrix*

$$\Gamma_c^+(\phi) = [\, \phi I - A \quad -B \,] \tag{2.210}$$

Note that an invariant zero of $\Gamma_c^+(\phi)$ is also an invariant zero of $\Gamma^+(\phi)$ if $\mu \le \kappa$. This is not usually the case, however, if $\mu > \kappa$.

2.7. SYSTEM REALIZATION

By system realization we mean finding a state-space representation for a system given in input-output form, by its transfer function or difference equation. While the opposite procedure, finding the input-output description for systems given in state-space form, is simple and straightforward (see Subsection 2.6.2), this is far from being true for the realization problem.

One of the difficulties is that the state-space representation is not unique, thus there is an infinite number of valid realizations for any system. Further, it may not be immediately obvious what the order of the state-space model should be. Usually, we are interested in the simplest possible realization. Since unobservable and uncontrollable modes do not contribute to the input-output model, there may be no point in having them in the realization. Beyond this, we will seek the minimum order (observable and controllable) state-space model(s) which realize the input-output system.

As seen earlier, the order of a state-space model is defined as the number of state variables. Since each state appears in a first-order difference equation, this is also the number of "delay elements" it takes to "realize" the set of state equations on a simulation device. The minimum realization of an input-output model thus can be thought of as (i) finding an algebraic conversion which results in the minimum number of delay elements and (ii) associating a state variable with each delay element. Note that graphic techniques (signal-flow graphs) may facilitate and/or replace the algebraic conversion.

We will start our discussion with single-input single-output systems where the issue of system order is clear, and will introduce a few standard realizations. These results will then be extended to multiple-input single-output systems; the zeroes of such systems will also be defined. Finally, the more complex problem of system order for multiple-input multiple-output systems will be addressed, together with the poles and zeroes of such systems.

2.7.1. Single-Input Single-Output Systems

Consider the system described by the difference equation (2.27). Though seemingly it contains a total of $\upsilon + d + \iota$ delays, we will show below that it can be rearranged so that it is realized with

$$\nu = max(\upsilon,\ d+\iota) \tag{2.211}$$

delays, which is therefore the order of the system (c.f. Eq. (2.37)).

Below we will present two standard ("canonical") realizations, one of which we will derive from the difference equation. Then we will introduce a powerful approach to realization utilizing the partial fraction expansion of the discrete transfer function.

Canonical realizations. Consider the difference equation (2.27) with $d=1$, and assume that $\upsilon = d + \iota = \nu$

$$y(t) + h^1 y(t-1) + h^2 y(t-2) + \ldots + h^{\nu} y(t-\nu) =$$

$$g^1 u(t-1) + g^2 u(t-2) + \ldots + g^{\nu} u(t-\nu) \qquad (2.212)$$

Express y(t) from this as

$$y(t) = -h^1 y(t-1) + g^1 u(t-1) - h^2 y(t-2) + g^2 u(t-2) - \ldots$$

$$-h^{\nu} y(t-\nu) + g^{\nu} u(t-\nu) \qquad (2.213)$$

Now choose the first state variable as

$$x_1(t) = y(t) \qquad (2.214)$$

and write it as

$$x_1(t) = -h^1 x_1(t-1) + g^1 u(t-1) + x_2(t-1) \qquad (2.215)$$

This implies that

$$x_2(t) = -h^2 y(t-1) + g^2 u(t-1) - \ldots - h^{\nu} y(t+1-\nu) + g^{\nu} u(t+1-\nu)$$

$$(2.216)$$

Now write this as

$$x_2(t) = -h^2 x_1(t-1) + g^2 u(t-1) + x_3(t-1) \qquad (2.217)$$

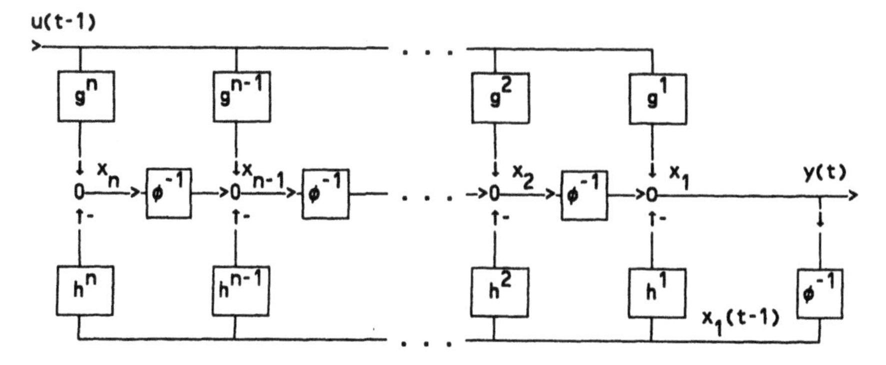

FIGURE 2.7. Observer canonical form ($n=\nu$).

and so on, until

$$x_\nu(t) = -h^\nu x_1(t-1) + g^\nu u(t-1) \tag{2.218}$$

This leads to the following scheme (Fig. 2.7):

$$\begin{bmatrix} x_1(t) \\ x_2(t) \\ \vdots \\ x_\nu(t) \end{bmatrix} = \begin{bmatrix} -h^1 & 1 & 0 \dots 0 \\ -h^2 & 0 & 1 \dots 0 \\ \vdots & \vdots & \vdots \\ -h^\nu & 0 & 0 \dots 0 \end{bmatrix} \begin{bmatrix} x_1(t-1) \\ x_2(t-1) \\ \vdots \\ x_\nu(t-1) \end{bmatrix} + \begin{bmatrix} g^1 \\ g^2 \\ \vdots \\ g^\nu \end{bmatrix} u(t-1)$$

$$y(t) = [\,1 \;\; 0 \;\; \dots \;\; 0\,]\, x(t) \tag{2.219}$$

which is known as the "*observer canonical form.*" Note that if $d>1$ or $\upsilon \neq d+\iota$ then some coefficients become zero in the scheme. Further, if $d=0$ then a $g^o u(t)$ term appears in the output equation and the g^j, $j=1\dots\nu$ coefficients are replaced with $g^j - g^o h^j$.

Another important canonical realization is the *controller form* (Kailath, 1980):

$$x(t) = \begin{bmatrix} -h^1 & -h^2 \dots & -h^{\nu-1} & -h^\nu \\ 1 & 0 \dots & 0 & 0 \\ \vdots & \vdots & & \\ 0 & 0 \dots & 1 & 0 \end{bmatrix} x(t-1) + \begin{bmatrix} 1 \\ 0 \\ \vdots \\ 0 \end{bmatrix} u(t-1)$$

$$y(t) = [\,g^1 \;\; g^2 \;\; \dots \;\; g^\nu\,]\, x(t) \tag{2.220}$$

Realization by partial fraction expansion. A diagonal or block-diagonal realization can be obtained by applying partial fraction expansion to the system's discrete transfer function and associating the state variables with the system modes.

Consider first a system which has only distinct poles $p_1 \dots p_\nu$. Obtain the partial fraction expansion of the transfer function $m(\phi)=g^+(\phi)/h^+(\phi)$ as

$$m(\phi) = \pi^o + \frac{\pi^1}{\phi - p_1} + \dots + \frac{\pi^\nu}{\phi - p_\nu} \tag{2.221}$$

(c.f. (2.91)), where the coefficients are computed as

$$\pi^o = g^o \qquad \pi^k = (\phi - p_k)\frac{g^+(\phi) - g^o h^+(\phi)}{h^+(\phi)}\Bigg|_{\phi=p_k} \quad k=1\ldots\nu \tag{2.222}$$

(c.f. (2.92)). Now define the state variables as

$$x_k(t) = \frac{\pi^k}{\phi - p_k}u(t) = \frac{\pi^k \phi^{-1}}{1 - p_k\phi^{-1}}u(t) \qquad k=1\ldots\nu \tag{2.223}$$

These can be realized as

$$x_k(t) = p_k x_k(t-1) + \pi^k u(t-1) \qquad k=1\ldots\nu \tag{2.224}$$

leading to the following diagonal system realization (Fig. 2.8):

$$x(t) = \begin{bmatrix} p_1 & 0 & \ldots & 0 \\ 0 & p_2 & \ldots & 0 \\ & \cdots & & \\ & \cdots & & \\ 0 & 0 & \ldots & p_\nu \end{bmatrix} x(t-1) + \begin{bmatrix} \pi^1 \\ \pi^2 \\ \cdot \\ \cdot \\ \pi^\nu \end{bmatrix} u(t-1)$$

$$y(t) = [\,1 \;\; 1 \;\; \ldots \;\; 1\,]\,x(t) + \pi^o u(t) \tag{2.225}$$

If a pair of poles is complex, so are their coefficients in the partial fraction expansion and also the corresponding state variables in the diagonal realization. The inconvenience of complex coefficients and variables may be avoided,

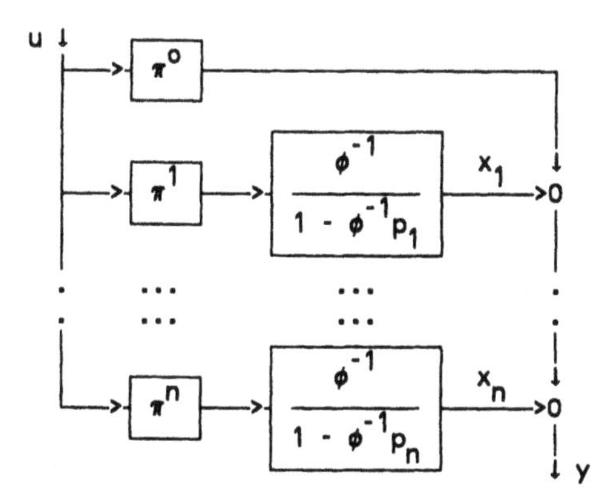

FIGURE 2.8. Realization by partial fractions ($n=\nu$).

at the expense of introducing a $2\cdot 2$ block in the A matrix. Consider the complex poles $p_{k,k+1} = \sigma e^{\pm i\varphi}$ with coefficients $\pi^{k,k+1} = \xi \pm i\eta$. Then the two partial fractions may be converted as

$$\frac{\xi + i\eta}{\phi - \sigma e^{i\varphi}} + \frac{\xi - i\eta}{\phi - \sigma e^{-i\varphi}} = \frac{2\xi(\phi - \sigma cos\varphi) - 2\eta\sigma sin\varphi}{\phi^2 - 2\phi\sigma cos\varphi + \sigma^2} \tag{2.226}$$

where all coefficients are real. Now define two state variables as

$$x_k(t) = 2\xi \frac{1 - \phi^{-1}\sigma cos\varphi}{1 - 2\phi^{-1}\sigma cos\varphi + \phi^{-2}\sigma^2}\phi^{-1}u(t) \tag{2.227a}$$

$$x_{k+1}(t) = -2\eta \frac{\phi^{-1}\sigma sin\varphi}{1 - 2\phi^{-1}\sigma cos\varphi + \phi^{-2}\sigma^2}\phi^{-1}u(t) \tag{2.227b}$$

which are also real. This leads to the following (partial) realization:

$$\begin{bmatrix} x_k(t) \\ x_{k+1}(t) \end{bmatrix} = \begin{bmatrix} \sigma cos\varphi & -\frac{\xi}{\eta}\sigma sin\varphi \\ -\frac{\eta}{\xi}\sigma sin\varphi & \sigma cos\varphi \end{bmatrix} \begin{bmatrix} x_k(t-1) \\ x_{k+1}(t-1) \end{bmatrix} + \begin{bmatrix} 2\xi \\ 0 \end{bmatrix} u(t-1)$$

$$y(t) = [\, \dots \underset{k}{1} \;\; \underset{k+1}{1} \dots]x(t) \tag{2.228}$$

Now consider a pole p_k with multiplicity ρ. The partial fraction expansion of $m(\phi)$ now contains the terms

$$m(\phi) = \frac{\pi^{k1}}{\phi - p_k} + \frac{\pi^{k2}}{(\phi - p_k)^2} + \dots + \frac{\pi^{k\rho}}{(\phi - p_k)^\rho} + \dots \tag{2.229}$$

Define the state variables as

$$x_{k1}(t) = \frac{\phi^{-1}}{1 - p_k\phi^{-1}}u(t) \quad \dots \quad x_{k\rho}(t) = \frac{\phi^{-\rho}}{(1 - p_k\phi^{-1})^\rho}u(t) \tag{2.230}$$

Since

$$x_{k,j+1}(t) = \frac{\phi^{-1}}{1 - p_k\phi^{-1}}x_{kj}(t) \tag{2.231}$$

this leads to the following scheme:

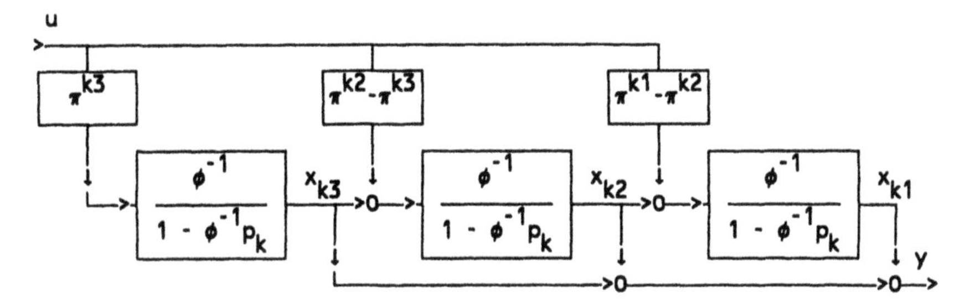

FIGURE 2.9. Realization of multiple pole, scheme (2.233) (ρ=3).

$$\begin{bmatrix} x_{k\rho}(t) \\ \cdots \\ \cdots \\ x_{k2}(t) \\ x_{k1}(t) \end{bmatrix} = \begin{bmatrix} p_k & 1 & \dots & 0 & 0 \\ & \cdots & & & \\ & \cdots & & & \\ 0 & 0 & \dots & p_k & 1 \\ 0 & 0 & \dots & 0 & p_k \end{bmatrix} \begin{bmatrix} x_{k\rho}(t-1) \\ \cdots \\ \cdots \\ x_{k2}(t-1) \\ x_{k1}(t-1) \end{bmatrix} + \begin{bmatrix} 0 \\ \cdot\cdot \\ \cdot\cdot \\ 0 \\ 1 \end{bmatrix} u(t-1)$$

$$y_k(t) = [\, \pi^{k\rho} \;\; \dots \;\; \pi^{k2} \;\; \pi^{k1} \,]\, \boldsymbol{x}_k(t) \tag{2.232}$$

where $y_k(t)$ is the partial output and $\boldsymbol{x}_k(t)$ is the partial state vector as defined above. This is clearly a Jordan block (c.f. (2.153)). An alternative realization, more amenable to extension to multiple-input systems is (Fig. 2.9)

$$\boldsymbol{x}_k(t) = \begin{bmatrix} p_k & 1 & \dots & 0 & 0 \\ & \cdots & & & \\ & \cdots & & & \\ 0 & 0 & \dots & p_k & 1 \\ 0 & 0 & \dots & 0 & p_k \end{bmatrix} \boldsymbol{x}_k(t-1) + \begin{bmatrix} \pi^{k1} - \pi^{k2} \\ \cdot\cdot \\ \cdot\cdot \\ \pi^{k,\rho-1} - \pi^{k\rho} \\ \pi^{k\rho} \end{bmatrix} u(t-1)$$

$$y_k(t) = [\, 1 \;\; \dots \;\; 1 \;\; 1 \,]\, \boldsymbol{x}_k(t) \tag{2.233}$$

2.7.2. Multiple-Input Single-Output Systems

Consider now the system described by the difference equation (2.41). As it will be seen below, some of the single-input single-output realization techniques, notably the observer canonical form and the partial fraction expansion technique, extend naturally to such systems. These extensions suggest that the number of delays needed for the realization, that is, the order of the system is (c.f. (2.42)):

$$\nu = max(d_j + \iota_j,\ \upsilon) \qquad j=1...\kappa \tag{2.234}$$

Since other realizations may be obtained by transformation, without changing the system order, (2.234) describes the order of multiple-input single output systems in general. We will discuss the extension of the mentioned realization techniques and will also revisit the poles and zeroes of multiple-input single-output systems, in the light of their state-space realization.

Observer canonical form. Write the difference equation (2.41) as

$$y(t) + h^1 y(t-1) + h^2 y(t-2) + \ldots + h^\nu y(t-\nu)$$

$$= \mathbf{g}'^1 \mathbf{u}(t-1) + \mathbf{g}'^2 \mathbf{u}(t-2) + \ldots + \mathbf{g}'^\nu \mathbf{u}(t-\nu) \tag{2.235}$$

where $\mathbf{g}'^k$, $k=1...\nu$, are row vectors composed of the k-th elements of the $g_j(\phi)$, $j=1...\kappa$ polynomials. Then the derivation (2.213) - (2.218) applies and the resulting realization scheme is identical with (2.219), except the scalar input $u(t)$ is replaced by the vector input $\mathbf{u}(t)$ and the scalar coefficients $g^1...g^\nu$ by the vectors $\mathbf{g}'^1...\mathbf{g}'^\nu$.

Partial fractions. Write the input-output relationship for the multiple-input single-output system, using transfer functions:

$$y(t) = \frac{g_1^+(\phi)}{h^+(\phi)} u_1(t) + \frac{g_2^+(\phi)}{h^+(\phi)} u_2(t) + \ldots + \frac{g_\kappa^+(\phi)}{h^+(\phi)} u_\kappa(t) \tag{2.236}$$

Recall that $h^+(\phi)$ is the least common multiple of the denominators of the single-input subsystem transfer functions.

Assume first that all roots $p_1...p_\nu$ of $h^+(\phi)$ are distinct and apply partial fraction expansion (2.221) separately to each subsystem:

$$\frac{g_j^+(\phi)}{h^+(\phi)} = \pi_j^0 + \frac{\pi_j^1}{\phi - p_1} + \ldots + \frac{\pi_j^\nu}{\phi - p_\nu} \qquad j=1...\kappa \tag{2.237}$$

Substitute this into (2.236) and collect the like terms:

$$y(t) = [\,\boldsymbol{\pi}'^0 + \frac{\boldsymbol{\pi}'^1}{\phi - p_1} + \ldots + \frac{\boldsymbol{\pi}'^\nu_j}{\phi - p_\nu}\,]\,\mathbf{u}(t) \tag{2.238}$$

where

$$\pi'^{k} = [\pi_1^k \ \ \pi_2^k \ \ldots \ \pi_K^k] \qquad k=1\ldots\nu \tag{2.239}$$

Now define the state variables as

$$x_k(t) = \frac{\pi'^{k}}{\phi - p_k} u(t) \qquad k=1\ldots\nu \tag{2.240}$$

Then the system can be realized by the direct extension of the realization (2.225), replacing the scalars $u(t)$ and $\pi^o \ldots \pi^\nu$ with their vector equivalents. One should notice that each system mode is realized only once, even if it appears in several single-input subsystems; superposition is applied at the input side and thus each system mode is "re-used" with several inputs.

If a pole p_k has a multiplicity $\rho > 1$ then an expansion of the kind of (2.229) yields coefficients $\pi_j^{k1} \ldots \pi_j^{k\rho}$ in each subsystem j which has this pole. Define the coefficient vectors

$$\pi'^{ki} = [\pi_1^{ki} \ \ \pi_2^{ki} \ \ldots \ \pi_K^{ki}] \qquad i=1\ldots\rho \tag{2.241}$$

Then the partial realization of the system is obtained by the extension of (2.233), the scalars $u(t)$ and $\pi^{k1} \ldots \pi^{k\rho}$ replaced by their vector equivalents. Notice that multiple poles, even if they appear in several single-input subsystems, need to be realized only once, with their highest multiplicity.

Poles and zeroes. The poles of the multiple-input single-output system are the roots of the common denominator polynomial $h^+(\phi)$. As seen above, a pole appearing singly in several single-input subsystems can be realized with a single state variable so it does not need to have a multiplicity in the system denominator. A multiple pole of any (or several) subsystem(s) will appear in the system denominator with the maximum multiplicity it has in any subsystem. Thus $\vartheta(\phi) = 1$.

As mentioned in Subsection 2.2.2, the various single-input subsystems have different sets of zeroes. We consider system zeroes those values ζ_i, if any, which are common zeroes of all subsystems. It will be shown below that these are the same as the invariant zeroes of the system matrix defined in subsection 2.6.3:

$$\Gamma^+(\phi) = \begin{bmatrix} \phi I - A & -B \\ c' & d' \end{bmatrix} \tag{2.242}$$

Consider the numerator of the j-th subsystem. With (2.174) and $\vartheta^+(\phi) = 1$:

$$g_j^+(\phi) = c' Adj(\phi I - A)\, b_{\cdot j} + d_j \, |\phi I - A| \tag{2.243}$$

It can be seen by expanding both sides that

$$g_j^+(\phi) = Det\,\Gamma_j^{\,+}(\phi) \tag{2.244}$$

where

$$\Gamma_j^{\,+}(\phi) = \begin{bmatrix} \phi I - A & -b_{.j} \\ c' & d_j \end{bmatrix} \tag{2.245}$$

So if $g_j^+(\zeta_i)=0$ then $|\Gamma_j^+(\zeta_i)|=0$ and vice versa. Also $g_j^+(\zeta_i)=0,\ j=1\ldots\kappa$ implies

$$|\Gamma_j^{\,+}(\zeta_i)| = 0 \quad for \;\; j=1\ldots\kappa \tag{2.246}$$

It remains to be seen that (2.246) implies that *all* the $(\nu+1)\cdot(\nu+1)$ submatrices of $\Gamma^+(\phi)$ drop rank at ζ_i. Write $\Gamma^+(\phi)$ as

$$\Gamma^+(\phi) = [\sigma_{.1}(\phi) \;\; \ldots \;\; \sigma_{.\nu}(\phi) \;\, \beta_{.1} \;\; \ldots \;\; \beta_{.\kappa}] \tag{2.247}$$

It follows from (2.246) that the columns of any $\Gamma_j^{\,+}(\phi)$ are linearly dependent at ζ_i, that is,

$$\alpha_{j1}\sigma_{.1}(\zeta_i) + \ldots + \alpha_{j\nu}\sigma_{.\nu}(\zeta_i) + \alpha_{j,\nu+1}\beta_{.j} = 0 \qquad j=1\ldots\kappa \tag{2.248}$$

Now any other $(\nu+1)\cdot(\nu+1)$ submatrix contains a combination of σ and β columns; by taking the appropriate linear combinations of the equations in (2.248), it can be seen that these, too, are linearly dependent at ζ_i. Thus the invariant zeroes of the system matrix $\Gamma^+(\phi)$ are, indeed, the same as the (common) zeroes of the system.

2.7.3. Realization of Multiple-Input Multiple-Output Systems

The minimum realization of multiple-input multiple-output systems is a rather complex problem which goes beyond the scope of this book. In the sequel, we are only summarizing some of the most fundamental results, based on the definitive work of Kailath (1980). For further details, the reader is referred to Kailath's book, and to other sources listed at the end of this chapter.

Gilbert's method (Gilbert, 1963). The technique of realization by partial fraction expansion can be extended to multiple-input multiple-output systems provided all single-input single-output subsystems have only simple poles. Find all the subsystem poles p_k, $k=1\ldots\tau$ (where $\tau=\tilde{\nu}$). Then the least common multiple of the denominators is

$$\bar{h}^{+}(\phi) = (\phi - p_1)(\phi - p_2) \dots (\phi - p_T) \tag{2.249}$$

Now perform partial fraction expansion for all subsystems, according to (2.221), and form the coefficient matrices

$$\Pi^k = \begin{bmatrix} \pi^k_{11} & \cdots & \pi^k_{1K} \\ & \cdots & \\ & \cdots & \\ \pi^k_{\mu 1} & \cdots & \pi^k_{\mu K} \end{bmatrix} \qquad k=1 \dots T \tag{2.250}$$

Define $\rho_k = Rank\ \Pi^k$. Then the system order is

$$\nu = \sum_{k=1}^{T} \rho_k \tag{2.251}$$

and the characteristic polynomial (c.f. (2.175)) is

$$|\phi I - A| = \vartheta^{+}(\phi)\bar{h}^{+}(\phi) = (\phi - p_1)^{\rho 1}(\phi - p_2)^{\rho 2} \dots (\phi - p_T)^{\rho T} \tag{2.252}$$

That is, each pole p_k appears with a multiplicity ρ_k.

Of course, the results we obtained for multiple-input single-output systems in the previous subsection are consistent with the above; there the rank of the Π^k matrices is one. Also one is the rank of these matrices in case of single-input multiple-output systems explicitly not discussed here. Gilbert's method is simple and straightforward, however, it breaks down if the single-input single-output subsystems have multiple poles.

The Smith-McMillan Form (McMillan, 1952). The following decomposition of the transfer function matrix $M(\phi)$ provides the order and the poles and zeroes of the minimum realization, with no restriction on the nature of subsystem poles.

Consider the transfer function matrix

$$M(\phi) = \bar{G}^{+}(\phi) / \bar{h}^{+}(\phi) \tag{2.253}$$

where $\bar{h}^{+}(\phi)$ is the *least* common multiple of the subsystem denominators. First expand the numerator matrix $\bar{G}^{+}(\phi)$ as

$$\bar{G}^{+}(\phi) = V^{+}(\phi)\, \Omega^{+}(\phi)\, W^{+}(\phi) \tag{2.254}$$

where $V^{+}(\phi)$ and $W^{+}(\phi)$ are unimodular matrices (see Subsection 2.5.1) and

$$\Omega^+(\phi) = \left[\begin{array}{cccc|c} \omega_1(\phi) & 0 & \dots & 0 & \\ 0 & \omega_2(\phi) & \dots & 0 & 0 \\ & & \dots & & \\ 0 & 0 & \dots & \omega_\rho(\phi) & \\ \hline & 0 & & & 0 \end{array}\right] \begin{array}{l} \rho \\ \\ \mu-\rho \end{array} \qquad \begin{array}{cc} \rho & \kappa-\rho \end{array} \tag{2.255}$$

Here ρ is the (normal) rank of $\bar{G}^+(\phi)$ and the polynomials $\omega_i(\phi)$ have the divisibility property

$$\omega_i(\phi) | \omega_{i+1}(\phi) \qquad (\omega_i \textit{ divides } \omega_{i+1}) \qquad i=1\dots\rho\text{-}1 \tag{2.256}$$

Equations (2.254) and (2.255) are the *Smith form* of the polynomial matrix $\bar{G}^+(\phi)$ and $\omega_1(\phi)\dots\omega_\rho(t)$ are its *invariant polynomials*.

Note that the matrices $V^+(\phi)$ and $W^+(\phi)$ are not unique but the invariant polynomials are. Further, if matrix $\bar{G}^+(\phi)$ has the Smith matrix $\Omega^+(\phi)$, any other matrix obtained from $\bar{G}^+(\phi)$ by multiplication with a unimodular matrix or by elementary matrix operations, has the same Smith matrix.

Now substitute (2.254) into (2.253):

$$M(\phi) = V^+(\phi)\frac{\Omega^+(\phi)}{\bar{h}^+(\phi)} W^+(\phi) \tag{2.257}$$

Convert the diagonal elements of $\Omega^+(\phi)/\bar{h}^+(\phi)$ as

$$\frac{\omega_i(\phi)}{\bar{h}^+(\phi)} = \frac{\varphi_i(\phi)}{\eta_i(\phi)} \qquad i=1\dots\rho \tag{2.258}$$

so that $\eta_i(\phi)$ are *monic*, $\varphi_i(\phi)$ and $\eta_i(\phi)$ are *relative prime*. Then they obey the divisibility conditions

$$\varphi_i(\phi)|\varphi_{i+1}(\phi) \qquad \eta_{i+1}(\phi)|\eta_i(\phi) \qquad i=1\dots\rho\text{-}1 \tag{2.259}$$

with $\eta_1(\phi)=\bar{h}^+(\phi)$. Equation (2.257), with (2.258), is the *Smith-McMillan form* of the transfer function matrix $M(\phi)$; the $\varphi_i(\phi)$ and $\eta_i(\phi)$ polynomials are unique properties of the system.

Kalman (1965) showed that the order of the minimal (observable and controllable) realization of $M(\phi)$ is

$$\nu = \sum_{i=1}^{\rho} Deg\ \eta_i(\phi) \tag{2.260}$$

and the characteristic polynomial is

$$|\phi I - A| = \vartheta^+(\phi)\bar{h}^+(\phi) = \eta_1(\phi)\,\eta_2(\phi)\ldots\,\eta_\rho(\phi) \tag{2.261}$$

Poles. We define the poles of the multiple-input multiple-output system as those of its minimal (observable and controllable) realizations. It follows from (2.261) that these are the roots of the denominator polynomial $\eta_1(\phi)\eta_2(\phi)\ldots\eta_\rho(\phi)$ of the Smith-McMillan matrix $\boldsymbol{\Omega}^+(\phi)/\bar{h}^+(\phi)$. These include all the poles of the multiple-input single-output subsystems, with multiplicities revealed by the Smith-McMillan form. If all poles are simple in the subsystems then this set of poles is the same as the one obtained in (2.252). Also, the set of poles derived from the Smith-McMillan form is identical with the roots of the determinant $|\phi I - A|$ (that is, the eigenvalues of the matrix A), where A is the state transition matrix of the minimum realization.

Zeroes. The zeroes of the multiple-input multiple-output system are defined as the roots of the invariant (numerator) polynomial $\varphi_1(\phi)\varphi_2(\phi)\ldots\varphi_\rho(\phi)$ of the Smith-McMillan matrix. Again, multiplicities are possible, if and as shown by the invariant polynomial. Note that a multiple-input multiple-output system may have a pole and a zero at the same value (the individual φ_i, η_i pairs are relative prime but their products may not be). It can be shown (MacFarlane and Karcanias, 1976) that the zeroes defined above are the same as the invariant zeroes defined in terms of the system matrix $\boldsymbol{\Gamma}^+(\phi)$ (see Subsection 2.5.5) of the minimum (observable and controllable) realizations. If the realization contains unobservable and/or uncontrollable modes, these result in extra zeroes and poles, but these are invisible in the transfer function due to cancellation (see Subsections 2.6.4 and 2.6.5).

2.8. NOTES AND REFERENCES

Most of the contents of this chapter is standard material of intermediate level textbooks on linear systems theory, though it has been somewhat condensed, re-emphasized and re-interpreted here. Perhaps the best modern coverage of the broad subject can be found in the seminal book of Thomas Kailath (1980). Other valuable sources include the books by C.T. Chen (1984) and W.L. Brogan (1985).

The matrix properties outlined in this chapter are described in detail in any matrix algebra textbook, see for example (Horn and Johnson, 1985). They are also well covered in Brogan's book referred to above. Many of the important results originate from the seminal work of F.R. Gantmakher (1959).

The shift-operator description of discrete dynamic systems has been primarily the contribution of Karl Johan Åström and coworkers, see for example (Åström and Wittenmark, 1989 and 1997). A fundamental understanding of the properties of linear systems, by means of their state space description, is due to Rudolf Kalman, see for example his papers (1960, 1963 and 1965). His work relied on and complemented results by, among others, B. McMillan (1952) and E. Gilbert (1963). System zeroes were first studied by H.H. Rosenbrock (1970), followed by A.G.J. MacFarlane and N. Karcanias (1976) and, more recently, C.B. Schrader and M.K. Sain (1989).

3
Random Variables

3.0. INTRODUCTION

Most engineering systems are subject to *noise*. Noise is an input variable, or a set thereof, which is not measurable and the exact behavior of which is not known. Such noise variables are usually the mathematical representations of a broad variety of effects acting on the physical plant and its measurement and control system. Noise variables are considered *random* functions of time and are characterized by their *average properties* and *probability distribution functions*. These concepts will be introduced in this chapter, together with the important notion of the *independence* of random variables.

In many cases, the type of probability distribution is known for a particular random variable, or at least is assumed to be known, but some parameters of the distribution are not. Determining (guessing) the missing parameters from experimental data may be an important task, which is referred to as *statistical estimation*. Another important task is making a decision on whether the experimental data supports certain presumptions about the distribution. This latter activity is referred to as *statistical (hypothesis) testing*. These concepts will also be addressed in the present chapter.

Just like in Chapter 2, the variables will be assumed to be *discrete functions of time*. This, however, does not imply any restriction on their magnitude, that is, they will generally be assumed to be *continuous-valued*.

3.1. DISTRIBUTION AND EXPECTATION OF A SINGLE VARIABLE

In this section, the fundamental concepts of probability distribution and density functions will be introduced, for a *single random variable*. This will be followed by the definitions of *mean, variance* and *autocorrelation*.

3.1.1. The Concept of Probability

The characterization of random variables is rooted in the properties of random experiments which have a finite number of possible outcomes. For example, casting a die may lead to one of six scores. Performing the experiment repeatedly, one may record the number of occurrences for each outcome, also called *events*, and compute their *relative frequency*. For an event A, occurring N_A times in the course of N trials, the relative frequency is

$$Freq[A] = \frac{N_A}{N} \tag{3.1}$$

The limit of the relative frequencies, as the number of trials tends to infinity, is referred to as the *probability* of the particular event

$$Prob[A] = \lim_{N \to \infty} \frac{N_A}{N} \tag{3.2}$$

A discrete-valued random variable, for example x, can be used to characterize the experiment, its allowable values x_i, $i=1...n$, being the possible outcomes. The probabilities are then associated with those values, leading to a *probability function* $P_x(x_i)$, which has finite values (the probabilities) at the discrete values of $x=x_i$ (the events). That is,

$$P_x(x_i) = Prob[x=x_i] \tag{3.3}$$

It follows from (3.1) and (3.2) that

$$\sum_{i=1}^{n} P_x(x_i) = 1 \tag{3.4}$$

For example, in the case of the die experiment, the random variable is the score obtained, its possible values are $1,2,3,4,5,6$, and each value has a probability of $1/6$.

Strictly speaking, the above set of trials could be performed in parallel (many identical dice thrown at the same time) or sequentially (the same die cast many times). The conclusions drawn from the parallel execution are called the *ensemble* properties of the experiment while those obtained from the sequential execution are its *time properties*. An experiment is called *stationary* if its ensemble properties are time invariant and *ergodic* if, beyond this, they are identical with its time properties. The random variables of our concern occur as time sequences, therefore the serial interpretation of their statistical properties follows naturally and will be implied throughout.

3.1.2. Distribution and Density of a Continuous-Valued Variable

For a continuous-valued variable, relative frequencies cannot be assigned to the occurrence of particular values, because a finite set of trials would never yield most of those values. However, it is possible to define the relative frequency, and thus the probability, of the variable being *smaller than a particular value*. This leads to the definition

$$F_x(\xi) = Prob[x < \xi] \tag{3.5}$$

Here $F_x(\xi)$ is the *probability distribution function* of the variable $x(t)$ evaluated at $x(t)=\xi$. This may also be denoted as $F_x(x)$ or simply $F(x)$. Obviously, $F_x(\xi)$ is a monotonously increasing function of ξ and

$$\lim_{\xi \to \infty} F_x(\xi) = 1 \tag{3.6}$$

The derivative of the distribution function is

$$f_x(\xi) = \frac{dF_x(\xi)}{d\xi} \tag{3.7}$$

$f_x(\xi)$ is referred to as the *probability density function* of the variable $x(t)$ at the value $x(t)=\xi$. The density function may also be denoted as $f_x(x)$ or simply $f(x)$. The density function itself is not a probability, only its integral is:

$$\int_{\xi=\alpha}^{\beta} f_x(\xi)\, d\xi = F_x(\beta) - F_x(\alpha) = Prob[\alpha \leq x < \beta] \tag{3.8}$$

$$\int_{\xi=-\infty}^{\beta} f_x(\xi)\, d\xi = F_x(\beta) = Prob[x < \beta] \tag{3.9}$$

Obviously,

$$\int_{\xi=-\infty}^{\infty} f_x(\xi)\, d\xi = 1 \tag{3.10}$$

If the distribution function is smooth, the density may be considered as a local measure of probability, in the sense that, for a small Δ,

$$2\Delta f_x(\xi) \approx Prob[\xi-\Delta<x<\xi+\Delta] \tag{3.11}$$

Eq. (3.11), which is a rectangular approximation of (3.9) for a smooth distribution, is very helpful in the visualization of the probability density function.

3.1.3. Mean, Mean-Square and Variance

The following are the average-type properties of the random variable.

The *mean* or *expectation* of the random variable $x(t)$ is

$$\mu_x = E\{x(t)\} = \lim_{N\to\infty} \frac{1}{N} \sum_{t=1}^{N} x(t) \tag{3.12}$$

The *mean square* of the variable $x(t)$ is

$$E\{x^2(t)\} = \lim_{N\to\infty} \frac{1}{N} \sum_{t=1}^{N} x^2(t) \tag{3.13}$$

The *variance* of the random variable $x(t)$ is

$$\sigma_x^2 = Var\{x(t)\} = E\{[x(t)-\mu_x]^2\} = \lim_{N\to\infty} \frac{1}{N} \sum_{t=1}^{N} [x(t)-\mu_x]^2 \tag{3.14}$$

Since E{...} is a linear operation, it follows that

$$\sigma_x^2 = E\{x^2(t)-2\mu_x x(t)+\mu_x^2\} = E\{x^2(t)\} - 2\mu_x E\{x(t)\} + \mu_x^2 = E\{x^2(t)\} - \mu_x^2 \tag{3.15}$$

Clearly, if $\mu_x=0$ then $E\{x^2(t)\}=\sigma_x^2$.

The mean, mean-square and variance can be expressed with the density function as follows:

$$\mu_x = \int_{\xi=-\infty}^{\infty} \xi f_x(\xi)\, d\xi \tag{3.16}$$

$$E\{x^2(t)\} = \int_{\xi=-\infty}^{\infty} \xi^2 f_x(\xi)\, d\xi \tag{3.17}$$

$$\sigma_x^2 = \int_{\xi=-\infty}^{\infty} [\xi-\mu_x]^2 f_x(\xi)\, d\xi \tag{3.18}$$

3.1.4. Autocorrelation and Autocovariance

An important characteristic of the random signal $x(t)$ is its *autocorrelation function*

$$\varphi_{xx}(\tau) = \mathrm{E}\{x(t)\, x(t-\tau)\} \tag{3.19}$$

Here τ is the (discrete) *time-shift* of the correlation. Similarly, one may define the *autocovariance* as

$$\psi_{xx}(\tau) = \mathrm{E}\{[x(t)-\mu_x][x(t-\tau)-\mu_x]\} \tag{3.20}$$

Clearly,

$$\psi_{xx}(\tau) = \varphi_{xx}(\tau) - \mu_x^2 \tag{3.21}$$

and, if $\mu_x = 0$, then $\varphi_{xx}(\tau) = \psi_{xx}(\tau)$. The autocorrelation and autocovariance functions are even, that is,

$$\varphi_{xx}(\tau) = \varphi_{xx}(-\tau) \qquad \psi_{xx}(\tau) = \psi_{xx}(-\tau) \tag{3.22}$$

Further

$$\varphi_{xx}(0) = \mathrm{E}\{x^2(t)\} \qquad \psi_{xx}(0) = \sigma_x^2 \tag{3.23}$$

Of particular interest are the random variables for which

$$\varphi_{xx}(\tau) = \mu_x^2 \qquad \psi_{xx}(\tau) = 0 \qquad \text{for all } \tau \neq 0 \tag{3.24}$$

Such random variables (signals) are referred to as uncorrelated in time, or *white*. If, further, they have zero mean, then

$$\varphi_{xx}(\tau) = \begin{cases} \sigma_x^2 & \text{for } \tau = 0 \\ 0 & \text{for } \tau \neq 0 \end{cases} \tag{3.25}$$

Note that the autocorrelation and autocovariance functions cannot be derived from the distribution function $f_x(\xi)$ because the latter carries no information on how the values $x(t)$ and $x(t-\tau)$ are related to each other. To describe this behavior, the *joint probability density function* of $x(t)$ and $x(t-\tau)$ is needed. Such joint density functions will be introduced in the next section where some implications of whiteness will also be explored.

3.2. BIVARIATE DISTRIBUTIONS

In this section, distributions describing the relationship between two random variables will be discussed. *Bivariate joint* and *conditional* distributions will be introduced, followed by the important concept of *statistical independence*. Finally, the *cross-correlation and covariance functions* and their behavior under independence will be explored.

3.2.1. Joint Distribution of Two Random Variables

As a background for this section, consider two random experiments, performed simultaneously, both with a finite number of outcomes. Select one each of the possible outcomes, e.g. *event A* for the first experiment and *event B* for the second. Then the relative frequency of *A and B* is

$$Freq[A \text{ and } B] = \frac{N_{AB}}{N} \tag{3.26}$$

where N_{AB} is the number of occurrences when, in the course of N trials, the first experiment yielded A and the second B. The pair of experiments can be characterized by the random variables x and y, which may take the discrete values x_i, $i=1...n$ and y_j, $j=1...m$, respectively. The *joint probability function* of the two variables is

$$P_{xy}(x_i, y_j) = Prob[x=x_i, y=y_j] \tag{3.27}$$

Consider now two continuous-valued random variables, $x(t)$ and $y(t)$. Their *joint probability distribution* is

$$F_{xy}(\xi, \eta) = Prob[x<\xi, y<\eta] \tag{3.28}$$

The *joint probability density function* of the two variables is

$$f_{xy}(\xi, \eta) = \frac{\partial^2 F_{xy}(\xi, \eta)}{\partial\xi \; \partial\eta} \tag{3.29}$$

so that

$$\int_{\eta=-\infty}^{\beta} \int_{\xi=-\infty}^{\alpha} f_{xy}(\xi, \eta)\, d\xi\, d\eta = F_{xy}(\alpha, \beta) \tag{3.30}$$

and

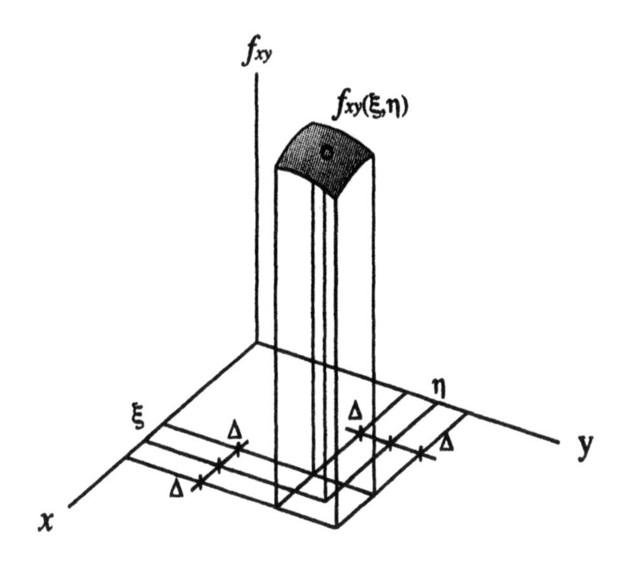

FIGURE 3.1. Joint local probability for two variables.

$$\int_{\eta=-\infty}^{\infty} \int_{\xi=-\infty}^{\infty} f_{xy}(\xi, \eta)\, d\xi\, d\eta = 1 \tag{3.31}$$

The bivariate density function, if it arises from a smooth distribution, can also be considered as a local measure of probability (Fig. 3.1). For a sufficiently small ,

$$4\Delta^2 f_{xy}(\xi, \eta) \approx Prob[\xi-\Delta<x<\xi+\Delta,\ \ \eta-\Delta<y<\eta+\Delta] \tag{3.32}$$

The single variable density functions $f_x(\xi)$, $f_y(\eta)$, which are now *marginal densities* of the bivariate distribution, can be computed as

$$f_x(\xi) = \int_{\eta=-\infty}^{\infty} f_{xy}(\xi, \eta)\, d\eta \qquad f_y(\eta) = \int_{\xi=-\infty}^{\infty} f_{xy}(\xi, \eta)\, d\xi \tag{3.33}$$

3.2.2. Conditional Distribution and Independence

Consider again the two simultaneous experiments with discrete outcomes. Count the occurrences of *event B* in the second experiment, with no respect to the outcome of the first experiment, and also the joint occurrences of *event A* in the first experiment and *event B* in the second. Then the *conditional frequency* of *A*, provided *B* happened, is

$$Freq[A|B] = \frac{N_{AB}}{N_B} \tag{3.34}$$

Notice that the reference now is not the total number of trials N, only the number of those trials when the condition B occurred. The respective probability relation for the discrete valued random variables x and y is

$$Prob[x=x_i \,|y=y_j] = \frac{Prob[x=x_i,\ y=y_j]}{Prob[y=y_j]} \tag{3.35}$$

that is,

$$P_{x|y}(x_i \,|y_j) = \frac{P_{xy}(x_i,\ y_j)}{P_y(y_j)} \tag{3.36}$$

where $P_{x|y}(x_i \,|y_j) = Prob[x=x_i \,|y=y_j]$ is the *conditional probability function* of x, relative to y.

The following *law of total probability* arises from (3.36)

$$P_x(x_i) = \sum_{j=1}^{m} P_{xy}(x_i,\ y_j) = \sum_{j=1}^{m} P_{x|y}(x_i \,|y_j)\, P_y(y_j) \tag{3.37}$$

The two events (experiments) are called *independent* if the conditional frequency $Freq[A|B]$ is the same as the unconditional frequency $Freq[A]$. In terms of the probability functions, the two random variables x and y are independent if

$$P_{x|y}(x_i|y_j) = P_x(x_i) \qquad for \quad i=1...n, \quad j=1...m \tag{3.38}$$

This implies that, if x and y are independent, then

$$P_{xy}(x_i,\ y_j) = P_x(x_i)\, P_y(y_j) \qquad for \quad i=1...n, \quad j=1...m \tag{3.39}$$

To apply the concept of conditional probability to continuous-valued random variables, consider the local probabilities expressed in (3.11) and (3.32) (Fig. 3.2). By (3.35)

$$Prob[\xi\text{-}\Delta<x<\xi+\Delta \mid \eta\text{-}\Delta<y<\eta+\Delta]$$

$$= Prob[\xi\text{-}\Delta<x<\xi+\Delta,\ \eta\text{-}\Delta<y<\eta+\Delta] \,/\, Prob[\eta\text{-}\Delta<y<\eta+\Delta] \tag{3.40}$$

With (3.11) and (3.32),

$$2\Delta f_{x|y}(\xi\,|\,\eta) = 4\Delta^2 f_{xy}(\xi,\ \eta) \,/\, 2\Delta f_y(\eta) \tag{3.41}$$

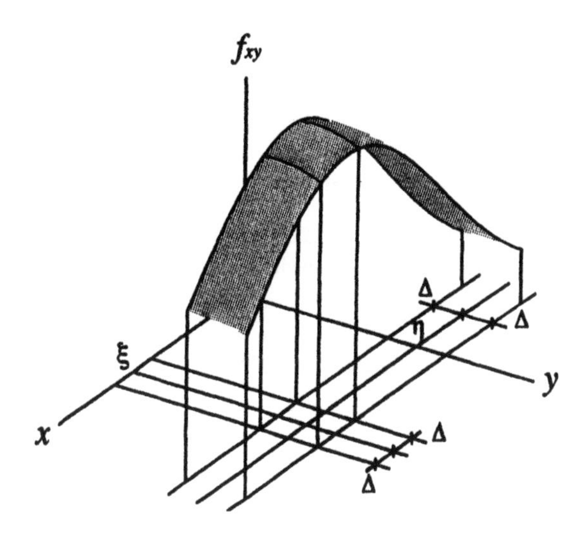

FIGURE 3.2. Local conditional probability for two variables.

where $f_{x|y}(\xi \mid \eta)$ is the *conditional probability density function* of x, relative to y. This clearly implies

$$f_{x|y}(\xi \mid \eta) = \frac{f_{xy}(\xi, \eta)}{f_y(\eta)} \tag{3.42}$$

Further, it follows from (3.33) and (3.42) that

$$f_x(\xi) = \int_{\eta=-\infty}^{\infty} f_{xy}(\xi, \eta)\, d\eta = \int_{\eta=-\infty}^{\infty} f_{x|y}(\xi \mid \eta) f_y(\eta)\, d\eta \tag{3.43}$$

which is the law of total probability for continuous-valued random variables. If

$$f_{x|y}(\xi \mid \eta) = f_x(\xi) \quad \textit{for all } \xi \textit{ and } \eta \tag{3.44}$$

then the two random variables x and y are independent. In this case,

$$f_{xy}(\xi, \eta) = f_x(\xi) f_y(\eta) \quad \textit{for all } \xi \textit{ and } \eta \tag{3.45}$$

Conditional mean. The *conditional mean* of the variable x, relative to y, is

$$\mu_{x|y}(\eta) = \int_{\xi=-\infty}^{\infty} \xi\, f_{x|y}(\xi \mid \eta)\, d\xi \tag{3.46}$$

In general, this depends on η, the value of y. However, if x and y are independent then

$$\mu_{x|y}(\eta) = \mu_x \tag{3.47}$$

3.2.3. Distributions with Time-Shift. Cross-Correlation and Covariance

Distributions with time-shift. If $x(t)$ and $y(t)$ are random variables, they are functions of time. In a bivariate distribution, the two variables do not necessarily belong to the same time value. Due to the assumption of stationarity, the absolute time of sampling is irrelevant, but the time-shift τ between the two variable samples becomes a parameter of the distribution. Thus

$$F_{xy}(\xi, \eta, \tau) = Prob[x(t) < \xi, y(t-\tau) < \eta] \tag{3.48}$$

is the joint probability distribution function of $x(t)$ and $y(t-\tau)$ and

$$f_{xy}(\xi, \eta, \tau) = \partial^2 F_{xy}(\xi, \eta, \tau) / \partial\xi \; \partial\eta \tag{3.49}$$

is their joint probability density function. Further,

$$f_{x|y}(\xi \,|\, \eta, \tau) = f_{xy}(\xi, \eta, \tau) / f_y(\eta) \tag{3.50}$$

is the conditional density function of $x(t)$ relative to $y(t-\tau)$. Clearly, $F_{xy}(\xi, \eta, \tau) = F_{yx}(\eta, \xi, -\tau)$, since they are just different forms of the same function. However, in general,

$$F_{xy}(\xi, \eta, \tau) \neq F_{xy}(\xi, \eta, -\tau) \tag{3.51}$$

The same properties hold for $f_{xy}(\xi, \eta, \tau)$ and $f_{x|y}(\xi \,|\, \eta, \tau)$.

The distributions linking the unlike samples of the same variable can be handled as a special case of the above. In particular,

$$F_{xx}(\xi, \eta, \tau) = Prob[x(t) < \xi, x(t-\tau) < \eta] \tag{3.52}$$

is the joint probability distribution function of $x(t)$ and $x(t-\tau)$ and

$$f_{xx}(\xi, \eta, \tau) = \partial^2 F_{xx}(\xi, \eta, \tau) / \partial\xi \; \partial\eta \tag{3.53}$$

is their joint probability density function. Further,

$$f_{x|x}(\xi \,|\, \eta, \tau) = f_{xx}(\xi, \eta, \tau) / f_x(\eta) \tag{3.54}$$

is the conditional density of $x(t)$ relative to $x(t-\tau)$.

Cross-correlation and covariance. The *cross-correlation* of the variables $x(t)$ and $y(t-\tau)$ is

$$\varphi_{xy}(\tau) = E\{x(t)\, y(t-\tau)\} \tag{3.55}$$

while their *covariance* is

$$\psi_{xy}(\tau) = E\{[x(t)-\mu_x]\, [y(t-\tau)-\mu_y]\} \tag{3.56}$$

Clearly (c.f. (3.21))

$$\psi_{xy}(\tau) = \varphi_{xy}(\tau) - \mu_x \mu_y \tag{3.57}$$

Cross-correlation and covariance may be expressed with the joint probability density function as

$$\varphi_{xy}(\tau) = \int_{\eta=-\infty}^{\infty} \int_{\xi=-\infty}^{\infty} \xi\, \eta\, f_{xy}(\xi, \eta, \tau)\, d\xi\, d\eta \tag{3.58}$$

$$\psi_{xy}(\tau) = \int_{\eta=-\infty}^{\infty} \int_{\xi=-\infty}^{\infty} [\xi-\mu_x]\, [\eta-\mu_y]\, f_{xy}(\xi, \eta, \tau)\, d\xi\, d\eta \tag{3.59}$$

If $x(t)$ and $y(t-\tau)$ are *independent* then

$$f_{xy}(\xi, \eta, \tau) = f_x(\xi)\, f_y(\eta) \tag{3.60}$$

With this

$$\varphi_{xy}(\tau) = \int_{\eta=-\infty}^{\infty} \xi\, f_x(\xi)\, d\xi \int_{\xi=-\infty}^{\infty} \eta\, f_y(\eta)\, d\eta = \mu_x \mu_y \tag{3.61}$$

$$\psi_{xy}(\tau) = \int_{\eta=-\infty}^{\infty} [\xi-\mu_x] f_x(\xi)\, d\xi \int_{\xi=-\infty}^{\infty} [\eta-\mu_y] f_y(\eta)\, d\eta = 0 \tag{3.62}$$

That is, the covariance of independent variables is zero. This is a very important result which has broad application. Note that the converse is not true in general, that is, zero covariance does not imply the independence of the two variables. However, if the joint distribution is normal, which is the most frequent case, then independence and zero covariance are equivalent properties.

The auto-correlation and auto-covariance, introduced in Subsection 3.1.4, can be expressed with the joint density function (3.53). For $\tau \neq 0$,

$$\varphi_{xx}(\tau) = \int_{\eta=-\infty}^{\infty} \int_{\xi=-\infty}^{\infty} \xi \, \eta \, f_{xx}(\xi, \eta, \tau) \, d\xi \, d\eta \tag{3.63}$$

$$\psi_{xx}(\tau) = \int_{\eta=-\infty}^{\infty} \int_{\xi=-\infty}^{\infty} [\xi-\mu_x] \, [\eta-\mu_x] f_{xx}(\xi, \eta, \tau) \, d\xi \, d\eta \tag{3.64}$$

If $x(t)$ and $x(t-\tau)$ are *independent* then

$$f_{xx}(\xi, \eta, \tau) = f_x(\xi) f_x(\eta) \tag{3.65}$$

and

$$\varphi_{xx}(\tau) = \mu_x^2 \qquad \psi_{xx}(\tau) = 0 \tag{3.66}$$

That is, if $x(t)$ and $x(t-\tau)$ are independent for all $\tau \neq 0$ then the *whiteness* of the signal $x(t)$ follows. (However, whiteness implies independence of the series only if the joint distributions are normal.)

3.2.4. The Distribution of the Sum

We will consider here the properties of a variable obtained as the sum of two random variables,

$$z(t) = x(t) + y(t) \tag{3.67}$$

The generalization of these results, in the next section, will lead to the properties of the sample mean, which plays a very important role in statistical analysis and decision making.

Assume first that the concerned variables are discrete-valued. Then a particular value z_k arises as any of the possible combinations of x_i and y_j for which $x_i + y_j = z_k$. Accordingly, the probability of the sum is

$$Prob[z = z_k] = \sum_{i,j} Prob[x = x_i, \ y = y_j, \ x_i + y_j = z_k] \tag{3.68}$$

that is

$$P_z(z_k) = \sum_i P_{xy}(x_i, \ y_j = z_k - x_i) \tag{3.69}$$

To apply this line of thought to continuous-valued variables, let us resort to the local probability interpretation of the density function. The probability that the sum $z(t)$ is in the $\pm\Delta$ vicinity of the value ζ is

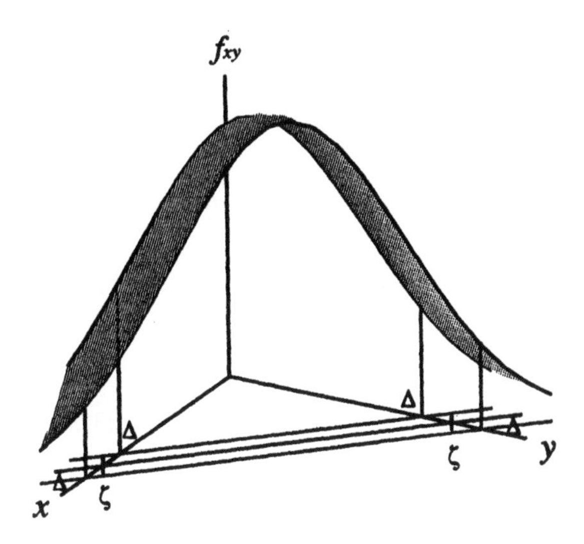

FIGURE 3.3. Local probability of the sum of two variables.

$$Prob[\zeta-\Delta<z<\zeta+\Delta] = 2\Delta f_z(\zeta) \tag{3.70}$$

Observe that (3.67) now implies the relationship $\zeta=\xi+\eta$ among the variable values, that is, $\eta=\zeta-\xi$, which (for a given ξ) is a straight line in the ξ - η plane (Fig. 3.3). For the sum to be in the $\pm\Delta$ vicinity of the value ζ, the components need to be between the lines $\eta=\zeta-\Delta-\xi$ and $\eta=\zeta+\Delta-\xi$. Thus

$$Prob[\zeta-\Delta<z<\zeta+\Delta]$$

$$= \int_{\xi=-\infty}^{\infty} \int_{\eta=\zeta-\Delta-\xi}^{\zeta+\Delta-\xi} f_{xy}(\xi,\eta)\,d\eta\,d\xi = 2\Delta \int_{\xi=-\infty}^{\infty} f_{xy}(\xi,\zeta-\xi)\,d\xi \tag{3.71}$$

This, with (3.70), implies

$$f_z(\zeta)= \int_{\xi=-\infty}^{\infty} f_{xy}(\xi,\zeta-\xi)\,d\xi \tag{3.72}$$

If x and y are independent then (3.72) becomes

$$f_z(\zeta)= \int_{\xi=-\infty}^{\infty} f_x(\xi) f_y(\zeta-\xi)\,d\xi \tag{3.73}$$

The mean of the sum is obtained simply as

$$\mu_z = \mathrm{E}\{x(t) + y(t)\} = \mu_x + \mu_y \tag{3.74}$$

while the mean-square and the variance are computed as

$$\mathrm{E}\{z^2(t)\} = \mathrm{E}\{[x(t)+y(t)]^2\} = \mathrm{E}\{x^2(t)\} + \mathrm{E}\{y^2(t)\} + 2\mathrm{E}\{x(t)y(t)\} \tag{3.75}$$

$$\sigma_z^2 = \mathrm{E}\{[z(t) - \mu_z]^2\} = \sigma_x^2 + \sigma_y^2 + 2\mathrm{E}\{[x(t)-\mu_x][y(t)-\mu_y]\} \tag{3.76}$$

If x and y are uncorrelated then

$$\mathrm{E}\{z^2(t)\} = \mathrm{E}\{x^2(t)\} + \mathrm{E}\{y^2(t)\} + 2\mu_x\mu_y \qquad \sigma_z^2 = \sigma_x^2 + \sigma_y^2 \tag{3.77}$$

3.3. MULTIVARIATE DISTRIBUTIONS

In this section, the concepts introduced previously for two variables will be generalized to sets (vectors) of random variables. This will also facilitate the analysis of the properties of time-series of random variables, both scalar and vector.

3.3.1. Single Vector Distributions

Consider a random vector

$$\boldsymbol{x}(t) = [x_1(t), x_2(t) \ldots x_n(t)]' \tag{3.78a}$$

and its value-set

$$\boldsymbol{\xi} = [\xi_1, \xi_2, \ldots \xi_n]' \tag{3.78b}$$

The joint probability distribution function is defined as

$$F_{\boldsymbol{x}}(\boldsymbol{\xi}) = Prob[x_i(t) < \xi_i,\ i=1\ldots n] \tag{3.79}$$

while the joint probability density function is

$$f_{\boldsymbol{x}}(\boldsymbol{\xi}) = \partial^n F_{\boldsymbol{x}}(\boldsymbol{\xi}) \,/\, \partial\xi_1\, \partial\xi_2 \ldots \partial\xi_n \tag{3.80}$$

If the elements of the random vector are independent then

$$f_{\boldsymbol{x}}(\boldsymbol{\xi}) = f_{x1}(\xi_1) f_{x2}(\xi_2) \ldots f_{xn}(\xi_n) \tag{3.81}$$

The independence of the full vector, of course, implies that the elements are pairwise independent.

The mean of the random vector $x(t)$ is

$$\mu_x = \mathrm{E}\{x(t)\} = [\mu_{x1}, \mu_{x2} \cdots \mu_{xn}]' \tag{3.82}$$

where $\mu_{xi} = \mathrm{E}\{x_i(t)\}$, $i=1 \ldots n$. The correlation matrix of the vector $x(t)$ is

$$\Phi_x = \mathrm{E}\{x(t)\, x'(t)\} \tag{3.83}$$

$$= \begin{bmatrix} \mathrm{E}\{x_1^2(t)\} & \mathrm{E}\{x_1(t)\, x_2(t)\} & \cdots & \mathrm{E}\{x_1(t)\, x_n(t)\} \\ \mathrm{E}\{x_2(t)\, x_1(t)\} & \mathrm{E}\{x_2^2(t)\} & \cdots & \mathrm{E}\{x_2(t)\, x_n(t)\} \\ \vdots & & & \\ \mathrm{E}\{x_n(t)\, x_1(t)\} & \mathrm{E}\{x_n(t)\, x_2(t)\} & \cdots & \mathrm{E}\{x_n^2(t)\} \end{bmatrix}$$

Note that this is basically an auto-correlation function; observe the symmetry of the matrix. The correlation matrix may be obtained with the joint density function as

$$\Phi_x = \int_{\xi_n=-\infty}^{\infty} \cdots \int_{\xi_2=-\infty}^{\infty} \int_{\xi_1=-\infty}^{\infty} \xi\, \xi' f_x(\xi)\, d\xi_1 d\xi_2 \ldots d\xi_n \tag{3.84}$$

This implies that the elements of the correlation matrix are computed as

$$[\Phi_x]_{ij} = \int_{\xi_j=-\infty}^{\infty} \int_{\xi_i=-\infty}^{\infty} \xi_i \xi_j f_{xi,xj}(\xi_i, \xi_j)\, d\xi_i d\xi_j \qquad for\ i \neq j \tag{3.85}$$

$$[\Phi_x]_{ii} = \int_{\xi_i=-\infty}^{\infty} \xi_i^2 f_{xi}(\xi_i)\, d\xi_i \tag{3.86}$$

where $f_{xi,xj}(\xi_i, \xi_j)$ is the bivariate marginal density function obtained by integrating $f_x(\xi)$ n-2 times, by all elements of ξ except the i-th and j-th, and $f_{xi}(\xi_i)$ is the univariate marginal density obtained by n-1 integrations. If the elements of the $x(t)$ vector are independent then, as it follows from (3.81) and (3.84),

$$\Phi_x = \mu_x \mu_x' + Diag[\sigma_{x1}^2, \sigma_{x2}^2 \cdots \sigma_{xn}^2] \tag{3.87}$$

The covariance matrix of the vector random variable $x(t)$ is defined as

$$\Psi_x = E\{[x(t)-\mu_x][x(t)-\mu_x]'\} = \Phi_x - \mu_x \mu_x' \tag{3.88}$$

The properties of the covariance matrix follow from those of the correlation matrix, as presented above, including $\Psi_x = Diag[\sigma^2_{x1}, \sigma^2_{x2} \ldots \sigma^2_{xn}]$ if the elements are independent.

Time-series. A special case of (3.78a) arises when the elements of the random vector are time samples of the same random variable, $x(t)$, $x(t-1)$ ... $x(t-m)$, where $m+1$ is the *window width*. Let us reperesent this series as

$$X(t) = [x(t), x(t-1) \ldots x(t-m)]' \tag{3.89}$$

Now the joint distribution function is

$$F_X(\xi) = Prob[x(t-\tau) < \xi_\tau,\ \tau=0 \ldots m] \tag{3.90}$$

The joint density function is obtained as in (3.80). The correlation matrix is

$$\Phi_X = E\{X(t)\, X'(t)\} \tag{3.91}$$

$$= \begin{bmatrix} E\{x^2(t)\} & E\{x(t)\, x(t-1)\} & \ldots & E\{x(t)\, x(t-m)\} \\ E\{x(t-1)\, x(t)\} & E\{x^2(t-1)\} & \ldots & E\{x(t-1)\, x(t-m)\} \\ \vdots & & & \\ E\{x(t-m)\, x(t)\} & E\{x(t-m)\, x(t-1)\} & \ldots & E\{x^2(t-m)\} \end{bmatrix}$$

Because of the stationarity of $X(t)$, the correlation matrix can also be written as

$$\Phi_X = \begin{bmatrix} \varphi_{xx}(0) & \varphi_{xx}(1) & \ldots & \varphi_{xx}(m) \\ \varphi_{xx}(1) & \varphi_{xx}(0) & \ldots & \varphi_{xx}(m-1) \\ \vdots & & & \\ \varphi_{xx}(m) & \varphi_{xx}(m-1) & \ldots & \varphi_{xx}(0) \end{bmatrix} \tag{3.92}$$

where $\varphi_{xx}(\tau)$ is the scalar auto-correlation function defined in (3.19). Clearly, if the time-series is independent (or at least uncorrelated) then, by (3.24), (3.23) and (3.15),

$$\mathbf{\Phi}_X = \mu_x^2 U + \sigma_x^2 I \tag{3.93}$$

where U is an $(m+1)\cdot(m+1)$ matrix having all 1 elements.

3.3.2. Bi-Vector Distributions and Time-Shifted Vectors

Consider now $x(t)$ as defined in (3.78a), together with another random vector $y(t)=[y_1(t), y_2(t) \dots y_k(t)]'$ and its value-set $\boldsymbol{\eta}$. The joint distribution of the two vector variables is

$$F_{xy}(\boldsymbol{\xi}, \boldsymbol{\eta}) = Prob[x_i(t) < \xi_i,\ y_j(t) < \eta_j,\ \ i=1\dots n,\ j=1\dots k] \tag{3.94}$$

and their joint density function is

$$f_{xy}(\boldsymbol{\xi}, \boldsymbol{\eta}) = \partial^{n+k} F_{xy}(\boldsymbol{\xi}, \boldsymbol{\eta}) \,/\, \partial\xi_1 \dots \partial\xi_n \partial\eta_1 \dots \partial\eta_k \tag{3.95}$$

The conditional density function of $x(t)$, relative to $y(t)$ is

$$f_{x|y}(\boldsymbol{\xi} \mid \boldsymbol{\eta}) = f_{xy}(\boldsymbol{\xi}, \boldsymbol{\eta}) \,/ f_y(\boldsymbol{\eta}) \tag{3.96}$$

If the two vector variables are independent then

$$f_{xy}(\boldsymbol{\xi}, \boldsymbol{\eta}) = f_x(\boldsymbol{\xi}) f_y(\boldsymbol{\eta}) \tag{3.97}$$

The cross-correlation matrix of the two vector variables is

$$\mathbf{\Phi}_{xy} = \mathrm{E}\{x(t)\, y'(t)\} \tag{3.98}$$

$$= \begin{bmatrix} \mathrm{E}\{x_1(t)\, y_1(t)\} & \mathrm{E}\{x_1(t)\, y_2(t)\} & \dots & \mathrm{E}\{x_1(t)\, y_k(t)\} \\ \mathrm{E}\{x_2(t)\, y_1(t)\} & \mathrm{E}\{x_2(t)\, y_2(t)\} & \dots & \mathrm{E}\{x_2(t)\, y_k(t)\} \\ \vdots & & & \\ \mathrm{E}\{x_n(t)\, y_1(t)\} & \mathrm{E}\{x_n(t)\, y_2(t)\} & \dots & \mathrm{E}\{x_n(t)\, y_k(t)\} \end{bmatrix}$$

Notice that this correlation matrix is not symmetrical and, in general, not even square. If the two vectors are independent (or at least uncorrelated) then

$$\mathbf{\Phi}_{xy} = \mu_x \mu'_y \tag{3.99}$$

Time-shifted vectors. The relationship between the time-shifted vectors $\boldsymbol{x}(t)$ and $\boldsymbol{x}(t-\tau)$, which is very important in the analysis of discrete dynamic systems, can be described as a special case of the bi-vector distributions. Let us first fix the time-shift τ. The joint distribution function is now

$$F_{\boldsymbol{xx}}(\boldsymbol{\xi}, \boldsymbol{\eta}, \tau) = Prob[x_i(t) < \xi_i,\ x_j(t-\tau) < \eta_j,\ \ i=1...n,\ j=1...n] \tag{3.100}$$

with the joint density function and conditional density function

$$f_{\boldsymbol{xx}}(\boldsymbol{\xi}, \boldsymbol{\eta}, \tau) = \partial^{2n} F_{\boldsymbol{xx}}(\boldsymbol{\xi}, \boldsymbol{\eta}, \tau) \,/\, \partial\xi_1 ... \partial\xi_n \partial\eta_1 ... \partial\eta_n \tag{3.101}$$

$$f_{\boldsymbol{x}|\boldsymbol{x}}(\boldsymbol{\xi} \mid \boldsymbol{\eta}, \tau) = f_{\boldsymbol{xx}}(\boldsymbol{\xi}, \boldsymbol{\eta}, \tau) \,/\, f_{\boldsymbol{x}}(\boldsymbol{\eta}) \tag{3.102}$$

If $\boldsymbol{x}(t)$ and $\boldsymbol{x}(t-\tau)$ are independent then

$$f_{\boldsymbol{xx}}(\boldsymbol{\xi}, \boldsymbol{\eta}, \tau) = f_{\boldsymbol{x}}(\boldsymbol{\xi}) f_{\boldsymbol{x}}(\boldsymbol{\eta}) \tag{3.103}$$

The correlation matrix becomes

$$\boldsymbol{\Phi}_{\boldsymbol{xx}}(\tau) = \mathrm{E}\{\boldsymbol{x}(t)\, \boldsymbol{x}'(t-\tau)\} \tag{3.104}$$

$$= \begin{bmatrix} \mathrm{E}\{x_1(t)\, x_1(t-\tau)\} & \mathrm{E}\{x_1(t)\, x_2(t-\tau)\} & \cdots & \mathrm{E}\{x_1(t)\, x_n(t-\tau)\} \\ \mathrm{E}\{x_2(t)\, x_1(t-\tau)\} & \mathrm{E}\{x_2(t)\, x_2(t-\tau)\} & \cdots & \mathrm{E}\{x_2(t)\, x_n(t-\tau)\} \\ \vdots & & & \\ \mathrm{E}\{x_n(t)\, x_1(t-\tau)\} & \mathrm{E}\{x_n(t)\, x_2(t-\tau)\} & \cdots & \mathrm{E}\{x_n(t)\, x_n(t-\tau)\} \end{bmatrix}$$

Now define

$$\varphi_{xi,xj}(\tau) = \mathrm{E}\{x_i(t)\, x_j(t-\tau)\} \tag{3.105}$$

Then (3.104) can be written as

$$\boldsymbol{\Phi}_{\boldsymbol{xx}}(\tau) = \begin{bmatrix} \varphi_{x1,x1}(\tau) & \varphi_{x1,x2}(\tau) & \cdots & \varphi_{x1,xn}(\tau) \\ \varphi_{x2,x1}(\tau) & \varphi_{x2,x2}(\tau) & \cdots & \varphi_{x2,xn}(\tau) \\ \vdots & & & \\ \varphi_{xn,x1}(\tau) & \varphi_{xn,x2}(\tau) & \cdots & \varphi_{xn,xn}(\tau) \end{bmatrix} \tag{3.106}$$

Notice that, in general,

$$\varphi_{xi,xj}(\tau) = \mathrm{E}\{x_i(t)\, x_j(t-\tau)\} \neq \mathrm{E}\{x_j(t)\, x_i(t-\tau)\} = \varphi_{xj,xi}(\tau) \tag{3.107}$$

thus $\mathbf{\Phi}_{xx}(\tau)$ is not symmetrical. However, for $\tau=0$, $\mathbf{\Phi}_{xx}(0)$ is symmetrical and (c.f. (3.83))

$$\mathbf{\Phi}_{xx}(0) = \mathbf{\Phi}_x \tag{3.108}$$

Finally, if $\boldsymbol{x}(t)$ and $\boldsymbol{x}(t-\tau)$, $\tau\neq 0$, are independent (or uncorrelated) then

$$\mathbf{\Phi}_{xx}(\tau) = \boldsymbol{\mu}_x \boldsymbol{\mu}_x' \tag{3.109}$$

Vector time-series. To conclude, consider now the series of random vectors $\boldsymbol{x}(t)$, $\boldsymbol{x}(t-1)$ ... $\boldsymbol{x}(t-m)$, where $m+1$ is the window width. Let us represent the series by the compound random vector

$$\boldsymbol{X}(t) = [\boldsymbol{x}'(t),\, \boldsymbol{x}'(t-1) \ldots \boldsymbol{x}'(t-m)]' \tag{3.110a}$$

Define the value-set

$$\boldsymbol{\Xi} = [\boldsymbol{\xi}'_{\cdot 0},\, \boldsymbol{\xi}'_{\cdot 1},\, \ldots,\, \boldsymbol{\xi}'_{\cdot m}]' \tag{3.110b}$$

where $\xi_{\cdot\tau}$, $\tau=0\ldots m$, are column vectors of n elements each. The properties of the $\boldsymbol{X}(t)$ series can be characterized by the joint distribution function

$$F_X(\Xi) = Prob[x_i(t-\tau) < \xi_{i\tau},\ i=1\ldots n,\ \tau=0\ldots m] \tag{3.111}$$

with the joint density function

$$f_X(\Xi) = \partial^{n(m+1)} F_X(\Xi) \,/\, \partial\xi_{10} \ldots \partial\xi_{nm} \tag{3.112}$$

Observe that this is a multivariate distribution with $n(m+1)$ variables. If $\boldsymbol{x}(t)$, $\boldsymbol{x}(t-1)$ ... $\boldsymbol{x}(t-m)$ are independent then

$$f_X(\Xi) = f_x(\xi_{\cdot 0}) \ldots f_x(\boldsymbol{\xi}_{\cdot m}) \tag{3.113}$$

The correlation function of the compound random vector $\boldsymbol{X}(t)$ is

$$\mathbf{\Phi}_X = \mathrm{E}\{\boldsymbol{X}(t)\, \boldsymbol{X}'(t)\}$$

$$= \begin{bmatrix} \Phi_{xx}(0) & \Phi_{xx}(1) & \cdots & \Phi_{xx}(m) \\ \Phi_{xx}(-1) & \Phi_{xx}(0) & \cdots & \Phi_{xx}(m-1) \\ \cdot & & & \\ \cdot & & & \\ \cdot & & & \\ \Phi_{xx}(-m) & \Phi_{xx}(-m+1) & \cdots & \Phi_{xx}(0) \end{bmatrix} \tag{3.114}$$

The size of this matrix is $[n(m+1)]\cdot[n(m+1)]$ which may be rather large, usually due to a long window. Since

$$\Phi_{xx}(-\tau) = \mathrm{E}\{x(t)\, x'(t+\tau)\} = \mathrm{E}\{x(t-\tau)\, x'(t)\} = \Phi'_{xx}(\tau) \tag{3.115}$$

the correlation matrix can also be written as

$$\Phi_X = \begin{bmatrix} \Phi_{xx}(0) & \Phi_{xx}(1) & \cdots & \Phi_{xx}(m) \\ \Phi'_{xx}(1) & \Phi_{xx}(0) & \cdots & \Phi_{xx}(m-1) \\ \cdot & & & \\ \cdot & & & \\ \cdot & & & \\ \Phi'_{xx}(m) & \Phi'_{xx}(m-1) & \cdots & \Phi_{xx}(0) \end{bmatrix} \tag{3.116}$$

which shows clear symmetry. If the vectors $x(t)$, $x(t-1)$ ... $x(t-m)$ are independent (or at least uncorrelated) then, by (3.109), all the off-diagonal matrices in (3.116) become $\mu_x \mu_x'$. If, further, $x(t)$ has zero mean then all off-diagonal matrices become zero.

3.3.3. The Distribution of the Sum and the Window Average

In this subsection, the results of Subsection 3.2.4 will be generalized to multivariate distributions and then applied to the window average of scalar and vector time-series.

The sum of multiple variables. Consider a variable arising as the sum of the elements in the vector $x(t)$:

$$z(t) = \sum_{i=1}^{n} x_i(t) \tag{3.117}$$

The density function of this sum may be obtained by the formal generalization of (3.72) as

$$f_z(\zeta) = \int_{\xi_{n-1}=-\infty}^{\infty} \cdots \int_{\xi_1=-\infty}^{\infty} f_x(\xi_1, \ldots, \xi_{n-1}, \zeta - \xi_1 - \ldots - \xi_{n-1})\, d\xi_1 \ldots d\xi_{n-1} \qquad (3.118)$$

It may, however, be more effective to generate the sum recursively, as $z_1 = x_1$, $z_2 = x_1 + x_2 = z_1 + x_2$, ... $z_j = z_{j-1} + x_j$, and apply the bivariate relation (3.72) in each step.

The mean of the sum is obtained as

$$\mu_z = \mathrm{E}\{z(t)\} = \mathrm{E}\{\sum_{i=1}^{n} x_i(t)\} = \sum_{i=1}^{n} \mu_{xi} \qquad (3.119)$$

The mean-square and variance are computed as

$$\mathrm{E}\{z^2(t)\} = \mathrm{E}\{[\sum_{i=1}^{n} x_i(t)]^2\} = \sum_{i=1}^{n} \mathrm{E}\{x_i^2(t)\} + 2\sum_{i=1}^{n-1} \sum_{j=i+1}^{n} \mathrm{E}\{x_i(t)\, x_j(t)\} \qquad (3.120)$$

$$\sigma_z^2 = \mathrm{E}\{[z(t)-\mu_z]^2\} = \mathrm{E}\{[\sum_{i=1}^{n} x_i(t) - \sum_{i=1}^{n} \mu_{xi}]^2\}$$

$$= \sum_{i=1}^{n} \sigma_{xi}^2 + 2\sum_{i=1}^{n-1} \sum_{j=i+1}^{n} \mathrm{E}\{[x_i(t)-\mu_{xi}][x_j(t)-\mu_{xj}]\} \qquad (3.121)$$

Observe that (3.120) contains the sum of all elements of the correlation matrix Φ_x (c.f. (3.83)). If the elements of $x(t)$ are uncorrelated then $\mathrm{E}\{x_i(t)x_j(t)\} = \mu_{xi}\,\mu_{xj}$ and $\mathrm{E}\{[x_i(t)-\mu_{xi}][x_j(t)-\mu_{xj}]\} = 0$ for $i \neq j$ and (3.120) and (3.121) simplify to

$$\mathrm{E}\{z^2(t)\} = \sum_{i=1}^{n} \mathrm{E}\{x_i^2(t)\} + 2\sum_{i=1}^{n-1} \sum_{j=i+1}^{n} \mu_{xi}\,\mu_{xj} \qquad \sigma_z^2 = \sum_{i=1}^{n} \sigma_{xi}^2 \qquad (3.122)$$

Window average of scalar time-series. Now the previous results will be used to compute the parameters of the average of the single random variable $x(t)$ over a time-window containing $m+1$ observations. The window average is obtained as

$$\bar{x}(t, m) = \frac{1}{m+1} \sum_{k=0}^{m} x(t-k) \qquad (3.123)$$

We will term this quantity window average rather than sample average (or sample mean), as it is usually referred to in the statistical literature, since in the systems context sample usually means a single observation of a variable.

The expectation of the window average is

$$\mu_{\bar{x}}(m) = E\{\bar{x}(t, m)\} = \frac{1}{m+1} \sum_{k=0}^{m} E\{x(t-k)\} = \mu_x \tag{3.124}$$

That is, the expectation of the window average is the same as the expectation of the observed variable. The mean-square and variance of the window average are computed as

$$E\{\bar{x}^2(t, m)\} = \frac{1}{(m+1)^2} E\{[\sum_{k=0}^{m} x(t-k)]^2\}$$

$$= \frac{1}{m+1} \varphi_{xx}(0) + \frac{2}{(m+1)^2} \sum_{\tau=1}^{m} (m+1-\tau)\, \varphi_{xx}(\tau) \tag{3.125}$$

$$\sigma_{\bar{x}}^2(m) = E\{[\bar{x}(t, m)-\mu_{\bar{x}}(m)]^2\} = \frac{1}{(m+1)^2} E\{[\sum_{k=0}^{m} (x(t-k)-\mu_x)]^2\}$$

$$= \frac{1}{m+1} \sigma_x^2 + \frac{2}{(m+1)^2} \sum_{\tau=1}^{m} (m+1-\tau)\, \psi_{xx}(\tau) \tag{3.126}$$

Notice that (3.125) is the sum of all elements in the correlation matrix Φ_X (c.f. (3.92)), divided by $(m+1)^2$. If the time-series is uncorrelated then $\varphi_{xx}(\tau)=\mu_x^2$ and $\psi_{xx}(\tau)=0$ for $\tau \neq 0$, yielding

$$E\{\bar{x}(t, m)\} = \frac{1}{m+1} \varphi_{xx}(0) + \frac{m}{m+1} \mu_x^2 = \frac{1}{m+1} \sigma_x^2 + \mu_x^2 \tag{3.127}$$

$$\sigma_{\bar{x}}^2(m) = \frac{1}{m+1} \sigma_x^2 \tag{3.128}$$

Window average of vector time-series. Consider now the vector variable $\boldsymbol{x}(t)$ and compute its average over a window:

$$\bar{\boldsymbol{x}}(t,m) = \frac{1}{m+1} \sum_{k=0}^{m} \boldsymbol{x}(t-k) \tag{3.129}$$

Obviously, $\mu_{\bar{x}}=\mu_x$. The autocorrelation and covariance matrices of the window average are obtained as

$$\Phi_{\bar{x}}(m) = E\{\bar{\boldsymbol{x}}(t,m)\, \bar{\boldsymbol{x}}'(t,m)\} = \frac{1}{(m+1)^2} \sum_{k=0}^{m} \sum_{l=0}^{m} E\{\boldsymbol{x}(t-k)\, \boldsymbol{x}'(t-l)\}$$

$$= \frac{1}{m+1} \Phi_{xx}(0) + \frac{1}{(m+1)^2} \sum_{\tau=1}^{m} (m+1-\tau)[\Phi_{xx}(\tau)+\Phi'_{xx}(\tau)] \tag{3.130}$$

$$\Psi_{\bar{x}}(m) = E\{[\bar{x}(t,m) - \mu_x][\bar{x}(t,m) - \mu_x]'\}$$

$$= \frac{1}{m+1}\Psi_{xx}(0) + \frac{1}{(m+1)^2}\sum_{\tau=1}^{m}(m+1-\tau)[\Psi_{xx}(\tau)+\Psi'_{xx}(\tau)] \qquad (3.131)$$

where $\Psi_{xx}(\tau)$ is defined following the pattern of (3.106). Observe that (3.130) contains the sum of all elements in the Φ_X matrix, c.f. (3.116). (Also, (3.131) contains all elements of a similarly defined Ψ_X matrix.) If the vector series is uncorrelated in time then (3.130) and (3.131) simplify to

$$\Phi_{\bar{x}}(m) = \frac{1}{m+1}\Phi_{xx}(0) + \frac{1}{m+1}\mu_x\mu_x' \qquad (3.132)$$

$$\Psi_{\bar{x}}(0) = \frac{1}{m+1}\Psi_{xx}(0) \qquad (3.133)$$

3.3.4. The Law of Large Numbers

The law of large numbers states that the window average of a time sequence approaches the theoretical mean of the random variable as the size of the window increases. Stated in terms of probabilities the law is

$$\lim_{m\to\infty} Prob[\,|\bar{x}(t,m) - \mu_x| > \epsilon] = 0 \qquad (3.134)$$

where ϵ is an arbitrarily small number. With less mathematical rigor, we may simply say

$$\lim_{m\to\infty} \bar{x}(t,m) = \mu_x \qquad (3.135)$$

The law of large numbers has numerous variants; for those the Reader is referred to the literature of mathematical statistics. Note that most variants assume the independence of the elements in the time-sequence in which case the relationship is quite straightforward.

Here we will explore, utilizing our previous results, how the variance of the window average behaves, with and without the independence assumption. Let us consider first the average of two variables, $z=(x+y)/2$. The variance of the sum, using (3.76), is

$$\sigma_z^2 = [\sigma_x^2 + \sigma_y^2 + 2E\{(x-\mu_x)(y-\mu_y)\}] / 4 \qquad (3.136)$$

where $E\{(x-\mu_x)(y-\mu_y)\}$ may be positive or negative. It can be shown that

$$|E\{(x-\mu_x)(y-\mu_y)\}| \leq \sigma_x\sigma_y \qquad (3.137)$$

To see this, observe that, for any pair of random variables v and w, and constant k, the following holds

$$E\{(v - kw)^2\} = E\{v^2\} + k^2E\{w^2\} - 2kE\{vw\} \geq 0 \tag{3.138}$$

Now substitute $v=x-\mu_x$, $w=y-\mu_y$, $k=E\{(x-\mu_x)(y-\mu_x)\}/E\{(y-\mu_y)^2$ into (3.138); after rearrangement, (3.137) follows. Thus, from (3.136),

$$(\sigma_x - \sigma_y)^2 / 4 \leq \sigma_z^2 \leq (\sigma_x + \sigma_y)^2 / 4 \tag{3.139}$$

The limits in (3.139) occur when there is linear dependence between the two variables, that is when (with $c>0$) $y=cx$ (upper limit) or $y=-cx$ (lower limit). If the two variables are independent then

$$\sigma_z^2 = (\sigma_x^2 + \sigma_y^2) / 4 \tag{3.140}$$

Now let us apply the above result to the window mean of a time series. Consider first $\bar{x}(t,1)=[x(t)+x(t-1)]/2$. By (3.139), it follows that

$$0 \leq \sigma_{\bar{x}}^2(1) \leq \sigma_x^2 \tag{3.141}$$

The upper limit would imply $x(t)=cx(t-1)$ while the lower limit $x(t)=-cx(t-1)$. The assumption of stationarity allows only $c=\pm 1$ (since with $|c|<1$, the signal would exponentially decrease while with $|c|>1$, it would exponentially increase). If $c=1$ then the signal is constant and its variance is zero. If $c=-1$ then the signal is alternating and the window average is zero. Thus (3.141) practically reduces to

$$0 < \sigma_{\bar{x}}^2(1) < \sigma_x^2 \tag{3.142}$$

If the two observations are uncorrelated then, from (3.140) (see also (3.128)),

$$\sigma_{\bar{x}}^2(1) = \sigma_x^2 / 2 \tag{3.143}$$

(3.142) and (3.143) generalize to windows of arbitrary length, so that

$$0 < \sigma_{\bar{x}}^2(m) < \sigma_x^2 \tag{3.144}$$

in general and

$$\sigma_{\bar{x}}^2(m) = \sigma_x^2 / (m+1) \tag{3.145}$$

for uncorrelated observations (see also (3.128)).

3.4. THE PROPAGATION OF RANDOM SIGNALS IN DISCRETE LINEAR DYNAMIC SYSTEMS

In this section, the behavior of discrete linear dynamic systems will be investigated when subjected to random inputs. In particular, the output mean, the input-output cross-correlation and the output autocorrelation will be computed. First a single-input single-output system will be considered, then the analysis will be extended to multiple-input multiple-output systems.

The reader should be reminded here that the shift operator is a linear operator (and the z-transformation is a linear transformation). Special care is needed when this technique is applied to variables which are subject to nonlinear manipulations, as it is the case with the correlation functions. In particular, one should keep in mind that the shift operator (z-transform variable) is not transferable in nonlinear operations, that is

$$x(t)\, y(t-1) = [x(t)]\, [\phi^{-1}y(t)] \neq [\phi^{-1}x(t)]\, [y(t)] = x(t-1)\, y(t) \tag{3.146}$$

3.4.1. Single-Input Single-Output Systems

Consider a linear system with the scalar input $u(t)$ and scalar output $y(t)$ (Fig. 3.4.a). Assume the system is described by the discrete transfer function

$$m(\phi) = \frac{g(\phi)}{h(\phi)} = \frac{g^o + g^1\phi^{-1} + \dots + g^\nu \phi^{-\nu}}{1 + h^1\phi^{-1} + \dots + h^\nu \phi^{-\nu}} \tag{3.147}$$

Obviously, this represents the input-output relationship

$$y(t) = g^o\, u(t) + g^1\, u(t-1) + \dots + g^\nu\, u(t-\nu) - h^1\, y(t-1) - \dots - h^\nu\, y(t-\nu) \tag{3.148}$$

The mean of the output. The mean of the output is obtained as

$$\mathrm{E}\{y(t)\} = [g^o+g^1+\dots+g^\nu]\, \mathrm{E}\{u(t)\} - [h^1+\dots+h^\nu]\, \mathrm{E}\{y(t)\} \tag{3.149}$$

yielding

$$\mathrm{E}\{y(t)\} = \frac{g^o+g^1+\dots+g^\nu}{1+h^1+\dots+h^\nu}\, \mathrm{E}\{u(t)\} \tag{3.150}$$

The gain between the means can, of course, be recognized as the steady state gain $m(1)$ of the system (Fig. 3.4.b).

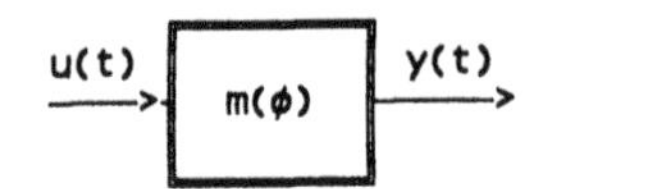

FIGURE 3.4.a. SISO system.

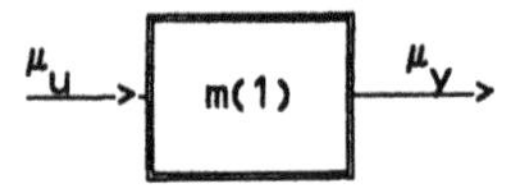

FIGURE 3.4.b. Propagation of mean.

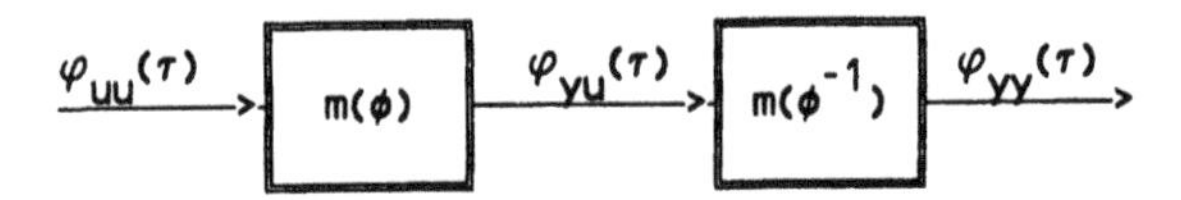

FIGURE 3.4.c. Propagation of correlations.

Input-output cross-correlation. Let us compute the cross-correlation

$$\varphi_{yu}(\tau) = \mathrm{E}\{y(t)\, u(t-\tau)\} \tag{3.151}$$

Expressing $y(t)$ with (3.148), this can be written as

$$\mathrm{E}\{[g^o u(t) + g^1 u(t-1) + \ldots + g^\nu u(t-\nu) - h^1 y(t-1) - \ldots - h^\nu y(t-\nu)]\, u(t-\tau)\}$$

$$= g^o \varphi_{uu}(\tau) + g^1 \varphi_{uu}(\tau-1) + \ldots + g^\nu \varphi_{uu}(\tau-\nu) - h^1 \varphi_{yu}(\tau-1) - \ldots - h^\nu \varphi_{yu}(\tau-\nu) \tag{3.152}$$

Notice that this is a linear relationship among the correlation functions. Thus the shift-operator may be applied as $\phi^{-i}\varphi_{..}(\tau)=\varphi_{..}(\tau-i)$ and (3.152) may be written as

$$\varphi_{yu}(\tau) = [g^o + g^1\phi^{-1} + \ldots + g^\nu \phi^{-\nu}]\varphi_{uu}(\tau) - [h^1\phi^{-1} + \ldots + h^\nu \phi^{-\nu}]\varphi_{yu}(\tau) \tag{3.153}$$

yielding (Fig. 3.4.c):

$$\boxed{\varphi_{yu}(\tau) = m(\phi)\, \varphi_{uu}(\tau)} \tag{3.154}$$

Output autocorrelation. Now compute

$$\varphi_{yy}(\tau) = \mathrm{E}\{y(t)\, y(t-\tau)\} \tag{3.155}$$

Expressing $y(t)$ from (3.148) and following the logic of (3.152) would lead to

$$\varphi_{yy}(\tau) = m(\phi)\, \varphi_{uy}(\tau) \tag{3.156}$$

Here $\varphi_{uy}(\tau) = \mathrm{E}\{u(t)y(t-\tau)\} = \varphi_{yu}(-\tau)$. To obtain a relationship in terms of $\varphi_{yu}(\tau)$, let us apply (3.148) to $y(t-\tau)$ in (3.155):

$$\mathrm{E}\{y(t)\,[g^o u(t-\tau) + g^1 u(t-\tau-1) + \ldots + g^\nu u(t-\tau-\nu) - h^1 y(t-\tau-1) - \ldots - h^\nu y(t-\tau-\nu)]\}$$

$$= g^o \varphi_{yu}(\tau) + g^1 \varphi_{yu}(\tau+1) + \ldots + g^\nu \varphi_{yu}(\tau+\nu) - h^1 \varphi_{yy}(\tau+1) - \ldots - h^\nu \varphi_{yy}(\tau+\nu) \tag{3.157}$$

With $\phi^i \varphi_{..}(\tau) = \varphi_{..}(\tau+i)$, this can be written as

$$\varphi_{yy}(\tau) = [g^o + g^1\phi + \ldots + g^\nu \phi^\nu]\varphi_{yu}(\tau) - [h^1\phi + \ldots + h^\nu \phi^\nu]\varphi_{yy}(\tau) \tag{3.158}$$

yielding (Fig. 3.4.c):

$$\boxed{\varphi_{yy}(\tau) = m(\phi^{-1})\, \varphi_{yu}(\tau) = m(\phi^{-1})\, m(\phi)\, \varphi_{uu}(\tau)} \tag{3.159}$$

3.4.2. Multiple-Input Multiple-Output Systems

Now the linear system has a vector input $\boldsymbol{u}(t)$ and vector output $\boldsymbol{y}(t)$ (Fig. 3.5.a) and is characterized by the discrete transfer function matrix

$$\boldsymbol{M}(\phi) = \frac{\boldsymbol{G}(\phi)}{h(\phi)} = \frac{\boldsymbol{G}^o + \boldsymbol{G}^1\phi^{-1} + \ldots + \boldsymbol{G}^\nu \phi^{-\nu}}{1 + h^1\phi^{-1} + \ldots + h^\nu \phi^{-\nu}} \tag{3.160}$$

The corresponding input-output relationship is

$$\boldsymbol{y}(t) = \boldsymbol{G}^o \boldsymbol{u}(t) + \boldsymbol{G}^1 \boldsymbol{u}(t-1) + \ldots + \boldsymbol{G}^\nu \boldsymbol{u}(t-\nu) - h^1 \boldsymbol{y}(t-1) - \ldots - h^\nu \boldsymbol{y}(t-\nu) \tag{3.161}$$

The mean of the output. The mean of the output is obtained as (Fig. 3.5.b):

$$\mathrm{E}\{\boldsymbol{y}(t)\} = \frac{\boldsymbol{G}^o + \boldsymbol{G}^1 + \ldots + \boldsymbol{G}^\nu}{1 + h^1 + \ldots + h^\nu}\, \mathrm{E}\{\boldsymbol{u}(t)\} \tag{3.162}$$

Input-output cross-correlation. The cross-correlation is defined as

$$\boldsymbol{\Phi}_{yu}(\tau) = \mathrm{E}\{\boldsymbol{y}(t)\, \boldsymbol{u}'(t-\tau)\} \tag{3.163}$$

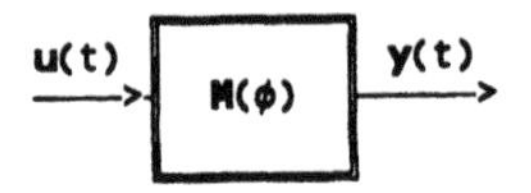

FIGURE 3.5.a. MIMO system.

FIGURE 3.5.b. Propagation of mean.

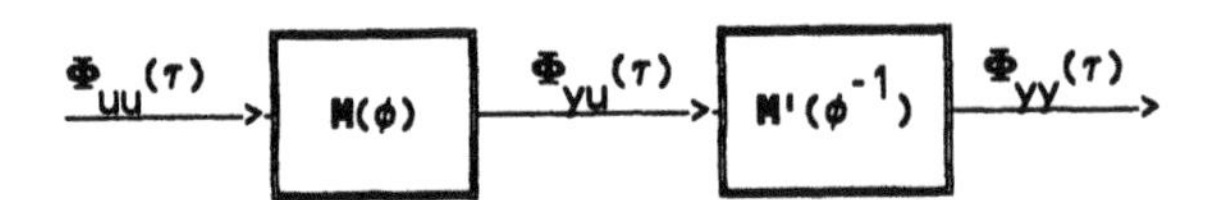

FIGURE 3.5.c. Propagation of correlations.

With (3.161) this can be written as

$$E\{[G^o u(t) + G^1 u(t-1) + \ldots + G^\nu u(t-\nu) - h^1 y(t-1) - \ldots - h^\nu y(t-\nu)]\, u'(t-\tau)\}$$

$$= G^o \Phi_{uu}(\tau) + G^1 \Phi_{uu}(\tau-1) + \ldots + G^\nu \Phi_{uu}(\tau-\nu) - h^1 \Phi_{yu}(\tau-1) - \ldots - h^\nu \Phi_{yu}(\tau-\nu)$$

$$= [G^o + G^1\phi^{-1} + \ldots + G^\nu \phi^{-\nu}]\Phi_{uu}(\tau) - [h^1\phi^{-1} + \ldots + h^\nu \phi^{-\nu}]\Phi_{yu}(\tau) \tag{3.164}$$

yielding (Fig. 3.5.c):

$$\boxed{\Phi_{yu}(\tau) = M(\phi)\, \Phi_{uu}(\tau)} \tag{3.165}$$

Output autocorrelation. The output autocorrelation is

$$\Phi_{yy}(\tau) = E\{y(t)\, y'(t-\tau)\} \tag{3.166}$$

Substituting $y(t)$ from (3.161), this would lead to

$$\Phi_{yy}(\tau) = M(\phi)\, \Phi_{uy}(\tau) \tag{3.167}$$

Here $\Phi_{uy}(\tau) = E\{u(t) y'(t-\tau)\} = \Phi'_{yu}(-\tau)$. To obtain an expression in terms of $\Phi_{yu}(\tau)$, let us write $y(t-\tau)$ in (3.166) with (3.161):

$$E\{y(t)\, [G^o u(t-\tau) + G^1 u(t-\tau-1) + \ldots + G^\nu u(t-\tau-\nu) - h^1 y(t-\tau-1) - \ldots - h^\nu y(t-\tau-\nu)]'\}$$

$$= \Phi_{yu}(\tau)\, [G^o + G^1\phi + \ldots + G^\nu \phi^\nu]' - \Phi_{yy}(\tau)[h^1\phi + \ldots + h^\nu \phi^\nu] \tag{3.168}$$

yielding (Fig. 3.5.c):

$$\Phi_{yy}(\tau) = \Phi_{yu}(\tau)\, M'(\phi^{-1}) = M(\phi)\, \Phi_{uu}(\tau)\, M'(\phi^{-1}) \tag{3.169}$$

3.5. THE NORMAL DISTRIBUTION

The normal (Gaussian) distribution plays an important role in the analysis of random signals, for a number of reasons

- many random physical processes can be rather accurately characterized by the normal distribution;
- the behavior of variables representing the combined effect of a large number of phenomena approaches the normal distribution (see the central limit theorem, Subsection 3.5.2);
- the normal distribution is relatively simple to handle mathematically.

3.5.1. Density Functions of the Normal Distribution

Univariate normal distribution. The probability density function of a single normally distributed variable is

$$f_x(\xi) = \frac{1}{\sqrt{(2\pi)}\sigma_x} \exp\left[- \frac{(\xi - \mu_x)^2}{2\sigma_x^2} \right] \tag{3.170}$$

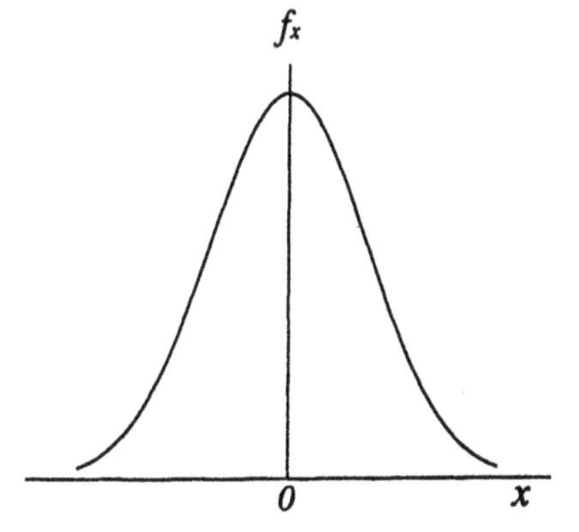

FIGURE 3.6. Univariate normal density function.

This is the equation of the well-known bell-shaped curve, shown in Figure 3.6. The usual shorthand notation for normal distribution is $N(\mu, \sigma)$. In many cases, the "standard" normal distribution $N(0, 1)$ is used, which is the special case of (3.170) with $\mu_x = 0$ and $\sigma_x = 1$:

$$f_x(\xi) = \frac{1}{\sqrt{(2\pi)}} \exp\left[-\frac{\xi^2}{2}\right] \tag{3.171}$$

Bivariate normal distribution. The joint probability density function of two, jointly normally distributed variables is

$$f_{xy}(\xi, \eta) = \frac{1}{2\pi\sigma_x\sigma_y\sqrt{(1-\rho_{xy}^2)}} \tag{3.172}$$

$$\exp\left[-\frac{1}{2(1-\rho_{xy}^2)}\left(\frac{(\xi-\mu_x)^2}{\sigma_x^2} - \frac{2\rho_{xy}(\xi-\mu_x)(\eta-\mu_y)}{\sigma_x\sigma_y} + \frac{(\eta-\mu_y)^2}{\sigma_y^2}\right)\right]$$

where

$$\rho_{xy} = \frac{\mathrm{E}\{xy\} - \mu_x\mu_y}{\sigma_x\sigma_y} \tag{3.173}$$

is the *correlation coefficient* of the pair x,y.

If the two variables are independent then $\rho_{xy} = 0$ and

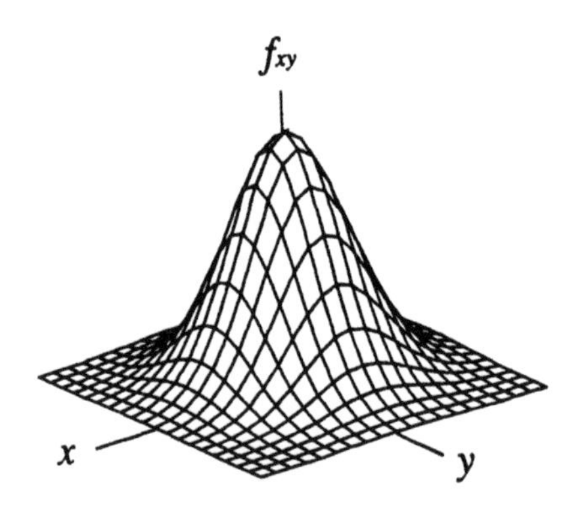

FIGURE 3.7. Bivariate normal density function.

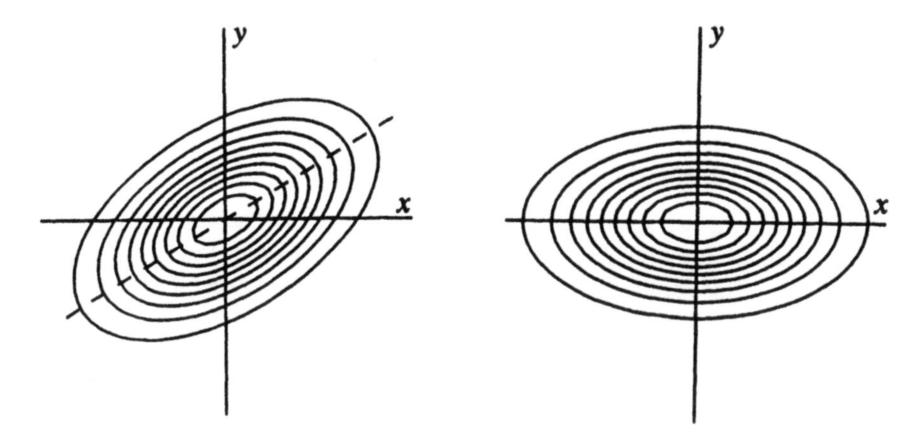

a. Dependent variables b. Independent variables

FIGURE 3.8. Constant density contours.

$$f_{xy}(\xi, \eta) = \frac{1}{2\pi\sigma_x\sigma_y} exp\,[-\frac{(\xi-\mu_x)^2}{2\sigma_x^2} - \frac{(\eta-\mu_y)^2}{2\sigma_y^2}] \tag{3.174}$$

Figure 3.7 shows a "three-dimensional" illustration of the bivariate normal density function. It may be more informative to represent the bivariate density function in the $\xi - \eta$ plane by its $f_{xy}(\xi, \eta)=const$ contours. These contours are ellipses, with axis directions depending on the correlation coefficient (Fig. 3.8.a). If the two variables are independent then the axes of the ellipse coincide with the coordinate axes (Fig. 3.8.b).

Note that the marginal distributions $f_x(\xi)$ and $f_y(\eta)$, derived from a normal joint distribution $f_{xy}(\xi, \eta)$, are also normal.

Multivariate (vector) normal density function. Consider now a random vector variable $\boldsymbol{x}(t)=[x_1(t), x_2(t),\ldots x_n(t)]'$ the elements of which are jointly normally distributed. The joint probability density function is

$$f_{\boldsymbol{x}}(\boldsymbol{\xi}) = \frac{1}{(2\pi)^{n/2}|\boldsymbol{\Psi}_x|^{1/2}} exp\,[-\tfrac{1}{2}(\boldsymbol{\xi}-\boldsymbol{\mu}_x)'\,\boldsymbol{\Psi}_x^{-1}(\boldsymbol{\xi}-\boldsymbol{\mu}_x)] \tag{3.175}$$

If the elements $x_1(t), x_2(t),\ldots x_n(t)$ are independent then $\boldsymbol{\Psi}_x = Diag[\sigma_{x1}^2, \sigma_{x2}^2,\ldots\sigma_{xn}^2]$ and (3.175) becomes

$$f_{\boldsymbol{x}}(\boldsymbol{\xi}) = \frac{1}{(2\pi)^{n/2}\sigma_{x1}\sigma_{x2}\ldots\sigma_{xn}} exp\,[-\sum_{i=1}^{n}\frac{(\xi_i-\mu_{xi})^2}{2\sigma_{xi}^2}] \tag{3.176}$$

The $f_x(\boldsymbol{\xi})=const$ contours are now hypersurfaces in the n-dimensional space of $\boldsymbol{\xi}$. Any marginal distribution derived from the joint normal distribution is also normal.

3.5.2. Normal Distribution and the Sum of Variables

In this subsection, two important results will be reviewed. The first states that the sum of an arbitrary number of independent normal variables is also normally distributed. This result, though not quite surprising, is very useful in dealing with, for example, the average of normal variables or the propagation of such variables through a discrete linear system. The other result, known as the *central limit theorem*, states that the distribution of the average of random variables tends to the normal distribution, as the number of observations goes to infinity, no matter what the distribution of the variables is (subject to some broad conditions). This result has far-reaching consequences, in that it justifies the use of normal distribution (i) when a variable represents a combination of a large number of random effects, and (ii) as an approximate distribution when the sum or average of a sufficiently large number of random variables is concerned.

The sum of independent normal variables. Consider two independent random variables $x(t)$ and $y(t)$ with distributions $N(\mu_x, \sigma_x)$ and $N(\mu_y, \sigma_y)$. The sum of the variables, $z(t)=x(t)+y(t)$, follows the distribution (c.f. (3.73) and (3.76)):

$$f_z(\zeta) = \frac{1}{2\pi\sigma_x\sigma_y} \int_{\xi=-\infty}^{\infty} exp\left[-\frac{(\xi-\mu_x)^2}{2\sigma_x^2}\right] exp\left[-\frac{(\zeta-\xi-\mu_y)^2}{2\sigma_y^2}\right] d\xi \tag{3.177}$$

This is a convolution type relationship which can be most conveniently handled by Fourier transformation. (The Fourier transform of the probability density is referred to as the *characteristic function* of the distribution - see e.g., Stuart and Ord, 1991). The analysis yields

$$f_z(\zeta) = \frac{1}{\sqrt{(2\pi)}\sigma_z} exp\left[-\frac{(\zeta-\mu_z)^2}{2\sigma_z^2}\right] \tag{3.178}$$

with $\mu_z=\mu_x+\mu_y$ and $\sigma_z=\sqrt{(\sigma_x^2+\sigma_y^2)}$ (c.f. (3.74) and (3.77)). That is, the sum of two independent, normally distributed variables is also normally distributed.

The above result easily extends to the sum of n independent, normally distributed variables, $z(t)=x_1(t)+x_2(t)+\ldots+x_n(t)$. Define the recursive partial sums $z_j(t)=z_{j-1}(t)+x_j(t)$, $z_1(t)=x_1(t)$. First, $z_2(t)$ is clearly normal by (3.178). Then, $z_2(t)$ and $x_3(t)$ are independent, thus $z_3(t)$ is also normal - and so on.

The central limit theorem. Consider the set of random variables $x_1(t), x_2(t), \ldots x_n(t)$. Assume that they are independent, identically distributed with mean μ_x and standard deviation σ_x. Define a new variable as $z_n(t)=x_1(t)+x_2(t)+\ldots+x_n(t)$. The basic form of the central limit theorem (e.g., Feller, 1950) states that

$$\lim_{n\to\infty} P\left(\frac{z_n - n\mu_x}{\sigma_x\sqrt{n}} < \beta\right) = \frac{1}{\sqrt{(2\pi)}} \int_{\xi=-\infty}^{\beta} exp\left[-\frac{\xi^2}{2}\right] d\xi \qquad (3.179)$$

That is, the distribution of the average z_n/n, normalized with its mean μ_x and standard deviation $\sigma_x/\sqrt{n}$, approaches the standard normal distribution $N(0, 1)$ as $n\to\infty$. Notice that nothing has been assumed about the distribution of the x variables, except their mutual independence and the existence of the mean and standard deviation. In an even more general form of the theorem, the assumption of independence is also removed (Feller, 1950).

Note that while the theorem states exact equality only in the limit, the standard normal distribution turns out to be a good approximation of the actual distribution of the mean even if the number of observations is limited. If the distribution of the x variables is known, the approximation error for various values of n can be computed.

3.5.3. The χ^2 Distribution

Consider again a set of independent random variables, $x_1(t), x_2(t), \ldots x_n(t)$, each with $N(0, 1)$ distribution. The distribution obeyed by the variable

$$w_n(t) = x_1^2(t) + x_2^2(t) + \ldots + x_n^2(t) \qquad (3.180)$$

is referred to as the (central) χ^2 *distribution with n degrees of freedom*. Any parameter computed from the observations acts as a constraint on the distribution and reduces the degrees of freedom accordingly. The distribution is called central because it arises from the zero-mean normal distribution; of course, the $w_n(t)$ variable can only take non-negative values and its distribution is increasingly skewed in the positive direction as n grows (Fig. 3.9). A non-central variant of the distribution, with μ_x as its parameter, also exists.

The closed analytical form of the distribution is rather complex; its values are usually taken from pre-computed tables or provided by a computer program. The χ^2 distribution plays an important role in statistical hypothesis testing involving independent normal variables.

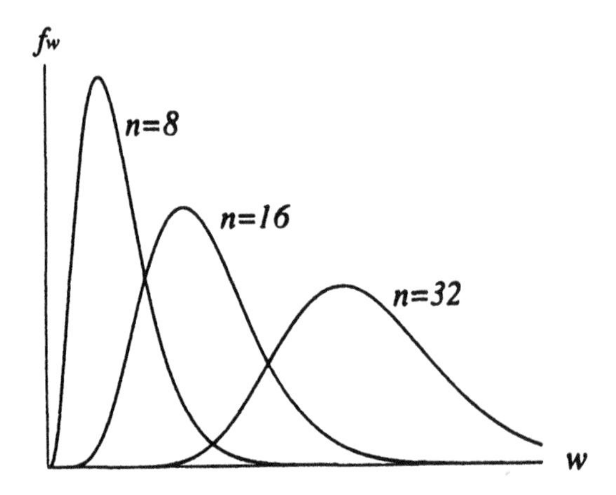

FIGURE 3.9. Central χ^2 distributions.

3.6. MAXIMUM LIKELIHOOD ESTIMATION

When the type of the distribution is known, the exact distribution (or density) function still depends on some parameters. The simplest examples of such parameters are the mean and the standard deviation of the distribution. Somewhat more complicated but basically the same is the situation when the random variable is the output of a dynamic system; now its distribution depends on the system parameters. Thus the probability density function may be written as $f_x(\xi, \theta)$, where θ, a scalar or a vector, represents the parameter(s) of the distribution which are assumed to be known.

Often some of the parameters are not known; then observations of the random variable, together with the knowledge of the type of the distribution (and its other parameters) may be used to obtain an *estimate* for the unknown parameters.

Sufficient statistic. It may be desirable to represent the information, embodied in a set of observations, in some condensed form. For example, one may wish to replace the time-series $X(t)$ with its window average $\bar{X}(t, m)$ (see (3.123)). In general, a function of a random variable is referred to as a *statistic*. For the purpose of estimating a parameter θ, a statistic may replace the original observations if it preserves all the information the latter carried concerning θ. Such statistics are referred to as *sufficient*.

Consider a random series $X(t)$, a parameter θ and a statistic $w(X)$. The statistic w is sufficient for θ if the density of X, given $w(\Xi)=\omega$ (where Ξ is a value-series), is independent of θ (see e.g., Scharf, 1991). According to the *Fisher-Neyman factorization theorem,* this condition is expressed as

$$f_X(\Xi, \theta) = \gamma(\Xi) f_w(\omega, \theta), \qquad \omega = w(\Xi) \tag{3.181}$$

where $\gamma(\Xi)$ is a scaling coefficient which does not depend on θ.

To see the implications of (3.181), let us apply the conditional density relationship (3.42) to X and w:

$$f_{Xw}(\Xi, \omega, \theta) = f_{X|w}(\Xi, \omega, \theta) f_w(\omega, \theta) \tag{3.182}$$

Because of the deterministic relationship $\omega = w(\Xi)$,

$$f_{Xw}(\Xi, \omega, \theta) = \begin{cases} f_X(\Xi, \theta) & \text{if } \omega = w(\Xi) \\ 0 & \text{otherwise} \end{cases} \tag{3.183}$$

Thus

$$f_X(\Xi, \theta) = f_{X|w}(\Xi, \omega, \theta) f_w(\omega, \theta) \qquad \omega = w(\Xi) \tag{3.184}$$

Comparing this with (3.181) reveals the condition

$$f_X(\Xi, \theta) = f_{X|w}(\Xi, \omega) f_w(\omega, \theta) \qquad \omega = w(\Xi) \tag{3.185}$$

That is, the statistic $w(X)$ is sufficient for the parameter θ if the conditional density function $f_{X|w}(\Xi, \omega)$ is independent of θ.

Note that the empirical window average is a sufficient statistic for the estimation of the mean if the time-series $X(t)$ is uncorrelated but it is not sufficient if the series is correlated (Scharf, 1991).

Maximum likelihood estimate. If the density function, together with the observation, is used to estimate the unknown parameters, it is usually reinterpreted as the *likelihood function*

$$L_x(\xi, \theta) = f_x(\xi, \theta) \tag{3.186}$$

in which ξ is known and θ is sought. Here ξ may represent a single (scalar or vector) observation or it may be a compound representation of a set of observations, such as a time-sequence vector or a window average.

Intuition suggests to seek the parameter estimates as the parameter values with which the distribution assigns the highest local probability to the actual observations. This concept is known as the *maximum likelihood* principle of parameter estimation. The parameter estimates are thus obtained by maximizing the likelihood function under the given observations, that is,

$$\hat{\theta} = arg\,[\max_{\theta} L_x(\xi,\ \theta)] \tag{3.187}$$

Technically, the estimates are sought by computing the partial derivative(s) of the likelihood function and equating it (them) with zero:

$$\hat{\theta} = \underset{\theta}{sol}\,[\partial L_x(\xi,\ \theta)/\partial\theta = 0] \tag{3.188}$$

with the additional requirement that the second derivative of the likelihood function, at the solution, has to be negative

$$\partial^2 L_x(\xi,\ \theta)\ /\ \partial\theta^2 \,\Big|_{\theta=\hat{\theta}} < 0 \tag{3.189}$$

In many cases, it is simpler to work with the natural logarithm of the likelihood function, the so called *log-likelihood function* $logL_x(\xi,\ \theta)$. Observe that

$$\frac{\partial logL(\xi,\ \theta)}{\partial\theta} = \frac{dlogL(\xi,\ \theta)}{dL(\xi,\ \theta)} \cdot \frac{\partial L(\xi,\ \theta)}{\partial\theta} = \frac{1}{L(\xi,\ \theta)} \cdot \frac{\partial L(\xi,\ \theta)}{\partial\theta} \tag{3.190}$$

and $1/L(\xi,\ \theta) \neq 0$ (provided the density function is bounded, that is, the distribution function has no jumps). Thus

$$\partial logL_x(\xi,\ \theta)/\partial\theta = 0 \tag{3.191}$$

has the same solution as (3.188). One advantageous property of the log-likelihood function is that it converts the product of likelihood functions, arising with independent observations, into a sum. Further, it replaces the exponential likelihood function of the normal distribution with the exponent, plus an additional term, so that for a univariate normal distribution

$$logL_x(\xi,\ \mu_x,\ \sigma_x) = -\ log\ \sqrt{(2\pi)}\sigma_x - (\xi - \mu_x)^2/2\sigma_x^2 \tag{3.192}$$

while for a multivariate normal distribution

$$logL_x(\boldsymbol{\xi},\ \mu_x,\ \boldsymbol{\Psi}_x^{-1}) = -\ ½\ log[(2\pi)^n|\boldsymbol{\Psi}_x|] - ½\ (\boldsymbol{\xi} - \mu_x)'\ \boldsymbol{\Psi}_x^{-1}(\boldsymbol{\xi} - \mu_x) \tag{3.193}$$

3.7. HYPOTHESIS TESTING

Hypothesis testing is one of the central subjects of mathematical statistics. The objective is to make a decision, based on empirical observations and the statistical properties of the ensemble (time-series) from which the observations are assumed to originate. The decision usually concerns one or more of the parameters of the underlying distribution.

3.7.1. Test Properties

The decision problem usually is posed as a choice between two or more *hypotheses*. The following definitions apply:

A hypothesis is *simple* if it involves a single specific parameter value. It is *composite* if it involves several specific values or one or more continuous range(s) of the parameter.

A decision problem is *binary* if it involves only two hypotheses, and it is *multiple* if it involves more.

One of the hypotheses is referred to as the *null-hypothesis*, and denoted as H°. The other hypothesis (or hypotheses) is (are) referred to as the *alternative hypothesis* (hypotheses) and denoted as H^1 ($H^j, j=1...k$).

A typical example of a binary decision between two simple hypotheses is the decision whether the mean of a time-series is zero or another known value. That is:

$$H^\circ: \mu=0; \qquad H^1: \mu=\mu_1 \tag{3.194}$$

where μ_1 is known. In contrast, H^1 is composite in the binary problem

$$H^\circ: \mu=0; \qquad H^1: \mu\neq 0 \tag{3.195}$$

Further, both hypotheses are composite in the binary problem

$$H^\circ: \mu\leq\mu_1; \qquad H^1: \mu>\mu_1 \tag{3.196}$$

A multiple decision problem arises if the alternative choice is among a set of known mean values $\mu_1, \mu_2, \ldots$:

$$H^\circ: \mu=0; \quad H^1: \mu=\mu_1; \quad H^2: \mu=\mu_2 \tag{3.197}$$

Statistical hypothesis tests are usually implemented as threshold tests on some statistic. That is, a binary test can be written as:

$$D = \begin{bmatrix} H^\circ & \text{if } w \leq \kappa \\ H^1 & \text{if } w > \kappa \end{bmatrix} \tag{3.198}$$

where D is the decision, w is the test statistic and κ is the threshold. The test effectively divides the space of the observation variable ξ into two regions, Ω_o, where the hypothesis H° is accepted, and Ω_1, where H^1 is accepted:

$$D = \begin{bmatrix} H^\circ & \text{if } \xi \text{ is in } \Omega_o \\ H^1 & \text{if } \xi \text{ is in } \Omega_1 \end{bmatrix} \tag{3.199}$$

(If there are multiple alternative hypotheses H^j, $j=1...k$, then each one is associated with a region Ω_j.)

Test quality measures. The quality of a test can be characterized by its *false alarm rate* and *missed detection rate*. Assume the choice is between H° and H^1. Then the false alarm rate α is the probability that, while in reality H° occurs, the decision is H^1:

$$\alpha = P[\, D=H^1 \mid H^\circ \,] \tag{3.200}$$

The false alarm rate is also called the *size* of the test. The missed detection rate $(1\text{-}\beta)$ is the probability that, while in reality H^1 occurs, the decision is H°:

$$1 - \beta = P[\, D=H^\circ \mid H^1 \,] \tag{3.201}$$

It should be emphasized that (3.200) and (3.201) are conditional probabilities. The actual probability of false alarm, P_F, or missed detection, P_M, in an experiment depends also on the probabilities that H° or H^1 happen:

$$P_F = P[\, D=H^1, H^\circ \,] = \alpha\, P[\, H^\circ \,] \tag{3.202}$$

$$P_M = P[\, D=H^\circ, H^1 \,] = (1 - \beta)\, P[\, H^1 \,] \tag{3.203}$$

The *detection power* of the test is the complement of the missed detection rate, that is, the probability of H^1 being declared when H^1 occurs:

$$\beta = P[\, D=H^1 \mid H^1 \,] \tag{3.204}$$

In a binary decision problem between two simple hypotheses, a test is termed *most powerful* of size α if, for the given α, there is no other test with greater detection power. In a binary problem when one or both hypotheses are composite, a test is termed *uniformly most powerful* of size α if, for the given α, it is most powerful for all parameter values under the composite hypothesi(e)s.

3.7.2. Simple Hypotheses: The Likelihood Ratio Test

Perhaps the most widely used statistical test is the so called likelihood ratio test. The main reason for its popularity is that it is the most powerful test if the decision is between two simple hypotheses. Also, it is easy to implement if the underlying distribution is normal.

The Neyman-Pearson Lemma. Consider a binary decision problem with the simple hypotheses

$$H^\circ: \boldsymbol{\theta} = \boldsymbol{\theta}_o; \qquad H^1: \boldsymbol{\theta} = \boldsymbol{\theta}_1 \tag{3.205}$$

Assume that $x=\xi$ is observed and the density functions $f_x(\xi, \boldsymbol{\theta}_j)$, $j=0,1$, are known. Select a test size α. Then, according to the Neyman-Pearson Lemma, the most powerful test arises from the following decision rule

$$D = \begin{cases} H^\circ & \text{if } f_x(\xi, \boldsymbol{\theta}_1) < \kappa f_x(\xi, \boldsymbol{\theta}_o) \quad (3.206a) \\ H^1 & \text{if } f_x(\xi, \boldsymbol{\theta}_1) > \kappa f_x(\xi, \boldsymbol{\theta}_o) \quad (3.206b) \end{cases}$$

(assuming that $P[f_x(\xi, \boldsymbol{\theta}_1)=\kappa f_x(\xi, \boldsymbol{\theta}_o)]=0$), where κ is a suitable threshold. That is, the statistic to test, against the threshold κ, is

$$w = \frac{f_x(\xi, \boldsymbol{\theta}_1)}{f_x(\xi, \boldsymbol{\theta}_o)} = \frac{L_x(\xi, \boldsymbol{\theta}_1)}{L_x(\xi, \boldsymbol{\theta}_o)} = \Lambda_x(\xi, \boldsymbol{\theta}_1, \boldsymbol{\theta}_o) \tag{3.207}$$

This statistic is known as the *likelihood ratio* and the test involving it as the *likelihood ratio test*.

In the following, we will outline a proof of the Neyman-Pearson Lemma. Define Ω_1 as the region of ξ in which $D=H^1$, according to (3.206). Then the size α and the power β of the test are obtained as (see Fig. 3.10):

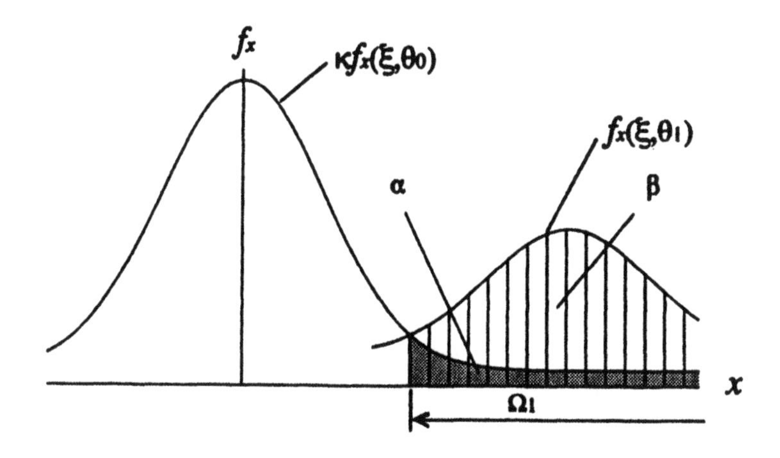

FIGURE 3.10. Size and power of binary test.

$$\alpha = \int_{\Omega_1} f_x(\xi, \theta_o)\, d\xi \tag{3.208}$$

$$\beta = \int_{\Omega_1} f_x(\xi, \theta_1)\, d\xi \tag{3.209}$$

Now let us assume that we want to deviate from the Neyman-Pearson strategy while keeping the size of the test constant. This implies replacing a part of the region inside Ω_1 with one outside. Let us represent the sub-region we discard with the volume $\Delta\xi_1$ and the one we introduce with the volume $\Delta\xi_2$, and assume that these are small enough so that the density functions within them may be characterized with the single values $f_x(\xi_1, \theta_j)$ and $f_x(\xi_2, \theta_j)$, $j=0,1$, where ξ_1 and ξ_2 are points within $\Delta\xi_1$ and $\Delta\xi_2$. Then clearly

$$f_x(\xi_1, \theta_1) > \kappa f_x(\xi_1, \theta_o); \qquad f_x(\xi_2, \theta_1) < \kappa f_x(\xi_2, \theta_o) \tag{3.210}$$

Further, since α must remain the same,

$$\Delta\alpha = -f_x(\xi_1, \theta_o)\, \Delta\xi_1 + f_x(\xi_2, \theta_o)\, \Delta\xi_2 = 0 \tag{3.211}$$

The change of β can be computed as

$$\Delta\beta = -f_x(\xi_1, \theta_1)\, \Delta\xi_1 + f_x(\xi_2, \theta_1)\, \Delta\xi_2 \tag{3.212}$$

Substituting (3.211) into (3.212) yields

$$\Delta\beta = \left[-\frac{f_x(\xi_1, \theta_1)}{f_x(\xi_1, \theta_o)} + \frac{f_x(\xi_2, \theta_1)}{f_x(\xi_2, \theta_o)} \right] f_x(\xi_2, \theta_o)\, \Delta\xi_2 \tag{3.213}$$

Now from (3.210), the first fraction in (3.213) is greater than κ while the second fraction is smaller than κ, thus $\Delta\beta < 0$. That is, any deviation from the Neyman-Pearson strategy results in a decrease of the test power.

The threshold κ remains to be determined. For any value of κ, a region Ω_1 follows. The size of the test may then be computed from (3.208). We need to choose κ so that the prescribed value of α results. Note that this procedure may not be easy to implement and in general requires numerical computation.

The test can also be performed on the *log-likelihood ratio* function $\log\Lambda_x(\xi, \theta_1, \theta_o)$, relative to the threshold $\log\kappa$.

Note that while a vector observation ξ has been assumed here, all the concepts and results carry over to scalar observations and time series.

Likelihood ratio for normal distribution. For the multivariate normal distribution (3.175), with parameters $\theta = \{\mu_x, \Psi_x\}$, the likelihood ratio is

$$\Lambda_x(\xi, \theta_1, \theta_0) = \frac{|\Psi_{xo}|^{1/2}}{|\Psi_{x1}|^{1/2}} exp\,[\,½\,(\xi - \mu_{xo})'\,\Psi_{xo}^{-1}(\xi - \mu_{xo})$$

$$- ½\,(\xi - \mu_{x1})'\,\Psi_{x1}^{-1}\,(\xi - \mu_{x1})] \qquad (3.214)$$

If $\Psi_{x1} = \Psi_{xo} = \Psi_x$ then (3.214) simplifies to

$$\Lambda_x(\xi, \mu_{x1}, \mu_{xo}) = exp\,[\,½\,(\mu_{xo}'\Psi_x^{-1}\mu_{xo} - \mu_{x1}'\Psi_x^{-1}\mu_{x1}) - \xi'\,\Psi_x^{-1}(\mu_{xo} - \mu_{x1})] \qquad (3.215)$$

The log likelihood functions for (3.214) and (3.215) are

$$log\Lambda_x(\xi, \theta_1, \theta_0) = \qquad (3.216)$$

$$½\,log\frac{|\Psi_{xo}|}{|\Psi_{x1}|} + ½\,(\xi - \mu_{xo})'\,\Psi_{xo}^{-1}(\xi - \mu_{xo}) - ½\,(\xi - \mu_{x1})'\,\Psi_{x1}^{-1}\,(\xi - \mu_{x1})$$

$$log\Lambda_x(\xi, \mu_{x1}, \mu_{xo}) = ½\,(\mu_{xo}'\Psi_x^{-1}\mu_{xo} - \mu_{x1}'\Psi_x^{-1}\mu_{x1}) - \xi'\,\Psi_x^{-1}(\mu_{xo} - \mu_{x1}) \qquad (3.217)$$

Multiple simple hypotheses. In many classification problems, we wish to determine which of a set of simple hypotheses is the most likely, given the observations. Now there are two or more simple alternative hypotheses, which are of identical nature (in contrast to fault versus no fault). Thus the concepts of false alarm and missed detection do not apply, and the threshold in the likelihood ratio test is simply $\kappa = 1$. Further, in the case of more than two hypotheses, the series of pairwise comparisons, represented by the likelihood ratio tests, may be replaced by a global comparison of the likelihoods. That is, if $H^j\!: \theta = \theta_j,\ j=1...k,$ then

$$D = H^i \quad if \quad L_x(\xi, \theta_i) = \max_j L_x(\xi, \theta_j), \quad j=1...k \qquad (3.218)$$

3.7.3. Testing Composite Hypotheses

If one or both of the hypotheses are composite then likelihood ratio tests cannot be directly constructed. The most typical occurrence of this situation is when the hypothesis $\theta = \theta_0 = 0$ is tested against the alternative $\theta \neq 0$. In such cases, the test may be constructed solely on the basis of $f_x(\xi, \theta_0)$, to satisfy a specified test size α. The decision rule is

$$D = \begin{cases} H^\circ & \text{if } f_x(\xi, \theta_o) > \kappa \\ H^1 & \text{if } f_x(\xi, \theta_o) < \kappa \end{cases} \tag{3.219}$$

The threshold κ is determined from (3.208) for the selected α, with the region Ω_o chosen as the inside of a constant likelihood hypercontour in the ξ space. Notice that now the test statistic w is the density function $f_x(\xi, \theta_o)$.

The χ^2 test. In connection with the multivariate normal distribution, the test (3.219) may frequently be converted into a χ^2 test; this is then much simpler to implement than the original test. Consider a multivariate normal distribution with known covariance and construct a test for the zero mean hypothesis. The density function is

$$f_x(\xi, \mu_o) = \frac{1}{(2\pi)^{n/2} |\Psi_x|^{1/2}} \exp[-½\, \xi' \Psi_x^{-1} \xi] \tag{3.220}$$

After cross-multiplying and taking logarithm, the test condition $f_x(\omega, \mu_o) > \kappa$ can be reformulated as

$$w^* = \xi' \Psi_x^{-1} \xi < \kappa^* \quad \text{where} \quad \kappa^* = -\log[(2\pi)^n |\Psi_x| \kappa^2] \tag{3.221}$$

and Ψ_x is usually assumed known. To see what distribution w^* obeys, write it as

$$\xi' \Psi_x^{-1} \xi = \xi^{*\prime} \xi^* \qquad \text{where} \quad \xi^* = \Psi_x^{-1/2} \xi \tag{3.222}$$

Under the H° hypothesis, both ξ and ξ^* have zero mean. Now compute the covariance matrix of ξ^*

$$\Psi_{x^*} = E[\xi^* \xi^{*\prime}] = \Psi_x^{-1/2} E[\xi\, \xi'] \Psi_x^{-1/2} = I \tag{3.223}$$

Thus the elements of ξ^* are uncorrelated, with zero mean and unit variance. This implies that the statistic w^*, which is the sum of squares of these elements, obeys the central χ^2 distribution with n degrees of freedom. The test may be performed on this statistic, and the threshold κ^* obtained from standard χ^2 tables, without having to deal with the multivariate normal distribution.

Note that similar results apply even if the mean μ_x of the underlying normal distribution is nonzero and known. In this case, the statistic

$$w^* = (\xi - \mu_x)' \Psi_x^{-1} (\xi - \mu_x) \tag{3.224}$$

may be used which follows the central χ^2 distribution. Alternatively, the statistic $w^* = \xi' \Psi_x^{-1} \xi$ may be tested against the appropriate non-central χ^2 distribution.

Generalized Likelihood Ratio Test. A problem involving composite hypotheses may be reduced to testing simple hypotheses by the following idea. Assume the subject of the decision is in which of the regions Θ_j, $j=0,1$, the parameter $\boldsymbol{\theta}$ is located, that is,

$$H^j: \ \boldsymbol{\theta} \ \text{is in} \ \Theta_j \qquad j=0,1 \tag{3.225}$$

Start by obtaining maximum likelihood estimates for the parameter under the conditions of each hypothesis:

$$\hat{\boldsymbol{\theta}}_j = \arg [\max_{\boldsymbol{\theta}} L_{\boldsymbol{x}}(\boldsymbol{\xi}, \boldsymbol{\theta} \mid H^j)] \qquad j=0, 1 \tag{3.226}$$

The computation of the conditional estimate requires the solution of a maximization problem under constraint. The likelihood ratio test is then performed with these estimates as the hypothesized single values of the parameter $\boldsymbol{\theta}$

$$\Lambda_{\boldsymbol{x}}(\boldsymbol{\xi}, \hat{\boldsymbol{\theta}}_1, \hat{\boldsymbol{\theta}}_0) = \frac{L_{\boldsymbol{x}}(\boldsymbol{\xi}, \hat{\boldsymbol{\theta}}_1)}{L_{\boldsymbol{x}}(\boldsymbol{\xi}, \hat{\boldsymbol{\theta}}_0)} \tag{3.227}$$

This technique is known as the *generalized likelihood ratio* approach. Naturally, it can also be applied to multiple composite hypotheses.

3.8. NOTES AND REFERENCES

Again, most of the contents of this chapter is standard material in intermediate level textbooks. We placed somewhat more than the usual emphasis on the correlation and covariance matrices of scalar and vector sequences.

Perhaps the most comprehensive book on the subject is *The Advanced Theory of Statistics*, originally by M. Kendall and A. Stuart (1969), revised later by A. Stuart and J.K. Ord (1991). The material on hypothesis testing originated from the classical British school of R.A. Fisher (collected papers, 1950), J. Neyman and E.S. Pearson (collected papers, 1967). Additional original contributions are due to W. Feller (1950) and A. Wald (collected papers, 1955), just to mention a few.

There is a great number of textbooks on the subject. We drew, at several places, from the excellent work of L. Scharf (1991). Another, more recent though more introductory text is due to H. Stark and J.W. Woods (1994). Finally, the author has made good use of a little old book by an old friend, András Prékopa (1962).

4
Parameter Estimation Fundamentals

4.0. INTRODUCTION

While most of our analysis and design activities are based on the mathematical model of physical systems, such models are seldom known a priori. A very important part of engineering work concerns the establishment of mathematical models. There are fundamentally two approaches to modeling:

- from physical considerations, by writing differential and algebraic equations characterizing the behavior of the system;
- empirically, by observing the existing system and fitting mathematical models to the experimental data.

The distinction between the two approaches may not be sharp: physically based models usually contain parameters which need to be determined experimentally and experimentally based models usually utilize some qualitative understanding of the system.

This chapter is devoted to the fundamental methods of empirical model building. The assumption is that the physical system can be observed, subjecting it to known and measured inputs and measuring its outputs. The task is then to construct mathematical models which represent the observed behavior. This happens either in a predetermined model structure or the selection of the structure is also part of the model building process. In either case, the parameters of the model need to be estimated.

The significance of model building in fault detection and diagnosis is two-fold:

- model-based diagnostic methods naturally rely on mathematical models
- parameter estimation is one of the major approaches to fault detection and diagnosis.

The fundamental technique of parameter estimation is the method of least squares. Most of this chapter deals with various aspects of this method. We will start, in Section 1, by introducing the least squares concept and deriving its central formula for simple static systems. The results will be extended to dynamic systems in Section 2, first to polynomial (MA) then to general rational (ARMA) structures. In addition to single-input single-output systems, systems with multiple inputs and with multiple outputs will be discussed. It will also be shown, in Section 3, how the basic off-line (batch) algorithm can be converted into on-line form, with various degrees of recursivity.

In Section 4, attention will be devoted to the existence criteria of the estimation algorithm. The concept of persistently exciting input signals will be introduced and applied to the various model structures. Other problems potentially threatening the existence of the solution will be discussed, including dependence of multiple inputs, over-parametrization of the model and data collection under feedback.

The effect of noise on the accuracy of the estimates will be investigated in Section 5. It will be seen what noise properties are required, with the various model structures, for accurate estimates provided that the dataset is sufficiently long. Also, the behavior of the estimates under finite (short) datasets will be explored.

Finally, Section 6 will deal with the problems and techniques of selecting the model structure.

4.1. STATIC MODEL IDENTIFICATION

In this section, we will derive the fundamental least squares (LS) parameter estimation algorithm, for a simple static linear system. This will be extended, still in this section, to nonlinear static systems which are linear in the parameters. In the next section, the fundamental LS algorithm will be applied to various cases of linear dynamic systems.

4.1.1. Linear Models

Consider a system with a single output $y(t)$ and a set of inputs $u_1(t), \dots u_K(t)$, connected by a linear relationship

$$y(t) = a_1 u_1(t) + \dots + a_K u_K(t) + \epsilon(t) \tag{4.1}$$

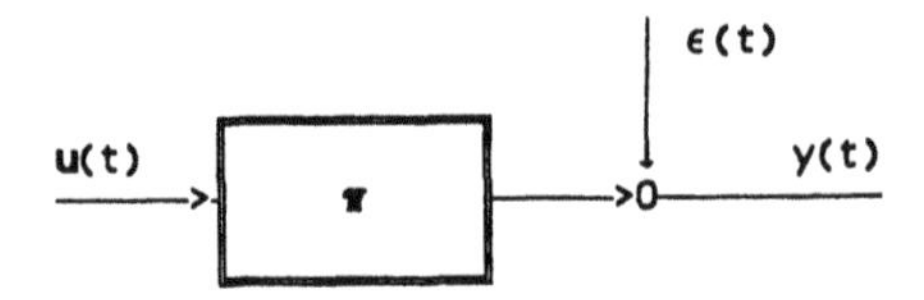

FIGURE 4.1.

Here $\epsilon(t)$ is an equation error term which usually represents the effects of noise and other unmodeled features, acting on (or reduced to) the output. Introduce the vector notation:

$$\boldsymbol{\varphi}'(t) = \boldsymbol{u}'(t) = [u_1(t) \ldots u_K(t)] \qquad \boldsymbol{\pi} = [a_1 \ldots a_K]' \tag{4.2}$$

where $\boldsymbol{\varphi}(t)=\boldsymbol{u}(t)$ is the observation or *regression vector* and $\boldsymbol{\pi}$ is the parameter vector. With these, (4.1) can be written as (see Fig. 4.1)

$$y(t) = \boldsymbol{\varphi}'(t)\,\boldsymbol{\pi} + \epsilon(t) \tag{4.3}$$

In (4.3), the inputs $\boldsymbol{\varphi}'(t)$ and the output $y(t)$ are measurable but the error $\epsilon(t)$ is not. Further, the parameters $\boldsymbol{\pi}$ are not known. We may predict the output, based on the inputs, as

$$\hat{y}(t) = \boldsymbol{\varphi}'(t)\,\hat{\boldsymbol{\pi}} \tag{4.4}$$

where $\hat{\boldsymbol{\pi}}$ is an *estimate* of the parameter vector. We will find this estimate so that the sum of the squared *prediction errors*, over a series of observations,

$$J = J(t, K) = \sum_{k=0}^{K-1} [\hat{y}(t-k) - y(t-k)]^2 \tag{4.5}$$

is minimal. This is the fundamental concept of least squares parameter estimation.

The basic least squares algorithm. To derive the basic LS algorithm, we will utilize some vector differentiation rules summarized in Appendix 4.A.

Substitute (4.4) into (4.5):

$$J = \sum_{k=0}^{K-1} [\boldsymbol{\varphi}'(t-k)\,\hat{\boldsymbol{\pi}} - y(t-k)]^2 \tag{4.6}$$

Now differentiate with respect to $\hat{\boldsymbol{\pi}}$:

$$\frac{\partial J}{\partial \hat{\boldsymbol{\pi}}} = \sum_{k=0}^{K-1} 2\,\boldsymbol{\varphi}(t-k)[\boldsymbol{\varphi}'(t-k)\,\hat{\boldsymbol{\pi}} - y(t-k)] \tag{4.7}$$

At the optimum, the derivative is zero, yielding

$$\sum_{k=0}^{K-1} \boldsymbol{\varphi}(t-k)\boldsymbol{\varphi}'(t-k)\,\hat{\boldsymbol{\pi}} = \sum_{k=0}^{K-1} \boldsymbol{\varphi}(t-k)y(t-k) \tag{4.8}$$

Notice that the solution of (4.8) is a minimum because the second derivative of the performance index J,

$$\frac{\partial^2 J}{\partial \hat{\boldsymbol{\pi}}\, \partial \hat{\boldsymbol{\pi}}'} = \sum_{k=0}^{K-1} 2\, \boldsymbol{\varphi}(t-k)\boldsymbol{\varphi}'(t-k) \tag{4.9}$$

is a positive definite matrix.

The solution of (4.8) yields the parameter estimates as

$$\hat{\boldsymbol{\pi}} = \left[\sum_{k=0}^{K-1} \boldsymbol{\varphi}(t-k)\boldsymbol{\varphi}'(t-k)\right]^{-1} \left[\sum_{k=0}^{K-1} \boldsymbol{\varphi}(t-k)y(t-k)\right] \tag{4.10}$$

(4.10) is the fundamental formula of least squares parameter estimation. The solution exists if the first bracketed matrix has full rank.

The following is a slightly different formulation of the fundamental LS algorithm. Define

$$\boldsymbol{\Phi} = \boldsymbol{\Phi}(t, K) = \begin{bmatrix} \boldsymbol{\varphi}'(t) \\ \cdot \\ \cdot \\ \cdot \\ \boldsymbol{\varphi}'(t-K+1) \end{bmatrix} \qquad Y = Y(t, K) = \begin{bmatrix} y(t) \\ \cdot \\ \cdot \\ y(t-K+1) \end{bmatrix} \tag{4.11}$$

Then

$$\sum_{k=0}^{K-1} \boldsymbol{\varphi}(t-k)\boldsymbol{\varphi}'(t-k) = \boldsymbol{\Phi}'\boldsymbol{\Phi} \qquad \sum_{k=0}^{K-1} \boldsymbol{\varphi}(t-k)y(t-k) = \boldsymbol{\Phi}' Y \tag{4.12}$$

With these, (4.10) becomes

$$\hat{\boldsymbol{\pi}} = [\boldsymbol{\Phi}'\boldsymbol{\Phi}]^{-1}\boldsymbol{\Phi}' Y \tag{4.13}$$

4.1.2. Nonlinear Models

While nonlinear models in general do not render themselves for easy treatment by the LS algorithm, there is a wide and important class which does. These are models in which the parameters appear in a linear way.

The general form of linear-in-the-parameters nonlinear model is

$$y(t) = a_1 f_1 [\mathbf{u}(t)] + \ldots + a_\rho f_\rho [\mathbf{u}(t)] + \epsilon(t) \tag{4.14}$$

where $f_1 [\mathbf{u}(t)], \ldots f_\rho [\mathbf{u}(t)]$ are *known* nonlinear functions of the input vector $\mathbf{u}(t)$, and $a_1 \ldots a_\rho$ are unknown parameters (and $\epsilon(t)$ is the equation error term). Now the regression and parameter vectors are defined as

$$\boldsymbol{\varphi}'(t) = \left[f_1 [\mathbf{u}(t)] \ldots f_\rho [\mathbf{u}(t)] \right] \qquad \boldsymbol{\pi} = [a_1 \ldots a_\rho]' \tag{4.15}$$

The output is predicted, just like in the linear case, as

$$\hat{y}(t) = \boldsymbol{\varphi}'(t)\, \hat{\boldsymbol{\pi}} \tag{4.16}$$

and the estimate $\hat{\boldsymbol{\pi}}$ is sought according to the LS criterion. The fundamental LS formulas (4.10) and (4.13) then apply without modification.

The most important case of the linear-in-the-parameters nonlinear model is the polynomial model. For example, for a model which contains quadratic, linear and constant terms, the regression vector is (for κ input variables and omitting the time argument for simplicity)

$$\boldsymbol{\varphi}' = [1;\, u_1, \ldots u_\kappa;\, u_1^2, u_1 u_2, \ldots u_1 u_\kappa;\, \ldots \ldots \ldots u_{\kappa-1}^2, u_{\kappa-1} u_\kappa;\, u_\kappa^2] \tag{4.17}$$

4.2. DYNAMIC MODELS

Models of linear discrete dynamic systems will be shown to allow the formal application of the same LS algorithm as static models, with some subtle but important differences. We will start with moving average (MA) systems, then extend the results to autoregressive moving average (ARMA) systems, proceeding from the single-input single-output case through multiple-input single-output to multiple-input multiple-output systems (see Section 2.2. for definitions).

4.2.1. Moving Average Systems

Consider first the following dynamic system:

$$y(t) = g(\phi)\, u(t) = g^0 u(t) + g^1 u(t-1) + \ldots + g^\nu u(t-\nu) + \epsilon(t) \tag{4.18}$$

This can be written as

$$y(t) = \boldsymbol{\varphi}'(t)\, \boldsymbol{\pi} + \epsilon(t) \tag{4.19}$$

where

$$\boldsymbol{\varphi}'(t) = [u(t) \quad u(t\text{-}1) \ldots u(t\text{-}\nu)] \tag{4.20}$$

$$\boldsymbol{\pi} = [g^o \quad g^1 \ldots g^\nu]' \tag{4.21}$$

Now the prediction of $y(t)$ is sought as

$$\hat{y}(t) = \boldsymbol{\varphi}'(t)\, \hat{\boldsymbol{\pi}} \tag{4.22}$$

With this, the basic LS formulas (4.10) and (4.13) apply. The only difference now is that the regression vector contains a series of time-shifted values of the input.

If there are multiple inputs in a MA system, so that

$$y(t) = g_1(\phi)\, u_1(t) + \ldots + g_K(\phi)\, u_K(t) + \epsilon(t) \tag{4.23}$$

then $\boldsymbol{\varphi}'(t)$ and $\boldsymbol{\pi}$ become

$$\boldsymbol{\varphi}'(t) = [u_1(t) \ldots u_1(t\text{-}\nu) \ldots \ldots u_K(t) \ldots u_K(t\text{-}\nu)] \tag{4.24}$$

$$\boldsymbol{\pi} = [g_1^o \ldots g_1^\nu \ldots \ldots g_K^o \ldots g_K^\nu]' \tag{4.25}$$

4.2.2. ARMA Systems: Single-Input Single-Output

Consider the system

$$y(t) = \frac{g(\phi)}{h(\phi)}\, u(t) + v(t) \tag{4.26}$$

where $y(t)$ is the scalar output and $u(t)$ is the scalar input. The transfer function polynomials are

$$g(\phi) = g^o + g^1 \phi^{-1} + \ldots + g^\nu \phi^{-\nu}$$

$$h(\phi) = 1 + h^1 \phi^{-1} + \ldots + h^\nu \phi^{-\nu} \tag{4.27}$$

Note that, in order to simplify our treatment of parameter estimation, the above representation does not show the deadtime d explicitly (see Subsection 2.2.1.); one may assume at this point that $d=0$, or that it is included in the polynomial $g(\phi)$ as the d leading coefficients being zero. Further, the degrees

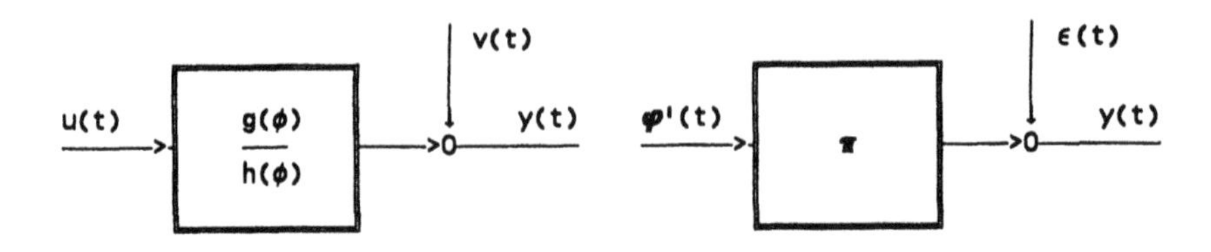

FIGURE 4.2.a

FIGURE 4.2.b

of the numerator and denominator are not necessarily equal. Finally, $v(t)$ in (4.26) is an error term representing the effect of noise and other unmodeled features, acting on (or reduced to) the output of the system (Fig. 4.2.a).

To convert (4.26) into recursive difference equation form, first cross-multiply with $h(\phi)$:

$$h(\phi)\, y(t) = g(\phi)\, u(t) + h(\phi)\, v(t) \tag{4.28}$$

then express $y(t)$ as

$$y(t) = g(\phi)\, u(t) - [h(\phi) - 1]\, y(t) + h(\phi)\, v(t) \tag{4.29}$$

This can then be written as

$$y(t) = \boldsymbol{\varphi}'(t)\, \boldsymbol{\pi} + \epsilon(t) \tag{4.30}$$

where

$$\boldsymbol{\varphi}'(t) = [u(t), u(t-1) \ldots u(t-\nu); -y(t-1) \ldots -y(t-\nu)] \tag{4.31}$$

$$\boldsymbol{\pi} = [g^{o}, g^{1} \ldots g^{\nu}; h^{1} \ldots h^{\nu}]' \tag{4.32}$$

$$\epsilon(t) = h(\phi)\, v(t) \tag{4.33}$$

(4.30) is formally the same as (4.3). However, there are some important differences, namely (see Fig. 4.2.b)

- the regression vector $\boldsymbol{\varphi}(t)$ contains present and past values;
- the regression vector contains past outputs, in addition to inputs;
- the equation error $\epsilon(t)$ is not identical with the output error $v(t)$.

Now the prediction may be sought as

$$\hat{y}(t) = \boldsymbol{\varphi}'(t)\,\hat{\boldsymbol{\pi}} \tag{4.34}$$

With this, the fundamental LS formulas (4.10) and (4.13) apply.

4.2.3. ARMA Systems: Multiple-Input Single-Output

The multiple-input single-output case is a simple extension of the previous results. Consider the system

$$y(t) = \frac{g_1(\phi)}{h(\phi)} u_1(t) + \frac{g_2(\phi)}{h(\phi)} u_2(t) + \ldots + \frac{g_K(\phi)}{h(\phi)} u_K(t) + v(t) \tag{4.35}$$

where

$$g_j(\phi) = g_j^o + g_j^1 \phi^{-1} + \ldots + g_j^{\nu} \phi^{-\nu} \qquad j=1 \ldots \kappa \tag{4.36}$$

and $h(\phi)$ is the common denominator of the transfer functions. This yields the recursive expression for $y(t)$

$$y(t) = g_1(\phi)\, u_1(t) + g_2(\phi)\, u_2(t) + \ldots + g_K(\phi)\, u_K(t)$$

$$- [h(\phi) - 1]\, y(t) + h(\phi)\, v(t) \tag{4.37}$$

This, in turn, can be described as

$$y(t) = \boldsymbol{\varphi}'(t)\,\boldsymbol{\pi} + \epsilon(t) \tag{4.38}$$

with

$$\boldsymbol{\varphi}'(t) = [u_1(t) \ldots u_1(t-\nu); \ldots \ldots; u_K(t) \ldots u_K(t-\nu); -y(t-1) \ldots -y(t-\nu)] \tag{4.39}$$

$$\boldsymbol{\pi} = [g_1^o \ldots g_1^{\nu}; \ldots \ldots; g_K^o \ldots g_K^{\nu}; h^1 \ldots h^{\nu}]' \tag{4.40}$$

$$\epsilon(t) = h(\phi)\, v(t) \tag{4.41}$$

With these, the standard results apply.

4.2.4. ARMA Systems: Multiple-Input Multiple-Output

If a multiple-input multiple-output system is modeled as

$$y_i(t) = \sum_{j=1}^{\kappa} \frac{g_{ij}(\phi)}{h_i(\phi)} u_j(t) + v_i(t) \qquad i=1\ldots\mu \tag{4.42}$$

where the denominator $h_i(\phi)$ is assumed to be different in each single-output subsystem, then the system can be identified as a set of μ unrelated single-output subsystems. This requires a relatively simple identification algorithm. However, in a real multiple-output system, the subsystem denominators are, at least partially, identical, and this remains hidden in the above approach. Alternatively, the system can be modeled in the structure

$$y_i(t) = \sum_{j=1}^{\kappa} \frac{\tilde{g}_{ij}(\phi)}{\tilde{h}(\phi)} u_j(t) + v_i(t) \qquad i=1\ldots\mu \tag{4.43}$$

where now all the subsystem denominators are identical ($\tilde{h}(\phi)$ is the least common multiple of $h_i(\phi)$, $i=1\ldots\mu$). This system needs to be identified as a single entity, which involves a more complex identification algorithm, but the estimated denominator is forced to be common throughout the system.

(4.43) leads to the recursive formulation

$$y_i(t) = \sum_{j=1}^{\kappa} \tilde{g}_{ij}(\phi)\, u_j(t) - [\tilde{h}(\phi) - 1]\, y_i(t) + \tilde{h}(\phi)\, v_i(t) \qquad i=1\ldots\mu \tag{4.44}$$

Define a single, common parameter vector

$$\pi = [[\tilde{g}_{11}]' \ldots [\tilde{g}_{1\kappa}]' \ldots\ldots\ldots [\tilde{g}_{\mu 1}]' \ldots [\tilde{g}_{\mu\kappa}]'; [\tilde{h}]']' \tag{4.45}$$

where

$$[\tilde{g}_{ij}]' = [\tilde{g}_{ij}^{0}, \tilde{g}_{ij}^{1} \ldots \tilde{g}_{ij}^{\nu}] \qquad i=1\ldots\mu,\; j=1\ldots\kappa \tag{4.46}$$

$$[\tilde{h}]' = [\tilde{h}^{1} \ldots \tilde{h}^{\nu}] \tag{4.47}$$

Now (4.44) can be written as

$$y_i(t) = \psi_i'(t)\, \pi + \epsilon_i(t) \qquad i=1\ldots\mu \tag{4.48}$$

where

$$\psi_i'(t) = [\underbrace{0' \ldots 0'}_{1} \ldots \ldots \underbrace{0' \ldots 0'}_{i-1} \quad \underbrace{u_1(t) \ldots u_1(t-\nu) \ldots \ldots u_\kappa(t) \ldots u_\kappa(t-\nu)}_{i}$$

$$\underbrace{0' \ldots 0'}_{i+1} \ldots \ldots \underbrace{0' \ldots 0'}_{\mu} \; -y_i(t-1) \ldots y_i(t-\nu)] \qquad i=1 \ldots \mu \tag{4.49}$$

(with $0'$ denoting zero row-vectors) and

$$\epsilon_i(t) = h(\phi)\, v_i(t) \qquad i=1 \ldots \mu \tag{4.50}$$

Observe that the parameter vector now may be quite long (contains $\mu \cdot \kappa \cdot (\nu+1)+\nu$ elements). The regression vectors are equally long but are packed with mostly zero elements.

Now write the prediction of the output vector $y(t)=[y_1(t) \ldots y_\mu(t)]'$ as

$$\hat{y}(t) = \Psi'(t)\, \hat{\pi} \tag{4.51}$$

where

$$\Psi'(t) = \begin{bmatrix} \psi_1'(t) \\ \cdot \\ \cdot \\ \cdot \\ \psi_\mu'(t) \end{bmatrix} \tag{4.52}$$

and define the LS performance index as

$$J = \sum_{k=0}^{K-1} \sum_{i=1}^{\mu} [\hat{y}_i(t-k) - y_i(t-k)]^2 = \sum_{k=0}^{K-1} [\hat{y}(t-k) - y(t-k)]' [\hat{y}(t-k) - y(t-k)] \tag{4.53}$$

Substituting (4.51) we obtain

$$J = \sum_{k=0}^{K-1} [\hat{\pi}'\, \Psi(t-k)\, \Psi'(t-k)\, \hat{\pi} - 2\, y'(t-k)\, \Psi'(t-k)\, \hat{\pi} + y'(t-k)\, y(t-k)] \tag{4.54}$$

which yields the derivative

$$\frac{\partial J}{\partial \hat{\pi}} = \sum_{k=0}^{K-1} 2\, [\Psi(t-k)\, \Psi'(t-k)\, \hat{\pi} - \Psi(t-k)\, y(t-k)] \tag{4.55}$$

From this, the parameter estimate is

$$\hat{\pi} = [\sum_{k=0}^{K-1} \Psi(t-k)\, \Psi'(t-k)]^{-1} [\sum_{k=0}^{K-1} \Psi(t-k)\, y(t-k)] = [\Phi'\, \Phi]^{-1}\, \Phi'\, Y \tag{4.56}$$

where

$$\boldsymbol{\Phi} = \begin{bmatrix} \boldsymbol{\varphi}'(t) \\ \cdot \\ \cdot \\ \cdot \\ \boldsymbol{\varphi}'(t - K+1) \end{bmatrix} \qquad Y = \begin{bmatrix} y(t) \\ \cdot \\ \cdot \\ \cdot \\ y(t - K+1) \end{bmatrix} \tag{4.57}$$

4.3. RECURSIVE ALGORITHMS

In this section, a number of recursive variants of the LS algorithm will be introduced. We will start from the standard, "batch" algorithm, then introduce recursive features gradually. Apart from some initialization aspects, which will be discussed at the end of the section, all the recursive variants are algebraically identical with the batch algorithm.

Refer to the basic LS algorithm (4.10) and (4.13) and introduce the shorthand notation

$$Q(t, K) = \boldsymbol{\Phi}'(t, K)\, \boldsymbol{\Phi}(t, K) = \sum_{k=0}^{K-1} \boldsymbol{\varphi}(t-k)\, \boldsymbol{\varphi}'(t-k) \tag{4.58}$$

$$p(t, K) = \boldsymbol{\Phi}'(t, K)\, Y(t, K) = \sum_{k=0}^{K-1} \boldsymbol{\varphi}(t-k)\, y(t-k) \tag{4.59}$$

Then the parameter estimate is obtained as

$$\hat{\boldsymbol{\pi}}(t, K) = Q^{-1}(t, K)\, p(t, K) \tag{4.60}$$

This notation indicates that, in general, the parameter estimate depends on the length of the observation dataset and on the time of the latest observation.

Batch algorithm. Under batch computation, the entire dataset is collected first and then the computation is performed off-line. This usually implies that the $\boldsymbol{\Phi}(t, K)$ and $Y(t, K)$ matrices are formed explicitly and then $Q(t, K)$ and $p(t, K)$ are computed from these. Finally, the parameter estimates are obtained from (4.60).

Observe that, with n parameters and K observations, $\boldsymbol{\Phi}(t, K)$ is $K{\cdot}n$ and $Y(t, K)$ is $K{\cdot}1$. Since the number of observations is usually large (in the order of magnitude of hundreds or thousands), these matrices may be quite extensive. Thus the batch algorithm may require significant memory.

Recursive data collection. Clearly, the size of $Q(t, K)$ is just $n{\cdot}n$ and of $p(t, K)$ is $n{\cdot}1$, no matter how many samples are taken. It follows from (4.58) and (4.59) that these matrices may be formed recursively, as

$$Q(t+1, K+1) = Q(t, K) + \varphi(t+1)\, \varphi'(t+1) \tag{4.61}$$

$$p(t+1, K+1) = p(t, K) + \varphi(t+1)\, y(t+1) \tag{4.62}$$

With this trick, it becomes unnecessary to actually form the large Φ and Y matrices. This is worth doing even if the computation is basically batch, that is, if the parameter estimates are computed only once, after all the data have been collected. Alternatively, new estimates may be computed on-line, using (4.60), following the arrival of each new sample.

Forgetting factor. In many practical situations, the plant changes (slowly) in time and it is desirable that the identification algorithm adapt the model to these changes. Such adaptation may be achieved by gradually suppressing old observations in the algorithm. Technically, this can be done by the inclusion of a "forgetting factor" λ in the updating formulas (4.61) and (4.62), as

$$Q(t+1, \lambda) = \lambda\, Q(t, \lambda) + \varphi(t+1)\, \varphi'(t+1) \tag{4.63}$$

$$p(t+1, \lambda) = \lambda\, p(t, \lambda) + \varphi(t+1)\, y(t+1) \tag{4.64}$$

With this forgetting mechanism, the data are filtered exponentially, that is, older observations are taken into account with exponentially decreasing weight. The length of the dataset becomes irrelevant, provided it is large enough compared to the time-constant of filtering. That is, Q and p are effectively formed as

$$Q(t, \lambda) = \sum_{k=0}^{\infty} \lambda^k \varphi(t-k)\, \varphi'(t-k) \tag{4.65}$$

$$p(t, \lambda) = \sum_{k=0}^{\infty} \lambda^k \varphi(t-k)\, y(t-k) \tag{4.66}$$

Recursive matrix inversion. If (4.60) is evaluated on-line, then it may be advantageous to compute the inverse matrix $Q^{-1}(t+1)$ recursively. This is possible since (4.58) and (4.63) represent a "rank-one update" for which a standard formula is available (e.g., Kailath, 1980):

$$[A + b\, b']^{-1} = A^{-1} - \frac{A^{-1} b\, b' A^{-1}}{1 + b' A^{-1} b} \tag{4.67}$$

If exponential forgetting is used, the recursive formula is

$$Q^{-1}(t+1, \lambda) = [\lambda\, Q(t, \lambda) + \varphi(t+1)\, \varphi'(t+1)]^{-1}$$

$$= \frac{1}{\lambda}\left[Q^{-1}(t,\lambda) - \frac{Q^{-1}(t,\lambda)\,\varphi(t+1)\,\varphi'(t+1)\,Q^{-1}(t,\lambda)}{\lambda + \varphi'(t+1)\,Q^{-1}(t,\lambda)\,\varphi(t+1)}\right] \tag{4.68}$$

With no forgetting, the formula is obtained by substituting $\lambda = 1$:

$$\begin{aligned} Q^{-1}(t+1, K+1) &= [Q(t,K) + \varphi(t+1)\,\varphi'(t+1)]^{-1} \\ &= Q^{-1}(t,K) - \frac{Q^{-1}(t,K)\,\varphi(t+1)\,\varphi'(t+1)\,Q^{-1}(t,K)}{1 + \varphi'(t+1)\,Q^{-1}(t,K)\,\varphi(t+1)} \end{aligned} \tag{4.69}$$

Full recursion. Finally, the parameter estimate $\hat{\pi}(t+1)$ may be expressed in terms of the previous estimate $\hat{\pi}(t)$, in a fully recursive algorithm. We will show this again first with forgetting factor. Following the pattern of (4.60)

$$\hat{\pi}(t+1, \lambda) = Q^{-1}(t+1, \lambda)\, p(t+1, \lambda) \tag{4.70}$$

The substitution of (4.68), and some algebra, yields

$$\hat{\pi}(t+1, \lambda) = \hat{\pi}(t, \lambda) + Q^{-1}(t+1, \lambda)\,\varphi(t+1)\, e(t+1\,|\,t) \tag{4.71}$$

where

$$e(t+1\,|\,t) = [y(t+1) - \varphi'(t+1)\,\hat{\pi}(t, \lambda)] \tag{4.72}$$

is the prediction error at time $t+1$, computed with the estimate $\hat{\pi}(t, \lambda)$. Without forgetting, the fully recursive algorithm becomes

$$\hat{\pi}(t+1, K+1) = \hat{\pi}(t, K) + Q^{-1}(t+1, K+1)\,\varphi(t+1)\, e(t+1\,|\,t) \tag{4.73}$$

where

$$e(t+1\,|\,t) = [y(t+1) - \varphi'(t+1)\,\hat{\pi}(t, K)] \tag{4.74}$$

The fully recursive algorithm looks very logical, in that it provides a correction to the parameter estimates which is proportional to the momentary prediction error. Computationally, it has to go together with (4.68) or (4.69) to calculate the updated inverses and thus it is not more effective than the direct application of (4.70) or (4.60), also with the recursive inverse computation.

Initialization. The recursive computations require initial values, in both cases for the inverse matrix and, with full recursion, also for the parameter estimates. The choice of the initial values affects the estimates, though this effect fades away if exponential forgetting is used. The literature contains various recommendations as to how to initialize the algorithms. In our experience, the best strategy is to use a short batch run to compute the initial values. In this case, the initial values are meaningful and the algebraic identity between the batch and recursive algorithms is maintained. As it will be shown in the next section, the minimum data-length for running the batch algorithm is $K=n$ samples (where n is the number of parameters to estimate).

4.4. EXISTENCE CONDITIONS

The solution of the least squares algorithm requires the inversion of the Q matrix, whether this is done explicitly (in the batch algorithm) or implicitly (in the recursive variants). Thus the condition for the parameter estimates to exist is that the Q matrix be invertible. Since with n parameters to estimate Q is an $n \cdot n$ matrix, this means

$$Rank\, Q(t, K) = n \tag{4.75}$$

Further, since

$$Rank\, Q(t, K) = Rank\, [\Phi'(t, K)\, \Phi(t, K)] = Rank\, \Phi(t, K) \tag{4.76}$$

this also implies

$$Rank\, \Phi(t, K) = n \tag{4.77}$$

Now $\Phi(t, K)$ is a $K \cdot n$ matrix, so what is required is that

$$K \geq n \tag{4.78}$$

and that *all n columns of* Φ *be linearly independent*. Usually, the number of samples K is much bigger than the number of parameters n. However, there may be several reasons why the columns of Φ may not be linearly independent.

It should be noted that outright singularity of the $\Phi' \Phi$ matrix seldom happens in practice. Even if in theory the matrix would be expected to be singular, it is usually invertible (though may be poorly conditioned), due to random effects such as noise and numerical errors. However, the parameter estimates obtained in such cases are very unreliable, and they reflect those random effects more than the true plant parameters.

4.4.1. Persistent Excitation

Perhaps the most important reason why the Φ matrix may have rank defect is if there are repetitions in the input sequence which establish relationships among the input columns. Such features of the input sequence are referred to as persistent excitation properties.

Persistently exciting input sequence: MA case. Let us consider a MA dynamic system and write out the Φ matrix for a single input in detail:

$$\Phi(t, K) = \begin{bmatrix} u(t) & u(t-1) & \dots & u(t-\nu) \\ u(t-1) & u(t-2) & \dots & u(t-1-\nu) \\ \cdots & & & \\ \cdots & & & \\ \cdots & & & \\ u(t-K+1) & u(t-K) & \dots & u(t-K+1-\nu) \end{bmatrix} \tag{4.79}$$

If, due to a repetition in the input sequence, for example $u(t) = u(t-2)$ for all t, then the entire third column is identical with the first column (and the fourth column with the second, etc.). Similarly, if due to a "regularity" in the input sequence, for example $u(t)=\alpha_1 u(t-1)+\dots+\alpha_\nu u(t-\nu)$ for all t then the first column is a linear combination of the second through the $(\nu+1)$th columns. In such cases, the input sequence is said to be *not persistently exciting*.

Strong and weak persistent excitation. The following is a more precise definition of the concept of persistent excitation. Consider the two identities:

$$\alpha_0 u(t) + \alpha_1 u(t-1) + \dots + \alpha_\rho u(t-\rho) = 0 \qquad \alpha_0 \alpha_1 \dots \alpha_\rho \neq 0 \tag{4.80a}$$

$$\alpha_0 u(t) + \alpha_1 u(t-1) + \dots + \alpha_{\rho-1} u(t-\rho+1) \neq 0 \tag{4.80b}$$

- If (4.80a) and (4.80b) are valid for all t then the $u(t)$ sequence is *strongly persistently exciting (SPE) of order* ρ.
- If (4.80a) is valid for all t and (4.80b) is valid for some t then the $u(t)$ sequence is *weakly persistently exciting (WPE) of order* ρ.

With strong persistent excitation, $\Phi(t, K)$ has full rank for any t, provided that

$$\rho \geq \nu + 1 \tag{4.81a}$$

$$K \geq \nu + 1 \tag{4.81b}$$

With weak persistent excitation, and (4.81) satisfied, $\Phi(t, K)$ has full rank for some t's, and certainly as $K \to \infty$.

Strong persistent excitation of the proper order is necessary in most cases, especially when exponential forgetting is involved. Weak persistent excitation, of proper order, may be sufficient for batch identification, if the dataset is long enough or if specific test signals are applied.

Persistent excitation properties of typical signals.

- *Impulse function* is SPE of order *0* and WPE of order ∞.
- *Step function* is SPE of order *1* and WPE of order ∞.
- *Sinusoidal functions* are SPE and WPE of order *2*. To see this, consider the samples taken from the function $\sin\beta\tau$ (where τ is the continuous time variable), at Δ sampling intervals:

$$u(t) = \sin \beta\tau_o$$

$$u(t-1) = \sin(\beta\tau_o - \Delta) = \cos\Delta \sin\beta\tau_o - \sin\Delta \cos\beta\tau_o$$

$$u(t-2) = \sin(\beta\tau_o - 2\Delta) = \cos2\Delta \sin\beta\tau_o - \sin2\Delta \cos\beta\tau_o \quad (4.82)$$

  Clearly, the pair $u(t)$ and $u(t-1)$ is independent (for every τ_o) but the triplet $u(t)$, $u(t-1)$ and $u(t-2)$ is not.

- *Periodic discrete sequences* are SPE and WPE of order M where M is the length of their period. However, sequences symmetrical within their period are SPE and WPE only of order $M/2$. Note that a sequence obtained by sampling a periodic continuous signal is not, in general, periodic. It is periodic only if the relationship $M \cdot \Delta = N \cdot T$ holds, where T is the period of the continuous signal and M and N are integers; in this case, the smallest M is the length of period of the discrete sequence.
- *Random sequences* are SPE of order ∞. The most frequently used random input sequences are the white (uncorrelated) sequence with normal or uniform distribution, and the (pseudo-)random binary sequence (a series of *0* and *1* signals, with random transition times). Computer realizations of random sequences are periodic but their period is usually very long.

Persistently exciting input sequence: ARMA case. Let us now write out the $\boldsymbol{\Phi}$ matrix for the single-input single-output ARMA case:

$$\boldsymbol{\Phi}(t, K) = \begin{bmatrix} u(t) & u(t-1) \; \dots \; u(t-\nu) & -y(t-1) \; \dots \; -y(t-\nu) \\ u(t-1) & u(t-2) \; \dots \; u(t-1-\nu) & -y(t-2) \; \dots \; -y(t-1-\nu) \\ \cdots & & \\ \cdots & & \\ \cdots & & \\ u(t-K+1) & u(t-K) \; \dots \; u(t-K+1-\nu) & -y(t-K) \; \dots \; -y(t-K+1-\nu) \end{bmatrix} \quad (4.83)$$

Clearly, any linear relations in the input columns cause a rank defect, calling for SPE of order $\nu+1$ or higher. However, the presence of the output columns requires more than this, as it will be shown below.

Let us express the outputs with convolution sums

$$y(t-k) = \sum_{l=0}^{\infty} m_l u(t-k-l) \qquad k=1\ldots\nu \tag{4.84}$$

where m_l are coefficients of the discrete impulse response. If (4.80a) and (4.80b) hold for all t then all samples of the input can be expressed in terms of the last ρ samples, yielding

$$y(t-k) = c_{ko}u(t) + c_{k1}u(t-1) + \ldots + c_{k,\rho-1}u(t-\rho+1) \qquad k=1\ldots\nu \tag{4.85}$$

Now $\Phi(t, K)$ has full rank if all columns are linearly independent, that is, if

$$\alpha_o u(t) + \alpha_1 u(t-1) + \ldots + \alpha_\nu u(t-\nu) + \alpha_{\nu+1} y(t-1) + \ldots + \alpha_{2\nu} y(t-\nu) = 0 \tag{4.86}$$

has only the trivial solution $\alpha_o = \alpha_1 = \ldots = \alpha_{2\nu} = 0$. Substituting $y(t-k)$, $k=1\ldots\nu$ from (4.85) and arranging the terms which contain $u(t)$, $u(t-1)$, etc. each into separate equations yields the formulation

$$Diag\,[u(t) \quad u(t-1) \ldots u(t-\rho+1)]\; C\,[\alpha_o\, \alpha_1 \ldots \alpha_{2\nu}]' = 0 \tag{4.87}$$

where C is a coefficient matrix and where we implied that $\rho \geq \nu+1$. Clearly, (4.87) represents ρ homogeneous conditions for $2\nu+1$ unknown coefficients. For only the trivial solution to exist, it is necessary that

$$\boxed{\rho \geq 2\,\nu + 1} \tag{4.88}$$

That is, *the input signal must be persistently exciting of order* $2\nu+1$ *or more.*

4.4.2. Other Reasons of Rank Defect

There are several other reasons why the Φ matrix may suffer rank defect. These include

- linear dependence among multiple inputs
- over-parametrization of a dynamic model
- data collection under feedback control.

These will be discussed below.

Linear independence of multiple inputs. Consider first the static linear case (4.1) and write out the Φ matrix in some detail:

$$\Phi(t, K) = \begin{bmatrix} u_1(t) & u_2(t) & \dots & u_K(t) \\ u_1(t-1) & u_2(t-1) & \dots & u_K(t-1) \\ \vdots & & & \\ u_1(t-K+1) & u_2(t-K+1) & \dots & u_K(t-K+1) \end{bmatrix} \tag{4.89}$$

Clearly, if any input is a multiple of another, $u_i(t)=cu_j(t)$, or there exists a general linear relationship among the inputs,

$$\alpha_1 u_1(t) + \alpha_2 u_2(t) + \dots + \alpha_K u_K(t) = 0, \qquad \alpha_1 \alpha_2 \dots \alpha_K \neq 0 \tag{4.90}$$

for all t, then $\Phi(t, K)$ has less than full rank. The same is true if multiple inputs in a dynamic system are linearly dependent.

Over-parametrization. A rank defect of the Φ matrix results also if the model of an ARMA system contains excess pole-zero pairs. Assume that we seek a model in the structure

$$\hat{y}(t) = \frac{\hat{g}^0 + \hat{g}^1 \phi^{-1} + \dots + \hat{g}^{\nu+1} \phi^{-\nu-1}}{1 + \hat{h}^1 \phi^{-1} + \dots + \hat{h}^{\nu+1} \phi^{-\nu-1}} u(t) \tag{4.91}$$

while the true system is

$$y(t) = \frac{g^0 + g^1 \phi^{-1} + \dots + g^{\nu} \phi^{-\nu}}{1 + h^1 \phi^{-1} + \dots + h^{\nu} \phi^{-\nu}} u(t) \tag{4.92}$$

where we assumed that there is no noise. Clearly, the model has an excess pole-zero pair which is undefined; the experimental data would provide no information to determine it.

To see how this plays out in terms of the Φ matrix, observe that, according to the model structure, the regression vector is

$$\varphi'(t) = [u(t) \dots u(t-\nu-1) \;\; -y(t-1) \dots -y(t-\nu-1)] \tag{4.93}$$

Now it follows from the true relationship (4.92) that

$$y(t-1) = g^0 u(t-1) + \dots + g^{\nu} u(t-\nu-1) - h^1 y(t-2) - \dots - h^{\nu} y(t-\nu-1) \tag{4.94}$$

That is, one of the elements in the regression vector is expressed completely in terms of the other elements. This implies that the Φ matrix has rank defect.

Note that if only the numerator or only the denominator is over-parametrized then rank defect does not occur. In these cases, the regression vectors are, respectively,

$$\varphi'(t) = [u(t) \ldots u(t-\nu-1) \quad -y(t-1) \ldots -y(t-\nu)] \tag{4.95}$$

$$\varphi'(t) = [u(t) \ldots u(t-\nu) \quad -y(t-1) \ldots -y(t-\nu-1)] \tag{4.96}$$

With these, (4.94) does not establish a linear relationship among the columns. Thus $\Phi'\Phi$ is invertible and the algorithm yields the estimate for the excess parameter (at least in the absence of noise and numerical errors) as zero.

Experiments under feedback. If the experimental data is collected while the plant operates under feedback control, then the plant inputs are linked to the outputs via the control algorithm (Fig. 4.3). The scope of the control algorithm (the variable samples involved) overlaps with that of the regression vector and it depends on the extent of this overlap whether or not rank defect results.

As an example, consider a PI control algorithm

$$u(t) = -\frac{b^o + b^1\phi^{-1}}{1 + a^1\phi^{-1}}\, y(t) \tag{4.97}$$

This means

$$u(t) = -b^o y(t) - b^1 y(t-1) - a^1 u(t-1) \tag{4.98}$$

but it also implies

$$u(t-1) = -b^o y(t-1) - b^1 y(t-2) - a^1 u(t-2) \tag{4.99}$$

Any model structure which places into the regression vector $\varphi'(t)$ all the four variable samples appearing in (4.99) runs into rank-defect with this control algorithm. However, if the model structure is such that some of the four variable samples are missing from $\varphi'(t)$ then rank defect does not occur.

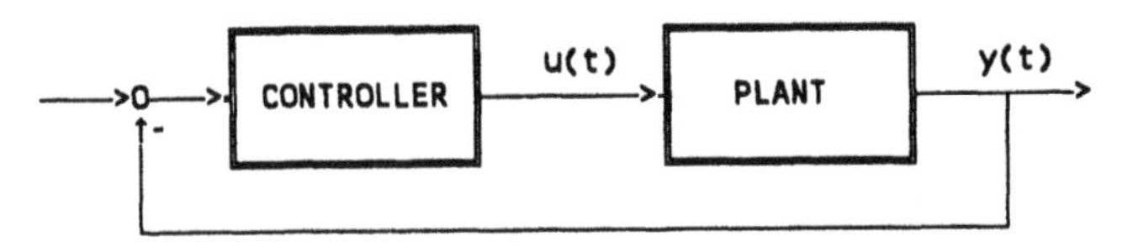

FIGURE 4.3. Closed-loop system.

4.5. THE EFFECT OF NOISE

If the observed outputs contain random noise, then they are random variables. Thus the parameter estimates computed from them also become random variables. In this situation, it is logical to ask two questions:

- How do the estimates behave if they are based on a large number of observations? The behavior of the estimates as the number of observations tends to infinity is referred to as the *consistency* properties of the estimator.
- How do the estimates behave if they are based on a smaller (finite) number of observations? In particular, how does the expectation of the estimates differ from the true parameters (what is their *bias*) and how do particular estimates differ from their expectation (what is their *variance*)?

These properties will be outlined in the following section. For a more detailed and more exact treatment of this rather complicated subject, the reader is referred to the specific literature of systems identification.

4.5.1. Consistency Properties

First we will discuss the conditions for exact parameter estimates from an infinite set of observations (zero consistency error), under various system models. Then we will give a formula for the consistency error in single-input single-output ARMA systems.

The conditions of zero consistency error. Consider the general estimation formula (4.13), with the time and set-length arguments written in:

$$\hat{\pi}(t, K) = [\Phi'(t, K)\, \Phi(t, K)]^{-1}\, \Phi'(t, K)\, Y(t, K) \tag{4.100}$$

It follows from (4.3) and (4.11) that

$$Y(t, K) = \Phi(t, K)\, \pi + E(t, K) \tag{4.101}$$

where

$$E(t, K) = \begin{bmatrix} \epsilon(t) \\ \cdot \\ \cdot \\ \cdot \\ \epsilon(t-K+1) \end{bmatrix} \tag{4.102}$$

Substituting (4.101) into (4.100) yields

$$\hat{\pi}(t, K) = \pi + [\Phi'(t, K)\, \Phi(t, K)]^{-1}\, \Phi'(t, K)\, E(t, K) \tag{4.103}$$

Thus, utilizing (4.12), the consistency error is

$$\hat{\pi}(t, \infty) - \pi = [\lim_{K\to\infty} \frac{1}{K} \sum_{k=0}^{\infty} \varphi(t-k)\, \varphi'(t-k)]^{-1} [\lim_{K\to\infty} \frac{1}{K} \sum_{k=0}^{\infty} \varphi(t-k)\, \epsilon(t-k)] \tag{4.104}$$

If the rank conditions discussed in the previous section are satisfied then the first bracketed matrix is invertible and the inverse is finite. Thus, it is the second bracketed matrix which determines whether or not the error is zero. We will examine the conditions of the second matrix disappearing for the various system models.

Multiple input static system. With (4.2)

$$\lim_{K\to\infty} \frac{1}{K} \sum_{k=0}^{\infty} \varphi(t-k)\, \epsilon(t-k) = \begin{bmatrix} \lim_{K\to\infty} \frac{1}{K} \sum_{k=0}^{\infty} u_1(t-k)\, \epsilon(t-k) \\ \vdots \\ \lim_{K\to\infty} \frac{1}{K} \sum_{k=0}^{\infty} u_\kappa(t-k)\, \epsilon(t-k) \end{bmatrix} \tag{4.105}$$

Thus the condition for zero consistency error is

$$\lim_{K\to\infty} \frac{1}{K} \sum_{k=0}^{\infty} u_j(t-k)\, \epsilon(t-k) = 0 \qquad j=1\ldots\kappa \tag{4.106}$$

The expression in (4.106) can be recognized as the cross-correlation between the inputs $u_j(t)$ and the equation error $\epsilon(t)$. That is, *for zero consistency error, the equation error must be independent of the inputs and have zero mean.*

Moving average dynamic system. With (4.20)

$$\lim_{K\to\infty} \frac{1}{K} \sum_{k=0}^{\infty} \varphi(t-k)\, \epsilon(t-k) = \begin{bmatrix} \lim_{K\to\infty} \frac{1}{K} \sum_{k=0}^{\infty} u(t-k)\, \epsilon(t-k) \\ \vdots \\ \lim_{K\to\infty} \frac{1}{K} \sum_{k=0}^{\infty} u(t-\nu-k)\, \epsilon(t-k) \end{bmatrix} \tag{4.107}$$

Thus the condition for zero consistency error is

$$\lim_{K\to\infty} \frac{1}{K} \sum_{k=0}^{\infty} u(t-l-k)\, \epsilon(t-k) = 0 \qquad l=0\ldots\nu \tag{4.108}$$

This is now the cross-correlation between the input samples $u(t-l)$ and the equation error $\epsilon(t)$. That is, *for zero consistency error, the equation error must be independent of the input sequence and have zero mean.*

ARMA dynamic system. With (4.31)

$$\lim_{K\to\infty} \frac{1}{K} \sum_{k=0}^{\infty} \varphi(t-k)\, \epsilon(t-k) = \begin{bmatrix} \lim_{K\to\infty} \frac{1}{K} \sum_{k=0}^{\infty} u(t-k)\, \epsilon(t-k) \\ \vdots \\ \lim_{K\to\infty} \frac{1}{K} \sum_{k=0}^{\infty} u(t-\nu-k)\, \epsilon(t-k) \\ \lim_{K\to\infty} \frac{1}{K} \sum_{k=0}^{\infty} -y(t-1-k)\, \epsilon(t-k) \\ \vdots \\ \lim_{K\to\infty} \frac{1}{K} \sum_{k=0}^{\infty} -y(t-\nu-k)\, \epsilon(t-k) \end{bmatrix} \tag{4.109}$$

Here the input rows are identical with (4.107) and lead to the requirement of (4.108). However, there is an additional set of requirements, arising from the output rows, namely

$$\lim_{K\to\infty} \frac{1}{K} \sum_{k=0}^{\infty} y(t-l-k)\, \epsilon(t-k) = 0 \qquad l=1\ldots\nu \tag{4.110}$$

These are cross-correlations between the output samples $y(t-l)$, $l=1\ldots\nu$, and the equation error $\epsilon(t)$. To see what it takes for these to disappear, let us rewrite (4.29) as

$$y(t) = g^{o} u(t) + \ldots + g^{\nu} u(t-\nu) - h^{1} y(t-1) - \ldots - h^{\nu} y(t-\nu) + \epsilon(t) \tag{4.111}$$

According to (4.111), the present value of the output depends on present and past values of the input, on the present value of the equation error ϵ and, via past values of the output, on past values of ϵ. Therefore, if the equation error sequence is uncorrelated, that is,

$$\lim_{K \to \infty} \frac{1}{K} \sum_{k=0}^{\infty} \epsilon(t-l-k)\, \epsilon(t-k) = 0 \qquad l \neq 0 \tag{4.112}$$

(in addition to (4.108)) then all cross-correlations between the present equation error and past output values in (4.110) become zero. Thus the condition for zero consistency error in the ARMA case is that

> *the equation error* $\epsilon(t)$
>
> - *has zero mean*
> - *is uncorrelated with the input sequence*
> - *is itself an uncorrelated (white) sequence.*

It should be emphasized that while in the case of static or MA dynamic systems the equation error $\epsilon(t)$ is identical with the noise acting on the plant output, this is not true for ARMA dynamic systems. With the latter (recall (4.41)), $\epsilon(t)=h(\phi)v(t)$, where $v(t)$ is the output noise. So *a white output noise does not make the equation error white*, rather, in order for $\epsilon(t)$ to be white, the output noise must be

$$v(t) = \epsilon_o(t) \,/\, h(\phi) \tag{4.113}$$

where $\epsilon_o(t)$ is some white source noise.

The size of the consistency error. In the following, we will derive a formula for the size of the consistency error in single-input single-output ARMA dynamic systems. Let us first rewrite (4.26) as

$$y(t) = \tilde{y}(t) + v(t) \tag{4.114}$$

where $\tilde{y}(t)$ is the noise-free output $[g(\phi)/h(\phi)]u(t)$ (Fig. 4.4). Define

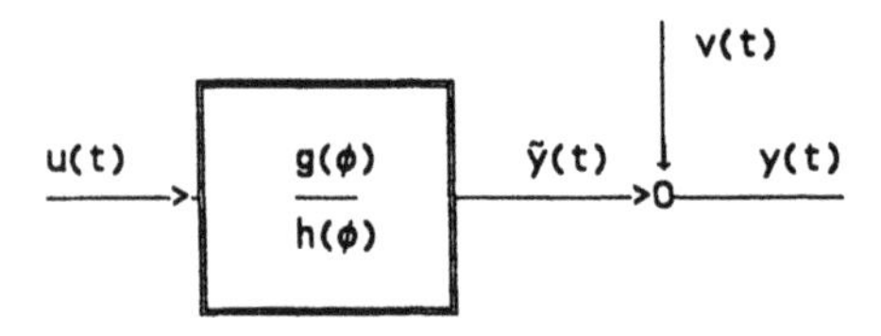

FIGURE 4.4. Noisy system.

$$\tilde{\boldsymbol{\varphi}}'(t) = [u(t) \ldots u(t-\nu) \quad -\tilde{y}(t-1) \ldots -\tilde{y}(t-\nu)] \tag{4.115}$$

$$\boldsymbol{\nu}'(t) = [0 \ldots 0 \quad -v(t-1) \ldots -v(t-\nu)] \tag{4.116}$$

With these, $\boldsymbol{\varphi}'(t)=\tilde{\boldsymbol{\varphi}}'(t)+\boldsymbol{\nu}'(t)$ and

$$\boldsymbol{\Phi}(t, K) = \tilde{\boldsymbol{\Phi}}(t, K) + \boldsymbol{\Omega}(t, K) \tag{4.117}$$

$$Y(t, K) = \tilde{Y}(t, K) + V(t, K) \tag{4.118}$$

where

$$\boldsymbol{\Omega}(t, K) = \begin{bmatrix} \boldsymbol{\nu}'(t) \\ \cdot \\ \cdot \\ \cdot \\ \boldsymbol{\nu}'(t-K+1) \end{bmatrix} \qquad V(t, K) = \begin{bmatrix} v(t) \\ \cdot \\ \\ \cdot \\ v(t-K+1) \end{bmatrix} \tag{4.119}$$

Now assume that the $v(t)$ sequence has zero mean and is uncorrelated with the $u(t)$ sequence and, therefore, with the $\tilde{y}(t)$ sequence (but is not, in general, a white sequence). Then

$$\lim_{K\to\infty} \frac{1}{K} \tilde{\boldsymbol{\Phi}}'(t, K)\, \boldsymbol{\Omega}(t, K) = 0 \tag{4.120}$$

$$\lim_{K\to\infty} \frac{1}{K} \tilde{\boldsymbol{\Phi}}'(t, K)\, V(t, K) = \lim_{K\to\infty} \frac{1}{K} \boldsymbol{\Omega}'(t, K)\, \tilde{Y}(t, K) = 0 \tag{4.121}$$

Therefore it follows that

$$\lim_{K\to\infty} \frac{1}{K} \boldsymbol{\Phi}'(t, K)\, \boldsymbol{\Phi}(t, K) = \lim_{K\to\infty} \frac{1}{K} \tilde{\boldsymbol{\Phi}}'(t, K)\, \tilde{\boldsymbol{\Phi}}(t, K) + \lim_{K\to\infty} \frac{1}{K} \boldsymbol{\Omega}'(t, K)\, \boldsymbol{\Omega}(t, K) \tag{4.122}$$

We will abbreviate this relationship as

$$Q^* = \tilde{Q}^* + \Delta Q^* \tag{4.123}$$

Similarly,

$$\lim_{K\to\infty} \frac{1}{K} \boldsymbol{\Phi}'(t, K)\, Y(t, K) = \lim_{K\to\infty} \frac{1}{K} \tilde{\boldsymbol{\Phi}}'(t, K)\, \tilde{Y}(t, K) + \lim_{K\to\infty} \frac{1}{K} \boldsymbol{\Omega}'(t, K)\, V(t, K) \tag{4.124}$$

which will be abbreviated as

$$p^* = \tilde{p}^* + \Delta p^* \tag{4.125}$$

Now clearly,

$$\hat{\pi}(t, \infty) = (Q^*)^{-1}p^* \qquad \pi = (\bar{Q}^*)^{-1}\bar{p}^* \tag{4.126}$$

Thus the consistency error can be described as

$$\hat{\pi}(t, \infty) - \pi = (Q^*)^{-1}[\bar{p}^* + \Delta p^* - (\bar{Q}^* + \Delta Q^*)(\bar{Q}^*)^{-1}\bar{p}^*] \tag{4.127}$$

which yields

$$\boxed{\hat{\pi}(t, \infty) - \pi = (Q^*)^{-1}[\Delta p^* - \Delta Q^* \pi]} \tag{4.128}$$

The matrices Δp^* and ΔQ^* are noise covariances while Q^* is the correlation matrix of the data. So *the consistency error is proportional to the noise covariances and inversely proportional to the data correlations.*

Further, the consistency error disappears when $\Delta p^* = \Delta Q^* \pi$. It can be shown that this happens when $v(t) = \epsilon_o(t)/h(\phi)$, where ϵ_o is a white sequence. This is in agreement with the results of the previous subsection.

4.5.2. Bias and Variance Under Finite Datasets

If the identification datasets are finite then every run will, in general, result in different parameter estimates. We will review here the expectation and the variance of the estimates under such conditions.

Expectation and bias of the estimates. The expectation of the parameter estimates, for a given data-length K, is

$$E\{\hat{\pi}(t, K)\} = E\{[\Phi'(t, K)\,\Phi(t, K)]^{-1}\,\Phi'(t, K)\,Y(t, K)\} \tag{4.129}$$

The bias is then obtained as

$$E\{\hat{\pi}(t, K)\} - \pi = E\{[\Phi'(t, K)\,\Phi(t, K)]^{-1}\,\Phi'(t, K)\,E(t, K)\} \tag{4.130}$$

where we utilized (4.103).

We may also define the expectation of the estimates as the values which minimize the expectation of the performance index $(1/K)J(t, K)$. Since, under rather broad assumptions of stationarity,

$$E\{\frac{1}{K}J(t, K)\} = \lim_{K \to \infty} \frac{1}{K} J(t, K) \tag{4.131}$$

we may conclude that

$$E\{\hat{\pi}(t, K)\} = \hat{\pi}(t, \infty) \tag{4.132}$$

Thus the bias and the consistency error may be considered the same.

With static and MA dynamic systems, $\Phi(t, K)$ is deterministic (it contains only input values) while $E(t, K)$ is random. If $E(t, K)$ is also independent of the inputs then (4.130) becomes

$$E\{\hat{\pi}(t, K)\} - \pi = [\Phi'(t, K)\,\Phi(t, K)]^{-1}\,\Phi'(t, K)\,E\{E(t, K)\} \tag{4.133}$$

This confirms the earlier results that, for such systems, the estimates are unbiased if the equation error has zero mean. It also provides an easy way to compute the bias if the mean of the equation error is nonzero.

With ARMA dynamic systems, $\Phi(t, K)$ is random, so the simple relationship described in (4.133) is not true. However, (4.132) still applies and the consistency error derived in 4.4.1. may be re-interpreted as bias.

The variance of the estimates. The variance of the parameter estimates is

$$\mathrm{Var}\{\hat{\pi}(t, K)\} = E\{[\hat{\pi}(t, K) - \hat{\pi}(t, \infty)]'\,[\hat{\pi}(t, K) - \hat{\pi}(t, \infty)]\} \tag{4.134}$$

where we utilized (4.132). If $\Phi(t, K)$ is deterministic and the bias is zero then this can be computed from (4.103) as

$$\mathrm{Var}\{\hat{\pi}(t, K)\} = \tag{4.135}$$

$$[\Phi'(t, K)\,\Phi(t, K)]^{-1}\,\Phi'(t, K)\,E\{E(t, K)\,E'(t, K)\}\,\Phi(t, K)\,[\Phi'(t, K)\,\Phi(t, K)]^{-1}$$

Here $E\{E(t, K)\,E'(t, K)\}$ is the autocorrelation of the equation error sequence. If this sequence is uncorrelated, with zero mean, then

$$E\{E(t, K)\,E'(t, K)\} = \sigma_{\epsilon}^{2}\,I \tag{4.136}$$

and (4.135) simplifies to

$$\mathrm{Var}\{\hat{\pi}(t, K)\} = \sigma_{\epsilon}^{2}\,[\Phi'(t, K)\,\Phi(t, K)]^{-1} = \frac{\sigma_{\epsilon}^{2}}{K}\left[\frac{1}{K}\sum_{k=0}^{K-1}\varphi(t-k)\,\varphi'(t-k)\right]^{-1} \tag{4.137}$$

That is, *the variance of the parameter estimates is roughly inversely proportional with the data-length* K (since the "size" of the bracketed matrix remains basically the same as K changes).

If $\Phi(t, K)$ is not deterministic then (4.135) is not valid anymore. It has been shown, however (see Ljung, 1987), that (4.137) still applies in most cases, at least asymptotically for large values of K, provided (4.136) holds.

The distribution of the estimates. It has also been shown (see Ljung, 1987) that the unbiased parameter estimates are asymptotically normally distributed if the equation error is a gaussian random variable (and even if it is not, in many cases, due to the central limit theorem). With the variance of the distribution estimated, according to the previous paragraphs, it is thus possible to construct confidence intervals and statistical tests for the parameter estimates.

4.5.3. Unbiased Methods

There is a great number of parameter estimation methods which aim at eliminating the consistency error (bias) in the estimates. These methods are more complex than the standard, prediction error least squares, they usually involve iterative procedures and the potential of multiple optima. Most of them belong to one of the following broad classes:

Extended or generalized least squares. These methods include an explicit dynamic model of the noise and estimate its parameters together with those of the plant. They are usually iterative by nature, in that the estimation of noise parameters and plant parameters alternate in an iterative scheme.

Output error least squares. The least squares performance index is formulated for the output error $y(t)$-$[\hat{g}(\phi)/\hat{h}(\phi)]u(t)$ (instead of the prediction error as in the standard LS). The resulting optimization problem is nonquadratic, it may have multiple minima and its solution requires numerical iteration.

Instrumental variable methods. The Φ' matrix in the LS algorithm is replaced with another matrix, a suitably chosen "instrumental variable," which is uncorrelated with the equation error (so that the consistency error becomes zero) and correlated with the measurements (so that the Q matrix remains invertible). Most instrumental variable methods require iteration, though there exist linear variants as well where added complexity arises from extended size matrices relative to the standard LS method.

Maximum likelihood methods. These rely on the statistical distribution of the noise which is assumed to be known and usually also normal. The parameters of the plant and noise model are estimated by maximizing the likelihood of the observed dataset. The procedure involves the numerical solution of nonlinear equations.

The detailed discussion of the unbiased methods goes far beyond the scope of this chapter. The reader is referred to the extensive literature of systems identification.

4.6. SELECTING THE MODEL STRUCTURE

We have assumed so far that the structure of the model which perfectly describes the physical system is known. This, of course, is seldom true in practice. We may have a general idea of the nature of the true system and structure the model accordingly. More often, though, the physical system is very complex and we just try to characterize it with a more or less simplified model.

Model structure selection covers the following issues:

Nonlinearities. One may want to determine what kind of nonlinearities are present in the system and how significant they are. If the nonlinearities are approximated by a polynomial expression, the question is what degree to select for this polynomial and which terms need to appear in it.

System order. It is necessary to estimate the dynamic order of the system. That is, one has to decide what the order of the numerator and denominator should be in a transfer function representation.

Pure time delay (deadtime). If the deadtime d is not characterized explicitly, then it manifests itself as a series of near-zero leading coefficients in the $g(\phi)$ polynomial. In this case, d becomes part of the transfer function order. Alternatively, the deadtime may be handled as an explicit structure parameter; then the regression vector contains the delayed input samples $u(t-d) \ldots u(t-d-\nu+1)$ and the leading coefficients in $g(\phi)$ are not zero. In the latter case, d needs to be determined as part of structure selection.

Unfortunately, there are no simple formulas for structure estimation. For polynomial degrees and transfer function orders, the following two approaches are generally used, combined with some "engineering judgment:"

- Start with a simple structure and compute the performance index J. Then increase the complexity of the structure (by adding a nonlinear term, one more pole/zero) and compute J again. The performance index will decrease monotonically but the improvement will gradually decrease. Choose the structure beyond which the improved performance does not justify the increased complexity. This decision may be placed on a more solid basis by constructing a statistical test for the performance index.
- Again, start with a simple structure and compute the prediction errors. Then compute the cross-correlation between the prediction error and the potential additional elements in the regression vector (nonlinear terms or previous input/output samples). Choose the candidates which have the highest correlation. Repeat this procedure with the new model structure, etc., until there is no candidate term left which has a notable correlation with the prediction error.

If the deadtime is estimated explicitly then the performance index needs to be computed for several hypothesized values of d and the value minimizing J chosen.

The structure selection procedures described above are well supported in today's identification software packages.

4.7. NOTES AND REFERENCES

Systems identification has been one of the most important and, in the past thirty years, most active fields of systems and control engineering. A triennial symposium series on Identification in Automatic Control Systems, under the aegis of IFAC (the International Federation of Automatic Control), was started in 1967 in Prague, Czechoslovakia (see Strejc, 1967), and is still running strong, with the latest event taking place in 1997 in Kitakyushu, Japan (see Sawaragi and Sagara, 1997).

Among the pioneers of modern systems identification, one should mention Karl Johan Åström (Åström and Bohlin, 1965), David Clarke (1967) and Pieter Eykhoff (1974). The early developments of the field were summarized in a landmark survey by Åström and Eykhoff at the 1970 IFAC Identification Symposium (Åström and Eykhoff, 1971). The triennial meetings have been producing between 200 and 300 papers each ever since. The field reached maturation in the eighties, marked by several excellent books from the Swedish school, by Lennart Ljung and Torsten Söderström (1983), L. Ljung (1987) and T. Söderström and P. Stoica (1987).

This chapter summarizes the standard foundations of systems identification methodology, somewhat re-emphasized and re-interpreted based on the author's teaching and research experience.

4.A. APPENDIX: VECTOR DERIVATIVES

The basic rules of differentiation with respect to vector quantities are summarized below.

We will define the derivative of a scalar with respect to a column vector as a column vector, and that with respect to a row vector as a row vector (the opposite definition would also be possible). Thus for a scalar f,

$$f = a' x = x' a \tag{4.A1}$$

the derivatives are

$$df / dx = a \qquad df / dx' = a' \tag{4.A2}$$

Further, for

$$f = x' A x \tag{4.A3}$$

the derivatives are

$$df / dx = (A+A')x \qquad df / dx' = x'(A+A') \tag{4.A4}$$

In either case,

$$\blacktriangle f = (df/dx') \blacktriangle x = \blacktriangle x' (df/dx) \tag{4.A5}$$

It follows that the derivative of a column vector with respect to a row vector is a matrix, and so is the derivative of a row vector with respect to a column vector. Thus for a column vector f

$$f = A x \qquad f' = x' A' \tag{4.A6}$$

the derivatives are

$$df / dx' = A \qquad df' / dx = A \tag{4.A7}$$

and

$$\blacktriangle f = (df/dx') \blacktriangle x \qquad \blacktriangle f' = x' (df'/x) \tag{4.A8}$$

5
Analytical Redundancy Concepts

5.0. INTRODUCTION

In this chapter, we are going to outline the fundamental concepts of fault detection and diagnosis using analytical redundancy.

The faults we are dealing with may arise in the *basic technological equipment* or in its *measurement and control instruments* (sensors and actuators). They may represent *performance deterioration* (such as surface fouling in a heat exchanger or bias in a sensor), *partial malfunctions* (such as leaks from a tank or a pipeline) or *total breakdowns* (such as the loss of a pump or a sensor). From the point of view of diagnosis, it is of interest how a particular fault affects the plant outputs (additive or multiplicative faults). This issue will be addressed in Section 5.1.

Fault detection and diagnosis in general include three functions:

1. Fault detection. To indicate the presence of fault(s).

2. Fault isolation. To determine the location of the fault(s).

3. Fault identification. To determine the size of the fault(s).

The detection function is indispensable and the isolation function is usually also required. However, the fault identification function can be omitted in most cases, except when the fault size is really important (for example in the case of a leak). The combination of isolation and identification (or isolation alone if the

the latter is missing) is also referred to as fault diagnosis. The detection and isolation functions may be performed *sequentially*, that is, by invoking the isolation function only once a fault has been detected, or *in parallel*, that is, simultaneously.

In most practical situations, fault diagnosis needs to be performed in the presence of *disturbances, noise* and *modeling errors*. These interfere with the diagnosis of faults and may lead to false alarms and mis-classification (mis-isolation). Therefore the diagnostic algorithm needs to be so designed that it

- is made *insensitive to (decoupled from) the disturbances*;
- includes mechanisms *to suppress the effects of noise*;
- is *robust with respect to modeling errors;*
- maintains sufficient *sensitivity with respect to faults*.

These important issues will be introduced in Subsection 5.2.2 and then revisited several times in later chapters of the book.

The basic idea of analytical redundancy is the comparison of the actual behavior of the monitored plant to the behavior predicted on the basis of a mathematical plant model. In other words, the *plant observations* (inputs and outputs) are *checked for consistency* with the mathematical model. The outcomes of the consistency checks are quantities called *residuals*. These residuals are nominally zero; they become nonzero as a result of faults, disturbances, noise and modeling errors. The residuals are then analyzed to arrive at a diagnostic decision (fault is/is not present; which component is failing). Thus any diagnostic algorithm which utilizes analytical redundancy consists of two blocks: the *residual generator* and the *decision maker* (Fig. 5.1). The fundamentals of residual generation will be discussed in Subsection 5.2.1; a more detailed treatment is the subject of later chapters where decision making will also be addressed.

While a single residual may be sufficient for fault detection, the isolation of faults requires a set of residuals. To facilitate the isolation function, the residuals are usually *enhanced*, that is, generated with specific isolation properties. The main residual enhancement techniques will be introduced in Subsection 5.2.3.

5.1. FAULTS AND DISTURBANCES

As pointed out in the Introduction above, faults are performance deteriorations, malfunctions or breakdowns in the monitored plant or in its instrumentation. Faults may be represented as unknown extra inputs acting on the system (*additive faults*) or as changes of some plant parameters (*multiplicative faults*). While in many cases the classification of a particular fault as additive or multiplicative follows naturally from its nature, sometimes it may also be arbitrary. As we will see, additive faults are significantly simpler to deal with so faults

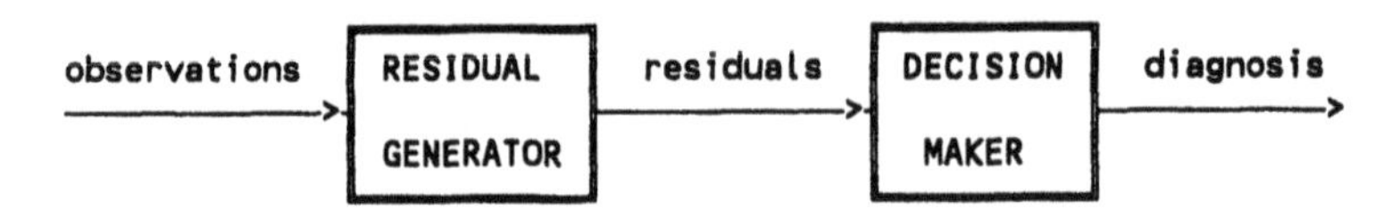

FIGURE 5.1. Analytical redundancy algorithm.

should be classified as additive whenever this can be reasonably justified.

A *disturbance* is also an unknown extra input acting on the plant. Thus physically there is no difference between a disturbance and certain additive faults. The distinction is, indeed, subjective; we consider as faults those extra inputs the presence of which we wish to detect while we consider as disturbances those which we want to ignore and be unaffected by. Some authors refer to the disturbances as "nuisance variables."

Additive faults and disturbances will be handled as unspecified deterministic functions of time. In general, no particular time-behavior will be assumed or utilized in the design of residual generators. In the analysis, however, it will be useful sometimes to refer to some typical time-functions, such as *drift-type, jump-type* and *intermittent* faults and disturbances (Fig. 5.2).

Noises are unknown extra inputs, just like the additive disturbances, but they are assumed to exhibit random behavior. They are also assumed to have zero mean; any nonzero mean can be handled as a separate disturbance. The noises are also nuisance variables the effects of which need to be suppressed.

Modeling errors are errors or uncertainties in the parameters of the monitored system. Just like the multiplicative faults, they are discrepancies between the true system and the model - but they represent an undesirable interference with fault diagnosis. Thus modeling errors can be considered as *multiplicative disturbances*. Note that in general it is difficult to distinguish parametric faults from certain modeling errors though their long-term behavior may provide some clue: parametric faults develop in the course of system operation while some model errors may have been there from the beginning.

The various types of faults and disturbances, and their relationship, are summarized in Table 5.1.

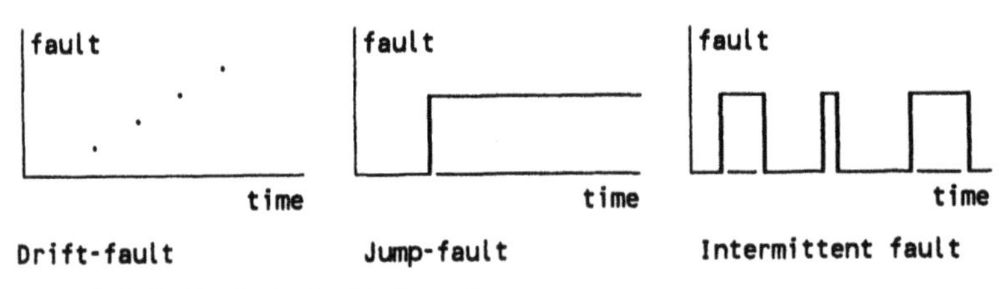

FIGURE 5.2. Typical fault functions.

TABLE 5.1

	ADDITIVE	MULTIPLICATIVE
FAULTS	sensor-fault actuator-fault plant (leak) fault	parametric fault
DISTURBANCES	plant disturbance	modeling error
NOISES	sensor-noise actuator-noise plant-noise	

5.1.1. Additive Faults and Disturbances

Consider a system with multiple inputs $\boldsymbol{u}(t)=[u_1(t)\ldots u_k(t)]'$ and multiple outputs $\boldsymbol{y}(t)=[y_1(t)\ldots y_\mu(t)]'$. For a linear discrete dynamic system, the nominal input-output relationship (without faults, disturbances and noise) is

$$\boldsymbol{y}(t) = \boldsymbol{M}(\phi)\boldsymbol{u}(t) \tag{5.1}$$

Assume that a subset $\boldsymbol{u}_C(t)$ of the inputs is controlled while the rest, $\boldsymbol{u}_M(t)$, are measured. (Note that if an input is neither controlled nor measured then it is unknown and has to be handled as a disturbance.) The observed variables are the *command values* for the controlled inputs and the *measurement values* for the measured inputs and the (measured) outputs.

Additive faults. Consider first the faults; in this generic system, the following additive faults are possible (Fig. 5.3):

* input actuator faults $\Delta\boldsymbol{u}_C(t)$
* input sensor faults $\Delta\boldsymbol{u}_M(t)$
* plant faults (leaks, etc) $\Delta\boldsymbol{u}_P(t)$
* output sensor faults $\Delta\boldsymbol{y}(t)$.

The observed variables $\boldsymbol{u}_C(t)$, $\boldsymbol{u}_M(t)$ and $\boldsymbol{y}(t)$ are related to the actual plant inputs $\boldsymbol{u}_C^\circ(t)$ and $\boldsymbol{u}_M^\circ(t)$ and the actual plant output $\boldsymbol{y}^\circ(t)$, as seen in Fig. 5.3, as

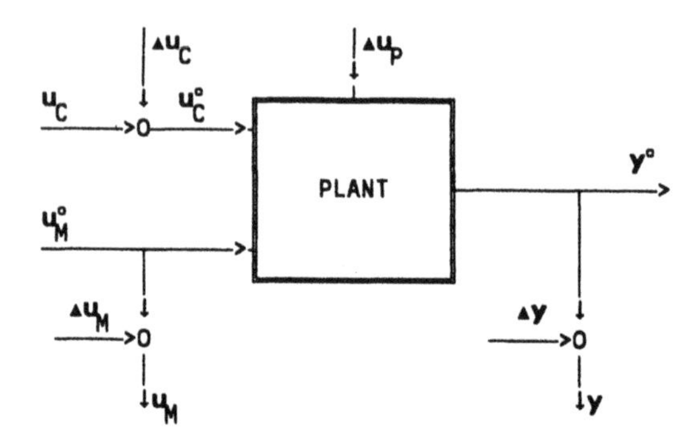

FIGURE 5.3. Additive faults.

$$u_C^\circ(t) = u_C(t) + \blacktriangle u_C(t) \qquad u_M^\circ(t) = u_M(t) - \blacktriangle u_M(t)$$

$$y^\circ(t) = y(t) - \blacktriangle y(t) \tag{5.2}$$

The nominal relationship (5.1) is valid between $u_C^\circ(t)$, $u_M^\circ(t)$ and $y^\circ(t)$, with the additional input $\blacktriangle u_P(t)$. Write

$$u(t) = \begin{bmatrix} u_C(t) \\ u_M(t) \end{bmatrix} \qquad M(\phi) = [M_C(\phi) \mid M_M(\phi)] \tag{5.3}$$

Thus the input-output relationship for the system with faults is

$$y(t) - \blacktriangle y(t) = [M_C(\phi) \mid M_M(\phi)] \begin{bmatrix} u_C(t) + \blacktriangle u_C(t) \\ u_M(t) - \blacktriangle u_M(t) \end{bmatrix} + S_{PF}(\phi)\blacktriangle u_P(t)$$

$$= M(\phi)u(t) + M_C(\phi)\blacktriangle u_C(t) - M_M(\phi)\blacktriangle u_M(t) + S_{PF}(\phi)\blacktriangle u_P(t) \tag{5.4}$$

where $S_{PF}(\phi)$ is the plant-fault transfer function. (5.4) can be written as

$$\boxed{y(t) = M(\phi)u(t) + S_F(\phi)p(t)} \tag{5.5}$$

where $p(t)$ is the combined vector of additive faults

$$p(t) = [\blacktriangle u_C'(t) \mid -\blacktriangle u_M'(t) \mid \blacktriangle u_P'(t) \mid \blacktriangle y'(t)]' \tag{5.6}$$

and $S_F(\phi)$ is the combined fault transfer function

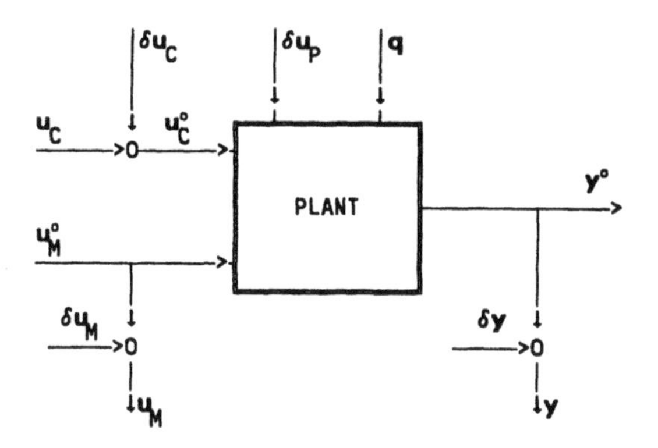

FIGURE 5.4. Additive disturbances and noises.

$$S_F(\phi) = [M_C(\phi) \mid M_M(\phi) \mid S_{PF}(\phi) \mid I] \tag{5.7}$$

Additive disturbances and noise. Now consider the additive disturbances and noises and temporarily ignore the faults (Fig. 5.4). The additive disturbances $q(t)$ act on the plant. The following noises are possible:

* input actuator noise $\delta u_C(t)$
* input sensor noise $\delta u_M(t)$
* plant noise $\delta u_p(t)$
* output sensor noise $\delta y(t)$.

Then

$$u_C^\circ(t) = u_C(t) + \delta u_C(t) \qquad u_M^\circ(t) = u_M(t) - \delta u_M(t)$$

$$y^\circ(t) = y(t) - \delta y(t) \tag{5.8}$$

and

$$\boxed{y(t) = M(\phi)u(t) + S_D(\phi)q(t) + S_N(\phi)v(t)} \tag{5.9}$$

where $v(t)$ is the combined vector of additive noises

$$v(t) = [\delta u_C'(t) \mid -\delta u_M'(t) \mid \delta u_P'(t) \mid \delta y'(t)]' \tag{5.10}$$

$S_D(\phi)$ is the disturbance transfer function and $S_N(\phi)$ is the combined noise transfer function (with $S_{PN}(\phi)$ denoting the plant-noise transfer function)

$$S_N(\phi) = [M_C(\phi) \mid M_M(\phi) \mid S_{PN}(\phi) \mid I] \tag{5.11}$$

Finally, if additive faults, disturbances and noise are present simultaneously, which is normally the case, then the input-output relationship becomes

$$\boxed{y(t) = M(\phi)u(t) + S_F(\phi)p(t) + S_D(\phi)q(t) + S_N(\phi)v(t)} \tag{5.12}$$

Example 5.1.
Consider a simple two-section pipeline system shown in Fig. 5.5. The fluid flow is controlled at the entry point (the variable u) and measured at the end of each section (variables y_1 and y_2). The system is static. The nominal input-output relations are

$$y_1 = u \qquad y_2 = u$$

Note that there is a third nominal relationship which, however, is not independent of the first two, namely $y_2 = y_1$.

The controlled input is subject to an actuator fault (bias) Δu, so that the true input is $u^\circ = u + \Delta u$, where u is now the command value. The measured outputs are subject to sensor faults (biases) Δy_1 and Δy_2, so that the measured values are $y_1 = y_1^\circ + \Delta y_1$ and $y_2 = y_2^\circ + \Delta y_2$, where y_1° and y_2° are the true flows at the measurement points. The nominal relationships are valid among the true plant values, that is

$$y_1^\circ = u^\circ \qquad y_2^\circ = u^\circ$$

Thus the complete equations for the observables (the command value of the input and the measured values of the outputs) are:

$$y_1 = u + \Delta u + \Delta y_1 \qquad y_2 = u + \Delta u + \Delta y_2$$

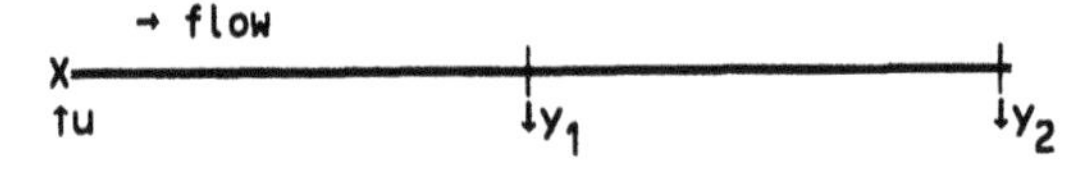

FIGURE 5.5. Pipeline system.

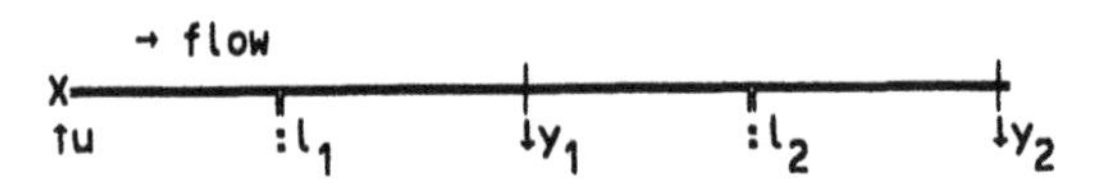

FIGURE 5.6. Pipeline system with leaks.

Now assume that the pipeline is subject to possible leaks along either section, as shown in Fig. 5.6. The leaks l_1 and l_2 change the relationships among the true flows to

$$y_1^\circ = u^\circ - l_1 \qquad y_2^\circ = u^\circ - l_1 - l_2$$

From the point of view of fault detection, the leaks may be considered as plant faults or as disturbances. In either case, they appear in the equations relating the observables as

$$y_1 = u + \blacktriangle u + \blacktriangle y_1 - l_1 \qquad y_2 = u + \blacktriangle u + \blacktriangle y_2 - l_1 - l_2$$

Example 5.2.

Consider now the second order dynamic system, with two inputs and two outputs, shown in Fig. 5.7. Assume that the parameter values are

$$\alpha_1 = -0.8 \quad \alpha_2 = -0.5 \quad \beta_{11} = 1 \quad \beta_{12} = 2 \quad \beta_{21} = 1 \quad \beta_{22} = 3$$

The nominal (fault-free) input-output equations of the two blocks (units) are

$$y_1(t) = \frac{\phi^{-1}}{1 - 0.8\phi^{-1}} u_1(t) + \frac{2\phi^{-1}}{1 - 0.8\phi^{-1}} u_2(t)$$

$$y_2(t) = \frac{\phi^{-1}}{1 - 0.5\phi^{-1}} y_1(t) + \frac{3\phi^{-1}}{1 - 0.5\phi^{-1}} u_2(t)$$

The first (nominal) unit equation is also a (nominal) system equation, describing the system output $y_1(t)$ in terms of the two system inputs $u_1(t)$ and $u_2(t)$. The second system equation, describing $y_2(t)$ in terms of the two system inputs, can be obtained by substituting the first unit equation into the second:

$$y_2(t) = \frac{\phi^{-1}}{1 - 0.5\phi^{-1}} \left[\frac{\phi^{-1}}{1 - 0.8\phi^{-1}} u_1(t) + \frac{2\phi^{-1}}{1 - 0.8\phi^{-1}} u_2(t) \right] + \frac{3\phi^{-1}}{1 - 0.5\phi^{-1}} u_2(t)$$

$$= \frac{\phi^{-2}}{1 - 1.3\phi^{-1} + 0.4\phi^{-2}} u_1(t) + \frac{3\phi^{-1} - 0.4\phi^{-2}}{1 - 1.3\phi^{-1} + 0.4\phi^{-2}} u_2(t)$$

Assume that the first input is controlled while the second is measured. They are subject to actuator fault (bias) $\blacktriangle u_1(t)$ and sensor fault (bias) $\blacktriangle u_2(t)$, so that the command $u_1(t)$ and measurement $u_2(t)$ are related to their respective true values as $u_1(t) = u_1^\circ(t) - \blacktriangle u_1(t)$ and $u_2(t) = u_2^\circ(t) + \blacktriangle u_2(t)$. The two outputs are measured; they are subject to sensor faults $\blacktriangle y_1(t)$ and $\blacktriangle y_2(t)$, so that $y_1(t) = y_1^\circ(t) + \blacktriangle y_1(t)$ and $y_2(t) = y_2^\circ(t) + \blacktriangle y_2(t)$. The nominal relations

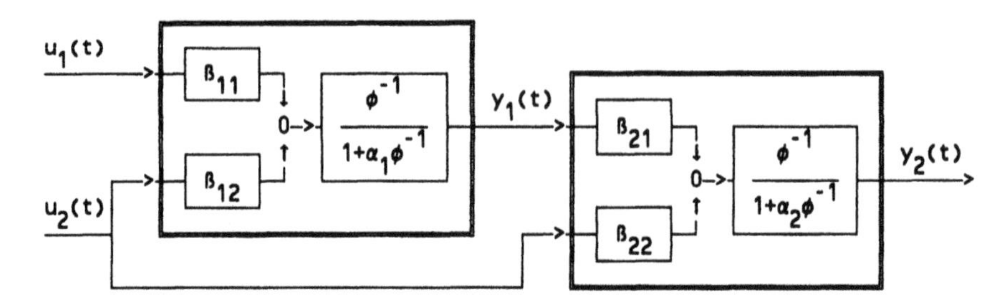

FIGURE 5.7. Second order system.

are valid for the true variables, thus for the observables the following equations apply:

$$y_1(t) = \frac{\phi^{-1}}{1 - 0.8\phi^{-1}} u_1(t) + \frac{2\phi^{-1}}{1 - 0.8\phi^{-1}} u_2(t)$$

$$+ \blacktriangle y_1(t) + \frac{\phi^{-1}}{1 - 0.8\phi^{-1}} \blacktriangle u_1(t) - \frac{2\phi^{-1}}{1 - 0.8\phi^{-1}} \blacktriangle u_2(t)$$

$$y_2(t) = \frac{\phi^{-2}}{1 - 1.3\phi^{-1} + 0.4\phi^{-2}} u_1(t) + \frac{3\phi^{-1} - 0.4\phi^{-2}}{1 - 1.3\phi^{-1} + 0.4\phi^{-2}} u_2(t)$$

$$+ \blacktriangle y_2(t) + \frac{\phi^{-2}}{1 - 1.3\phi^{-1} + 0.4\phi^{-2}} \blacktriangle u_1(t) - \frac{3\phi^{-1} - 0.4\phi^{-2}}{1 - 1.3\phi^{-1} + 0.4\phi^{-2}} \blacktriangle u_2(t)$$

5.1.2. Multiplicative Faults and Disturbances

Multiplicative faults and disturbances affect the model of the system; with the input-output description (5.1) the model is the transfer function matrix $M(\phi)$. Denote by $M^{\circ}(\phi)$ the actual ("true") transfer function of the physical system, then

$$M^{\circ}(\phi) = M(\phi) + \blacktriangle M(\phi) \tag{5.13}$$

where $\blacktriangle M(\phi)$ is the *discrepancy* between the model and the true system. Eq. (5.13) represents two conceptually different situations:

1. The discrepancy may reflect a *parametric fault*. In this case, it is the plant that has deviated from its earlier normal behavior, which was properly represented by the model.

2. The discrepancy may reflect a modeling error. This may be constant and present ever since the implementation of the algorithm. The error may simply be the inaccuracy of some parameters, due perhaps to identification inaccura-

cies, or may be the result of the approximation of a higher order plant with a lower order model. Another possible source of model error is the approximation of a nonlinear plant with a linear model; then the inaccuracy depends on the operating point and thus may vary with time.

We will assume, for the sake of simplicity, that the model discrepancy $\blacktriangle M(\phi)$ is not a function of time. Though this may not be completely true, certainly the variations of the model, if any, are much slower than those of the variables.

The input-output relationship (5.1) is valid for the actual transfer function $M^{\circ}(\phi)$ but this is not known. Applying it to the model $M(\phi)$ introduces the model discrepancy into the equation

$$y(t) = M^{\circ}(\phi)u(t) = M(\phi)u(t) + \blacktriangle M(\phi)u(t) \tag{5.14}$$

Here the last term is the effect of the model discrepancy. If $\blacktriangle M(\phi)$ can be decomposed as $\blacktriangle M_F(\phi)+\blacktriangle M_D(\phi)$, so that the first matrix represents the parametric faults and the second the modeling errors, then (5.14) can be further written as

$$y(t) = M(\phi)u(t) + \blacktriangle M_F(\phi)u(t) + \blacktriangle M_D(\phi)u(t) \tag{5.15}$$

Though the model discrepancy terms appear additively, they differ from the additive fault and disturbance terms in Eq. (5.12) in an important way. The additive faults and disturbances were *variables* and their coefficients in the input-output equation were *time-invariant transfer functions*. In contrast, the parametric faults and model errors are *parameter matrices* and their coefficients in the input-output equation are *variables*. This is the reason why we consider the model discrepancies as multiplicative. The above difference between model discrepancies and additive faults and disturbances will have important consequences in the treatment of the two groups.

Underlying parameters. One of the difficulties in handling model discrepancies is that the transfer function matrix usually has a large number of elements and each one of these may be accompanied by a separate (scalar) discrepancy. A significant simplification may be achieved if all model elements can be deduced from a small set of *"underlying parameters."* Such underlying parameters are the parameters of a first principle characterization of the plant, and usually have direct physical meaning (such as resistance, heat-transfer coefficient, etc). Faults and modeling errors/uncertainties may primarily concern these underlying parameters and then propagate from these to the transfer function (or state-space) model.

Consider a set of underlying parameters $\boldsymbol{\theta}=[\theta_1 \ \ldots \ \theta_U]'$, with uncertainty $\blacktriangle\boldsymbol{\theta}=[\blacktriangle\theta_1 \ \ldots \ \blacktriangle\theta_U]'$. The transfer function is a function of the $\boldsymbol{\theta}$ vector, that is, $M(\phi, \boldsymbol{\theta})$. Denote the underlying parameter values which yield the model $M(\phi)$ actually used as $\boldsymbol{\theta}^{\#}$. Then $\blacktriangle M(\phi)$ can be approximated as

$$\blacktriangle M(\phi) = \sum_{k=1}^{\upsilon} M_{\theta k}(\phi) \blacktriangle \theta_k \qquad (5.16)$$

where

$$M_{\theta k}(\phi) = \partial M(\phi, \boldsymbol{\theta}) / \partial \theta_k \Big|_{\boldsymbol{\theta} = \boldsymbol{\theta}^{\#}} \qquad k=1 \ldots \upsilon \qquad (5.17)$$

Note that, in general, the transfer function parameters are nonlinear functions of the underlying parameters. Therefore the accuracy of the approximation (5.16) deteriorates as the deviation from the model values grows.

With the above approximation, the model discrepancy term in (5.14) can be written as

$$\blacktriangle M(\phi)u(t) = [\sum_{k=1}^{\upsilon} M_{\theta k}(\phi) \blacktriangle \theta_k]\, u(t) = \sum_{k=1}^{\upsilon} [M_{\theta k}(\phi)\, u(t)] \blacktriangle \theta_k \qquad (5.18)$$

Define

$$N(t) = [n_{\cdot 1}(t) \ldots n_{\cdot \upsilon}(t)] \qquad n_{\cdot k}(t) = M_{\theta k}(\phi)u(t), \quad k=1 \ldots \upsilon \qquad (5.19)$$

then

$$\boxed{\blacktriangle M(\phi)u(t) = N(t) \blacktriangle \boldsymbol{\theta}} \qquad (5.20)$$

Now assume that the underlying parameter vector can be decomposed as $\boldsymbol{\theta}=[\ \boldsymbol{\theta}_F' \ |\ \boldsymbol{\theta}_D'\ ]'$, so that discrepancies in the first group represent parametric faults while those in the second represent modeling errors. Then, with the appropriate decomposition $N(t)=[\ N_F(t)\ |\ N_D(t)\ ]$,

$$\blacktriangle M_F(\phi)u(t) + \blacktriangle M_D(\phi)u(t) = N_F(t) \blacktriangle \boldsymbol{\theta}_F + N_D(t) \blacktriangle \boldsymbol{\theta}_D \qquad (5.21)$$

Eq. (5.21) can be incorporated into the input-output relationship (5.12):

$$\boxed{y(t) = M(\phi)u(t)+S_F(\phi)p(t)+N_F(t)\blacktriangle \boldsymbol{\theta}_F+S_D(\phi)q(t)+N_D(t)\blacktriangle \boldsymbol{\theta}_D+S_N(\phi)v(t)}$$

$$(5.22)$$

This latter equation integrates the model discrepancy effects with the additive fault, disturbance and noise effects. By expressing them in terms of the underlying parameters, the model discrepancy effects become formally similar to the effects of additive faults and disturbances which act via entry matrices. However, the multiplicative faults $\Delta\boldsymbol{\theta}_F$ and disturbances $\Delta\boldsymbol{\theta}_D$ are still constants with time-varying coefficients $N_F(t)$ and $N_D(t)$ while the additive faults $p(t)$ and disturbances $q(t)$ are variables, with time-invariant transfer function coefficients $S_F(\phi)$ and $S_D(\phi)$.

Note also that, strictly speaking, the model discrepancies should affect the $S(\phi)$ matrices as well, resulting in bilinear fault and/or disturbance terms like $\Delta S_F(\phi)p(t)$, but these usually can be ignored.

Example 5.3.

The idea of representing model uncertainties and parametric faults with underlying parameter uncertainties/faults will be illustrated on the two-input two-output system of Example 5.2 (Fig. 5.7). The results will appear somewhat cumbersome; techniques to obtain the same in simpler forms will be explored in Chapter 9, devoted entirely to multiplicative faults/disturbances.

Assume that the poles of the two units, α_1 and α_2, are the source of model uncertainty, that is, these are the underlying parameters. To see how their uncertainty propagates to the parameters of the system transfer function, write the latter in terms of the unit parameters:

$$M(\phi) = \begin{bmatrix} \dfrac{\beta_{11}\phi^{-1}}{1+\alpha_1\phi^{-1}} & \dfrac{\beta_{12}\phi^{-1}}{1+\alpha_1\phi^{-1}} \\[2ex] \dfrac{\beta_{11}\beta_{21}\phi^{-2}}{(1+\alpha_1\phi^{-1})(1+\alpha_2\phi^{-1})} & \dfrac{\beta_{12}\beta_{21}\phi^{-2}+\beta_{22}\phi^{-1}(1+\alpha_1\phi^{-1})}{(1+\alpha_1\phi^{-1})(1+\alpha_2\phi^{-1})} \end{bmatrix}$$

The partial derivatives of the transfer function with respect to the underlying parameters can be obtained by formal differentiation, yielding

$$M_{\alpha 1}(\phi) = \frac{1}{(1+\alpha_1\phi^{-1})^2} \begin{bmatrix} -\beta_{11}\phi^{-2} & -\beta_{12}\phi^{-2} \\[2ex] \dfrac{-\beta_{11}\beta_{21}\phi^{-3}}{1+\alpha_2\phi^{-1}} & \dfrac{-\beta_{12}\beta_{21}\phi^{-3}}{1+\alpha_2\phi^{-1}} \end{bmatrix}$$

$$M_{\alpha 2}(\phi) = \frac{1}{(1+\alpha_2\phi^{-1})^2}\begin{bmatrix} 0 & 0 \\ \dfrac{-\beta_{11}\beta_{21}\phi^{-3}}{1+\alpha_1\phi^{-1}} & \dfrac{-\beta_{12}\beta_{21}\phi^{-3}}{1+\alpha_1\phi^{-1}} - \beta_{22}\phi^{-2} \end{bmatrix}$$

The effect of model uncertainty is then computed as

$$\blacktriangle M(\phi)u(t) = [M_{\alpha 1}(\phi)u(t)]\blacktriangle\alpha_1 + [M_{\alpha 2}(\phi)u(t)]\blacktriangle\alpha_2$$

5.2. RESIDUAL GENERATION

As we pointed out in the Introduction, residuals are quantities which are nominally zero; they become nonzero in response to faults (and also to disturbances, noise and modeling error). Residuals are generated from the observables of the monitored plant, that is, from the command values of the controlled inputs and the measured values of the measured inputs and outputs.

5.2.1. Generic Residual Generator

The *residual generator* is a linear discrete dynamic algorithm (a computational "system") acting on the observables. Its *generic form* is

$$\boxed{r(t) = V(\phi)u(t) + W(\phi)y(t)} \tag{5.23}$$

where $r(t)$ is the vector of residuals and $V(\phi)$ and $W(\phi)$ are transfer function matrices. However, Eq. (5.23) is not necessarily a residual generator; to qualify it has to return zero residuals when all unknown inputs are zero, that is, when Eq. (5.1) holds. Thus

$$V(\phi)u(t) + W(\phi)M(\phi)u(t) = 0 \tag{5.24}$$

has to be satisfied for all $u(t)$, requiring

$$\boxed{V(\phi) = -W(\phi)M(\phi)} \tag{5.25}$$

Thus the generic residual generator can also be written as

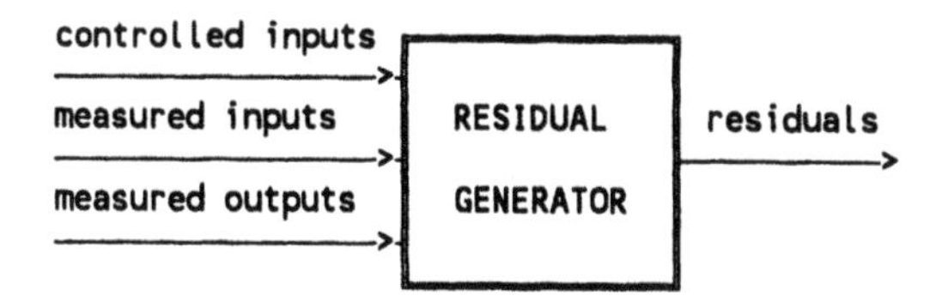

FIGURE 5.8.a. Residual generator: computational form.

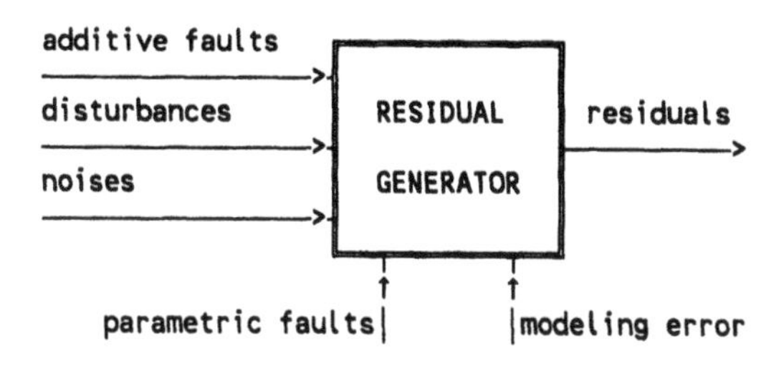

FIGURE 5.8.b. Residual generator: internal form.

$$\boxed{\mathbf{r}(t) = \mathbf{W}(\phi)\,[\mathbf{y}(t) - \mathbf{M}(\phi)\mathbf{u}(t)]} \tag{5.26}$$

Equations (5.23) and (5.26) are the *computational form* of the generic residual generator (Fig. 5.8.a).

Now express $\mathbf{y}(t) - \mathbf{M}(\phi)\mathbf{u}(t)$ from Eq. (5.22) and substitute into (5.26):

$$\mathbf{y}(t) - \mathbf{M}(\phi)\mathbf{u}(t) \tag{5.27}$$

$$= \mathbf{S}_F(\phi)\mathbf{p}(t) + \mathbf{N}_F(t)\blacktriangle\boldsymbol{\theta}_F + \mathbf{S}_D(\phi)\mathbf{q}(t) + \mathbf{N}_D(t)\blacktriangle\boldsymbol{\theta}_D + \mathbf{S}_N(\phi)\mathbf{v}(t)$$

$$\boxed{\mathbf{r}(t) = \mathbf{W}(\phi)[\mathbf{S}_F(\phi)\mathbf{p}(t) + \mathbf{N}_F(t)\blacktriangle\boldsymbol{\theta}_F + \mathbf{S}_D(\phi)\mathbf{q}(t) + \mathbf{N}_D(t)\blacktriangle\boldsymbol{\theta}_D + \mathbf{S}_N(\phi)\mathbf{v}(t)]} \tag{5.28}$$

Equation (5.28) is the *internal* or *unknown-input-effect* form of the generic residual generator; it shows how the residuals depend on the faults, disturbances and noises (Fig. 5.8.b).

Example 5.4.

We will first demonstrate the computational and internal forms of the generic residual generator using the pipeline system of Example 5.1. The system has a single input u; the output vector and the transfer matrix are

$$y(t) = [\, y_1(t) \quad y_2(t) \,]' \qquad M = [\, 1 \quad 1 \,]'$$

(Since the system is static, the "transfer function" is actually a gain matrix.) Thus the computational form (5.26) of the residual generator is

$$r(t) = W \left[\begin{bmatrix} y_1(t) \\ y_2(t) \end{bmatrix} - \begin{bmatrix} 1 \\ 1 \end{bmatrix} u(t) \right]$$

Among the unknown inputs, consider first the actuator and sensor faults. These form the fault vector $p(t)$, with (static) entry matrix S_F:

$$p(t) = [\, \Delta u(t) \quad \Delta y_1(t) \quad \Delta y_2(t) \,]' \qquad S_F = \begin{bmatrix} 1 & 1 & 0 \\ 1 & 0 & 1 \end{bmatrix}$$

If the leaks are considered as disturbances then

$$q(t) = [\, l_1(t) \quad l_2(t) \,]' \qquad S_D = \begin{bmatrix} -1 & 0 \\ -1 & -1 \end{bmatrix}$$

Alternatively, the leaks may be included among the faults. Then

$$p(t) = [\Delta u(t) \;\; \Delta y_1(t) \;\; \Delta y_2(t) \;\; l_1 \;\; l_2]' \qquad S_F = \begin{bmatrix} 1 & 1 & 0 & -1 & 0 \\ 1 & 0 & 1 & -1 & -1 \end{bmatrix}$$

The internal form of the residual generator is then

$$r(t) = W [\, S_F\, p(t) + S_D\, q(t) \,] \quad \text{or} \quad r(t) = W S_F\, p(t)$$

Note that such a flow system is special in the sense that the input-output equations are material balances. For this reason, the gains are known exactly and their value is usually 1 or -1 (or 0).

Example 5.5.

As a second example, let us revisit the dynamic system introduced in Example 5.2. Recall that the first input was assumed to be controlled while the second measured. The actuator fault $\Delta u_1(t)$, input measurement fault $\Delta u_2(t)$ and the two output measurement faults $\Delta y_1(t)$ and $\Delta y_2(t)$ form the fault vector

$$p(t) = [\ \Delta u_1(t) \quad -\Delta u_2(t) \quad \Delta y_1(t) \quad \Delta y_2(t)\ ]'$$

The system transfer matrix $M(\phi)$ and the fault transfer matrix $S_F(t)$ are

$$M(\phi) = \begin{bmatrix} \dfrac{\phi^{-1}}{1-0.8\phi^{-1}} & \dfrac{2\phi^{-1}}{1-0.8\phi^{-1}} \\ \dfrac{\phi^{-2}}{1-1.3\phi^{-1}+0.4\phi^{-2}} & \dfrac{3\phi^{-1}-0.4\phi^{-2}}{1-1.3\phi^{-1}+0.4\phi^{-2}} \end{bmatrix}$$

$$S_F(\phi) = \left[\begin{array}{c|cc} M(\phi) & 1 & 0 \\ & 0 & 1 \end{array} \right]$$

With these, the computational and internal forms are obtained according to (5.26) and (5.28).

5.2.2. Detection Properties of the Residual Generator

Ideally, the residuals should only be affected by the faults. However, the presence of disturbances, noise and modeling errors also causes the residuals to become nonzero and thus interferes with the detection of faults. Therefore the residual generator needs to be designed so that it is maximally unaffected by these nuisance inputs, that is, it is *robust* in the face of disturbances, noise and model errors. Robustness is perhaps the most important requirement in residual generation. Much of the effort in designing residual generators goes into achieving sufficiently robust residual performance.

The approaches to the three classes of nuisance inputs are somewhat different. This is due primarily to the differences in their temporal behavior. Figure 5.9 shows a rough qualitative picture of the relative frequency ranges of the various unknown inputs. While these frequency properties are characteristic of most applications, they of course may vary from case to case. Another reason for the differences in their treatment is that while usually nothing is assumed about disturbances, noises are assumed to have known statistical properties.

Additive disturbances are usually "slow," their temporal behavior (or frequency spectrum) is similar to those of the additive faults. Therefore robust residual generation in their presence can only be achieved by explicitly

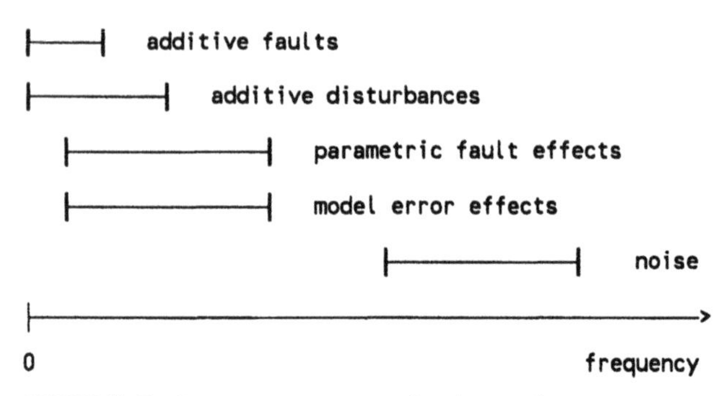

FIGURE 5.9. Frequency ranges of unknown inputs.

decoupling the residuals from such disturbances. That is, the residual generators need to be so designed that the residuals are *unaffected* by the disturbances. As we will show later, this can be done *exactly* if the number of disturbances is small while only *approximate decoupling* is possible if this is not the case. Usually decoupling is designed by considering the disturbances completely unknown, that is, no particular temporal behavior is assumed. The coefficients of additive disturbances are *time-invariant* transfer function matrices, therefore the residual generator may also be time-invariant. That is, the design may be performed completely *off-line*.

The **multiplicative disturbances (model errors)** are also usually slow. The model error itself is either permanently present, or it arises as a result of (normally slow) variations of the operating point. The temporal behavior of the model error effect is thus primarily determined by the coefficient of the model error which, in fact, is the plant input. The frequency range of the model error effects is partially overlapping with that of the faults thus it is desirable to achieve explicit decoupling from the multiplicative disturbances as well. In fact, robustness in the face of modeling errors is the most significant of the robustness problems.

If the model errors are expressed directly in terms of the transfer function (or state-space) model, the great number of affected parameters usually precludes any effective decoupling. If it is possible to represent the model uncertainty in terms of a limited number of underlying parameters then robust design, exactly or approximately, may be feasible. The linear approximation of the error propagation may somewhat compromise the accuracy of decoupling. The coefficients of the multiplicative faults are *time-varying* thus the residual generator designed for model error robustness has to be time-varying as well. We may say that the residual generator "design" is performed partially *on-line*.

Noises have zero mean and are usually in a higher frequency range than faults and disturbances. Also their statistical distribution is at least partially known (or assumed). Therefore, instead of explicit decoupling (which is expensive in terms of design restrictions), one or both of the following techniques are applied:

- The residuals are *filtered*, usually by low-pass algorithms, to reduce the effects of noise without significantly altering those of the faults.
- The residuals are *threshold tested*, instead of simply checking for nonzero values.

The threshold values may be determined by statistical considerations, on the basis of the known or assumed noise statistics. Alternatively, they may be obtained experimentally. In the latter case, they may cover not only the noise effects but also the modeling errors which are otherwise unaccounted for. Ideally, the selection of thresholds is guided by a pre-specified *trade-off between false alarms and missed detections*.

The ease or difficulty of the implementation of statistical tests depends on the way the noises propagate to the residuals. It is desirable that the noise-to-residual transfer function be of moving average type. It is even more advantageous if the residual sequence in response to a white noise input is also white. These issues will be addressed in detail in Chapter 11.

The **fault sensitivity** of the residuals is an important performance characteristic of the residual generator. While other measures of sensitivity are also possible, we will introduce here the *triggering limit*, the value of a particular fault which brings a particular residual to its threshold, provided no other faults and nuisance inputs are present. Assume that κ_i is the threshold for the residual $r_i(t)$. The response of the same residual to a fault $p_j(t)$, with no other fault or nuisance input present (see (5.28)), is

$$r_i(t|p_j) = w'_{i\cdot}(\phi) s_{F\cdot j}(\phi) p_j(t) \tag{5.29}$$

where $w'_{i\cdot}(\phi)$ is the i-th row of $W(\phi)$ and $s_{F\cdot j}(\phi)$ is the j-th column of $S_F(\phi)$. Obviously, $r_i(t|p_j)$ is a time-function which depends on $p_j(t)$. To be more specific, choose a unit-step function $\epsilon(t)$ for $p_j(t)$ and consider the response in steady state (another possibility would be the maximal value of the response). Thus

$$\lim_{t\to\infty} r_i[t|p_j]_{step} = [w'_{i\cdot}(\phi) s_{F\cdot j}(\phi)]_{\phi=1} \tag{5.30}$$

From this, the triggering limit η_{ij} is

$$\eta_{ij} = \frac{\kappa_i}{[w'_{i\cdot}(\phi) s_{F\cdot j}(\phi)]_{\phi=1}} \tag{5.31}$$

Obviously, a smaller triggering limit signifies greater fault sensitivity.

In many cases, the nominal value of the faults is known (is part of the specification). Then sensitivity may be characterized by the ratio of the nominal-fault residual response to the threshold, assuming steady state (or other well defined) gain of the fault-to-residual transfer. Denoting the nominal fault sizes as p_j°, and working with the steady state gain, this ratio is

$$\zeta_{ij} = \frac{p_j^{\circ}[w'_{i.}(\phi)s_{F.j}(\phi)]_{\phi=1}}{\kappa_i} \tag{5.32}$$

It is desirable that the above ratio be slightly above one for all faults in all equations. Clearly, a ratio smaller than one signifies that the nominal fault does not bring the residual to its threshold while a ratio much larger than one indicates that even a very small fault may result in threshold crossing.

While the fault sensitivities may be influenced by the filtering of the residuals, their ratio within a particular equation is fixed. Thus the spread of the ζ_{ij} ratios within an equation is an important measure of the detection quality of that equation. This spread will be characterized by the *sensitivity condition* of the equation, defined as

$$\xi_i = \max_j \zeta_{ij} \, / \min_j \zeta_{ij} \tag{5.33}$$

Example 5.6.

a. To demonstrate the concept of triggering limit, let us continue Example 5.5. Consider the steady-state responses to step-type faults. For the sake of simplicity, assume that the transformation is $W(\phi)=I$. The steady-state gains for the various faults in the two residual equations are, from the $S_F(\phi)$ matrix in Example 5.5:

$$[S_F(\phi)]_{\phi=1} = \begin{bmatrix} 5 & 10 & 1 & 0 \\ 10 & 26 & 0 & 1 \end{bmatrix}$$

Assume that the thresholds for the two residuals are chosen as $\kappa_1=\kappa_2=10$. Then the steady-state triggering limits are

$$[\eta_{ij}] = \begin{bmatrix} 2 & 1 & 10 & - \\ 1 & 0.385 & - & 10 \end{bmatrix}$$

That is, it takes $\blacktriangle u_1=2$ or $\blacktriangle u_2=1$ or $\blacktriangle y_1=10$ to bring the first residual to its threshold, and $\blacktriangle u_1=1$ or $\blacktriangle u_2=0.385$ or $\blacktriangle y_2=10$ to bring the second residual to its threshold.

b. Now assume that the nominal fault sizes are

$$\blacktriangle u_1^\circ = \blacktriangle u_2^\circ = 2 \qquad \blacktriangle y_1^\circ = \blacktriangle y_2^\circ = 10$$

Then the sensitivity ratios and the sensitivity conditions are obtained as

$$[\zeta_{ij}] = \begin{bmatrix} 1 & 2 & 1 & - \\ 2 & 5.2 & - & 1 \end{bmatrix} \qquad [\xi_i] = \begin{bmatrix} 2 \\ 5.2 \end{bmatrix}$$

Observe that while the first residual has reasonably balanced fault sensitivities, the second residual does much worse. ⁝⁝⁝

Modeling error robustness may be quantified in a way similar to fault sensitivity, provided the model error is expressed in terms of underlying parameters. Define the *limit model error* as the error in a particular underlying parameter which brings a particular residual to its threshold, with no other model error or other unknown input present. The response of $r_i(t)$ to a model error $\blacktriangle\theta_j$ is

$$r_i(t|\blacktriangle\theta_j) = \mathbf{w}'_{i.}(\phi)\,\mathbf{n}_{.j}(t)\,\blacktriangle\theta_j \tag{5.34}$$

Thus the limit error ϑ_{ij} is

$$\vartheta_{ij}(t) = \frac{\kappa_i}{\mathbf{w}'_{i.}(\phi)\mathbf{n}_{.j}(t)} \tag{5.35}$$

Now higher limit error signifies lower sensitivity to model errors, that is, better robustness. Notice that the denominator in (5.35), and thus the limit error, are functions of time (depend on the plant input).

5.2.3. Isolation Properties of the Residual Generator

In addition to having robust detection properties, the residual generator needs to be designed to support the isolation of faults. As pointed out earlier, isolation always requires a *set (vector) of residuals*. To facilitate fault isolation, the residual set needs to have distinctive properties, uniquely characteristic of particular faults. Residual sets designed with this objective in mind are referred to as *enhanced residuals*. There are two fundamental residual enhancement approaches: *structured residuals* and *directional residuals*.

Structured residuals. Structured residuals are so designed that each residual responds to a different subset of faults and is insensitive to the others (Fig. 5.10). When a particular fault occurs, some of the residuals do respond and others do not. Then the pattern of the response set, the *fault signature* or *fault code*, is characteristic of the fault.

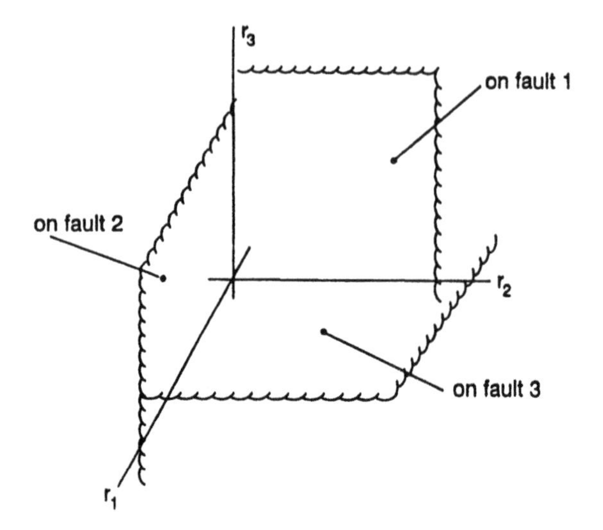

FIGURE 5.10. Structured residuals.

With structured residuals, threshold tests are applied separately to each element of the residual vector. The outcome of the test applied to residual $r_i(t)$ is a binary variable $\epsilon_i(t)$ so that

$$\epsilon_i(t) = \begin{bmatrix} 0 & \text{if } |r_i(t)| < \kappa_i \\ 1 & \text{if } |r_i(t)| \geq \kappa_i \end{bmatrix} \qquad i=1...n \tag{5.36}$$

The vector $\epsilon=[\epsilon_1 \ldots \epsilon_n]'$ is the fault code or signature. Now fault isolation amounts to comparing the actually obtained fault code to a pre-defined set of codes.

The fault codes are determined by the structure of the fault transfer matrix $W(\phi)S_F(\phi)$. A minimum requirement for structured isolation of single faults is that for each fault (which is large enough to exceed its triggering limit for each residual, and with no other fault and nuisance input present) the fault code returned be different and nonzero. A residual set possessing the above property will be referred to as *weakly isolating*. A more rigorous and complete treatment of structured residual sets will be given in Chapter 7.

Example 5.7.

To obtain a preliminary glimpse of structured residuals, let us return to the pipe-line system of Examples 5.1. and 5.4. Consider only the actuator and sensor faults and assume that no transformation is applied (that is, $W=I$). The structure of the two residuals is

	$\blacktriangle u$	$\blacktriangle y_1$	$\blacktriangle y_2$
r_1	1	1	0
r_2	1	0	1

Here a "*1*" in an intersection signifies that the fault of the column affects the residual of the row, while a "*0*" means that it does not. Each column is a fault code, obtained in response to the fault at the top of the column. Since the three columns are all different, the two residuals allow the distinction of the three faults. (Note that the above structure matrix is identical with the gain matrix S_F in Example 5.4; this may happen in a flow system where the gains are static and usually of value *1*.)

Example 5.8.

Now consider the dynamic system of Example 5.2. Assume again that $W(\phi)=I$. The structure of the two residuals, as can be seen from the $S_F(\phi)$ matrix in Example 5.5, is

	$\blacktriangle u_1$	$\blacktriangle u_2$	$\blacktriangle y_1$	$\blacktriangle y_2$
r_1	1	1	1	0
r_2	1	1	0	1

Now the first two columns are identical, so $\blacktriangle u_1$ and $\blacktriangle u_2$ would return identical fault codes. Thus the two residuals, in their "raw" form, are not suitable for fault isolation; they have to be enhanced by a transformation $W(\phi)\neq I$. ⁝⁝⁝

Directional residuals. Directional residuals are so designed that, in response to a particular fault, the residual vector is confined to a fault-specific straight line, at all times including transients (Fig. 5.11). That is

$$r(t|p_j) = \boldsymbol{\beta}_j \gamma_j(\phi)\, p_j(t) \tag{5.37}$$

where the vector $\boldsymbol{\beta}_j$ is the direction of the *j*-th fault response and the scalar transfer function $\gamma_j(\phi)$ is its dynamic. With directional residuals, fault isolation amounts to determining to which pre-defined fault direction the observed residual vector (or time-series of residual vectors) lies the closest. Directional residuals will be discussed in more detail in Chapter 8.

A residual set in which each residual responds to only one of the faults is both structured and directional. Such sets play an important role in the isolation of multiple faults and can serve as a common basis for any structured or directional design. These sets, referred to as *diagonal* or *basis residuals*, will also be discussed in Chapters 7 and 8.

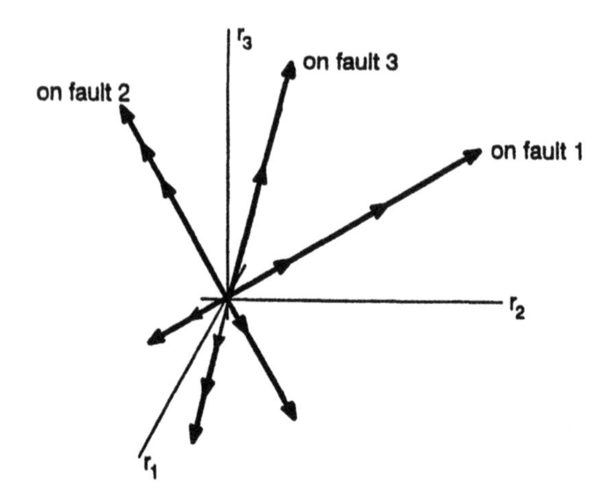

FIGURE 5.11. Directional residuals.

Example 5.9.
Recall the fault gain matrix of the pipeline system, for the actuator and sensor faults, from Example 5.4:

$$S_F = \begin{bmatrix} 1 & 1 & 0 \\ 1 & 0 & 1 \end{bmatrix}$$

Assuming again that the residuals are not transformed, the gain matrix reveals the following geometric properties in the space of the residuals r_1 and r_2 (Fig. 5.12):

- in response to a fault $\blacktriangle u$, the residual vector is confined to the +45° (or -135°) line;
- in response to a fault $\blacktriangle y_1$, the residual vector is confined to the r_1 axis;
- in response to a fault $\blacktriangle y_2$, the residual vector is confined to the r_2 axis.

Clearly, the residuals exhibit a directional behavior, without any explicit directional design. This is always the case with static systems; dynamic systems, however, require residual manipulations to achieve directional properties.

5.2.4. Computational Properties of the Generator

Another issue in the design of residual generators is the on-line computational procedure they involve. Recall that the on-line computations are always carried out on samples of the observable plant inputs and outputs.

The residual generator in general is computationally autoregressive-moving average (ARMA), meaning that the actual on-line computations are finite, but the residuals effectively reflect an infinite series of plant input and output

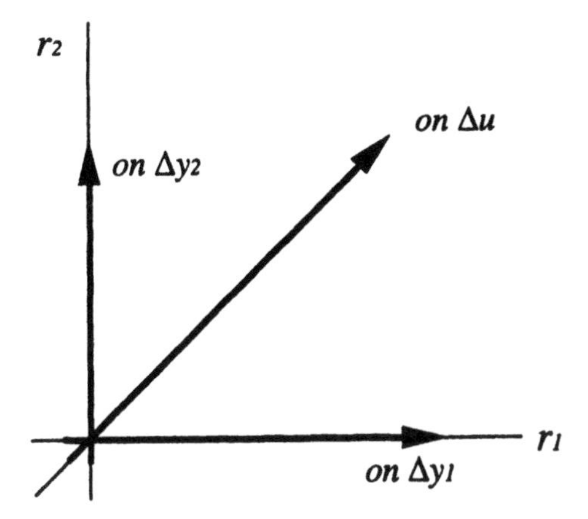

FIGURE 5.12. Directional behavior of the pipeline residuals.

values. It may be advantageous to design the generator to be computationally moving average (MA). According to the generic equations (5.23) and (5.26), this requires that both $W(\phi)$ and $V(\phi) = -W(\phi)M(\phi)$ be moving average transfer functions. With such design, the residuals are obtained from the inputs and outputs through a finite sliding window.

From the point of view of on-line computations, dealing with multiplicative faults and disturbances is more expensive than handling additive ones. Exact decoupling techniques, if at all feasible, are usually computationally less expensive than approximate ones which involve some kind of optimization at each on-line sample. Also, the particular implementation method (see the next section) affects the computational complexity of the algorithm.

5.2.5. The Stability of the Residual Generator

Obviously, the residuals obtained from the generator need to be bounded at all times. It is reasonable to assume, as described in the next paragraph, that the signals arising from the physical plant are bounded. Thus bounded residuals require that the residual generator be *bounded-input bounded-output (BIBO) stable.*

Figure 5.13 shows a typical setup of the physical plant, with controller and residual generator. As far as the control system is concerned, both the measured and the unknown inputs of the plant act as disturbances. The controller is so designed that it guarantees a finite closed-loop response to both the reference input and the disturbances, even if the plant itself is unstable. So the bounded signal assumption is reasonable in any physically functional system.

For the generic residual generator (5.26), the BIBO stability requirement has to be satisfied for both $W(\phi)$ and $W(\phi)M(\phi)$. If the generator is compu-

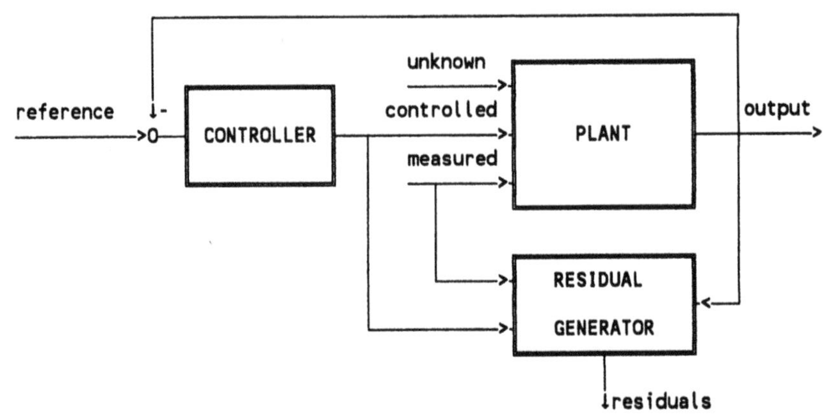

FIGURE 5.13. Physical setup.

tationally moving average, BIBO stability follows naturally. For ARMA generators, $W(\phi)$ has to be so chosen that it is stable itself and stabilizes $W(\phi)M(\phi)$. The latter requires the cancellation of any unstable poles $M(\phi)$ may have. This can be done without the usual risks pole-zero cancellation carries in controller design because the residual generator is not in the loop and because it is designed with the plant *model* $M(\phi)$ which is known exactly (in contrast to the real plant transfer $M^{\circ}(\phi)$ which is not known). The stabilization of the residual generator will be addressed in detail in Chapter 6.

5.3. THE DESIGN OF RESIDUAL GENERATORS

The design of a residual generator comprises two stages (Fig. 5.14):
- *specification* of residual properties
- *implementation* of the residual generator.

5.3.1. Residual Specification

The specification phase may cover the following aspects:
* *Detection properties*, including disturbance decoupling (exact or approximate), model error decoupling/robustness, noise propagation and fault sensitivity, as outlined in Subsection 5.2.2.
* *Isolation properties*, structured or directional residuals, as outlined in Subsection 5.2.3.
* *Computational properties*, in particular ARMA versus MA residual generation and off-line versus on-line design.

The detection and isolation requirements are formulated in terms of *residual responses* to various faults and disturbances. An important point here is that, by doing so, we pose the residual generation problem in a fundamentally deterministic framework. An alternative approach, fundamentally stochastic,

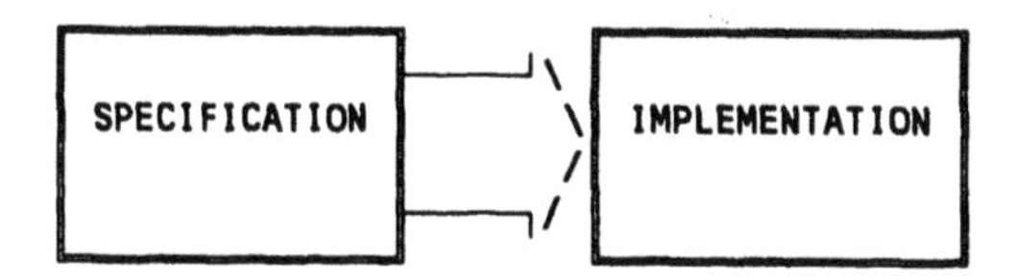

FIGURE 5.14. Residual generator design.

would be to specify the desired residual behavior in relation to the (random) noise input.

As introduced in Subsection 5.2.2, the response of the (scalar) residual $r_i(t)$ to the faults $p_j(t)$ and $\blacktriangle\theta_{Fj}$, respectively, is denoted as $r_i(t|p_j)$ and $r_i(t|\blacktriangle\theta_{Fj})$. Similarly, the response to disturbances $q_j(t)$ and $\blacktriangle\theta_{Dj}$, respectively, is written as $r_i(t|q_j)$ and $r_i(t|\blacktriangle\theta_{Dj})$. We can write the response of a particular scalar residual to the set of faults and disturbances under consideration as a vector, for example for additive faults $p'(t) = [p_1(t),\ p_2(t),\ \ldots]$ and disturbances $q'(t) = [q_1(t),\ q_2(t),\ \ldots\ ]$,

$$r_i(t|p',\ q') = [r_i(t|p_1),\ r_i(t|p_2),\ \ldots\ ,\ r_i(t|q_1),\ r_i(t|q_2),\ \ldots\ ] \qquad (5.38)$$

Note that $r_i(t|p',\ q')$ is just a symbolic enumeration of requirements and not a signal in the "physical" sense.

For additive faults and disturbances, the response specifications are naturally given in the form of transfer functions, such as

$$r_i(t|p_j) = z_{Fij}(\phi)p_j(t) \qquad r_i(t|q_j) = z_{Dij}(\phi)q_j(t) \qquad (5.39)$$

where $z_{Fij}(\phi)$ and $z_{Dij}(\phi)$ are scalar transfer functions. Thus the response specification (5.38) is equivalent to

$$\begin{aligned} r_i(t) &= [\ z_{Fi1}(\phi) \quad z_{Fi2}(\phi) \ \ldots\ ]\ p(t) + [\ z_{Di1}(\phi) \quad z_{Di2}(\phi)\ \ldots\ ]\ q(t) \\ &= z_{Fi.}(\phi)p(t) + z_{Di.}(\phi)q(t) \end{aligned} \qquad (5.40)$$

(where $z_{Fi.}(\phi)$ and $z_{Di.}(\phi)$ are collections (vectors) of the individual transfer functions) since (5.40) includes (5.39) for all the concerned faults and disturbances. Thus the response specification means nothing else but specifying the fault-to-residual and disturbance-to-residual transfer functions in the internal form of the residual generator. For multiplicative faults and disturbances, the response specifications are expressed as time-varying gains (rather than time-invariant transfer functions) but otherwise the mechanism is the same.

The response of the full residual vector $r(t)$ to a particular fault or disturbance, for example the additive fault $p_j(t)$, is described as

$$r(t|p_j) = [r_1(t|p_j),\ r_2(t|p_j),\ \dots]' \tag{5.41}$$

In terms of the desired fault-to-residual transfer functions, this is equivalent to

$$r(t|p_j) = z_{F\cdot j}(\phi)p_j(t) \tag{5.42}$$

Similarly, the full response specification for the vector $r(t)$ is

$$r(t|p', q') = [r(t|p_1),\ r(t|p_2),\ \dots,\ r(t|q_1),\ r(t|q_2),\ \dots] \tag{5.43}$$

which can then be expressed with transfer functions as

$$r(t) = \mathbf{Z}_F(\phi)\mathbf{p}(t) + \mathbf{Z}_D(\phi)\mathbf{q}(t) \tag{5.44}$$

What exactly the residual response specifications should be will be the subject of the forthcoming chapters. At this point, we can only summarize some general facts.

- For disturbance decoupling, the response to the disturbances is specified as (exactly or approximately) zero, that is, $r_i(t|q_j)=0$ or $z_{Dij}(\phi)=0$. For the faults, either specific nonzero responses are specified, such as $r_i(t|p_j)=z_{Fij}(\phi)p_j(t)$, or the responses are required to be nonzero but otherwise are allowed to "float" (are unspecified), that is, $r_i(t|p_j)\neq 0$.
- For structured residuals, some fault responses are specified as zero, exactly as $r_i(t|p_j)=0$ or $z_{Fij}(\phi)=0$, or approximately; while others are required to be nonzero, either exactly specified, such as $r_i(t|p_k)=z_{Fik}(\phi)p_k(t)$, or floating, that is, $r_i(t|p_k)\neq 0$.
- For directional residuals, all fault responses are given definite specifications, such as $r_i(t|p_j)=z_{Fij}(\phi)p_j(t)$.

Note that disturbance decoupling may be viewed as an extension of structured residuals; while each residual is to be decoupled from a different subset of faults, they are to be decoupled from all disturbances.

The computational requirements are described, explicitly or implicitly, in terms of the transfer functions $V(\phi)$ and $W(\phi)$. Moving average computations imply that $V(\phi)$ and $W(\phi)$ are polynomial while completely off-line design implies that they are time-invariant.

5.3.2. Residual Implementation

The implementation of the residual generator may rely on the input-output system description (5.12), or on the equivalent state-space description

$$x(t+1) = Ax(t) + Bu(t) + E_F p(t) + E_D q(t) + E_N \nu(t)$$

$$y(t) = Cx(t) + Du(t) + F_F p(t) + F_D q(t) + F_N \nu(t) \tag{5.45}$$

In the implementation phase, a particular method needs to be selected. The usual *implementation methods* may be classified into two groups:

1. direct implementation of the *generic form* (various *parity relation* techniques)
2. indirect implementation (*diagnostic observers*).

Parity relation implementation. This arises from the combination of the internal residual expression (5.28) (ignoring the noise term) and the specification (5.44). If only additive faults are considered then the implementation equation is

$$W(\phi)[S_F(\phi) \quad S_D(\phi)] = [Z_F(\phi) \quad Z_D(\phi)] \tag{5.46}$$

The residual generator is obtained by solving this relationship for $W(\phi)$. The solution is subject to some existence conditions which will be discussed in the next chapter. The generator needs to be realizable (causal) and stable (or polynomial if so specified); this may require a modification of the response relative to the original specification. In simple cases, the solution may be sought by direct inversion, a straightforward but computationally troublesome procedure. Alternatively, the fault system matrix, relying on the state space description of the plant, may be utilized; this leads to a family of systematic implementation procedures which will be described in detail in Chapter 6.

Parity equations may also be constructed for multiplicative faults and disturbances; this will be the subject of Chapter 9.

Observer implementation. The observer implementation relies entirely on the state space description and utilizes an estimate of the state as an intermediate vehicle of residual generation. The fundamental observer equations are

$$\hat{x}(t+1) = A\hat{x}(t) + Bu(t) + Ko(t) \tag{5.47}$$

$$o(t) = y(t) - C\hat{x}(t) - Du(t) \tag{5.48}$$

where K is the observer gain and $o(t)$ is the innovations vector. Expressing the innovations in terms of the input and output yields

$$o(t) = [I + C(\phi I - A)^{-1} K]^{-1} [y(t) - (C(\phi I - A)^{-1} B+D)\, u(t)] \tag{5.49}$$

Utilizing a well-known matrix equality (see e.g., Kailath, 1980, p. 656), this may be simplified to

$$o(t) = [I - C(\phi I - A+KC)^{-1} K]\, [y(t) - (C(\phi I - A)^{-1} B+D)\, u(t)] \tag{5.50}$$

Since $C(\phi I - A)^{-1} B+D=M(\phi)$, condition (5.25) is clearly satisfied, thus the innovations qualify as a primary residual vector. A residual can then be obtained as

$$r(t) = H\, o(t) \tag{5.51}$$

where H is a static transformation.

Implementing the generator would amount to computing K and H so that the response specifications are satisfied and the observer is stable. Expressing the innovations in terms of the faults and disturbances yields the internal form of the residuals as

$$\begin{aligned} r(t) = {} & H[C(\phi I - A+KC)^{-1} (E_F - KF_F) + F_F]\, p(t) \\ & + H[C(\phi I - A+KC)^{-1} (E_D - KF_D) + F_D]\, q(t) \end{aligned} \tag{5.52}$$

Clearly, (5.52) does not facilitate a direct solution for K and H. Instead, a great wealth of indirect design methods has been developed which rely on eigenvalue/eigenvector assignment or geometric considerations.

Equivalence and comparison. As it is revealed by (5.50) and (5.51), the equivalent generic form for the observer based residual generator is

$$W(\phi) = H[I - C(\phi I - A+KC)^{-1} K] \tag{5.53}$$

Thus any observer implementation can be replaced by a direct implementation, in accordance with (5.53). Then the residuals obtained from the direct implementation are identical, under any circumstances, with those generated by the observer.

It also follows from the above argument that a given design specification can be implemented by any of several methods. The detection and isolation performance of generators designed *for the same specification* and implemented by different methods is identical. That is, the detection and isolation performance are determined by the specification and not by the implementation method.

On the other hand, whether a residual generator with a given specification can be obtained for a particular system depends primarily on the *properties of the system*. However, the selected implementation method may impose some additional design constraints.

Conceptually, the direct approach is more straightforward than the observer approach. Also, its extension to parametric faults and relationship to parameter estimation follow naturally. Further, since it is formulated in terms of transfer functions, it may be more appealing to the practitioner.

The *design computations* with the direct approach require manipulations with polynomial matrices, whether they are transfer functions or the system matrix. For this numerical packages are available but not yet widespread. In many practically important "simple" cases, design with the direct approach may not require conversion into the state space. The observer design is always done in the state space, utilizing standard state-space computational procedures.

Beyond and above these considerations, *cultural aspects*, especially familiarity with one approach or another through experience gained in fields outside fault diagnosis, may play a role in one's preference for a particular implementation method.

The focus of this work is on the direct approach to residual generator design. We feel that, for the reasons explained above, this provides a complete exposure (and solution, to the extent of our present knowledge) to the model based fault detection and isolation problem. The reader interested in the observer approach is referred to the numerous papers on the subject; a selection of these is mentioned in the next section and listed in the References.

5.4. NOTES AND REFERENCES

Various classifications of additive faults and disturbances can be found in a number of early papers, survey and other (Willsky, 1976; Viswanadham et al, 1987a; Gertler, 1988). Multiplicative faults were addressed in the parameter estimation framework in (Isermann, 1984), and integrated with additive faults in (Gertler et al, 1985); see also (Patton and Chen, 1993). The classification in Section 5.1 has been expanded from (Gertler, 1991).

The concept of analytical redundancy arose from the aerospace fault detection community (Deckert et al, 1977; Potter and Suman, 1977). Direct residual generation approaches have been proposed by various authors (Willsky, 1976; Mironovskii, 1979; Ben-Haim, 1980; Gertler and Singer, 1985; Viswanadham et al, 1987a; Massoumnia and Vander Velde, 1989). The generic formulation of Subsection 5.2.1 has been taken from (Gertler, 1991); see also (Viswanadham et al, 1987a; Ding and Frank, 1990; Patton and Chen, 1991).

The fault sensitivity definitions of Subsection 5.2.2 originated from (Gertler and Luo, 1989); see the related work by Staroswiecki and colleagues (1993). Threshold selection on the basis of statistical considerations has been treated extensively in the literature; for the most exhaustive coverage see the book by Basseville and Nikiforov (1993). Threshold selection has been combined with robustness considerations by Horak (1988), Emami-Naeini and colleagues (1988) and Frank (1993).

Fault isolation by directional residual properties appeared in the early work by Beard (1971) and Jones (1973). Structured residuals were proposed and studied in (Ben-Haim, 1980, 1983; Gertler and Singer, 1985; 1990; Massoumnia et al, 1989; Staroswiecki et al, 1993). For a survey of isolation methods, see (Gertler, 1991, 1993).

Residual generation for unstable systems was explored by Kinnaert and coworkers (1995).

The explicit specification of residual responses was first proposed in (Gertler, 1995a), see also (Gertler, 1997). The equivalence of diagnostic observers and generic residual generators (parity relations) was shown in (Gertler, 1991); see also Viswanadham et al (1987a) and Frank (1992).

Fault detection and diagnosis using observers has a very extensive literature, though no comprehensive book has been published on the subject. The interested reader is referred to several survey papers by Paul Frank (1990, 1991, 1992, 1993) and their lists of references. The edited book by Patton, Frank and Clark (1989) contains several chapters on the subject (Clark, 1989; Frank and Wünnenberg, 1989; Patton and Kangethe, 1989). Other individual contributions we wish to mention include (Massoumnia, 1986; White and Speyer, 1987; Viswanadham and Srichander, 1987; Ge and Fang, 1988; Hou and Müller, 1992; Park and Rizzoni, 1994; Chen et al, 1996).

6
Parity Equation Implementation of Residual Generators

6.0. INTRODUCTION

In this chapter, the implementation of diagnostic residual generators by *parity equation* (or *consistency relation*) methods will be discussed. The system to be considered here is the one described by Eq. (5.12), which we repeat here without the noise term:

$$\mathbf{y}(t) = \mathbf{M}(\phi)\mathbf{u}(t) + \mathbf{S}_F(\phi)\mathbf{p}(t) + \mathbf{S}_D(\phi)\mathbf{q}(t) \tag{6.1}$$

Multiplicative faults and disturbances will also be ignored temporarily; they will be addressed in Chapter 9.

Several of the implementation methods discussed in this chapter rely on the state-space representation of the system. A realization of (6.1) can be written as

$$\begin{aligned}
\mathbf{x}(t) &= \mathbf{A}\mathbf{x}(t\text{-}1) + \mathbf{B}\mathbf{u}(t\text{-}1) + \mathbf{E}_F\mathbf{p}(t\text{-}1) + \mathbf{E}_D\mathbf{q}(t\text{-}1) \\
\mathbf{y}(t) &= \mathbf{C}\mathbf{x}(t) + \mathbf{D}\mathbf{u}(t) + \mathbf{F}_F\mathbf{p}(t) + \mathbf{F}_D\mathbf{q}(t)
\end{aligned} \tag{6.2}$$

The transfer function matrices are related to the coefficients in the state-space representation as

$$M(\phi) = C(I - \phi^{-1}A)^{-1}\phi^{-1}B + D$$

$$S_F(\phi) = C(I - \phi^{-1}A)^{-1}\phi^{-1}E_F + F_F$$

$$S_D(\phi) = C(I - \phi^{-1}A)^{-1}\phi^{-1}E_D + F_D \tag{6.3}$$

Parity equations are residual generators which generate the residuals by the direct manipulation of the plant observables. They can be derived from the input-output description of the plant or from a state-space representation. Computationally they may be autoregressive - moving average (ARMA) or moving average (MA). In this chapter, first the different parity equation formulations will be introduced, then it will be shown how they can be applied to implement *a single residual specification*. This will later serve as the basis for the parity equation implementation of structured and directional residual sets, to be discussed in Chapters 7 and 8.

6.1. PARITY EQUATION FORMULATIONS

Let us start by recalling the *generic residual generator* introduced in Subsection 5.2.1. Remember that a residual generator is a computational algorithm operating on the plant observables, that is, on the command values of the controlled inputs and on the measured values of the measured inputs and outputs. Assuming that the residual generator is linear and discrete dynamic, its generic form is

$$\boxed{r(t) = V(\phi)u(t) + W(\phi)y(t)} \tag{6.4}$$

as shown in Subsection 5.2.1. The residuals have to be zero in the absence of unknown inputs, that is, when the input-output equation

$$y(t) = M(\phi)u(t) \tag{6.5}$$

holds explicitly. Therefore $V(\phi)$ and $W(\phi)$ must obey

$$V(\phi) = - W(\phi)M(\phi) \tag{6.6}$$

and the generic residual generator (6.4) becomes

$$r(t) = W(\phi)[y(t) - M(\phi)u(t)] \tag{6.7}$$

The generic equation (6.7) may be used directly to implement the residual generator. Implementation now amounts to finding $W(\phi)$ so that the design specifications are met; $V(\phi)$ then follows from (6.6). The detection and isolation aspects of the specification are given in terms of the internal form (5.28) (without the multiplicative components) while the computational specifications are formulated on (6.4).

6.1.1. ARMA Parity Equations from the Input-Output Description

Consider the input-output equation (6.1), with additive faults and disturbances. Recall that $M(\phi)$ is a transfer function matrix, the elements of which are ARMA functions of the shift operator ϕ. Similarly, $S_F(\phi)$ and $S_D(\phi)$ are transfer function matrices.

Now move the $u(t)$ term in (6.1) to the left hand side:

$$o(t) = y(t) - M(\phi)u(t) = S_F(\phi)p(t) + S_D(\phi)q(t) \tag{6.8}$$

This is clearly a set (vector) of residuals which are computationally autoregressive - moving average. We will refer to Eq. (6.8) as the *ARMA primary parity equations* and to $o(t)$ as the *ARMA primary residuals*. The middle expression in (6.8) gives the computational form of these residuals while the right-hand side provides their internal (unknown-input-effect) form.

Example 6.1.

Consider a second order system with two outputs, two observed inputs and two fault-inputs. Assume the system is described by the following transfer function matrices:

$$M(\phi) = \frac{1}{1 - 1.3\phi^{-1} + 0.4\phi^{-2}} \begin{bmatrix} 2\phi^{-1} - \phi^{-2} & 0 \\ 3\phi^{-1} - 0.4\phi^{-2} & \phi^{-1} - 0.8\phi^{-2} \end{bmatrix}$$

$$S(\phi) = \frac{1}{1 - 1.3\phi^{-1} + 0.4\phi^{-2}} \begin{bmatrix} 1 - 0.3\phi^{-1} - 0.1\phi^{-2} & 2\phi^{-1} - \phi^{-2} \\ 1 - 1.3\phi^{-1} + 1.4\phi^{-2} & 3\phi^{-1} - 0.4\phi^{-2} \end{bmatrix}$$

The ARMA primary residuals for this system are (omitting the time argument for the variables):

$$\begin{bmatrix} o_1 \\ o_2 \end{bmatrix} = \begin{bmatrix} y_1 \\ y_2 \end{bmatrix} - \frac{1}{1 - 1.3\phi^{-1} + 0.4\phi^{-2}} \begin{bmatrix} 2\phi^{-1} - \phi^{-2} & 0 \\ 3\phi^{-1} - 0.4\phi^{-2} & \phi^{-1} - 0.8\phi^{-2} \end{bmatrix} \begin{bmatrix} u_1 \\ u_2 \end{bmatrix}$$

$$= \frac{1}{1 - 1.3\phi^{-1} + 0.4\phi^{-2}} \begin{bmatrix} 1 - 0.3\phi^{-1} - 0.1\phi^{-2} & 2\phi^{-1} - \phi^{-2} \\ 1 - 1.3\phi^{-1} + 1.4\phi^{-2} & 3\phi^{-1} - 0.4\phi^{-2} \end{bmatrix} \begin{bmatrix} p_1 \\ p_2 \end{bmatrix}$$

The primary residuals depend completely on the plant and thus allow no design flexibility. However, additional residuals can be generated from them by the transformation

$$\boxed{r(t) = W(\phi)o(t)} \tag{6.9}$$

where $W(\phi)$ is a transfer function matrix. Then $W(\phi)$ needs to be chosen so that the design specifications are satisfied.

The inspection of Eqs. (6.8) and (6.9) reveals that they yield exactly the generic equation (6.7). Thus the transformation applied to the ARMA primary parity equations is just a re-interpretation of the generic residual generator.

6.1.2. MA Parity Equations from the Input-Output Description

Write the transfer functions appearing in (6.1) as

$$M(\phi) = \check{G}(\phi)/\check{h}(\phi)$$

$$S_F(\phi) = \check{N}_F(\phi)/\check{h}(\phi) \qquad S_D(\phi) = \check{N}_D(\phi)/\check{h}(\phi) \tag{6.10}$$

where $\check{G}(\phi)$, $\check{N}_F(\phi)$ and $\check{N}_D(\phi)$ are polynomial matrices and the scalar polynomial $\check{h}(\phi)$ is the *least common multiple* of the denominators of all elements in $M(\phi)$, $S_F(\phi)$ and $S_D(\phi)$. Then (6.1) can be replaced with the input-output relationship

$$\check{H}(\phi)y(t) = \check{G}(\phi)u(t) + \check{N}_F(\phi)p(t) + \check{N}_D(\phi)q(t) \tag{6.11}$$

where $\check{H}(\phi) = \check{h}(\phi)I$. Further, cancel all common factors separately in each row of (6.11), leaving

$$H(\phi)y(t) = G(\phi)u(t) + N_F(\phi)p(t) + N_D(\phi)q(t) \qquad (6.12)$$

where $H(\phi)=Diag[h_1(\phi) \ \ldots \ h_\mu(\phi)]$ and $h_i(\phi)$, $i=1\ldots\mu$, is the common denominator in the i-th row of Eq. (6.1). Thus $h_i(\phi)$, and the rows $g_{i.}(\phi)$, $n_{Fi.}(\phi)$ and $n_{Di.}(\phi)$ are relative prime (have no common factor), separately for each $i=1\ldots\mu$.

Eq. (6.12) is a polynomial input-output relationship in which each row has been reduced to minimum degree. This leads to a set of *moving average (MA) primary parity equations* as

$$o^*(t) = H(\phi)y(t) - G(\phi)u(t) = N_F(\phi)p(t) + N_D(\phi)q(t) \qquad (6.13)$$

These equations are moving average both computationally (relative to the observables) and internally (relative to the unknown inputs). Also, the i-th residual $o_i^*(\phi)$ is computed from the i-th output $y_i(t)$ and all inputs. Note, further, that $o_i^*(t)$ is related to the i-th ARMA primary residual $o_i(t)$ as

$$o_i^*(t) = h_i(\phi)o_i(t) \qquad (6.14)$$

Example 6.2.

For the system introduced in Example 6.1, the MA primary residuals are:

$$\begin{bmatrix} o_1^* \\ o_2^* \end{bmatrix} = (1 - 1.3\phi^{-1}+0.4\phi^{-2}) \begin{bmatrix} y_1 \\ y_2 \end{bmatrix} - \begin{bmatrix} 2\phi^{-1} - \phi^{-2} & 0 \\ 3\phi^{-1} - 0.4\phi^{-2} & \phi^{-1} - 0.8\phi^{-2} \end{bmatrix} \begin{bmatrix} u_1 \\ u_2 \end{bmatrix}$$

$$= \begin{bmatrix} 1 - 0.3\phi^{-1} - 0.1\phi^{-2} & 2\phi^{-1} - \phi^{-2} \\ 1 - 1.3\phi^{-1}+1.4\phi^{-2} & 3\phi^{-1} - 0.4\phi^{-2} \end{bmatrix} \begin{bmatrix} p_1 \\ p_2 \end{bmatrix}$$

The MA primary parity equations, just like their ARMA counterparts, are completely determined by the system and allow no design freedom. However, additional residuals may be obtained by transformation as

$$r(t) = W^*(\phi)o^*(t) \qquad (6.15)$$

where $W^*(\phi)$ is usually a polynomial matrix; this is to be chosen so that the design specifications are satisfied.

This generator is easy to re-write in a form consistent with the generic equation (6.7). Substitute (6.13) into (6.15) and recall that $\boldsymbol{H}^{-1}(\phi)\boldsymbol{G}(\phi)=\boldsymbol{M}(\phi)$:

$$r(t) = \boldsymbol{W}^*(\phi)\boldsymbol{H}(\phi)[y(t) - \boldsymbol{M}(\phi)\boldsymbol{u}(t)]$$

$$= \boldsymbol{W}^*(\phi)[\boldsymbol{N}_F(\phi)\boldsymbol{p}(t) + \boldsymbol{N}_D(\phi)\boldsymbol{q}(t)] \tag{6.16}$$

If the transformation $\boldsymbol{W}^*(\phi)$ is restricted to polynomial matrices then the generator is moving average, both computationally and as far as the responses to the various unknown inputs are concerned.

6.1.3. Parity Equations from the State-Space Model

The following procedure is due originally to Chow and Willsky (1984).

We will first ignore the disturbance, to keep the derivation manageable; it will be re-introduced later. Consider Eq. (6.2) and express $y(t - \sigma)$ from the output equation (where σ is a delay to be specified later):

$$y(t - \sigma) = \boldsymbol{C}x(t - \sigma) + \boldsymbol{D}\boldsymbol{u}(t - \sigma) + \boldsymbol{F}_F\boldsymbol{p}(t - \sigma) \tag{6.17a}$$

Now write the same equation for $y(t - \sigma+1)$ and express $\boldsymbol{x}(t - \sigma+1)$ with the state-equation:

$$y(t - \sigma+1) = \boldsymbol{C}x(t - \sigma+1) + \boldsymbol{D}\boldsymbol{u}(t - \sigma+1) + \boldsymbol{F}_F\boldsymbol{p}(t - \sigma+1)$$

$$= \boldsymbol{C}\boldsymbol{A}x(t - \sigma) + \boldsymbol{C}\boldsymbol{B}\boldsymbol{u}(t - \sigma) + \boldsymbol{C}\boldsymbol{E}_F\boldsymbol{p}(t - \sigma) + \boldsymbol{D}\boldsymbol{u}(t - \sigma+1) + \boldsymbol{F}_F\boldsymbol{p}(t - \sigma+1) \tag{6.17b}$$

Similarly

$$y(t - \sigma+2) = \boldsymbol{C}\boldsymbol{A}^2x(t - \sigma) + \boldsymbol{C}\boldsymbol{A}\boldsymbol{B}\boldsymbol{u}(t - \sigma) + \boldsymbol{C}\boldsymbol{B}\boldsymbol{u}(t - \sigma+1) + \boldsymbol{D}\boldsymbol{u}(t - \sigma+2)$$

$$+ \boldsymbol{C}\boldsymbol{A}\boldsymbol{E}_F\boldsymbol{p}(t - \sigma) + \boldsymbol{C}\boldsymbol{E}_F\boldsymbol{p}(t - \sigma+1) + \boldsymbol{F}_F\boldsymbol{p}(t - \sigma+2)$$

$$\vdots$$

$$y(t) = \boldsymbol{C}\boldsymbol{A}^\sigma x(t - \sigma)$$

$$+ \boldsymbol{C}\boldsymbol{A}^{\sigma-1}\boldsymbol{B}\boldsymbol{u}(t - \sigma) + \boldsymbol{C}\boldsymbol{A}^{\sigma-2}\boldsymbol{B}\boldsymbol{u}(t - \sigma+1) + \ldots + \boldsymbol{C}\boldsymbol{B}\boldsymbol{u}(t - 1) + \boldsymbol{D}\boldsymbol{u}(t)$$

$$+ \boldsymbol{C}\boldsymbol{A}^{\sigma-1}\boldsymbol{E}_F\boldsymbol{p}(t - \sigma) + \boldsymbol{C}\boldsymbol{A}^{\sigma-2}\boldsymbol{E}_F\boldsymbol{p}(t - \sigma+1) + \ldots + \boldsymbol{C}\boldsymbol{E}_F\boldsymbol{p}(t - 1) + \boldsymbol{F}_F\boldsymbol{p}(t) \tag{6.17c}$$

Define now

$$Y(t)=\begin{bmatrix} y(t-\sigma) \\ y(t-\sigma+1) \\ y(t-\sigma+2) \\ \vdots \\ y(t) \end{bmatrix} \qquad U(t)=\begin{bmatrix} u(t-\sigma) \\ u(t-\sigma+1) \\ \vdots \\ u(t-1) \\ u(t) \end{bmatrix} \qquad P(t)=\begin{bmatrix} p(t-\sigma) \\ p(t-\sigma+1) \\ \vdots \\ p(t-1) \\ p(t) \end{bmatrix} \tag{6.18}$$

With this, the set of equations (6.17) can be written as

$$\boxed{Y(t) = Jx(t-\sigma) + KU(t) + L_F P(t)} \tag{6.19}$$

where

$$J=\begin{bmatrix} C \\ CA \\ CA^2 \\ \vdots \\ CA^{\sigma} \end{bmatrix} \qquad K=\begin{bmatrix} D & & & & \\ CB & D & & & \\ CAB & CB & & & \\ \vdots & & & & \\ CA^{\sigma-1}B & CA^{\sigma-2}B & \dots & CB & D \end{bmatrix}$$

$$L_F=\begin{bmatrix} F_F & & & & \\ CE_F & F_F & & & \\ CAE_F & CE_F & & & \\ \vdots & & & & \\ CA^{\sigma-1}E_F & CA^{\sigma-2}E_F & \dots & CE_F & F_F \end{bmatrix} \tag{6.20}$$

Eq. (6.19) is just another way of describing the input-output relationship of the system with additive faults. Note that the size of the vectors and matrices in the equation is quite large; with κ inputs, μ outputs and ρ faults, the vectors Y, U and P are $(\sigma+1)\cdot\mu$, $(\sigma+1)\cdot\kappa$ and $(\sigma+1)\cdot\rho$ long, respectively, and the size of the matrices J, K and L_F is $[(\sigma+1)\mu]\cdot\nu$, $[(\sigma+1)\mu]\cdot[(\sigma+1)\kappa]$ and $[(\sigma+1)\mu]\cdot[(\sigma+1)\rho]$, where ν is the order of the system realization.

By moving the $\boldsymbol{KU}(t)$ term to the left hand-side in Eq. (6.19), a vector

$$\boldsymbol{Y}(t) - \boldsymbol{KU}(t) = \boldsymbol{J}\boldsymbol{x}(t - \sigma) + \boldsymbol{L}_F \boldsymbol{P}(t) \tag{6.21}$$

is obtained which resembles a set of moving average primary residuals except that the "internal form" contains the state vector $\boldsymbol{x}(t - \sigma)$. This dependence on the state can be eliminated in a transformation; obtain a (scalar) transformed residual as

$$\boxed{r_i(t) = \boldsymbol{w}_i' [\boldsymbol{Y}(t) - \boldsymbol{KU}(t)]} \tag{6.22}$$

where the transformation $\boldsymbol{w}_i'$ is subject to the zeroing constraint

$$\boxed{\boldsymbol{w}_i' \boldsymbol{J} = \boldsymbol{0}'} \tag{6.23}$$

By satisfying (6.23), the transformation *decouples* the residual from the state.

Note that (6.23) represents $Rank\boldsymbol{J} \leq \nu$ scalar constraints for the $(\sigma+1)\cdot\mu$ elements of the transforming vector $\boldsymbol{w}_i'$. The matrix $\boldsymbol{J}$ can be recognized as the observability matrix of the system, truncated or extended depending on the value of σ. Normally, $Rank\boldsymbol{J} = \nu$. However, $Rank\boldsymbol{J} < \nu$ if the realization is unobservable, or if $\sigma < \nu - 1$.

The internal form of the transformed residual, with (6.23) satisfied, is

$$r_i(t) = \boldsymbol{w}_i' \boldsymbol{L}_F \boldsymbol{P}(t) \tag{6.24}$$

If there are also additive disturbances in the system then the vector $\boldsymbol{Q}(t)$ can be defined, the way $\boldsymbol{P}(t)$ has been defined in (6.18), and its accompanying matrix $\boldsymbol{L}_D$ can be formed, following the pattern of $\boldsymbol{L}_F$ in (6.20). These are then incorporated in the internal form (6.24) yielding

$$\boxed{r_i(t) = \boldsymbol{w}_i' [\boldsymbol{L}_F \boldsymbol{P}(t) + \boldsymbol{L}_D \boldsymbol{Q}(t)]} \tag{6.25}$$

The implementation of the residual generator now amounts to finding the transformation $\boldsymbol{w}_i'$ which, while satisfying the constraint (6.23), meets the detection and isolation specifications. Finding (choosing) the window-length $\sigma+1$ is part of the implementation; it depends on the system (strictly proper or proper) and on the design specification. This and other design issues will be addressed in Subsection 6.4.

Example 6.3.

For the system introduced in Example 6.1, a minimal state-space representation is obtained as:

$$\boldsymbol{A} = \begin{bmatrix} 0.8 & 0 \\ 1 & 0.5 \end{bmatrix} \qquad \boldsymbol{B} = \begin{bmatrix} 2 & 0 \\ 3 & 1 \end{bmatrix} \qquad \boldsymbol{C} = \begin{bmatrix} 1 & 0 \\ 0 & 1 \end{bmatrix}$$

$$\boldsymbol{E} = \begin{bmatrix} 1 & 2 \\ 0 & 3 \end{bmatrix} \qquad \boldsymbol{F} = \begin{bmatrix} 1 & 0 \\ 1 & 0 \end{bmatrix} \qquad \boldsymbol{D} = \boldsymbol{0}$$

This leads to the following $\boldsymbol{J}$, $\boldsymbol{K}$ and $\boldsymbol{L}$ matrices, for the window width $\sigma=2$:

$$\boldsymbol{J} = \begin{bmatrix} 1 & 0 \\ 0 & 1 \\ 0.8 & 0 \\ 1 & 0.5 \\ 0.64 & 0 \\ 1.3 & 0.25 \end{bmatrix} \qquad \boldsymbol{K} = \begin{bmatrix} 0 & 0 & 0 & 0 & 0 & 0 \\ 0 & 0 & 0 & 0 & 0 & 0 \\ 2 & 0 & 0 & 0 & 0 & 0 \\ 3 & 1 & 0 & 0 & 0 & 0 \\ 1.6 & 0 & 2 & 0 & 0 & 0 \\ 3.5 & 0.5 & 3 & 1 & 0 & 0 \end{bmatrix}$$

$$\boldsymbol{L} = \begin{bmatrix} 1 & 0 & 0 & 0 & 0 & 0 \\ 1 & 0 & 0 & 0 & 0 & 0 \\ 1 & 2 & 1 & 0 & 0 & 0 \\ 0 & 3 & 1 & 0 & 0 & 0 \\ 0.8 & 1.6 & 1 & 2 & 1 & 0 \\ 1 & 3.5 & 0 & 3 & 1 & 0 \end{bmatrix}$$

Equivalent generic formulation. In order to derive the generic formulation for the Chow-Willsky scheme, first we will convert it into shift-operator form (Fang, 1993). Write the transformation as

$$w'_i = [w'_{i\sigma} \quad w'_{i,\sigma-1} \quad \dots \quad w'_{i1} \quad w'_{io}] \tag{6.26}$$

where each sub-vector is μ elements long. Also, define

$$v'_i = - w'_i K = [v'_{i\sigma} \quad v'_{i,\sigma-1} \quad \dots \quad v'_{i1} \quad v'_{io}]$$

$$z'_{Fi} = w'_i L_F = [z'_{Fi\sigma} \quad z'_{Fi,\sigma-1} \quad \dots \quad z'_{Fi1} \quad z'_{Fio}] \tag{6.27}$$

where the v'_{ij} and z'_{ij} sub-vectors are κ and ρ elements long, respectively. Now introduce the polynomial vectors

$$w'_i(\phi) = w'_{io} + w'_{i1}\phi^{-1} + \dots + w'_{i\sigma}\phi^{-\sigma}$$

$$v'_i(\phi) = v'_{io} + v'_{i1}\phi^{-1} + \dots + v'_{i\sigma}\phi^{-\sigma}$$

$$z'_{Fi}(\phi) = z'_{Fio} + z'_{Fi1}\phi^{-1} + \dots + z'_{Fi\sigma}\phi^{-\sigma} \tag{6.28}$$

With these, the residual generator (6.22), (6.24) can be written as

$$\boxed{r_i(t) = w'_i(\phi)y(t) + v'_i(\phi)u(t) = z'_{Fi}(\phi)p(t)} \tag{6.29}$$

Eq. (6.29) is the Chow-Willsky generator expressed in terms of the shift operator. Recall that this equation is valid only if the transformation $w'_i(\phi)$ (or w'_i) has been so chosen that it satisfies the constraint (6.23).

Further insight can be gained by considering the fault-free situation. Now $y(t)=M(\phi)u(t)$ and $p(t)=0$. Thus (6.29) becomes

$$w'_i(\phi)M(\phi)u(t) + v'_i(\phi)u(t) = 0 \tag{6.30}$$

This reveals that, for any $u(t)$,

$$v'_i(\phi)u(t) = - w'_i(\phi)M(\phi)u(t) \tag{6.31}$$

as long as (6.23) is satisfied. Substituting this into (6.29) and bringing back the fault term yields another form of the Chow-Willsky generator

$$\boxed{r_i(t) = w'_i(\phi)[y(t) - M(\phi)u(t)] = z'_{Fi}(\phi)p(t)} \tag{6.32}$$

Eq. (6.32) is clearly in the form of the generic residual generator (6.7).

Eq. (6.29) reveals that the generator is computationally moving average. This, with (6.32), implies that $\mathbf{w}_i'(\phi)\mathbf{M}(\phi)$ is polynomial (but it does not imply that the denominator of $\mathbf{M}(\phi)$ divides $\mathbf{w}_i'(\phi)$). Also, from (6.29), the fault-response of the generator is moving average. Eq. (6.32) is somewhat more restrictive than the generic residual generator (6.7) because here the transformation needs to be polynomial (in (6.7) it can be rational) and it has to satisfy the constraint (6.23). This is the price paid for the moving average behavior.

6.2. IMPLEMENTATION OF A SINGLE RESIDUAL

In the rest of this chapter, the *exact* implementation of a single residual, using the various parity relation formulations, will be investigated. First, however, we need to introduce a few important definitions.

Strictly input and strictly output faults/disturbances

- We will refer to a fault $p_j(t)$ or disturbance $q_j(t)$ as a *strictly input fault or disturbance* if its column $\mathbf{f}_{F\cdot j}$ or $\mathbf{f}_{D\cdot j}$ in the state-space description (6.2) is zero. Equivalently, the transfer function columns $\mathbf{s}_{F\cdot j}(\phi)$ or $\mathbf{s}_{D\cdot j}(\phi)$, in Eq. (6.1), of a stricly input fault/disturbance are strictly proper, that is, ϕ^{-1} can be factored out from their respective numerators $\mathbf{n}_{F\cdot j}(\phi)$ and $\mathbf{n}_{D\cdot j}(\phi)$.

- We will refer to a fault $p_j(t)$ or disturbance $q_j(t)$ as a *strictly output fault or disturbance* if its column $\mathbf{e}_{F\cdot j}$ or $\mathbf{e}_{D\cdot j}$ in the state-space description (6.2) is zero. Equivalently, the transfer function columns $\mathbf{s}_{F\cdot j}(\phi)$ or $\mathbf{s}_{D\cdot j}(\phi)$, in Eq. (6.1), of a stricly output fault/disturbance are constant, that is, the elements $n_{Fij}(\phi)$ and $n_{Dij}(\phi)$ in their numerators are constant multiples of the respective denominator $h_i(\phi)$.

Homogeneous and non-homogeneous specifications. The response specification will be given, in accordance with Eq. (5.40), as

$$r_i(t) = \mathbf{z}_{Fi}'(\phi)\mathbf{p}(t) + \mathbf{z}_{Di}'(\phi)\mathbf{q}(t) \tag{6.33}$$

where $\mathbf{z}_{Fi}'(\phi)=[\,z_{Fi1}(\phi) \quad z_{Fi2}(\phi) \;\dots\,]$ and $\mathbf{z}_{Di}'(\phi)=[\,z_{Di1}(\phi) \quad z_{Di2}(\phi) \;\dots\,]$. The individual transfer function specifications may be homogeneous (i.e., $z_{\cdot ij}(\phi)=0$) or non-homogeneous (i.e., $z_{\cdot ij}(\phi)\neq 0$). A non-homogeneous specification may be a constant, a polynomial or a rational function in the shift operator. We will refer to a full specification-set

$$\mathbf{z}_i'(\phi) = [\mathbf{z}_{Fi}'(\phi) \quad \mathbf{z}_{Di}'(\phi)] \tag{6.34}$$

as *homogeneous* if *all* the scalar transfer functions it contains are zero. Otherwise, we will call the specification *non-homogeneous*.

Full and almost full specifications. When seeking an implementation, the mathematical objective will be to find an appropriate transformation

$$w_i'(\phi) = [\ w_{i1}(\phi)\ \ldots\ w_{i\mu}(\phi)\ ] \tag{6.35}$$

where the elements $w_{i1}(\phi)\ \ldots\ w_{i\mu}(\phi)$ are rational functions or polynomials in the shift operator. The length of the vector $w_i'(\phi)$ is always μ, the number of system outputs. Obviously, it is not possible to find a transformation which satisfies more than μ independent (rational or polynomial) conditions. Denote by ρ the number of scalar responses specified. We will refer to a specification as *full* if $\rho = \mu$ and *almost full* if $\rho = \mu - 1$.

An important special case is the specification which contains μ-1 zero elements and one nonzero element. We will refer to this case as *full almost homogeneous specification*.

Full non-homogeneous specifications arise naturally in the design of directional residual sets. Structured residuals, including diagonal sets, may be implemented either as full almost homogeneous or almost full homogeneous specifications.

Similar transformations. We will say that the transformations $[w_i'(\phi)]_1$ and $[w_i'(\phi)]_2$ are *similar* (have the same polynomial direction) if they are linearly dependent in the polynomial sense, that is, if

$$\alpha_1(\phi)[w_i'(\phi)]_1 + \alpha_2(\phi)[w_i'(\phi)]_2 = 0' \tag{6.36}$$

where $\alpha_1(\phi)$ and $\alpha_2(\phi)$ are polynomials. If a transformation is subject to almost full homogeneous specifications, that is, if

$$w_i'(\phi)S(\phi) = 0' \tag{6.37}$$

where $S(\phi)$ has μ-1 independent columns (and μ is the number of elements in $w_i'(\phi)$), then all solutions of the transformation have the same polynomial direction. This follows as the polynomial generalization of the fact that μ-1 independent orthogonality conditions determine the direction of a vector in the μ-dimensional space.

There are several issues associated with the implementation procedure:

A. *Does the solution exist?* Existence requires certain properties of the monitored plant but may also be affected by the specification.

B. *Is the solution realizable*? To obtain realizable solutions, the specification may need to be matched to the properties of the plant (in particular, to strictly input faults only delayed responses are feasible).

C. *Is the solution stable?* Moving average residual generators guarantee stability naturally but with ARMA generators it needs to be assured explicitly.

D. *Is the solution unique?* A solution may be unique or have a unique direction or neither, depending on the specification and, to some extent, the system.

E. *Is the solution of minimal complexity?* Minimal complexity means minimal polynomial degree in the generator; with non-unique solutions, this does not follow automatically.

In the sequel, first the implementation of a single residual generator by the direct utilization of the input-output system models (ARMA and MA) will be explored. This "naive" approach does not provide any clue to such properties as the stability of ARMA generators or minimum complexity of non-unique implementations; in this framework, the concerned properties can only be attained empirically. Then implementation by the Chow-Willsky scheme (state-space approach) will be investigated, where minimum complexity design will be seen to be straightforward. Finally, the fault system matrix will be introduced and analyzed, leading to a deeper insight into several internal properties of the system which will allow placing the transfer function implementations on a solid foundation.

The various approaches will be applied to the same design example, for illustration and comparison.

6.3. NAIVE IMPLEMENTATION WITH INPUT-OUTPUT RELATIONS

In this section, we will see what can be done with parity relations within the strict input-output transfer function framework. The techniques introduced here will be re-interpreted and substantially refined in Section 6.6.

6.3.1. Implementation with ARMA Parity Relations

Write Eq. (6.9) for a single residual and substitute the primary residuals from (6.8):

$$r_i(t) = \mathbf{w}_i'(\phi)[\mathbf{y}(t) - \mathbf{M}(\phi)\mathbf{u}(t)] = \mathbf{w}_i'(\phi)[\mathbf{S}_F(\phi)\mathbf{p}(t) + \mathbf{S}_D(\phi)\mathbf{q}(t)] \qquad (6.38)$$

The right-hand side of this equation is the internal form of the residual generator. Compare this with the specification (6.33):

$$\mathbf{w}_i'(\phi)[\mathbf{S}_F(\phi)\mathbf{p}(t) + \mathbf{S}_D(\phi)\mathbf{q}(t)] = \mathbf{z}_{Fi}'(\phi)\mathbf{p}(t) + \mathbf{z}_{Di}'(\phi)\mathbf{q}(t) \qquad (6.39)$$

This yields the implementation equation

$$w_i'(\phi)[S_F(\phi) \quad S_D(\phi)] = [z_{Fi}'(\phi) \quad z_{Di}'(\phi)] \tag{6.40a}$$

We will use this equation in the more compact form

$$\boxed{w_i'(\phi)S(\phi) = z_i'(\phi)} \tag{6.40b}$$

where

$$S(\phi) = [S_F(\phi) \quad S_D(\phi)] \qquad z_i'(\phi) = [z_{Fi}'(\phi) \quad z_{Di}'(\phi)] \tag{6.40c}$$

In this equation, $S(\phi)$ describes properties of the monitored plant and is assumed to be known while $z_i'(\phi)$ represents the design specification, also known. The equation needs to be solved for the transformation $w_i'(\phi)$.

The solution depends on
- the number of the specified responses ρ
- the nature (homogeneous/non-homogeneous) of the specification-set
- the rank of the matrix $S(\phi)$.

A1. Full non-homogeneous specification. Now $\rho = \mu$. Assuming that

$$Rank\, S(\phi) = \mu \tag{6.41}$$

Eq. (6.40b) can be solved, uniquely, as

$$\boxed{w_i'(\phi) = z_i'(\phi)S^{-1}(\phi)} \tag{6.42}$$

Example 6.4.
We will demonstrate the design on the system introduced in Example 6.1. First let us compute the inverse of the fault transfer matrix:

$$S^{-1}(\phi) = \frac{1 - 1.3\phi^{-1} + 0.4\phi^{-2}}{\phi^{-1} + 2.3\phi^{-2} - 4.28\phi^{-3} + 1.44\phi^{-4}} \begin{bmatrix} 3\phi^{-1} - 0.4\phi^{-2} & -2\phi^{-1} + \phi^{-2} \\ -1 + 1.3\phi^{-1} - 1.4\phi^{-2} & 1 - 0.3\phi^{-1} - 0.1\phi^{-2} \end{bmatrix}$$

which is of rather high degree. However, the fraction can be simplified to

$$\frac{1}{\phi^{-1}(1 + 3.6\phi^{-1})}$$

Notice the ϕ^{-1} factor in the denominator. This is canceled with the first row of the adjoint matrix but not with the second. Now let us specify the fault responses as

$$z_i'(\phi) = [\ 0.5 \quad \phi^{-1}\ ]$$

Note the one step delay in the second specification; this is needed for realizability in connection with the second fault which is strictly input. With this, the transformation is obtained as

$$w_i'(\phi) = [\ 0.5+1.1\phi^{-1} - 1.4\phi^{-2} \qquad 0.2\phi^{-1} - 0.1\phi^{-2}\ ] \,/\, (1+3.6\phi^{-1})$$

Finally, the input multipliers can be computed as $v_i'(\phi) = -\ w_i'(\phi)M(\phi)$. After identifying and removing the common factors again, these are found as

$$v_i'(\phi) = [\ -\ \phi^{-1} - 3.6\phi^{-2} \qquad -\ 0.2\phi^{-2}\ ] \,/\, (1+3.6\phi^{-1})$$

Notice that the generator designed in Example 6.4 is unstable. In fact, this naive implementation procedure does not guarantee the *stability* of the generator; it must be checked in a separate design step and, if necessary, the specification modified in order to obtain a stable generator. The results in Section 6.6 will lead to a systematic procedure of stable implementation.

Example 6.4. continued

To attain stability, change the specification to

$$z_i'(\phi) = [\ 0.5(1+3.6\phi^{-1}) \qquad \phi^{-1}(1+3.6\phi^{-1})]$$

Then the generator becomes

$$w_i'(\phi) = [\ 0.5+1.1\phi^{-1} - 1.4\phi^{-2} \qquad 0.2\phi^{-1} - 0.1\phi^{-2}\ ]$$

$$v_i'(\phi) = [\ -\ \phi^{-1} - 3.6\phi^{-2} \qquad -\ 0.2\phi^{-2}\ ]$$

AB. Full almost homogeneous specification. Consider a full specification with only one nonzero element $z_{ig}(\phi)$. Then (6.42) simplifies to

$$w_i'(\phi) = [0 \ldots 0\ z_{ig}(\phi)\ 0 \ldots 0]S^{-1}(\phi) = z_{ig}(\phi)[S^{-1}(\phi)]_g. \qquad (6.43)$$

B1. Full homogeneous specification. With a full homogeneous specification-set and *Rank* $S(\phi)=\mu$, only a trivial solution $w_i'(\phi)=0'$ could be obtained. This would not allow any fault detection and would thus be useless.

B2. Almost full homogeneous specification. Now $\rho=\mu-1$, with

$$Rank\ S(\phi) = \mu - 1 \tag{6.44}$$

Non-trivial solution(s) exist but are not unique. To obtain a solution, we may resort to either of the following tricks:

a. Add a non-homogeneous scalar fault specification $z_{i\mu}(\phi) \neq 0$, with the respective column $s_{\cdot\mu}(\phi)$, so that the rank of the augmented matrix $S^A(\phi)$ is μ. Then the problem is reduced technically to Case AB. The solution will not be unique and of minimal complexity but all solutions will be similar, independent of the choice of the auxiliary specification. The residual generator may be simplified empirically, by identifying and canceling common factors. This procedure will maintain the specified homogeneous response while violating the auxiliary specification - which is usually arbitrary anyway.

b. Choose one of the elements in $w_i'(\phi)$ arbitrarily. For example, choose $w_{i\mu}(\phi) = \gamma(\phi)$. Then (6.40b) can be rearranged as

$$[\, w_{i1}(\phi) \dots w_{i,\mu-1}(\phi)\,] S^{\mu}(\phi) = -\, \gamma(\phi) s_{\mu\cdot}(\phi) \tag{6.45}$$

where

$$S^{\mu}(\phi) = \begin{bmatrix} s_{1\cdot}(\phi) \\ \cdot \\ \cdot \\ \cdot \\ s_{\mu-1\cdot}(\phi) \end{bmatrix} \tag{6.46}$$

This is now a non-homogeneous equation which can be solved, provided that $S^{\mu}(\phi)$ has full rank, as

$$[\, w_{i1}(\phi) \dots w_{i,\mu-1}(\phi)\,] = -\, \gamma(\phi) s_{\mu\cdot}(\phi) \cdot [S^{\mu}(\phi)]^{-1} \tag{6.47}$$

The solution will vary depending on which element was arbitrarily assigned and to what value or function, but its direction is unique. The transformation may be simplified empirically, by identifying and canceling common factors.

Example 6.5.

Let us design an ARMA residual generator for our system so that it produces zero response to the first fault. The fault transfer matrix for this single-fault system is

$$S(\phi) = \frac{1}{1 - 1.3\phi^{-1} + 0.4\phi^{-2}} \begin{bmatrix} 1 - 0.3\phi^{-1} - 0.1\phi^{-2} \\ 1 - 1.3\phi^{-1} + 1.4\phi^{-2} \end{bmatrix}$$

Now select $w_{i2}(\phi)=1$. By this choice, the first row of the fault transfer function becomes $S^{\mu}(\phi)$ and the second row $s_{\mu.}(\phi)$. The remaining element of the transformation is then obtained as

$$w_{i1}(\phi) = -\frac{1 - 1.3\phi^{-1} + 1.4\phi^{-2}}{1 - 0.3\phi^{-1} - 0.1\phi^{-2}}$$

Again, stability of the solution is not automatic. If the resulting generator is unstable, the respective pole factors need to be included in $\gamma(\phi)$.

C. Less than full non-homogeneous or almost full homogeneous specification. If the matrix $S(\phi)$ has full (column) rank, then solutions to (6.40b) exist but are not unique. To obtain a solution, one may introduce additional constraints or assume some of the elements of $w_i'(\phi)$.

Technically, the *existence condition* for a (non-trivial) solution is that $S(\phi)$ must have full column rank. This guarantees that Eq. (6.40b) can be solved under *any* specification-set. However, particular specification-sets may allow meaningful solutions even if $S(\phi)$ has a rank defect. Most notably, if the rank defect is restricted to columns which are all accompanied by zero response specifications, then the problem can be solved by omitting some of the columns until the remaining matrix has full rank.

The *computational algorithm* arising from the above implementation procedure is, in general, ARMA.

6.3.2. Implementation with MA Parity Relations

Now write (6.15) for a single residual and substitute the primary residuals from (6.13)

$$r_i(t) = w_i^{*\prime}(\phi)[H(\phi)y(t) - G(\phi)u(t)] = w_i^{*\prime}(\phi)[N_F(\phi)p(t) + N_D(\phi)q(t)] \quad (6.48)$$

Here $H(\phi)$, $G(\phi)$, $N_F(\phi)$ and $N_D(\phi)$ are polynomial matrices. When using this form, it is reasonable to aim at a completely moving average design. This implies that the specification, given in accordance with (6.33), contains all polynomial responses and the transformation $w_i^{*\prime}(\phi)$ is also sought to be polynomial.

Combine the right-hand side of (6.48) with (6.33):

$$w_i^{*\prime}(\phi)[N_F(\phi)p(t) + N_D(\phi)q(t)] = z_{Fi}'(\phi)p(t) + z_{Di}'(\phi)q(t) \quad (6.49)$$

which yields the implementation equation

$$w_i^{*\prime}(\phi)[N_F(\phi) \quad N_D(\phi)] = [z_{Fi}'(\phi) \quad z_{Di}'(\phi)] \quad (6.50a)$$

This will be used in the more compact form

$$w_i^{*\prime}(\phi)N(\phi) = z_i'(\phi) \tag{6.50b}$$

where

$$N(\phi) = [N_F(\phi) \quad N_D(\phi)] \tag{6.51}$$

A1. Full non-homogeneous specification. There is a unique solution, assuming that $N(\phi)$ has full rank. However, this is not polynomial for arbitrary specifications, even if they are polynomial. It will be shown in Section 6.6 that, with some constraint on the specified responses, polynomial implementation is still possible.

A2. Almost full non-homogeneous specification. Now $\rho=\mu - 1$. Assume that *Rank* $N(\phi) = \mu - 1$. Now one of the elements of $w_i^{*\prime}(\phi)$ can be assigned. Select $w_{i\mu}^*(\phi)$ for assignment and rearrange (6.50b) as

$$[w_{i1}^*(\phi) \ldots w_{i,\mu-1}^*(\phi)]N^\mu(\phi) = z_i'(\phi) - w_{i\mu}^*(\phi)n_{\mu\cdot}(\phi) \tag{6.52}$$

where $N^\mu(\phi)$ is defined following the pattern of (6.46). This can be solved as

$$[w_{i1}^*(\phi) \ldots w_{i,\mu-1}^*(\phi)] = [z_i'(\phi) - w_{i\mu}^*(\phi)n_{\mu\cdot}(\phi)]\,[N^\mu(\phi)]^{-1} \tag{6.53}$$

In general, this solution is not moving average. In some cases, it may be made such by the appropriate selection of $w_{i\mu}^*(\phi)$. At this stage, without further analysis, we can only propose to select it so that

$$z_i'(\phi) - w_{i\mu}^*(\phi)n_{\mu\cdot}(\phi) = \boldsymbol{\beta}'(\phi)\cdot Det N^\mu(\phi) \tag{6.54}$$

where $\boldsymbol{\beta}'(\phi)$ is a polynomial vector. However, there is no guarantee that any choice of $w_{i\mu}^*(\phi)$ will indeed lead to a polynomial solution; it may not be possible to obtain an arbitrarily specified response in the polynomial framework. This problem will be approached in a systematic way in Section 6.6.

Example 6.6.

Let us design an MA residual generator for our example system so that it produces a response of $z_{i1}(\phi)=0.5$ to the first fault. For this single fault system, now

$$N(\phi) = \begin{bmatrix} 1 - 0.3\phi^{-1} - 0.1\phi^{-2} \\ 1 - 1.3\phi^{-1} + 1.4\phi^{-2} \end{bmatrix}$$

Obtain the second element of the transformation from (6.54); now the first row of $N(\phi)$ will be $N^{\mu}(\phi)$ and the second row will be $n_{\mu\cdot}(\phi)$. Thus $Det[N^{\mu}(\phi)] = 1 - 0.3\phi^{-1} - 0.1\phi^{-2}$ and the equation to solve is

$$0.5 - w^*_{i2}(\phi)(1 - 1.3\phi^{-1} + 1.4\phi^{-2}) = \beta_i(\phi)(1 - 0.3\phi^{-1} - 0.1\phi^{-2})$$

The minimum degree solution is found as

$$w^*_{i2}(\phi) = 0.0926 + 0.0162\phi^{-1} \qquad\qquad \beta_i(\phi) = 0.4074 + 0.2265\phi^{-1}$$

Then from (6.53) and (6.54), $w^*_{i1}(\phi) = \beta_i(\phi)$. ▩

B1. Full homogeneous specification. With full homogeneous specification, and $N(\phi)$ having full rank, only the trivial solution of $w_i^{*\prime}(\phi) = 0'$ would be possible.

B2. Almost full homogeneous specification. Now $\rho = \mu - 1$; with $Rank N(\phi) = \mu - 1$, this can be posed as a special case of (6.52), with $z_i'(\phi) = 0'$:

$$[w^*_{i1}(\phi) \ \ldots \ w^*_{i,\mu-1}(\phi)] N^{\mu}(\phi) = - w^*_{i\mu}(\phi) n_{\mu\cdot}(\phi) \tag{6.55}$$

Now choose

$$- w^*_{i\mu}(\phi) = Det N^{\mu}(\phi) \tag{6.56}$$

Then a polynomial solution is obtained as

$$[w^*_{i1}(\phi) \ \ldots \ w^*_{i,\mu-1}(\phi)] = n_{\mu\cdot}(\phi) \cdot Adj N^{\mu}(\phi) \tag{6.57}$$

The algorithm (6.56)-(6.57) is simple and straightforward. However, it usually produces transformations of excessive polynomial degree (though of unique direction). The minimum complexity residual generator can be found empirically, by locating and canceling the common factors. A direct procedure to implement the minimum complexity generator will be introduced later in Section 6.6.

Example 6.7.

Now let us solve the problem of Example 6.6 so that the specified response is $z_{i1}(\phi) = 0$. With

$$w^*_{i2}(\phi) = - Det N^{\mu}(\phi) = - 1 + 0.3\phi^{-1} + 0.1\phi^{-2}$$

and $AdjN^{\mu}(\phi)=1$, (6.57) yields

$$w_{i1}^{*}(\phi) = 1 - 1.3\phi^{-1} + 1.4\phi^{-2}$$

A comparison with the result of Example 6.5 reveals that the two transformations, as expected, are of the same direction; the present transformation $[w_{i1}^{*} \quad w_{i2}^{*}]$ can be obtained from the previous $[w_{i1} \quad w_{i2}]$ by multiplication with $-(1 - 0.3\phi^{-1} - 0.1\phi^{-2})$.

Note that the excess complexity (cancelable common factors), usually arising with these naive methods, could not be demonstrated in Examples 6.6 and 6.7 because the system itself is trivially simple.

6.4. IMPLEMENTATION WITH THE CHOW-WILLSKY SCHEME

When the Chow-Willsky scheme is used to implement a residual generator, the specifications are to be given, in accordance with (6.27), in the form

$$z'_{Fi} = [z'_{Fi\sigma} \quad z'_{Fi,\sigma-1} \quad \cdots \quad z'_{Fi1} \quad z'_{Fio}]$$

$$z'_{Di} = [z'_{Di\sigma} \quad z'_{Di,\sigma-1} \quad \cdots \quad z'_{Di1} \quad z'_{Dio}] \tag{6.58}$$

If necessary, z'_{Fi} and z'_{Di} can easily be obtained from the polynomial form $z'_{Fi}(\phi)$ and $z'_{Di}(\phi)$, see Eq. (6.28). According to the internal form (6.25) of the residual, the implementation calls for

$$w'_i[L_F \quad L_D] = [z'_{Fi} \quad z'_{Di}] \tag{6.59}$$

In addition, the state decoupling condition (6.23) needs to be satisfied, that is,

$$w'_i J = 0' \tag{6.60}$$

Combine the two conditions as

$$\boxed{w'_i [J \quad L] = [0' \quad z'_i]} \tag{6.61}$$

where

$$L = [L_F \quad L_D] \qquad z'_i = [z'_{Fi} \quad z'_{Di}] \tag{6.62}$$

Eq. (6.61) needs to be solved for the transformation

$$w_i' = [w_{i\sigma}' \quad w_{i,\sigma-1}' \quad \cdots \quad w_{i1}' \quad w_{io}'] \tag{6.63}$$

(c.f. (6.26)), with J and L (the system) and z_i' (the response specification) given. Recall that the number of rows in J and L is $\mu(\sigma+1)$, where $(\sigma+1)$ is the width of the time-window. The solution, in general, is not unique because σ is not specified. We will seek the transformation with the minimum complexity, that is the shortest possible time window.

To find the minimum complexity solution, we need to consider the number of unknowns versus the number of conditions in (6.61). The unknowns are the elements of w_i'; since each sub-vector w_{ij}' is μ elements long, this is

$$\text{\# of unknowns} = \mu(\sigma+1) \tag{6.64}$$

As for the number of conditions, recall that J has ν columns and L has $\rho(\sigma+1)$ columns, where ρ is the combined number of faults and disturbances considered. However, as seen in Eq. (6.20), one of the columns in the L matrix is zero for each strictly input fault/disturbance (since the respective column in F_F or F_D is zero). Denote by L° the matrix obtained from L by omitting the zero columns and assume that $[\,J \quad L^\circ\,]$ has full rank. Then the number of conditions in (6.61) is the number of columns in $[\,J \quad L^\circ\,]$, which is

$$\text{\# of conditions} = \nu + \rho(\sigma+1) - \rho_I \tag{6.65}$$

where ρ_I is the number of strictly input faults and disturbances. With this:

A. For non-homogeneous specifications, the minimum complexity non-trivial solution arises when the *# of unknowns* = *# of conditions*, that is, when

$$\mu(\sigma+1) = \nu + \rho(\sigma+1) - \rho_I \tag{6.66}$$

B. For homogeneous specifications, the minimum complexity solution arises when the *# of unknowns* = *# of conditions*+*1*, that is, when

$$\mu(\sigma+1) = \nu + \rho(\sigma+1) - \rho_I + 1 \tag{6.67}$$

We will investigate the implications of (6.66) and (6.67) under full and almost full specifications.

A1. Full non-homogeneous specification. With $\rho=\mu$, (6.66) yields $\rho_I=\nu$; this is a special case which implies $\mu=\nu$ and $\rho_I=\rho$. In this case, (6.66) is satisfied for any σ and the transformation may be obtained as $w_i' = [0 \quad z_i'] \, [J \quad L^\circ]^{-1}$. Since $[J \quad L^\circ]$ is singular for $\sigma=0$ (because $F=0$), the minimum complexity solution arises with $\sigma=1$. However, the $\rho_I=\nu$ condition also reveals that, apart from this special case, it is not possible to find a polynomial residual generator which satisfies an arbitrary full non-homogeneous specification.

A2. Almost full non-homogeneous specification. Now $\rho=\mu-1$ and (6.66) yields

$$\boxed{\sigma = \nu - \rho_I - 1} \tag{6.68}$$

The matrix $[J \quad L^\circ]$ is square and, provided it has full rank, a unique transformation is obtained as

$$\boxed{w_i' = [\, 0' \quad z_i' \,][\, J \quad L^\circ \,]^{-1}} \tag{6.69}$$

Note that in the special case of $\mu=\nu$, $\rho_I=\rho$, (6.68) yields $\sigma=0$. However, now $[J \quad L^\circ]$ is singular ($F=0$) thus the solution does not exist. Therefore $\sigma=1$ is the minimum complexity in this case as well, with which, of course, the transformation is not unique.

Example 6.8.

Let us design a residual generator for our example system so that the response to the first fault is $z_{i1}(\phi)=0.5$. This fault is not strictly input so, for the single fault system, $\sigma=2-0-1=1$. We have constructed the J, L and K matrices for the full system with $\sigma=2$ in Example 6.3; the matrices applicable to the present example are obtained from those by keeping the first four rows of each, and the odd numbered columns of L. Note that $L^\circ=L$. Thus

$$J = \begin{bmatrix} 1 & 0 \\ 0 & 1 \\ 0.8 & 0 \\ 1 & 0.5 \end{bmatrix} \qquad L^\circ = \begin{bmatrix} 1 & 0 \\ 1 & 0 \\ 1 & 1 \\ 0 & 1 \end{bmatrix} \qquad K = \begin{bmatrix} 0 & 0 & 0 & 0 \\ 0 & 0 & 0 & 0 \\ 2 & 0 & 0 & 0 \\ 3 & 1 & 0 & 0 \end{bmatrix}$$

The specification, in the format consistent with the Ch-W. design is $z_i'=[0 \quad 0.5]$. The transformation is obtained from (6.69) as

$$w_i' = [\, -0.412 \quad -0.029 \quad 0.441 \quad 0.059 \,]$$

Then the input multiplier by (6.27) is

$$v_i' = [\,-1.059 \quad -0.059 \quad 0 \quad 0\,]$$

Finally, w_i' and v_i' can be converted into polynomial form to yield

$$w_i'(\phi) = [\,0.441 - 0.412\phi^{-1} \qquad 0.059 - 0.029\phi^{-1}\,]$$

$$v_i'(\phi) = [\,-1.059\phi^{-1} \qquad -0.059\phi^{-1}\,]$$

▓

B1. Full homogeneous specification. With $\rho=\mu$, (6.67) yields $\rho_1=\nu+1$ which is a contradiction (since $\rho_1 \leq \rho=\mu \leq \nu$). Thus non-trivial solution for full homogeneous specification is not possible.

B2. Almost full homogeneous specification. Now $\rho=\mu-1$ and (6.67) yields

$$\boxed{\sigma = \nu - \rho_1} \qquad (6.70)$$

To find a transformation which satisfies (6.61) with $z_i'=0'$, one of the elements of the w_i' vector needs to be assumed. Choose for example $w_{i\sigma 1}=c$. Then (6.61) can be re-arranged as

$$[(w_{i\sigma}')^1 \quad w_{i,\sigma-1}' \quad \cdots \quad w_{i1}' \quad w_{io}'\,]\,[\,J^1 \quad L^{o1}] = -\,c\,[\,j_{1.} \quad l_{1.}^{o}] \qquad (6.71)$$

where $(w_{i\sigma}')^1$ denotes $w_{i\sigma}'$ without its first element, J^1 and L^{o1} denote J and L^o without their first row and $[\,j_{1.} \quad l_{1.}^{o}]$ is the missing first row. A solution is obtained as

$$\boxed{[(w_{i\sigma}')^1 \quad w_{i,\sigma-1}' \quad \cdots \quad w_{i1}' \quad w_{io}'\,] = -\,c\,[\,j_{1.} \quad l_{1.}^{o}]\,[\,J^1 \quad L^{o1}]^{-1}} \qquad (6.72)$$

Since orthogonality to $\mu(\sigma+1)-1$ vectors determines the direction of the $\mu(\sigma+1)$ dimensional w_i' vector, the choice of the arbitrarily assigned element, and the value assigned to it, affects only its size. Thus, under almost full homogeneous response specification, the *direction* of the transformation is unique.

Example 6.9.

Now let us design the residual generator so that the response to the first fault is zero. The window width by (6.70) is $\sigma=2-0=2$. We will use the $\boldsymbol{J}$, $\boldsymbol{L}$ and $\boldsymbol{K}$ matrices, constructed in Example 6.3, with all six rows, but only the odd numbered columns in $\boldsymbol{L}=\boldsymbol{L}^o$. Choose $w_{i21}=1$. Then by solving (6.71)

$$w'_i = [\,1 \quad 0.071 \quad -0.929 \quad 0.215 \quad 0.714 \quad -0.714\,]$$

and from (6.27)

$$v'_i = [\,2.570 \quad 0.142 \quad 0.714 \quad 0.714 \quad 0 \quad 0\,]$$

The polynomial forms are

$$w'_i(\phi) = 0.714[\,1 - 1.3\phi^{-1} + 1.4\phi^{-2} \qquad -1 + 0.3\phi^{-1} + 0.1\phi^{-2}\,]$$

$$v'_i(\phi) = 0.714[\,\phi^{-1} + 3.6\phi^{-2} \qquad \phi^{-1} + 0.2\phi^{-2}\,]$$

Clearly, this transformation has the same direction as the one found in Example 6.7 with input-output implementation. ▩

Technically, the full column rank of the matrix $[J \quad L^{\circ}]$ is a precondition of the solution (6.69) and (6.72). However, solutions can be obtained even in case of rank defects, if the linearly related columns are accompanied by zero response specifications. In such cases, the solution is sought after eliminating the redundant columns and specifications.

6.5. THE FAULT SYSTEM MATRIX

In this section, the properties of the fault/disturbance system matrix will be explored. These will provide valuable insight which will then be utilized in a re-examination of transfer function based implementation of residual generators.

First recall a few fundamental relationships from Chapter 2.

$$M(\phi) = H^{-1}(\phi)G(\phi) = \bar{G}(\phi)/\bar{h}(\phi) \tag{6.73}$$

$$S(\phi) = H^{-1}(\phi)N(\phi) = \bar{N}(\phi)/\bar{h}(\phi) \tag{6.74}$$

where

$H(\phi) = Diag[h_1(\phi) \ldots h_\mu(\phi)]$ and $h_i(\phi)$, $g_{i.}(\phi)$ and $n_{i.}(\phi)$ (the rows of $G(\phi)$ and $N(\phi)$), $i = 1 \ldots \mu$, are row-wise relative prime;

$\bar{h}(\phi)$ is the least common multiple of $h_1(\phi) \ldots h_\mu(\phi)$;

$S(\phi) = [S_F(\phi) \quad S_D(\phi)]$ and $N(\phi) = [N_F(\phi) \quad N_D(\phi)]$.

Further, with $E=[E_F \;\; E_D]$ and $F=[F_F \;\; F_D]$,

$$M^+(\phi) = C(\phi I - A)^{-1}B + D \tag{6.75}$$

$$S^+(\phi) = C(\phi I - A)^{-1}E + F \tag{6.76}$$

and

$$\vartheta^+(\phi)\bar{h}^+(\phi) = Det(\phi I - A) \tag{6.77}$$

$$\vartheta^+(\phi)\bar{G}^+(\phi) = C\,Adj(\phi I - A)\,B + Det(\phi I - A)\,D \tag{6.78}$$

$$\vartheta^+(\phi)\bar{N}^+(\phi) = C\,Adj(\phi I - A)\,E + Det(\phi I - A)\,F \tag{6.79}$$

where $\vartheta^+(\phi)$ contains, with the proper multiplicities, the poles of $\bar{h}^+(\phi)$ which are repeated in the realization.

The system matrix for the combined fault-disturbance system (to be referred to as *the fault system matrix*) is

$$\Gamma^+(\phi) = \begin{bmatrix} \phi I - A & -E \\ C & F \end{bmatrix} \begin{matrix} \nu \\ \mu \end{matrix} \qquad \begin{matrix} \nu & \rho \end{matrix} \tag{6.80}$$

Square systems. Recall that with full specification $\rho=\mu$ thus the system matrix is square. With almost full specifications, $\rho=\mu-1$; such problems will be approached by converting them into square systems, either by adding a column or by separating a row. With square systems, we may utilize the ordinary inverse of the fault system matrix which greatly simplifies the implementation procedure. Therefore, we will focus our attention on *square fault systems*, in which the fault system matrix is $(\nu+\rho)\cdot(\nu+\rho)$ and $\rho\leq\mu$.

The rank of the fault system matrix. As it was shown in Chapter 2 (c.f. Eq. (2.183)), the normal rank of the fault system matrix is related to the normal rank of the fault transfer function as

$$Rank\,\Gamma^+(\phi) = Rank\,\bar{N}^+(\phi) + \nu \qquad for\ all\ \phi \neq \lambda_j \tag{6.81}$$

where ν is the system order (the order of the minimum realization) and λ_j, $j=1\ldots\nu$, are the system poles.

Obviously, the fault system matrix has a rank defect if the last ρ columns (that is, the columns of the $[-\,\boldsymbol{E}' \quad \boldsymbol{F}']'$ matrix) are linearly dependent. Though it is less obvious, rank defect may also arise from (polynomial) linear relations among the columns of the full fault system matrix.

The inverse of the fault system matrix. Let us write the inverse of the fault system matrix for a square system as

$$[\boldsymbol{\Gamma}^{+}(\phi)]^{-1} = \frac{\boldsymbol{\Omega}^{+}(\phi)}{\pi^{+}(\phi)} = \frac{1}{\pi^{+}(\phi)} \begin{bmatrix} \boldsymbol{\Omega}_{A}^{+}(\phi) & \boldsymbol{\Omega}_{E}^{+}(\phi) \\ \boldsymbol{\Omega}_{C}^{+}(\phi) & \boldsymbol{\Omega}_{F}^{+}(\phi) \end{bmatrix} \begin{matrix} \nu \\ \rho \end{matrix} \qquad \begin{matrix} \nu & \rho \end{matrix} \tag{6.82}$$

where

$$\boldsymbol{\Omega}^{+}(\phi) = Adj\,\boldsymbol{\Gamma}^{+}(\phi) \qquad \pi^{+}(\phi) = Det\,\boldsymbol{\Gamma}^{+}(\phi) \tag{6.83}$$

It can be verified by cross-multiplication that the following relationships hold [Kailath, 1980]:

$$\frac{\boldsymbol{\Omega}_{A}^{+}(\phi)}{\pi^{+}(\phi)} = (\phi\boldsymbol{I} - \boldsymbol{A})^{-1} - (\phi\boldsymbol{I} - \boldsymbol{A})^{-1}\boldsymbol{E}[\boldsymbol{C}(\phi\boldsymbol{I} - \boldsymbol{A})^{-1}\boldsymbol{E}+\boldsymbol{F}]^{-1}\boldsymbol{C}(\phi\boldsymbol{I} - \boldsymbol{A})^{-1} \tag{6.84}$$

$$\frac{\boldsymbol{\Omega}_{E}^{+}(\phi)}{\pi^{+}(\phi)} = (\phi\boldsymbol{I} - \boldsymbol{A})^{-1}\boldsymbol{E}[\boldsymbol{C}(\phi\boldsymbol{I} - \boldsymbol{A})^{-1}\boldsymbol{E}+\boldsymbol{F}]^{-1} \tag{6.85}$$

$$\frac{\boldsymbol{\Omega}_{C}^{+}(\phi)}{\pi^{+}(\phi)} = -\,[\boldsymbol{C}(\phi\boldsymbol{I} - \boldsymbol{A})^{-1}\boldsymbol{E}+\boldsymbol{F}]^{-1}\boldsymbol{C}(\phi\boldsymbol{I} - \boldsymbol{A})^{-1} \tag{6.86}$$

$$\frac{\boldsymbol{\Omega}_{F}^{+}(\phi)}{\pi^{+}(\phi)} = [\boldsymbol{C}(\phi\boldsymbol{I} - \boldsymbol{A})^{-1}\boldsymbol{E}+\boldsymbol{F}]^{-1} \tag{6.87}$$

Invariant zeroes. The roots of the polynomial $\pi^{+}(\phi)$ are the invariant zeroes of the fault system matrix. Denoting them as ζ_1, ζ_2, ..., the denominator polynomial (also called the invariant zero polynomial) can be written as

$$\pi^{+}(\phi) = \alpha(\phi - \zeta_1)(\phi - \zeta_2) \cdots \tag{6.88}$$

where α is a constant. These invariant zeroes will play a crucial role in the systematic implementation of residual generators.

Now define the j-th single-fault subsystem as

$$\Gamma_j^+(\phi) = \begin{bmatrix} \phi I - A & -e_{\cdot j} \\ C & f_{\cdot j} \end{bmatrix} \begin{matrix} \nu \\ \rho \end{matrix} \qquad (6.89)$$
$$\qquad\qquad \nu \qquad\quad 1$$

The value ζ_k is an invariant zero of this single-fault subsystem if it makes all $(\nu+1)\cdot(\nu+1)$ subdeterminants of $\Gamma_j^+(\phi)$ zero. Such subsystem zero(es) may or may not exist, but if they do they are also the invariant zeroes of the full fault system $\Gamma^+(\phi)$. This is because it is always possible to expand $Det\Gamma^+(\phi)$ along the $(\nu+1)\cdot(\nu+1)$ subdeterminants of $\Gamma_j^+(\phi)$. Similarly, if we build up the fault system matrix gradually, starting with a single fault/disturbance column and then adding more such columns then the set of invariant zeroes of any expanded (sub)system will contain all the invariant zeroes of the embedded subsystems.

Observe that the elements of $\Omega_C^+(\phi)$ and $\Omega_F^+(\phi)$ are determinants belonging to elements of $-E$ and F in the $\Gamma^+(\phi)$ matrix. Therefore it follows that if ζ_k is an invariant zero of the j-th single fault subsystem then it is a root of all rows of $\Omega_C^+(\phi)$ and $\Omega_F^+(\phi)$, except the j-th row. Similarly, if ζ_k is a zero of the $j1$, $j2$ two-fault subsystem then it will be a root of all rows of $\Omega_C^+(\phi)$ and $\Omega_F^+(\phi)$, except the $j1$-th and $j2$-th rows - and so on.

The polynomial degrees of the inverse fault system matrix. Next we will investigate the degree of the polynomial $\pi^+(\phi)$ and of the polynomials appearing in the various sub-matrices of $\Omega^+(\phi)$. While $\pi^+(\phi)$ is the $(\nu+\rho)\cdot(\nu+\rho)$ determinant of the matrix $\Gamma^+(\phi)$, the elements of $\Omega^+(\phi)$ are $(\nu+\rho-1)\cdot(\nu+\rho-1)$ subdeterminants of the same. The polynomial degrees will be established by the analysis of these determinants. In general, the maximum polynomial degree of any of the concerned determinants is the size of the largest diagonal minor of $\phi I-A$ they contain because $\phi I-A$ is the only submatrix of $\Gamma^+(\phi)$ which contributes any powers of the shift operator ϕ. As it will be seen, the polynomial degrees depend on the system order ν and on the number ρ_I of strictly input faults/disturbances. Note that by the degree of a matrix (or row/column thereof) we mean the highest element degree.

1. The degree of $\pi^+(\phi)$. The maximum degree of this determinant is ν. Expand the determinant by a column $[-e'_{\cdot j}\ \ f'_{\cdot j}]'$. Any subdeterminant belonging to an element of $e_{\cdot j}$ loses a row of $\phi I-A$, and thus a degree, while the subdeterminants belonging to an element of $f_{\cdot j}$ do not. Each strictly input fault/disturbance is accompanied by a column $f_{.j}=0$, so only the *e*-related subdeterminants remain in the expansion, resulting in a one degree loss. Thus

$$Deg\ \pi^+(\phi) \leq \nu - \rho_I \tag{6.90}$$

2. *The degree of* $\Omega_F^+(\phi)$. The elements of $\Omega_F^+(\phi)$ are subdeterminants of $\Gamma^+(\phi)$ obtained by deleting the *j*-th column of the $[-E' \quad F']'$ submatrix and the *l*-th row of the $[C \quad F]$ submatrix of the fault system matrix. Since this does not cut into the ϕI-A submatrix, the maximum possible degree of the determinant is ν. The degree loss caused by $f_{.j}=0$ columns can be traced the same way as in Point 1 above. There are two sub-cases, though:

a. In the eliminated *j*-th column, $f_{.j}\neq 0$. Then all the ρ_I zero columns contribute to the degree loss, that is, in the *j*-th row of $\Omega_F^+(\phi)$,

$$Deg\ \omega_{Fj\cdot}^+(\phi) \leq \nu - \rho_I \qquad if\ f_{.j}\neq 0 \tag{6.91a}$$

b. In the eliminated *j*-th column, $f_{.j}=0$. Then only the remaining ρ_I-1 zero columns contribute to the degree loss, that is,

$$Deg\ \omega_{Fj\cdot}^+(\phi) \leq \nu - \rho_I + 1 \qquad if\ f_{.j}=0 \tag{6.91b}$$

Thus for the entire $\Omega_F^+(\phi)$ matrix, provided $\rho_I\neq 0$:

$$Deg\ \Omega_F^+(\phi) \leq \nu - \rho_I + 1 \qquad for\ \rho_I\neq 0 \tag{6.92}$$

For $\rho_I=0$, only case a. applies thus

$$Deg\ \Omega_F^+(\phi) \leq \nu \qquad for\ \rho_I=0 \tag{6.93}$$

3. *The degree of* $\Omega_C^+(\phi)$. Now a row of the $[\phi I$-$A \quad -E]$ submatrix is eliminated, together with the *j*-th column of $[-E' \quad F']$. This reduces the maximum possible degree to ν-1. Otherwise, the logic of Point 2 above applies, leading to:

$$Deg\ \omega_{Cj\cdot}^+(\phi) \leq \nu - \rho_I - 1 \qquad if\ f_{.j}\neq 0 \tag{6.94a}$$

$$Deg\ \omega_{Cj\cdot}^+(\phi) \leq \nu - \rho_I \qquad if\ f_{.j}=0 \tag{6.94b}$$

$$Deg\ \Omega_C^+(\phi) \leq \nu - \rho_I \qquad for\ \rho_I\neq 0 \tag{6.95a}$$

$$Deg\ \Omega_C^+(\phi) \leq \nu - 1 \qquad for\ \rho_I=0 \tag{6.95b}$$

4. *The degree of* $\Omega_E^+(\phi)$. The initial elimination involves a column of the submatrix $[(\phi I\text{-}A)' \quad C']'$ and a row of $[C \quad F]$. Since $\phi I\text{-}A$ is affected, the maximum possible degree is $\nu\text{-}1$. Expand the determinant by a column of $[-E' \quad F']'$. Then the subdeterminants belonging to an element in the $f_{\cdot j}$ column will not suffer any degree loss, but neither will the subdeterminant accompanying *one* of the elements in the $e_{\cdot j}$ column. If $\rho_I > 0$, choose a column with $f_{\cdot j} = 0$ to start the expansion. Then the presence of the zero column will not cause a degree reduction in this step. From here on, the logic of Point 1 applies, that is, all but the first one of the $f_{\cdot j} = 0$ columns result in degree reduction. Thus

$$Deg\ \Omega_E^+(\phi) \le \nu - \rho_I \qquad for\ \rho_I > 0 \tag{6.96}$$

$$Deg\ \Omega_E^+(\phi) \le \nu - 1 \qquad for\ \rho_I = 0 \tag{6.97}$$

5. *The degree of* $\Omega_A^+(\phi)$. The initial elimination now involves a column of the submatrix $[(\phi I\text{-}A)' \quad C']'$ and a row of $[(\phi I\text{-}A) \quad -E]$. The maximum possible degree is limited to $\nu\text{-}1$ if the concerned element is on the main diagonal and to $\nu\text{-}2$ if it is off the main diagonal. In the first case, any $f_{\cdot j} = 0$ column then results in one degree reduction, up to a maximal reduction of $\nu\text{-}1$. In the second case, the first such column has no effect while the rest reduce the degree by one each, up to a maximal reduction of $\nu\text{-}2$. Thus

$$Deg\ \Omega_A^+(\phi) \le \nu - \rho_I - 1 \qquad for\ \rho_I < \nu \tag{6.98}$$

$$Deg\ \Omega_A^+(\phi) = 0 \qquad for\ \rho_I = \nu \tag{6.99}$$

and $Deg\,\omega_{Aij}^+(\phi) = \nu\text{-}2$ if $\rho_I = 0$ *and* $i \neq j$.

Let us summarize the degree properties of $\Omega^+(\phi)$ below:

$$\begin{bmatrix} Deg\ \Omega_A^+(\phi) & Deg\ \Omega_E^+(\phi) \\ Deg\ \Omega_C^+(\phi) & Deg\ \Omega_F^+(\phi) \end{bmatrix} \le \begin{bmatrix} \nu - \rho_I - 1^{\#} & \nu - \rho_I \\ \nu - \rho_I^{*} & \nu - \rho_I + 1^{*} \end{bmatrix} \qquad \rho_I > 0 \tag{6.100}$$

$^{\#}0$ *for* $\rho_I = \nu$

* *deduct 1 for rows arising from columns of* $\Gamma^+(\phi)$ *with* $f_{\cdot j} \neq 0$

$$\begin{bmatrix} Deg\ \Omega_A^+(\phi) & Deg\ \Omega_E^+(\phi) \\ Deg\ \Omega_C^+(\phi) & Deg\ \Omega_F^+(\phi) \end{bmatrix} \leq \begin{bmatrix} \nu-1 & \nu-1 \\ \nu-1 & \nu \end{bmatrix} \qquad \rho_I=0 \tag{6.101}$$

Example 6.10.

Consider the system studied in the previous examples. As seen in Example 6.3, the system parameters are $\nu=2$, $\mu=2$, $\rho=2$, $\rho_I=1$. The fault system matrix can be constructed as:

$$\Gamma^+(\phi) = \left[\begin{array}{cc|cc} \phi-0.8 & 0 & -1 & -2 \\ -1 & \phi-0.5 & 0 & -3 \\ \hline 1 & 0 & 1 & 0 \\ 0 & 1 & 1 & 0 \end{array}\right]$$

The predicted degrees, based on (6.90) and (6.100), are $Deg\ \pi^+(\phi)=1$ and

$$Deg\ \Omega^+(\phi) = \left[\begin{array}{c|c} 0 & 1 \\ \hline 1 & 2 \end{array}\right]$$

with $1-1=0$ and $2-1=1$ applying to the third row. The inverse is computed as

$$\pi^+(\phi) = \phi+3.6$$

$$\Omega^+(\phi) = \left[\begin{array}{cc|cc} 3 & -2 & -2\phi+4 & 2\phi-1 \\ 3 & -2 & -3\phi+0.4 & 3\phi+2.6 \\ \hline -3 & 2 & 3\phi-0.4 & -2\phi+1 \\ \phi-1.5 & -\phi-0.2 & -\phi^2+1.3\phi-1.4 & \phi^2-0.3\phi-0.1 \end{array}\right]$$

The degrees, of course, are as expected. ▩

The rank of the adjoint matrix at the invariant zeroes. It follows from a result in the theory of determinants [Gantmakher, 1959, p. 21] that any *2x2* determinant formed of the elements of the adjoint matrix $\mathbf{\Omega}^+(\phi)$ can be expressed as

$$\omega^+_{ij}(\phi)\omega^+_{lm}(\phi) - \omega^+_{im}(\phi)\omega^+_{lj}(\phi) = \pm\, \theta^+_{jm,il}(\phi)\pi^+(\phi) \tag{6.102}$$

where $\omega^+_{ij}(\phi)$, $\omega^+_{lm}(\phi)$, $\omega^+_{im}(\phi)$, $\omega^+_{lj}(\phi)$ is any foursome of the elements of $\mathbf{\Omega}^+(\phi)$ and $\theta^+_{jm,il}(\phi)$ is the $(\nu+\rho\text{-}2)\cdot(\nu+\rho\text{-}2)$ subdeterminant of $\mathbf{\Gamma}^+(\phi)$ with the j-th and m-th rows and i-th and l-th columns missing. The symbol $\pm$ signifies that the sign may be + or - , depending on the position of the elements. Eq. (6.102) reveals the important fact that $\pi^+(\phi)$ can be factored out from any of the 2·2 determinants. This implies that at $\phi=\zeta_k$, where ζ_k is any root of $\pi^+(\phi)$,

$$Rank\,\mathbf{\Omega}^+(\zeta_k) = 1 \tag{6.103}$$

In other words, all columns (or rows) of $\mathbf{\Omega}^+(\phi)$, or of any of its submatrices, are linearly dependent on a single column (or row) at $\phi=\zeta_k$.

Example 6.11.

In the previous example, the only invariant zero is $\zeta_1=-3.6$. Substituting this into $\mathbf{\Omega}^+(\phi)$ yields

$$\mathbf{\Omega}^+(-3.6) = \begin{bmatrix} 3 & -2 & 11.2 & -8.2 \\ 3 & -2 & 11.2 & -8.2 \\ -3 & 2 & -11.2 & 8.2 \\ -5.1 & 3.4 & 19.04 & 13.94 \end{bmatrix}$$

The complete linear dependence is obvious. 

6.6. SYSTEMATIC IMPLEMENTATION WITH TRANSFER FUNCTIONS

In this section, the transfer function based implementation techniques of Section 6.3 will be revisited. Some of the results of that section will be reinterpreted and the naive implementation methods replaced with systematic ones, utilizing the system properties explored in Section 6.5.

The results of Section 6.4, concerning state-space implementation, will also be recalled for reference. As it will be seen, the Chow-Willsky implementation can be exactly reproduced by systematic transfer function implementation. Further, the insight gained in the previous section will lead to implementations even in such situations which are outside the obvious scope of the Chow-Willsky technique (see summary in Subsection 6.6.5).

Note that while the properties utilized here have been obtained from the fault system matrix, which relies on the state-space description of the system, the knowledge of a state-space representation in general is not necessary for systematic implementation. When dealing with simple systems, one may apply a systematic procedure entirely in the framework of transfer functions. Such procedure, though, utilizes some ***structural*** system properties and requires the knowledge of the ***true*** order of the system (that is, the order of the minimum state-space realization).

6.6.1. ARMA Implementation Revisited

Here we are re-considering the results of Subsection 6.3.1.

Response modifier. As we have seen in Subsection 6.3.1, it may be necessary to revise the original response specification, in order to make the residual generator causal and stable (or polynomial). This response modification, which has been handled in an informal way, will be formalized here. Define the response modifier matrix for the i-th residual as

$$\mathbf{T}_i(\phi) = \mathbf{T}_{Pi}(\phi)\mathbf{T}_{Ri}(\phi) \tag{6.104}$$

where

$$\mathbf{T}_{Pi}(\phi) = Diag[\upsilon_{Pi1}(\phi) \ldots \upsilon_{Pip}(\phi)]$$

$$\mathbf{T}_{Ri}(\phi) = Diag[\upsilon_{Ri1}(\phi) \ldots \upsilon_{Rip}(\phi)] \tag{6.105}$$

and where the $\mathbf{T}_R(\phi)$ matrix will be so chosen that the generator is causal (realizable) while $\mathbf{T}_P(\phi)$ will be selected so that it is stable (or polynomial if so desired). With the modifier matrix, the implementation equation for a single residual, (6.40b), becomes

$$\boxed{\mathbf{w}_i'(\phi)\mathbf{S}(\phi) = \mathbf{z}_i'(\phi)\mathbf{T}_i(\phi)} \tag{6.106}$$

The implementation will be discussed first by assuming *full non-homogeneous specification*; implementation under other typical specifications will be deferred to Subsections 6.6.2 and 6.6.3.

(6.106) gives the transformation as

$$w_i'(\phi) = z_i'(\phi)\mathbf{T}_i(\phi)S^{-1}(\phi) \tag{6.107}$$

Now by (6.76) and (6.87)

$$S^{-1}(\phi) = [S^+(\phi)]^{-1} = [C(\phi I - A)^{-1}E+F]^{-1} = \Omega_F^+(\phi)/\pi^+(\phi) \tag{6.108}$$

Thus

$$\boxed{w_i'(\phi) = z_i'(\phi)\,\mathbf{T}_i(\phi)\,\Omega_F^+(\phi)/\pi^+(\phi)} \tag{6.109}$$

It will be useful to also see the generic form of the residual generator

$$r_i(\phi) = w_i'(\phi)y(t) + v_i'(\phi)u(t) \tag{6.110}$$

where $v_i'(\phi)=-w_i'(\phi)M(\phi)$. By (6.75), (6.76), (6.86) and (6.87)

$$S^{-1}(\phi)M(\phi) = [S^+(\phi)]^{-1}M^+(\phi) = [C(\phi I - A)^{-1}E+F]^{-1}\,[C(\phi I - A)^{-1}B+D]$$

$$= [\,-\Omega_C^+(\phi)B + \Omega_F^+(\phi)D]/\pi^+(\phi) \tag{6.111}$$

Thus

$$\boxed{v_i'(\phi) = -\,w_i'(\phi)M(\phi) = z_i'(\phi)\,\mathbf{T}_i(\phi)\,[\Omega_C^+(\phi)B - \Omega_F^+(\phi)D]/\pi^+(\phi)} \tag{6.112}$$

The realizability of the residual generator. As it is seen from (6.90) and (6.100), the degree of $\Omega_F^+(\phi)$ may exceed the degree of $\pi^+(\phi)$. If this is the case then $\Omega_F^+(\phi)/\pi^+(\phi)$ is not realizable. Eq. (6.91) also reveals that, normally, this situation arises only in those rows $\omega_{Fj.}(\phi)$ which belong to strictly input faults. We will select the elements of the $\mathbf{T}_{Ri}(\phi)$ matrix so that the generator is guaranteed to be causal, with minimal complexity.

Define

$$\delta_o = Deg\,\pi^+(\phi) \leq \nu - \rho_I \tag{6.113a}$$

$$\delta_j = Deg\,[\omega_{Cj.}^+(\phi) \quad \omega_{Fj.}^+(\phi)] \leq \begin{cases} \nu - \rho_I & \text{if } f_{.j} \neq 0 \\ & j=1 \dots \rho \\ \nu - \rho_I + 1 & \text{if } f_{.j} = 0 \end{cases} \tag{6.113b}$$

and note that $\delta_j \geq \delta_o$, $j=1 \ldots p$. Further, define

$$\pi(\phi) = \phi^{-\delta o} \pi^+(\phi) \tag{6.114a}$$

$$[\boldsymbol{\Omega}_C(\phi) \quad \boldsymbol{\Omega}_F(\phi)] = Diag\,[\phi^{-\delta 1} \ldots \phi^{-\delta p}]\,[\boldsymbol{\Omega}_C^+(\phi) \quad \boldsymbol{\Omega}_F^+(\phi)] \tag{6.114b}$$

Now choose the elements of $\boldsymbol{\Upsilon}_{Ri}(\phi)$ as

$$\boxed{\upsilon_{Rij}(\phi) = \phi^{\delta 0 - \delta j}} \tag{6.115}$$

Then

$$\boldsymbol{\Upsilon}_{Ri}(\phi)\,[\boldsymbol{\Omega}_C^+(\phi) \quad \boldsymbol{\Omega}_F^+(\phi)] \,/\, \pi^+(\phi) = [\boldsymbol{\Omega}_C(\phi) \quad \boldsymbol{\Omega}_F(\phi)] \,/\, \pi(\phi) \tag{6.116}$$

which is clearly causal, with maximal degrees given in (6.113). Notice that if $\delta_j > \delta_o$ then the response to the j-th fault is delayed; this is natural if the fault is strictly input.

With (6.116) and (6.104), the implementation equations (6.109) and (6.112) become

$$w_i'(\phi) = z_i'(\phi)\,\boldsymbol{\Upsilon}_{Pi}(\phi)\,\boldsymbol{\Omega}_F(\phi) \,/\, \pi(\phi) \tag{6.117}$$

$$v_i'(\phi) = z_i'(\phi)\,\boldsymbol{\Upsilon}_{Pi}(\phi)\,[\boldsymbol{\Omega}_C(\phi)\boldsymbol{B} - \boldsymbol{\Omega}_F(\phi)\boldsymbol{D}] \,/\, \pi(\phi) \tag{6.118}$$

Example 6.12.

Let us continue Example 6.10. There $\nu=2$ and $\rho_1=1$ and, as seen from the results there,

$$\delta_o = 1 \qquad \delta_1 = 1 \qquad \delta_2 = 2$$

Then

$$\boldsymbol{\Upsilon}_R(\phi) = Diag[1 \quad \phi^{-1}]$$

and $\pi(\phi)$, $\boldsymbol{\Omega}_F(\phi)$ and $\boldsymbol{\Omega}_C(\phi)$ are obtained as

$$\pi(\phi) = 1 + 3.6\phi^{-1}$$

$$\Omega_F(\phi) = \begin{bmatrix} 3(1 - 0.133\phi^{-1}) & -2(1 - 0.5\phi^{-1}) \\ -(1 - 1.3\phi^{-1} + 1.4\phi^{-2}) & (1 - 0.3\phi^{-1} - 0.1\phi^{-2}) \end{bmatrix}$$

$$\Omega_C(\phi) = \begin{bmatrix} -3\phi^{-1} & 2\phi^{-1} \\ \phi^{-1}(1 - 1.5\phi^{-1}) & -\phi^{-1}(1 + 0.2\phi^{-1}) \end{bmatrix}$$

This is clearly realizable. ▩

The stability of the residual generator. Eqs. (6.117) and (6.118) reveal that the stability of the residual generator depends only on the invariant zero polynomial $\pi(\phi)$. This implies the important observation that the stability of the plant transfer function $M(\phi)$ is irrelevant for the stability of the residual generator.

If any of the invariant zeroes ζ_k is on or outside the unit circle, that is, if $|\zeta_k| \geq 1$ for any k, then $\Omega_C(\phi)/\pi(\phi)$ and $\Omega_F(\phi)/\pi(\phi)$ are unstable. The generator will be stabilized by choosing $\mathbf{T}_{Pi}(\phi)$ so that the unstable invariant zeroes are canceled. Factor the invariant zero polynomial as

$$\pi(\phi) = \pi_U(\phi) \cdot \pi_S(\phi) \tag{6.119a}$$

where

$$\pi_U(\phi) = (1 - \zeta_1 \phi^{-1}) \ldots (1 - \zeta_\tau \phi^{-1}) \tag{6.119b}$$

contains all the unstable invariant zeroes. Further, denote by $\pi_{Uj}(\phi)$ the factors of $\pi_U(\phi)$ which are present in $[\boldsymbol{\omega}_{Cj.}(\phi) \quad \boldsymbol{\omega}_{Fj.}(\phi)]$ due to subsystem zeroes; these result in partial cancellation with $\pi_U(\phi)$. Then $\mathbf{T}_{Pi}(\phi)$ needs to be so chosen that it leads to complete cancellation of $\pi_U(\phi)$ from $\Omega_F(\phi)/\pi(\phi)$ and $\Omega_C(\phi)/\pi(\phi)$. Two approaches will be considered:

Algorithm I. Choose

$$\boxed{\upsilon_{Pij}(\phi) = \pi_U(\phi) \,/\, \pi_{Uj}(\phi) \qquad j = 1 \ldots \rho} \tag{6.120}$$

Then

$$\mathbf{T}_{Pi}(\phi)\,[\Omega_C(\phi) \quad \Omega_F(\phi)] \,/\, \pi(\phi) = [\ \ldots\ ] \,/\, \pi_S(\phi) \tag{6.121}$$

where $[\ \ldots\ ]$ is a polynomial matrix. Clearly, $\mathbf{w}_i'(\phi)$ and $\mathbf{v}_i'(\phi)$ are stable. Further, with (6.113) and (6.119b), and assuming that $\mathbf{z}_i'(\phi)$ is of minimum complexity (has no dynamics), the complexity of the generator is

$$Deg\ [num\ w_i'(\phi)] = Deg\ [num\ v_i'(\phi)] \leq \begin{bmatrix} \nu - \rho_I + 1 & if\ \ \rho_I > 0 \\ \nu & if\ \ \rho_I = 0 \end{bmatrix} \tag{6.122a}$$

$$Deg\ [denom\ w_i'(\phi)] = Deg\ [denom\ v_i'(\phi)] \leq \nu - \rho_I - \tau \tag{6.122b}$$

Algorithm II. Modify only one element of the specification, for example as

$$\mathbf{T}_{Pi}(\phi) = Diag[1 \ldots 1 \quad \upsilon_{Pig}(\phi) \quad 1 \ldots 1] \tag{6.123}$$

and choose the $\upsilon_{Pig}(\phi)$ polynomial so that

$$\boxed{\mathbf{z}_i'(\phi)\ \mathbf{T}_{Pi}(\phi)\ \boldsymbol{\omega}_{F\cdot l}(\phi) = \beta_{il}(\phi)\pi_U(\phi)} \tag{6.124}$$

where $\boldsymbol{\omega}_{F\cdot l}(\phi)$ is an *almost arbitrary* column of $\boldsymbol{\Omega}_F(\phi)$ and $\beta_{il}(\phi)$ is a polynomial. If $\pi_U(\phi)$ has only simple roots ζ_k, $k=1\ldots\tau$, then this can be solved by requiring

$$\boxed{\mathbf{z}_i'(\zeta_k)\ \mathbf{T}_{Pi}(\zeta_k)\ \boldsymbol{\omega}_{F\cdot l}(\zeta_k) = 0 \qquad k=1\ldots\tau} \tag{6.125}$$

A detailed treatment of the algorithm will be given in Subsection 6.6.2. As it will be shown there, it follows from the rank property (6.103) that *satisfying (6.125) for a suitable column of* $\boldsymbol{\Omega}_F(\phi)$ *implies that it is satisfied for all the other columns, and also for all columns of* $\boldsymbol{\Omega}_C(\phi)$. Thus the unstable zeroes are effectively canceled out.

The polynomial $\upsilon_{Pig}(\phi)$ has to satisfy τ conditions in (6.125), thus it needs to have (at least) τ coefficients. Minimum complexity is achieved with $Deg\ \upsilon_{Pig}(\phi)=\tau-1$. If $\mathbf{z}_i'(\phi)$ has no dynamics then (6.122b) applies to the denominator of the computational algorithm while (6.122a) changes to

$$Deg\ [num\ w_i'(\phi)] = Deg\ [num\ v_i'(\phi)] \leq \begin{bmatrix} \nu - \rho_I & if\ \ \rho_I > 0 \\ \nu - 1 & if\ \ \rho_I = 0 \end{bmatrix} \tag{6.126}$$

Example 6.13.

$[\boldsymbol{\Omega}_C(\phi) \quad \boldsymbol{\Omega}_F(\phi)]/\pi(\phi)$ as obtained in Example 6.12 is clearly unstable and needs to be stabilized.

Algorithm I. Choose

$$z_i'(\phi) = [\,0.5 \qquad 1\,]$$

and make

$$\boldsymbol{\Upsilon}_{Pi}(\phi) = Diag\,[(1+3.6\phi^{-1}) \qquad (1+3.6\phi^{-1})\,]$$

Then

$$w_i'(\phi) = [\,0.5+1.1\phi^{-1} - 1.4\phi^{-2} \qquad 0.2\phi^{-1} - 0.1\phi^{-2}\,]$$

$$v_i'(\phi) = [\,-\phi^{-1} - 3.6\phi^{-2} \qquad -0.2\phi^{-2}\,]$$

which is, of course, stable.

Algorithm II. Choose

$$z_i'(\phi) = [\,0.5 \qquad 1\,] \quad \text{and} \quad \boldsymbol{\Upsilon}_{Pi}(\phi) = Diag\,[1 \qquad \alpha]$$

and seek α so that $(1+3.6\phi^{-1})$ is canceled out. Use for example the first column of $\boldsymbol{\Omega}_F(\phi)$ for the calculation, so that

$$[0.5 \quad \alpha][3(1-0.133\phi^{-1}) \qquad -(1-1.3\phi^{-1}+1.4\phi^{-2})]'\,\Big|_{\phi=-3.6} = 0$$

This is

$$[0.5 \quad -0.2778\alpha][3.111 \quad 5.289]' = 0$$

which yields $\alpha = 1.059$. With this, we obtain

$$z_i'(\phi)\;\boldsymbol{\Upsilon}_{Pi}(\phi)\;\boldsymbol{\Omega}_F(\phi)$$
$$= [0.441(1+3.6\phi^{-1})(1-0.933\phi^{-1}) \quad 0.059(1+3.6\phi^{-1})(1-0.5\phi^{-1})]$$

$$z_i'(\phi)\,\boldsymbol{\Upsilon}_{Pi}(\phi)\,\boldsymbol{\Omega}_C(\phi) = [\,-0.441\phi^{-1}(1+3.6\phi^{-1}) \qquad -0.059\phi^{-1}(1+3.6\phi^{-1})]$$

so the cancellation takes place as expected. Then the computational form is

$$w_i'(\phi) = [0.441(1-0.933\phi^{-1}) \qquad 0.059(1-0.5\phi^{-1})]$$

$$v_i'(\phi) = [\,-1.059\phi^{-1} \qquad -0.059\phi^{-1}\,]$$

This result is the same as the one obtained in Example 6.8, using the Chow-Willsky scheme. Note that here we obtained a moving average residual generator because the only invariant zero was unstable and has been canceled out. In general, canceling just the unstable zeroes does not lead to a MA generator. ❖

Whether Algorithm I or II is employed, it is important that the cancellations be performed *analytically*. That is, the analytically simplified transfer functions are to be applied to the variables $y(t)$ and $u(t)$. Alternatively, by performing the cancellation numerically (first applying $\mathbf{\Omega}_F(\phi)/\pi(\phi)$, etc. to the variables and then multiplying with $\mathbf{z}_i'(\phi)\mathbf{T}_{Pi}(\phi)$), the resulting numerical inaccuracies may cause the residual to explode.

6.6.2. Polynomial Implementation from the ARMA Representation: Full Specifications

Here we will show how polynomial (MA) residual generators can be implemented starting from the ARMA system representation. As we pointed out earlier, transformations applied to the ARMA representation lead to the generic form of residual generation. The generation of MA residuals in this ARMA (generic) framework was not addressed in Section 6.3 where we were dealing with various "naive" implementation ideas. We will discuss full specifications here; almost full specifications will be addressed in the next subsection.

Full non-homogeneous specification. The polynomial (MA) implementation of full non-homogeneous specifications is obtained easily as an extension of the stabilization procedures introduced above. For polynomial implementation, the specification has to be moving average. Further, Algorithms I and II need to be modified so that the entire denominator is canceled out.

Algorithm I. Choose

$$\boxed{\upsilon_{Pij}(\phi) = \pi(\phi) \,/\, \pi_j(\phi) \qquad j = 1 \ldots p} \tag{6.127}$$

where $\pi_j(\phi)$ is the factor of $\pi(\phi)$ contained in the j-th row of $[\mathbf{\Omega}_C(\phi) \quad \mathbf{\Omega}_F(\phi)]$. Then

$$\mathbf{T}_{Pi}(\phi)\,[\mathbf{\Omega}_C(\phi) \quad \mathbf{\Omega}_F(\phi)] \,/\, \pi(\phi) = [\ \ldots\ ] \tag{6.128}$$

where $[\ \ldots\ ]$ is a polynomial matrix. Clearly, $\mathbf{w}_i'(\phi)$ and $\mathbf{v}_i'(\phi)$ are polynomial. If $\mathbf{z}_i'(\phi)$ has no dynamics the computational window length σ is the numerator degree shown in (6.122a). On the other hand, the polynomial degree of the fault response is the denominator degree shown in (6.122b), plus the pure delay, if any, needed for realizability.

Algorithm II. We will cancel the entire denominator $\pi(\phi)$ by the appropriate choice of one element $\upsilon_{Pig}(\phi)$ of the response modifier (6.123). This element will be determined so that the equation

$$z_i'(\phi)\,\mathbf{T}_{Pi}(\phi)\,\boldsymbol{\omega}_{F\cdot l}(\phi) = \beta_{il}(\phi)\pi(\phi) \tag{6.129}$$

for the (almost) arbitrary l-th column of $\boldsymbol{\Omega}_F(\phi)$ is satisfied, where $\beta_{il}(\phi)$ is an unspecified polynomial. It will be shown then that by this choice (6.129) is satisfied for all other columns of $\boldsymbol{\Omega}_F(\phi)$ and $\boldsymbol{\Omega}_C(\phi)$ as well.

Write (6.129) as

$$\varphi_{il}(\phi) + z_{ig}(\phi)\,\upsilon_{Pig}(\phi)\,\omega_{Fgl}(\phi) = \beta_{il}(\phi)\pi(\phi) \tag{6.130}$$

where

$$\varphi_{il}(\phi) = \sum_{j=1}^{p}{}^{(g)}\, z_{ij}(\phi)\omega_{Fjl}(\phi) \tag{6.131}$$

and where everything except $\upsilon_{Pig}(\phi)$ and $\beta_{il}(\phi)$ is given. Two solution techniques will be considered:

Technique 1. Assume that all roots ζ_k of $\pi(\phi)$ are simple. Then (6.130) is equivalent to

$$\varphi_{il}(\zeta_k) + z_{ig}(\zeta_k)\,\upsilon_{Pig}(\zeta_g)\,\omega_{Fgl}(\zeta_k) = 0 \qquad k=1\ldots\nu-\rho_l \tag{6.132}$$

leading to

$$\upsilon_{Pig}(\zeta_k) = -\varphi_{il}(\zeta_k)/z_{ig}(\zeta_k)\,\omega_{Fgl}(\zeta_k) \qquad k=1\ldots\nu-\rho_l \tag{6.133}$$

Then a unique polynomial

$$\upsilon_{Pig}(\phi) = \upsilon_{Pig}^{o} + \upsilon_{Pig}^{1}\phi^{-1} + \ldots + \upsilon_{Pig}^{\nu-\rho l-1}\phi^{-\nu+\rho l+1} \tag{6.134}$$

can be found which satisfies (6.133). If $\pi(\phi)$ has multiple roots then (6.132) has to be supplemented with the appropriate derivative equations; for example for a double root ζ_k

$$\varphi_{il}^{'}(\zeta_k) + [z_{ig}(\zeta_k)\omega_{Fgl}(\zeta_k)]^{'}\upsilon_{Pig}(\zeta_k) + z_{ig}(\zeta_k)\omega_{Fgl}(\zeta_k)\upsilon_{Pig}^{'}(\zeta_k) = 0 \quad (6.135)$$

where $\varphi_{il}^{'}(\zeta_k) = d\varphi_{il}(\phi)/d\phi^{-1}$ at $\phi = \zeta_k$, etc.

Technique 2. Consider (6.130) as a polynomial Diophantine equation. Solution equations are obtained separately for each power of ϕ^{-1}. The minimum degree solution arises when the number of unknown parameters equals the number of solution equations. In general (except when $\nu - \rho_l = 1$),

$$Deg[z_{ig}(\phi)\upsilon_{Pig}(\phi)\omega_{Fgl}(\phi)] = Deg[\beta_{il}(\phi)\pi(\phi)] \geq Deg\varphi_{il}(\phi) \quad (6.136)$$

Thus

$$\#\ of\ equations = Degz_{ig} + Deg\upsilon_{Pig}(\phi) + Deg\omega_{Fgl} + 1 = Deg\beta_{il} + Deg\pi + 1 \quad (6.137)$$

On the other hand,

$$\#\ of\ unknowns = Deg\upsilon_{Pig} + Deg\beta_{il} + 2 \quad (6.138)$$

This leads to

$$Deg\upsilon_{Pig} = Deg\pi - 1 \qquad Deg\beta_{il} = Degz_{ig} + Deg\omega_{Fgl} - 1 \quad (6.139)$$

Usually, $Degz_{ig} = 0$. Then from (6.139) with (6.91)

$$\sigma = Deg\beta_{il} \leq \begin{cases} \nu - \rho_l - 1 & if\ f_{\cdot g} \neq 0 \\ \nu - \rho_l & if\ f_{\cdot g} = 0 \end{cases} \quad (6.140)$$

It follows from the rank property (6.102)-(6.103) that, in general, a response modifier $\upsilon_{Pig}(\phi)$ designed to satisfy (6.130) for the l-th column of the $\mathbf{\Omega}_F(\phi)$ matrix will satisfy the same for all other columns of $\mathbf{\Omega}_F(\phi)$ and also for all columns of $\mathbf{\Omega}_C(\phi)$. If Technique 1, Eq. (6.132) is considered, it is obvious from (6.103) that any other column would yield exactly the same set of conditions and thus the same response, with the exception of Case 1 below. For Technique 2, assume that the pair $\upsilon_{Pig}(\phi)$, $\beta_{il}(\phi)$ has been obtained for column l. Then for column m with the same $\upsilon_{Pig}(\phi)$, expressed from (6.130)

$$z_i^{'}(\phi)\mathbf{T}_{Pi}(\phi)\boldsymbol{\omega}_{F\cdot m}(\phi) = \varphi_{im}(\phi) + z_{ig}(\phi)\upsilon_{Pig}(\phi)\omega_{Fgm}(\phi) \quad (6.141a)$$

$$= \frac{1}{\omega_{Fgl}(\phi)} [\omega_{Fgl}(\phi)\varphi_{im}(\phi) - \omega_{Fgm}(\phi)\varphi_{il}(\phi) + \omega_{Fgm}(\phi)\beta_{il}(\phi)\pi(\phi)]$$

Utilizing (6.102), this becomes

$$\varphi_{im}(\phi) + z_{ig}(\phi)\upsilon_{Pig}(\phi)\omega_{Fgm}(\phi)$$

$$= \frac{\pi(\phi)}{\omega_{Fgl}(\phi)} [\beta_{il}(\phi)\omega_{Fgm}(\phi) - \sum_{j=1}^{\rho^{(g)}} z_{ij}(\phi)\theta_{lm,gj}(\phi)] \qquad (6.141b)$$

Since the left hand-side is polynomial, the right hand side must be polynomial as well. That is, with the exception of Case 1 below, β_{im} is also polynomial in

$$z_i'(\phi)\mathbf{T}_{Pi}(\phi)\boldsymbol{\omega}_{F\cdot m}(\phi) = \beta_{im}(\phi)\pi(\phi) \qquad (6.142)$$

The polynomial $\omega_{Fgl}(\phi)$ plays a pivotal role in the algorithm. Obviously, an element $\omega_{Fgl}(\phi)=0$ cannot serve as the pivotal element. Also of concern is the situation when $\omega_{Fgl}(\zeta_k)=0$ for any k. It follows from (6.103) that this latter can not happen in isolation; if an element of $\boldsymbol{\Omega}_F(\phi)$ contains an invariant zero then either the full row or the full column of $\boldsymbol{\Omega}(\phi)$ has to contain it (that is, ζ_k is a subsystem zero). Now the following two cases may occur:

1. The l-th column contains one or more of the invariant zeroes. Then (6.130) can be solved for this column, after eliminating the common factors, but this solution will not completely satisfy (6.130) for the other columns (the subsystem zeroes cancel out between $\pi(\phi)$ and $\omega_{Fgl}(\phi)$ in (6.141)-(6.142)). Therefore another column should be chosen for the implementation. This, however, does not affect how the implementation complies with the specification.

2. The g-th row contains one or more of the invariant zeroes. Then (6.130) can not be solved for any of the columns. Now another row has to be chosen, that is, another element of the response selected for adjustment. In particular, if the g-th single-input subsystem has an invariant zero ζ_k then only the g-th row can be chosen (and $(1 - \zeta_k\phi^{-1})$ needs to be included in the response modifier $\upsilon_{Pig}(\phi)$). Further, if there are more than one single-input subsystems with invariant zero(es) (or in any case when subsystems with zeroes overlap) then more than one element of the response needs to be modified to achieve cancellation.

Example 6.14.

We will demonstrate the use of the polynomial Diophantine equation in Algorithm II. Implement the specification $z_i'(\phi)=[0.5 \quad 1]$ with the response modifier $[1 \quad \upsilon_{Pi2}]$. By (6.139), $Deg\,\upsilon_{Pi2}=0$ so $\upsilon_{Pi2}=\alpha$. Also $Deg\beta=1$ so it is sought as $\beta(\phi)=\beta^o+\beta^1\phi^{-1}$. Let us choose the first column of Ω_F for the cancellation computation. Then (6.130) is

$$0.5(3-0.4\phi^{-1})+\alpha(-1+1.3\phi^{-1}-1.4\phi^{-2})=(\beta^o+\beta^1\phi^{-1})(1+3.6\phi^{-1})$$

Three coefficient equations can be written for the three powers of ϕ^{-1} occurring in the equation:

$$\phi^o: \quad 1.5-\alpha=\beta^o$$

$$\phi^{-1}: \quad -0.2+1.3\alpha=3.6\beta^o+\beta^1$$

$$\phi^{-2}: \quad -1.4\alpha=3.6\beta^1$$

The solution is

$$\alpha=1.059 \qquad \beta(\phi)=0.441-0.412\phi^{-1}$$

Then from (6.117) and (6.118):

$$w'_i(\phi)=[0.441-0.412\phi^{-1} \qquad 0.059-0.029\phi^{-1}]$$

$$v'_i(\phi)=[-1.059\phi^{-1} \qquad -0.059\phi^{-1}]$$

This, of course, is identical with the results of Example 6.13, Algorithm II. ▓

Full almost homogeneous specification. Recall that this is a special case of the full non-homogeneous specification, with

$$z'_i(\phi)=[0\ldots 0 \quad z_{ig}(\phi) \quad 0\ldots 0] \tag{6.143}$$

Clearly, now it is only the g-th response which may require modification for causality or polynomial solution. By (6.115),

$$\upsilon_{Rig}(\phi)=\phi^{\delta o-\delta g} \tag{6.144}$$

Further, by (6.117),

$$w'_i(\phi)=z_{ig}(\phi)\upsilon_{Pig}(\phi)\boldsymbol{\omega}_{Fg\cdot}(\phi)/\pi(\phi) \tag{6.145}$$

For (minimum complexity) polynomial solution,

$$\upsilon_{Pig}(\phi) = \pi(\phi) / \pi_g(\phi) \tag{6.146}$$

where $\pi_g(\phi)$ is the factor of $\pi(\phi)$ included in $\boldsymbol{\omega}_{Fg\cdot}(\phi)$. The input component of the computational algorithm is, by (6.118),

$$\boldsymbol{v}_i'(\phi) = z_{ig}(\phi)\upsilon_{Pig}(\phi)[\boldsymbol{\omega}_{Cg\cdot}(\phi)\boldsymbol{B} - \boldsymbol{\omega}_{Fg\cdot}(\phi)\boldsymbol{D}]/\pi(\phi) \tag{6.147}$$

which is also polynomial by virtue of (6.146). Note that the above is a special case of Algorithm I. The window width is the maximal polynomial degree arising from (6.145) and (6.147). It depends on the nature of the g-th fault and can be described as

$$\sigma \leq \begin{cases} \nu - \rho_I & \text{if } f_{\cdot g} \neq 0 \\ \nu - \rho_I + 1 & \text{if } f_{\cdot g} = 0 \end{cases} \tag{6.148}$$

Example 6.15.

Design a polynomial residual generator for the example system so that $z_i'(\phi) = [0 \quad 1]$. By (6.144) and (6.146):

$$\upsilon_{Ri2}(\phi) = \phi^{-1} \qquad \upsilon_{Pi2}(\phi) = 1 + 3.6\phi^{-1}$$

Then from (6.145) and (6.147):

$$\boldsymbol{w}_i'(\phi) = [\,-1 + 1.3\phi^{-1} - 1.4\phi^{-2} \qquad 1 - 0.3\phi^{-1} - 0.1\phi^{-2}\,]$$

$$\boldsymbol{v}_i'(\phi) = [\,-\phi^{-1} - 3.6\phi^{-2} \qquad -\phi^{-1} - 0.2\phi^{-2}\,]$$

6.6.3. Polynomial Implementation from the ARMA Representation: Almost Full Specifications

The polynomial implementation of a full non-homogeneous specification, and also the stabilization of an unstable realization, required some deviation from the specified responses. If the specification is only *almost full*, then the additional design freedom can be utilized to remove this constraint. The basic idea is to convert the non-square system, arising from an almost full specification, into a square system. We will explore this avenue here, first in connection with non-homogeneous specifications. Homogeneous specifications will then be addressed as a special case.

Almost full non-homogeneous specification. One possible approach to converting the implementation of an almost full specification into a square problem involves a modification of Algorithm II, resulting in a $\mu \cdot \mu$ system. Another approach leads to a third algorithm, first outlined in Subsection 6.3.1, and involves a $(\mu-1)\cdot(\mu-1)$ system.

Algorithm II. The specification contains $\rho=\mu-1$ elements $z_{i1}(\phi) \ldots z_{i,\mu-1}(\phi)$. Add an auxiliary specification $z_{i\mu}(\phi)$, with an entry column $s_{\cdot\mu}(\phi)$ independent of the original $\mu-1$ columns. The response modifier associated with this auxiliary input can be used to tune the system so that it is computationally polynomial, that is, (6.129) is satisfied, without affecting the original responses. Also, with this the otherwise non-square system becomes square so that the convenient techniques developed above can be readily applied. The auxiliary specification may be an additional fault response or the response to one of the observed inputs, as long as the augmented $S^A(\phi)$ matrix has full rank and the last row of the augmented $\Omega^A(\phi)$ matrix (that is, the original system) has no invariant zeroes.

Now the degree relation (6.140) applies to the augmented system. Define ρ_I^A as the number of strictly input faults in the latter. If the auxiliary fault is not strictly input then $\rho_I^A=\rho_I$ and the first line of (6.140) holds. If it is strictly input then $\rho_I^A=\rho_I+1$ and the second line holds. In either case

$$\boxed{\sigma \leq \nu - \rho_I - 1} \tag{6.149}$$

This is the same window width as the one obtained, for the same type of specification, with the Chow-Willsky scheme, see (6.68).

Algorithm III. This algorithm makes the otherwise non-square system square by separating one of the rows, for example the last one, and assigning the respective element of the transformation $w_i'(\phi)$. Define

$$w_i^\mu(\phi)=[w_{i1}(\phi) \ldots w_{i,\mu-1}(\phi)] \tag{6.150}$$

and partition the implementation equation (6.106) as

$$[w_i^\mu(\phi) \quad w_{i\mu}(\phi)] \begin{bmatrix} S^\mu(\phi) \\ s_{\mu\cdot}(\phi) \end{bmatrix} = z_i'(\phi)\mathbf{T}_i(\phi) \tag{6.151}$$

where $S^{\mu}(\phi)$ is the fault transfer matrix $S(\phi)$ without its last row. If this has full rank, (6.151) can be solved for $w_i^{\mu}(\phi)$, with $w_{i\mu}(\phi)$ as a parameter, as

$$w_i^{\mu}(\phi) = [z_i'(\phi)\mathbf{T}_i(\phi) - w_{i\mu}(\phi)s_{\mu\cdot}(\phi)][S^{\mu}(\phi)]^{-1} \tag{6.152}$$

Now $w_{i\mu}(\phi)$ and $\mathbf{T}_i(\phi)$ will be so chosen that the solution is causal and stable.

Express $[S^{\mu}(\phi)]^{-1}$ and $s_{\mu\cdot}(\phi)[S^{\mu}(\phi)]^{-1}$ with the determinant $\pi^{\mu}(\phi)$ and adjoint $\Omega^{\mu}(\phi)$ of the fault system matrix

$$\Gamma^{\mu+}(\phi) = \begin{bmatrix} \phi I - A & -E \\ C^{\mu} & F^{\mu} \end{bmatrix} \tag{6.153}$$

Since

$$S^{\mu}(\phi) = C^{\mu}(\phi I - A)^{-1}E + F^{\mu} \tag{6.154a}$$

$$s_{\mu\cdot}(\phi) = c_{\mu\cdot}(\phi I - A)^{-1}B + f_{\mu\cdot} \tag{6.154b}$$

it follows from (6.87) and (6.85) that

$$[S^{\mu}(\phi)]^{-1} = \Omega_F^{\mu+}(\phi)/\pi^{\mu+}(\phi) \tag{6.155a}$$

$$s_{\mu\cdot}(\phi)[S^{\mu}(\phi)]^{-1} = [c_{\mu\cdot}\Omega_E^{\mu+}(\phi) + f_{\mu\cdot}\Omega_F^{\mu+}(\phi)]/\pi^{\mu+}(\phi) \tag{6.155b}$$

With these, (6.152) becomes

$$w_i^{\mu}(\phi) = [-w_{i\mu}(\phi)c_{\mu\cdot} \quad z_i'(\phi)\mathbf{T}_i(\phi) - w_{i\mu}(\phi)f_{\mu\cdot}] \begin{bmatrix} \Omega_E^{\mu+}(\phi) \\ \Omega_F^{\mu+}(\phi) \end{bmatrix} \frac{1}{\pi^{\mu+}(\phi)} \tag{6.156}$$

The response modifier $\mathbf{T}_i(\phi)$ will be so chosen that the term $z_i'(\phi)\mathbf{T}_i(\phi)\Omega_F^{\mu+}(\phi)/\pi^{\mu+}(\phi)$ is causal. This requires

$$\mathbf{T}_i(\phi) = Diag[\phi^{\delta o - \delta j}], \qquad j=1\ldots p \tag{6.157}$$

$$\text{with} \quad \delta_o = Deg\pi^{\mu+}(\phi) \qquad \delta_j = Deg[\omega_{C\,j\cdot}^{\mu+}(\phi) \quad \omega_{F\,j\cdot}^{\mu+}(\phi)]$$

$w_{i\mu}(\phi)$ will be sought as

$$w_{i\mu}(\phi) = \phi^{\delta o - \delta}\gamma_{i\mu}(\phi) \tag{6.158a}$$

where

$$\delta = Deg[c_{\mu\cdot} \quad f_{\mu\cdot}]\, \Omega^{\mu+}(\phi) \tag{6.158b}$$

The factor $\phi^{\delta o - \delta}$ is to make the remaining terms in (6.156) causal. Here we took into account that (i) all rows need to be made causal with a single modifier, and, (ii) the rows which belong to strictly input faults (and thus would contribute the highest degrees) are multiplied with zero elements in the $f_{\mu\cdot}$ vector. Note that normally $\delta_o = \delta$; the exceptional case $\delta_o < \delta$ only occurs when the degree of $\pi^{\mu+}(\phi)$ drops due to cancellation.

With the above assignments, (6.156) becomes

$$w_i^{\mu}(\phi) = \frac{\kappa'_{i\mu}(\phi)}{\pi^{\mu}(\phi)} \begin{bmatrix} \Omega_E^{\mu}(\phi) \\ \Omega_F^{\mu}(\phi) \end{bmatrix} \tag{6.159}$$

where $\Omega_E^{\mu}(\phi)$, $\Omega_F^{\mu}(\phi)$ and $\pi^{\mu}(\phi)$ are causal polynomials defined in accordance with (6.114). Further,

$$\kappa'_{i\mu}(\phi) = [\, -\gamma_{i\mu}(\phi) c_{\mu\cdot} \mathbf{T}_{Ei}(\phi) \qquad z'_i(\phi) - \gamma_{i\mu}(\phi) f_{\mu\cdot} \mathbf{T}_{Fi}(\phi)] \tag{6.160}$$

where degree adjustments are applied to reconcile (6.114) and (6.158), as

$$\mathbf{T}_{Ei}(\phi) = Diag[\phi^{\delta j - \delta}] \qquad j=1 \ldots \nu$$

$$\text{with} \quad \delta_j = Deg[\omega_{A\,j\cdot}^{\mu+}(\phi) \quad \omega_{E\,j\cdot}^{\mu+}(\phi)] \tag{6.161a}$$

$$\mathbf{T}_{Fi}(\phi) = Diag[\phi^{\delta j - \delta}] \qquad j=1 \ldots \mu\text{-}1$$

$$\text{with} \quad \delta_j = Deg[\omega_{C\,j\cdot}^{\mu+}(\phi) \quad \omega_{F\,j\cdot}^{\mu+}(\phi)] \tag{6.161b}$$

The factor $\gamma(\phi)$ will be used to make the generator polynomial. For the cancellation of the denominator, it suffices to choose $\gamma_{i\mu}(\phi)$ so that

$$z'_i(\phi)\omega_{F\cdot l}^{\mu}(\phi) - \gamma_{i\mu}(\phi)[c_{\mu\cdot} \mathbf{T}_{Ei}(\phi) \qquad f_{\mu\cdot} \mathbf{T}_{Fi}(\phi)] \begin{bmatrix} \omega_{E\cdot l}^{\mu}(\phi) \\ \omega_{F\cdot l}^{\mu}(\phi) \end{bmatrix} = \pi^{\mu}(\phi)\beta_{il}^{\mu}(\phi) \tag{6.162}$$

is satisfied for an (almost) arbitrary column of $[\Omega_E^{\mu\prime}(\phi) \quad \Omega_F^{\mu\prime}(\phi)]'$ (as long as the selected column has no invariant zeroes); by the rank condition (6.103), this implies the same for all the other columns. If all the system zeroes are simple, (6.162) may be implemented as

$$z_i'(\zeta_k)\omega^{\mu}_{F\cdot l}(\zeta_k) - \gamma_{i\mu}(\zeta_k)[c_{\mu\cdot}\mathbf{T}_{Ei}(\zeta_k) \quad f_{\mu\cdot}\mathbf{T}_{Fi}(\zeta_k)] \begin{bmatrix} \omega^{\mu}_{E\cdot l}(\zeta_k) \\ \omega^{\mu}_{F\cdot l}(\zeta_k) \end{bmatrix} = 0$$

$$\text{for } k=1\ldots\nu - \rho_I \tag{6.163}$$

If some of the invariant zeroes have multiplicities, then (6.163) needs to be supplemented with the appropriate derivative equations. Alternatively, (6.162) may be solved directly as a polynomial Diophantine equation. As in Algorithm II, the selected column only qualifies for computing the cancellation if it does not contain invariant zeroes. Note that by omitting the last (or any other) row, the resulting reduced system may become *unobservable*. This, however, will lead to the appearance of invariant zeroes in the Ω^{μ}_{C} and Ω^{μ}_{F} matrices (c.f. Subsection 2.6.4) and thus will not interfere with the cancellation procedure.

Now let us investigate the behavior of $v_i'(\phi)$. It can be written as

$$v_i'(\phi) = - w_i'(\phi)M(\phi) = - [\, w^{\mu}_i(\phi) \quad w_{i\mu}(\phi)\,] \begin{bmatrix} M^{\mu}(\phi) \\ m_{\mu\cdot}(\phi) \end{bmatrix} \tag{6.164}$$

where

$$M^{\mu}(\phi) = C^{\mu}(\phi I - A)^{-1}B + D^{\mu} \tag{6.165a}$$

$$m_{\mu\cdot}(\phi) = c_{\mu\cdot}(\phi I - A)^{-1}B + d_{\mu\cdot} \tag{6.165b}$$

Substituting $w^{\mu}_i(\phi)$ from (6.152) and utilizing (6.154), (6.84)-(6.87) and (6.159)-(6.161), we obtain

$$v^{\mu}_i(\phi) = \frac{\kappa'_{i\mu}(\phi)}{\pi^{\mu}(\phi)} \underbrace{\begin{bmatrix} \Omega^{\mu}_A(\phi) & \Omega^{\mu}_E(\phi) \\ \Omega^{\mu}_C(\phi) & \Omega^{\mu}_F(\phi) \end{bmatrix}}_{\Omega^{\mu}(\phi)} \begin{bmatrix} B \\ -D^{\mu} \end{bmatrix} - w_{i\mu}(\phi)d_{\mu\cdot} \tag{6.166}$$

This is clearly causal, due to the way $\mathbf{T}_{Ei}(\phi)$ and $\mathbf{T}_{Fi}(\phi)$ have been chosen. It is also polynomial since (i) $\gamma_{i\mu}(\phi)$ has been designed so that, if combined with $[\Omega^{\mu\prime}_E(\phi) \;\; \Omega^{\mu\prime}_F(\phi)]'$, the denominator $\pi^{\mu}(\phi)$ is canceled, and (ii) by virtue of the rank condition (6.103), this implies cancellation with $[\Omega^{\mu}_A{}'(\phi) \;\; \Omega^{\mu}_C{}'(\phi)]'$ as well.

Complexity. Finally, let us explore the attainable minimum complexity of the solution. Since there are $\nu\text{-}\rho_I$ invariant zeroes, the solution of (6.163) requires $Deg\gamma_{i\mu}(\phi)=\nu-\rho_I-1$. Consider now (6.159). $Deg\ \Omega_E^{\mu}(\phi)\leq\nu-\rho_I$; also $Deg\, f_{\mu\cdot}\Omega_F^{\mu}\leq\nu-\rho_I$ because the zero elements in $f_{\mu\cdot}$ eliminate the rows of $\Omega_F^{\mu}(\phi)$ which would contribute higher degree. Now assume that $z_i'(\phi)$ is non-dynamic. Then, with the exception of the special case of $\nu-\rho_I=1$, the following holds:

$$Deg\ z_i'(\phi)\Omega_F^{\mu}(\phi) \leq Deg\ \gamma_{i\mu}(\phi)[c_{\mu\cdot}\mathbf{T}_{Ei}(\phi)\Omega_E^{\mu}(\phi)+f_{\mu\cdot}\mathbf{T}_{Fi}(\phi)\Omega_F^{\mu}(\phi)] \qquad (6.167)$$

This leads to

$$Deg\,\beta_i^{\mu}(\phi) \qquad (6.168)$$

$$= Deg\ \gamma_{i\mu}(\phi) + Deg\ [c_{\mu\cdot}\mathbf{T}_{Ei}(\phi)\Omega_E^{\mu}(\phi)+f_{\mu\cdot}\mathbf{T}_{Fi}(\phi)\Omega_F^{\mu}(\phi)] - Deg\ \pi^{\mu}(\phi)$$

Observe that the presence of $\mathbf{T}_{Ei}(\phi)$ and $\mathbf{T}_{Fi}(\phi)$ in the bracketed expression does not increase its degree relative to that of $\Omega_E^{\mu}(\phi)$ and $f_{\mu\cdot}\Omega_F^{\mu}(\phi)$. Thus the window length is obtained as

$$\boxed{\sigma \leq \nu - \rho_I - 1 \qquad for\ \ \nu - \rho_I > 1} \qquad (6.169)$$

In the special case of $\nu-\rho_I=1$, (6.167) does not hold and the window length is obtained as $\sigma=1$. The analysis of (6.166) leads to the same result. Note that (6.169), including the special case, is *identical with the minimum complexity attainable with the Chow-Willsky scheme*, compare with (6.68).

Example 6.16.

Let us design a polynomial generator for our example system, so that the response to the first fault is $z_{i1}=0.5$, using Algorithm III. The fault system matrix for this single-fault system, after removing the second output, is

$$\Gamma^{2+}(\phi) = \left[\begin{array}{cc|c} \phi - 0.8 & 0 & -1 \\ -1 & \phi - 0.5 & 0 \\ \hline 1 & 0 & 1 \end{array}\right]$$

with $c_{2\cdot}=[0\ \ 1]$ and $f_{2\cdot}=1$. Note that this reduced system is unobservable. The adjoint matrix and the invariant zero polynomial are

$$\Omega^{2+}(\phi) = \left[\begin{array}{cc|c} \phi - 0.5 & 0 & \phi - 0.5 \\ 1 & \phi+0.2 & 1 \\ \hline -(\phi - 0.5) & 0 & (\phi - 0.5)(\phi - 0.8) \end{array}\right]$$

$$\pi^{2+}(\phi) = (\phi - 0.5)(\phi+0.2)$$

Clearly, $\delta=\delta_o=2$, thus $w_{i\mu}(\phi)=\gamma_{i\mu}(\phi)$. Further, $\tau_i(\phi)=1$, $\tau_{Fi}(\phi)=1$ and $\tau_{Ei}(\phi)=Diag[\phi^{-1} \;\; \phi^{-1}]$. The causal form of the inverse matrix is

$$\Omega^{2}(\phi) = \left[\begin{array}{cc|c} 1 - 0.5\phi^{-1} & 0 & 1 - 0.5\phi^{-1} \\ \phi^{-1} & 1+0.2\phi^{-1} & \phi^{-1} \\ \hline -\phi^{-1}(1 - 0.5\phi^{-1}) & 0 & (1 - 0.5\phi^{-1})(1 - 0.8\phi^{-1}) \end{array}\right]$$

$$\pi^{2}(\phi) = (1 - 0.5\phi^{-1})(1+0.2\phi^{-1})$$

The system pole 0.5 is an invariant zero, due to unobservability, but it does not appear in the entire ω_E column so the implementation is feasible. The polynomial $\gamma_{i2}(\phi)$ is designed so that (6.163) is satisfied for $\zeta_1=0.5$ and $\zeta_2=-0.2$, yielding $\gamma_{i2}(0.5)=0$, $\gamma_{i2}(-0.2)=0.2059$. The minimum degree polynomial fitting these points is

$$w_{i2}(\phi) = \gamma_{i2}(\phi) = 0.059 - 0.029\phi^{-1}$$

which is well known from the previous examples. The other element of the transformation is obtained from (6.159), also in harmony with past results, as

$$w_{i1}(\phi) = 0.441 - 0.412\phi^{-1}$$

Almost full homogeneous specification. The polynomial implementation of an almost full homogeneous specification can be reduced to the full almost homogeneous specification (Algorithm I) or handled as a special case of the polynomial implementation of an almost full non-homogeneous specification (Algorithm III). As we will see, the homogeneous case is significantly simpler than the non-homogeneous one.

Algorithm I. Add a nonzero auxiliary specification $z_{i\mu}(\phi)$, with the respective transfer function column $s_{\cdot\mu}(\phi)$, so that the augmented fault transfer function matrix $S^A(\phi)$ has full rank. Then the augmented specification is

$$z_i^{A\prime}(\phi) = [0 \ldots 0 \quad z_{i\mu}(\phi)] \tag{6.170}$$

This is clearly a special case of (6.143), with $g=\mu$, so the equations (6.145) and (6.147) apply, with $\boldsymbol{\omega}^A_{C\mu\cdot}(\phi)$ and $\boldsymbol{\omega}^A_{F\mu\cdot}(\phi)$ replacing $\boldsymbol{\omega}_{Cg\cdot}(\phi)$ and $\boldsymbol{\omega}_{Fg\cdot}(\phi)$, and so does the result (6.146). Note that the specification constraint now concerns an auxiliary input and thus may be practically irrelevant. To compute the window width, recall that ρ_I^A in the augmented system is not necessarily the same as ρ_I in the original system. If the auxiliary fault is not strictly input then $\rho_I^A=\rho_I$ and $Deg\ \boldsymbol{\omega}^A_{F\mu\cdot}(\phi)=\nu-\rho_I^A=\nu-\rho_I$. If it is strictly input then $\rho_I^A=\rho_I+1$ and $Deg\ \boldsymbol{\omega}^A_{F\mu\cdot}(\phi)=\nu-\rho_I^A+1=\nu-\rho_I$. So in either case

$$\sigma \leq \nu - \rho_I \tag{6.171}$$

This is identical with the minimum complexity attainable, in the same situation, with the Chow-Willsky scheme, see (6.70).

Note that $\boldsymbol{\omega}^A_{C\mu\cdot}(\phi)$ and $\boldsymbol{\omega}^A_{F\mu\cdot}(\phi)$ depend only on the original system and do not depend on what the augmenting column $\boldsymbol{s}_{\cdot\mu}(\phi)$ is. Thus, given the original system, *all solutions are similar* and $z_{i\mu}(\phi)$ only acts as a scaling factor.

Algorithm III. Now $\boldsymbol{z}_i'(\phi)=\boldsymbol{0}$. From (6.159) with (6.160) and (6.158)

$$\begin{aligned} \boldsymbol{w}_i^{\mu}(\phi) &= - w_{i\mu}(\phi)\boldsymbol{s}_{\mu\cdot}(\phi)[\boldsymbol{S}^{\mu}(\phi)]^{-1} \\ &= - \frac{\gamma_{i\mu}(\phi)}{\pi^{\mu}(\phi)} [\boldsymbol{c}_{\mu\cdot}\mathbf{T}_{Ei}(\phi) \quad \boldsymbol{f}_{\mu\cdot}\mathbf{T}_{Fi}(\phi)] \begin{bmatrix} \boldsymbol{\Omega}^{\mu}_E(\phi) \\ \boldsymbol{\Omega}^{\mu}_F(\phi) \end{bmatrix} \end{aligned} \tag{6.172}$$

Clearly, the minimal complexity polynomial transformation is obtained with

$$\boxed{w_{i\mu}(\phi) = \phi^{\delta o - \delta}\gamma_{i\mu}(\phi) = \alpha\, \phi^{\delta o - \delta}\pi^{\mu}(\phi)} \tag{6.173}$$

where α is any constant. Further, $\boldsymbol{v}_i'(\phi)$ is, from (6.166),

$$\boldsymbol{v}_i^{\mu}(\phi) = \frac{\gamma_{i\mu}(\phi)}{\pi^{\mu}(\phi)} [\boldsymbol{c}_{\mu\cdot}\mathbf{T}_{Ei}(\phi) \quad \boldsymbol{f}_{\mu\cdot}\mathbf{T}_{Fi}(\phi)]\, \boldsymbol{\Omega}^{\mu}(\phi) \begin{bmatrix} -\boldsymbol{B} \\ \boldsymbol{D}^{\mu} \end{bmatrix} - w_{i\mu}(\phi)\boldsymbol{d}_{\mu\cdot} \tag{6.174}$$

which is also made polynomial by (6.173). Observe that all solutions are similar, scaled by $\gamma_{i\mu}(\phi)$.

The polynomial degree of the algorithm is the highest degree appearing in the numerators of (6.172) and (6.174). Since the zero elements in $f_{\mu \cdot}$ eliminate the rows of $\Omega^{\mu}_{F}(\phi)$ with degree $\nu-\rho_I+1$, and since the $\mathbf{T}_{Ei}(\phi)$ and $\mathbf{T}_{Fi}(\phi)$ matrices do not interfere with the highest degree, this is $\nu-\rho_I$ throughout, yielding again the minimum window width (c.f. (6.171) and (6.70))

$$\boxed{\sigma \leq \nu - \rho_I} \tag{6.175}$$

Example 6.17.

Let us obtain the generator by Algorithm III. Refer to $\pi^2(\phi)$ and $\Omega^2(\phi)$ for the single-fault reduced system from Example 6.16. By (6.173),

$$w_{i2}(\phi) = 1 - 0.3\phi^{-1} - 0.1\phi^{-2}$$

Then by (6.172) and (6.174):

$$w_{i1}(\phi) = -1 + 1.3\phi^{-1} - 1.4\phi^{-2}$$

$$\mathbf{v}_i'(\phi) = [\phi^{-1} + 3.6\phi^{-2} \qquad \phi^{-1} + 0.2\phi^{-2}]$$

▩

6.6.4. Direct MA Implementation Revisited

We will revisit here the direct moving average implementation, introduced in Subsection 6.1.2 and elaborated on in 6.3.2. Recall the residual generator equation (6.48) and write it in terms of the common denominator matrices in (6.10) and (6.11):

$$r_i(t) = \mathbf{w}^{*\prime}_i(\phi)\mathbf{o}^*(t) \tag{6.176}$$

where

$$\mathbf{o}^*(t) = \bar{h}(\phi)\mathbf{y}(t) - \bar{\mathbf{G}}(\phi)\mathbf{u}(t) = \bar{\mathbf{N}}_F(\phi)\mathbf{p}(t) + \bar{\mathbf{N}}_D(\phi)\mathbf{q}(t) \tag{6.177}$$

is the vector of moving average primary residuals. Thus the design equation is

$$\boxed{\mathbf{w}^{*\prime}_i(\phi)\bar{\mathbf{N}}(\phi) = \mathbf{z}_i'(\phi)\mathbf{T}_i(\phi)} \tag{6.178}$$

Here $\tilde{N}(\phi)=[\tilde{N}_F(\phi)\ \ \tilde{N}_D(\phi)]$, while $z_i'(\phi)=[z_{Fi}'(\phi)\ \ z_{Di}'(\phi)]$ is the design specification and $\mathbf{T}_i(\phi)=\mathbf{T}_{Pi}(\phi)\mathbf{T}_{Ri}(\phi)$ is the response modifier. Now two objectives are possible in terms of the polynomial behavior of the algorithm:

1. The transformation $w^*{}_i'(\phi)$, applied to the MA primary residuals, is polynomial;
2. The generic form $r_i(t)=w_i'(\phi)y(t)+v_i'(\phi)u(t)$ of the residual generator is polynomial.

From (6.178), the transformation is obtained as

$$w^*{}_i'(\phi) = z_i'(\phi)\mathbf{T}_i(\phi)[\tilde{N}(\phi)]^{-1} \tag{6.179}$$

while the coefficients in the generic form are

$$w_i'(\phi) = w^*{}_i'(\phi)\bar{h}(\phi) = z_i'(\phi)\mathbf{T}_i(\phi)[\tilde{N}(\phi)]^{-1}\bar{h}(\phi) \tag{6.180}$$

$$v_i'(\phi) = -\,w^*{}_i'(\phi)\bar{G}(\phi) = -\,z_i'(\phi)\mathbf{T}_i(\phi)[\tilde{N}(\phi)]^{-1}\bar{G}(\phi) \tag{6.181}$$

Now it follows from (6.10) (6.76), (6.87) and (6.116) that

$$\mathbf{T}_{Ri}(\phi)[\tilde{N}(\phi)]^{-1} = \mathbf{T}_{Ri}(\phi)[S(\phi)\bar{h}(\phi)]^{-1} = \mathbf{\Omega}_F(\phi)\ /\ [\bar{h}(\phi)\pi(\phi)] \tag{6.182}$$

With this,

$$\boxed{w^*{}_i'(\phi) = z_i'(\phi)\mathbf{T}_{Pi}(\phi)\mathbf{\Omega}_F(\phi)\ /\ [\bar{h}(\phi)\pi(\phi)]} \tag{6.183}$$

and

$$\boxed{w_i'(\phi) = z_i'(\phi)\mathbf{T}_{Pi}(\phi)\mathbf{\Omega}_F(\phi)\ /\ \pi(\phi)} \tag{6.184}$$

Further, with (6.10), (6.75), (6.86) and (6.116),

$$\mathbf{T}_{Ri}(\phi)[\tilde{N}(\phi)]^{-1}\bar{G}(\phi) = \mathbf{T}_{Ri}(\phi)[S(\phi)\bar{h}(\phi)]^{-1}[M(\phi)\bar{h}(\phi)]$$

$$= \mathbf{T}_{Ri}(\phi)S^{-1}(\phi)M(\phi) = [-\,\mathbf{\Omega}_C(\phi)\boldsymbol{B}+\mathbf{\Omega}_F(\phi)\boldsymbol{D}]\ /\ \pi(\phi) \tag{6.185}$$

thus

$$v_i'(\phi) = z_i'(\phi)\mathbf{T}_{Pi}(\phi)[\boldsymbol{\Omega}_C(\phi)\boldsymbol{B} - \boldsymbol{\Omega}_F(\phi)\boldsymbol{D}] \,/\, \pi(\phi) \tag{6.186}$$

As it is seen in (6.184) and (6.186), the generic form of the MA residual generator is identical with the form obtained with ARMA design, compare to (6.117) and (6.118). To make this generator polynomial, only the invariant zero polynomial $\pi(\phi)$ needs to be canceled out. The various implementation techniques discussed in the previous subsections may be applied.

As revealed by (6.183), the MA transformation $w^*{}_i'(\phi)$ contains the transfer function denominator $\bar{h}(\phi)$ in its denominator. To make this transformation polynomial, $\bar{h}(\phi)$ also needs to be canceled out, together with the invariant zero polynomial. This will certainly increase the polynomial degree of the generator. Otherwise, the design techniques developed in the previous subsection are applicable. This includes the fact that it is sufficient to assure the cancellation for one column of the $\boldsymbol{\Omega}_F(\phi)$ matrix. To see this, consider (6.102) and notice that, if the $2{\cdot}2$ determinant on the right comes entirely from the $\boldsymbol{\Omega}_F(\phi)$ matrix, then $\theta^+_{jm,il}(\phi)$ always includes the determinant $|\phi\boldsymbol{I} - \boldsymbol{A}|$. Thus, with (2.130),

$$\theta^+_{jm,il}(\phi) = \alpha^+_{jm,il}(\phi)\, |\phi\boldsymbol{I} - \boldsymbol{A}| \tag{6.187}$$

where $\alpha^+_{jm,il}(\phi)$ is a determinant coefficient, and therefore, with (6.77),

$$\omega^+_{Fij}(\phi)\omega^+_{Flm}(\phi) - \omega^+_{Fim}(\phi)\omega^+_{Flj}(\phi) = \pm\alpha^+_{jm,il}(\phi)\vartheta^+(\phi)\bar{h}^+(\phi)\pi^+(\phi) \tag{6.188}$$

That is,

$$Rank\, \boldsymbol{\Omega}_F(\lambda_k) = 1 \qquad k=1\ldots\bar{\nu} \tag{6.189}$$

where λ_k, $k=1\ldots\bar{\nu}$ are the roots of the system denominator $\bar{h}(\phi)$.

Note that if the specification is homogeneous then it is not necessary to cancel $\bar{h}(\phi)$ in order to make $w_i^{*\prime}(\phi)$ polynomial. With $z_i'(\phi)=\boldsymbol{0}'$ and following the logic of Algorithm III, (6.151)-(6.152) and (6.172):

$$w_i^{*\prime}(\phi)\bar{N}(\phi) = [w_i^{*\mu}(\phi) \quad w^*_{i\mu}(\phi)] \begin{bmatrix} \bar{N}^{\mu}(\phi) \\ \bar{n}_{\mu}(\phi) \end{bmatrix} = z_i'(\phi) = \boldsymbol{0}' \tag{6.190}$$

$$\begin{aligned} w_i^{*\mu}(\phi) &= -\, w^*_{i\mu}(\phi)\bar{n}_{\mu}(\phi)[\bar{N}^{\mu}(\phi)]^{-1} = -\, w^*_{i\mu}(\phi)s_{\mu}(\phi)[S^{\mu}(\phi)]^{-1} \\ &= -\, \gamma^*_{i\mu}(\phi)[c_{\mu\cdot}\mathbf{T}_{Ei}(\phi)\boldsymbol{\Omega}^{\mu}_E(\phi) + f_{\mu\cdot}\mathbf{T}_{Fi}(\phi)\boldsymbol{\Omega}^{\mu}_F(\phi)] \,/\, \pi^{\mu}(\phi) \end{aligned} \tag{6.191}$$

The generic coefficients $w_i'(\phi)$ and $v_i'(\phi)$ are then computed by (6.180) and (6.181) and do contain $\bar{h}(\phi)$.

6.6.5. Summary and Comparison

We are going to summarize the main ideas of the input-output parity equation implementation techniques, review the applicability of the various algorithms under different types of specifications and compare their computational complexity. We will also compare the input-output techniques to the state-space based scheme of Chow and Willsky.

The essence of the input-output implementation techniques discussed in this section is the twin equations

$$\boxed{\begin{aligned} w_i'(\phi) &= z_i'(\phi)\mathbf{T}_i(\phi)S^{-1}(\phi) \\ v_i'(\phi) &= -\,z_i'(\phi)\mathbf{T}_i(\phi)S^{-1}(\phi)M(\phi) \end{aligned}} \tag{6.192}$$

With the help of the inverse of the fault/disturbance system matrix, these expressions have been re-stated as

$$\boxed{\begin{aligned} w_i'(\phi) &= z_i'(\phi)\mathbf{T}_i(\phi)\boldsymbol{\Omega}_F^+(\phi)\,/\,\pi^+(\phi) \\ v_i'(\phi) &= z_i'(\phi)\mathbf{T}_i(\phi)[\boldsymbol{\Omega}_C^+(\phi)B - \boldsymbol{\Omega}_F^+(\phi)D]\,/\,\pi^+(\phi) \end{aligned}} \tag{6.193}$$

The various algorithms are meant to achieve the following objectives:

1. *To convert the problem into a "square" one*, so that (6.192) and (6.193) should apply, in case the original system has fewer fault/disturbance inputs than outputs. This is done
   a. either by adding a column (or more) to the system, in the form of auxiliary response specification(s) (Algorithm I or II),
   b. or by removing a row (or more), and assigning values to one (or more) element(s) of the transformation (Algorithm III).
2. *To assure that the residual generator is realizable*,
   a. by introducing time delays into the response modifier whenever necessary (Algorithm I and II),
   b. or by including time-delays in the assigned elements of the transformation (Algorithm III).

3. *To cancel out the denominator $\pi^{+}(\phi)$ partially*, in order to attain stability, *or completely*, in order to attain a polynomial generator. This is achieved

a. either by including the denominator, partially or fully, in all elements of the response modifier $\mathbf{T}_i(\phi)$ (Algorithm I),

b. or by assigning one element of the response modifier so that the cancellation takes place (Algorithm II),

c. or by assigning the selected element of the transformation vector so that the cancellation takes place (Algorithm III).

The basic condition for the *existence of solutions* under any of the input-output algorithms is, in general, that the fault/disturbance transfer matrix $S(\phi)$ must have full normal rank. However, meaningful solutions may be obtained even in the case of rank defect, provided the specification is, or can be made to be, in harmony with the special restrictions of the system (e.g., zero response associated with linearly dependent columns).

Of the above implementation techniques, 3a. (Algorithm I) achieves stability or polynomial behavior by effectively modifying, relative to their original specifications, all the fault responses. Technique 3b. (Algorithm II) does the same by modifying a single response. If Algorithm II is employed to implement an almost full specification, by adding a dummy fault, then the response modification affects only the latter.

The presence of subsystem zeroes places restrictions on Algorithms II and III, in that the choice of the response to be modified (Alg. II) or the transformation element to be assigned (Alg. III) is not arbitrary. In extreme cases, subsystem zeroes may cause either algorithm to fail. In contrast, Algorithm I always works (and subsystem zeroes even make the solution simpler).

Note that the cancellation of the invariant zeroes is formally a pole-zero cancellation procedure. However, it is completely algebraic and does not involve the true poles and zeroes of the plant. Therefore, the stability problems encountered with such algorithms in controller design *do not* arise here.

Controllability and observability are normally not an issue with the parity equation implementation of residual generators. We start from transfer function descriptions and seek minimum state-space realizations which are controllable and observable by definition. Non-observability may arise in Algorithm III, due to the elimination of an output; this may affect choices in the application of the algorithm but not the feasibility of the implementation.

For *minimal computational complexity* (polynomial degrees), the specified responses need to be of minimal complexity as well. This means a constant response specification, which then may be expanded with minimum delay if and as needed for realizability and with further minimal expansion for stabilization or polynomial generation. In general. Algorithms II and III, if feasible, result in lower polynomial degrees than Algorithm I.

The minimum complexity solution of the basic equation (6.192) is, in general, *unique*. With homogeneous specifications, the solution is not unique but all solutions are similar.

The Chow-Willsky scheme implements almost full specifications by polynomial transformation. It results in the same polynomial complexity as one would obtain if solving the same problem by Algorithm III, or by Algorithm II with a dummy fault. The scheme fails if the $[J \quad L^{\circ}]$ matrix is singular. Solution in this case may be attained by increasing the window length (that is, the complexity of the transformation). The Chow-Willsky scheme represents a special case in the sense that it does not provide for the implementation of the broader and more flexible ARMA transformation.

Table 6.1. summarizes the applicability of the various algorithms for polynomial residual generation, under different types of specification. The table also shows the polynomial degrees of the minimum complexity solutions.

An important practical issue is how the design computations are actually performed. The derivations in this section rely heavily on the inverse fault/disturbance system matrix. The actual computations are conveniently executed if the system matrix is available; this, however, requires the state-space representation of the system. As an alternative, the computations can be performed directly on the transfer matrices. This latter amounts basically to the "naive" approach where common factors need to be recognized numerically

TABLE 6.1

	Full	Almost full
Non-homogeneous	Alg.I (+1) Alg.II (0)	Alg.II (-1) Alg.III (-1) Ch-W (-1)
Almost homogen.	Alg.I (+1)	
Homogeneous		Alg.I (0) Alg.III (0) Ch-W (0)

Degree key: (-1): $\nu+\rho_I-1$ (0): $\nu+\rho_I$ (+1): $\nu+\rho_I+1$

and eliminated to obtain minimum complexity. The understanding gained through the systematic procedures can be utilized, though, at least in terms of the polynomial degrees of the minimum solutions sought. This, in turn, requires the knowledge of the true system order, that is, the order of the minimum system realization (which takes half the effort of actually finding such a realization).

Whether we operate on the fault system matrix or the transfer matrices, the computations involve manipulations with polynomial matrices. This non-trivial task can be approached symbolically, or numerically by interpolation. The latter approach, which involves the numerical solution of the polynomial equations for a set of numerical values assigned to the polynomial variable, has been found to be quite effective. Additional numerical techniques may be found in the literature, see for example (Van Dooren, 1981; Stefanidis et al, 1992).

6.6.6. Design Example

We include here a somewhat more extensive example to demonstrate the design procedure, especially the application of Algorithms I and II.

System description. The system is described by the following transfer functions:

$$M(\phi) = \frac{\phi^{-1}}{(1-0.5\phi^{-1})(1-0.6\phi^{-1})(1-0.8\phi^{-1})} \cdot \begin{bmatrix} 2.2(1-0.6\phi^{-1})(1-0.8\phi^{-1}) & 3(1-0.567\phi^{-1})(1-0.8\phi^{-1}) \\ (1-0.5\phi^{-1})(1-0.6\phi^{-1}) & (1-0.5\phi^{-1})(1-0.8\phi^{-1}) \end{bmatrix}$$

$$S(\phi) = \frac{1}{(1-0.5\phi^{-1})(1-0.6\phi^{-1})(1-0.8\phi^{-1})} \cdot \begin{bmatrix} 10(1-0.4\phi^{-1})(1-0.7\phi^{-1})(1-0.8\phi^{-1}) & 2.2\phi^{-1}(1-0.6\phi^{-1})(1-0.8\phi^{-1}) \\ 10(1-0.5\phi^{-1})(1-0.7\phi^{-1})(1-\phi^{-1}) & \phi^{-1}(1-0.5\phi^{-1})(1-0.6\phi^{-1}) \end{bmatrix}$$

A minimal state-space representation is obtained as:

$$A = \begin{bmatrix} 0.5 & 0 & 0 \\ 0 & 0.6 & 0 \\ 0 & 0 & 0.8 \end{bmatrix} \qquad B = \begin{bmatrix} 2.2 & 2 \\ 0 & 1 \\ 1 & 0 \end{bmatrix} \qquad E = \begin{bmatrix} 2 & 2.2 \\ -2 & 0 \\ -1 & 1 \end{bmatrix}$$

$$C = \begin{bmatrix} 1 & 1 & 0 \\ 0 & 1 & 1 \end{bmatrix} \qquad D = \begin{bmatrix} 0 & 0 \\ 0 & 0 \end{bmatrix} \qquad F = \begin{bmatrix} 10 & 0 \\ 10 & 0 \end{bmatrix}$$

Clearly, the system parameters are:

$$\nu = 3 \qquad \mu = 2 \qquad \rho = 2 \qquad \rho_I = 1$$

where we noted that the second fault is strictly input.

The fault system matrix (6.80) can be constructed as

$$\Gamma^{+}(\phi) = \left[\begin{array}{ccc|cc} \phi - 0.5 & 0 & 0 & -2 & -2.2 \\ 0 & \phi - 0.6 & 0 & 2 & 0 \\ 0 & 0 & \phi - 0.8 & 1 & 1 \\ \hline 1 & 1 & 0 & 10 & 0 \\ 0 & 1 & 1 & 10 & 0 \end{array}\right]$$

We include here only those parts of the inverse which will be used in the sequel:

$$\pi^{+}(\phi) = -12(\phi - 0.7)(\phi - 1.5)$$

$$\Omega_C^{+}(\phi) = \begin{bmatrix} -(\phi-0.6) & 1.2(\phi-1.05) & 2.2(\phi-0.6) \\ 10(\phi-0.7)(\phi-1) & -3(\phi-0.7) & -10(\phi-0.4)(\phi-0.7) \end{bmatrix}$$

$$\mathbf{\Omega}_F^+(\phi) = \begin{bmatrix} (\phi-0.5)(\phi-0.6) & -2.2(\phi-0.6)(\phi-0.8) \\ -10(\phi-0.5)(\phi-0.7)(\phi-1) & 10(\phi-0.4)(\phi-0.7)(\phi-0.8) \end{bmatrix}$$

The degrees are as predicted by (6.90) and (6.100). There are two invariant zeroes, $\zeta_1=0.7$ and $\zeta_2=1.5$. Observe that the factor $\phi-0.7$ appears in all elements of the second rows of $\mathbf{\Omega}_C^+(\phi)$ and $\mathbf{\Omega}_F^+(\phi)$, indicating that $\zeta_1=0.7$ is a zero also of the first single-fault subsystem, see (6.89).

Adjustment for causality. As seen above, the row-degrees are

$$\delta_o = 2 \qquad \delta_1 = 2 \qquad \delta_2 = 3$$

Then by (6.115)

$$\mathbf{T}_R(\phi) = Diag[1 \quad \phi^{-1}]$$

and the realizable forms $\pi(\phi)$, $\mathbf{\Omega}_C(\phi)$ and $\mathbf{\Omega}_F(\phi)$ are obtained, according to (6.114), as

$$\pi(\phi) = -12(1-0.7\phi^{-1})(1-1.5\phi^{-1})$$

$$\mathbf{\Omega}_C(\phi) =$$

$$\begin{bmatrix} \phi^{-1}(1-0.6\phi^{-1}) & 1.2\phi^{-1}(1-1.05\phi^{-1}) & 2.2\phi^{-1}(1-0.6\phi^{-1}) \\ 10\phi^{-1}(1-0.7\phi^{-1})(1-\phi^{-1}) & -3\phi^{-2}(1-0.7\phi^{-1}) & -10(1-0.4\phi^{-1})(1-0.7\phi^{-1}) \end{bmatrix}$$

$$\mathbf{\Omega}_F(\phi) =$$

$$\begin{bmatrix} (1-0.5\phi^{-1})(1-0.6\phi^{-1}) & -2.2(1-0.6\phi^{-1})(1-0.8\phi^{-1}) \\ -10(1-0.5\phi^{-1})(1-0.7\phi^{-1})(1-\phi^{-1}) & -10(1-0.4\phi^{-1})(1-0.7\phi^{-1})(1-0.8\phi^{-1}) \end{bmatrix}$$

Stabilization. The transfer function $[\mathbf{\Omega}_C(\phi) \quad \mathbf{\Omega}_F(\phi)]/\pi(\phi)$ as obtained above is clearly unstable and needs to be stabilized. The unstable invariant zero is $\zeta_2=1.5$ and it is not a subsystem zero.

We will show the stabilization procedure by both Algorithm I and II. The specified response in both cases will be chosen as

$$z_i'(\phi) = [\,1 \quad 2\,]$$

Algorithm I. To cancel the unstable zero, apply (6.120) as

$$\mathbf{T}_{Pi}(\phi) = Diag\,[(1-1.5\phi^{-1}) \qquad (1-1.5\phi^{-1})]$$

Then the transformation is obtained from (6.117) and (6.118) as

$$w_i'(\phi) = \frac{1}{(1-0.7\phi^{-1})} \cdot [1.583(1-2.258\phi^{-1}+1.616\phi^{-2}-0.368\phi^{-3})$$
$$-1.483(1-1.962\phi^{-1}+1.244\phi^{-2}-0.252\phi^{-3})]$$

$$v_i'(\phi) = \frac{1}{(1-0.7\phi^{-1})} \cdot [-2\phi^{-1}(1-0.7\phi^{-1})(1-1.5\phi^{-1})$$
$$-\,3.267\phi^{-1}(1-1.890\phi^{-1}+0.821\phi^{-2})]$$

which is, of course, stable. The actual fault response becomes, by (6.106),

$$z_i'(\phi)\,\mathbf{T}_{Ri}(\phi)\,\mathbf{T}_{Pi}(\phi) = [(1-1.5\phi^{-1}) \qquad 2\phi^{-1}(1-1.5\phi^{-1})]$$

Algorithm II. Since the second row of $[\boldsymbol{\Omega}_C(\phi) \quad \boldsymbol{\Omega}_F(\phi)]$ contains an invariant zero, only the first response may be adjusted. Thus by (6.123)

$$\mathbf{T}_{Pi}(\phi) = Diag\,[\alpha \quad 1]$$

and we will seek α so that the $(1-1.5\phi^{-1})$ factor is canceled out. Use for example the first column of $\boldsymbol{\Omega}_F(\phi)$ for the calculation, so that according to (6.129) and (6.132)

$$[\alpha \quad 2][(1-0.5\phi^{-1})(1-0.6\phi^{-1}) \qquad -10(1-0.5\phi^{-1})(1-0.7\phi^{-1})(1-\phi^{-1})]'\,\Big|_{\phi=1.5} = 0$$

This is, after dropping the $1-0.5\phi^{-1}$ factor,

$$[\alpha \quad 2][0.6 \quad -3.555]' = 0$$

which yields $\alpha = 5.926$. With this, we obtain

$$z_i'(\phi)\,\mathbf{T}_{Pi}(\phi)\,\boldsymbol{\Omega}_F(\phi) =$$
$$(1-1.5\phi^{-1})[-14.074(1-0.5\phi^{-1})(1-0.663\phi^{-1}) \qquad 6.963(1-0.8\phi^{-1})(1-0.536\phi^{-1})]$$

$$z_i'(\phi)\ \mathbf{T}_{Pi}(\phi)\ \mathbf{\Omega}_C(\phi) =$$

$$(1\text{-}1.5\phi^{-1})\phi^{-1}\,[14.074(1\text{-}0.663\phi^{-1}) \quad 7.111(1\text{-}0.394\phi^{-1}) \quad -6.963(1\text{-}0.536\phi^{-1})]$$

so the cancellation takes place as expected. From this, the computational form is found as

$$w_i'(\phi) = \frac{1}{(1\text{-}0.7\phi^{-1})}\,[1.173(1\text{-}0.5\phi^{-1})(1\text{-}0.663\phi^{-1})$$

$$-0.580(1\text{-}0.8\phi^{-1})(1\text{-}0.536\phi^{-1})]$$

$$v_i'(\phi) = \frac{1}{(1\text{-}0.7\phi^{-1})}\,[-2\phi^{-1}(1\text{-}0.7\phi^{-1}) \quad -2.938(1\text{-}0.609\phi^{-1})]$$

Finally, the actual fault responses are

$$z_i'(\phi)\ \mathbf{T}_{Ri}(\phi)\ \mathbf{T}_{Pi}(\phi) = [\,5.926 \quad 2\phi^{-1}\,]$$

Design for polynomial transformation. We will demonstrate polynomial implementation with both algorithms. The target response in either case is again $z_i'(\phi)=[1 \quad 2]$.

Algorithm I. Since the invariant zero factor $(1\text{-}0.7\phi^{-1})$ appears in the full second row of the $[\mathbf{\Omega}_C(\phi) \quad \mathbf{\Omega}_F(\phi)]$ matrix, it needs to be included only in the first response modifier, that is, by (6.127),

$$\mathbf{T}_{Pi}(\phi) = Diag[(1\text{-}0.7\phi^{-1})(1\text{-}1.5\phi^{-1}) \quad (1\text{-}1.5\phi^{-1})]$$

With this, the transformation is obtained from (6.117) and (6.118) as

$$w_i'(\phi) = [1.583(1\text{-}0.5\phi^{-1})(1\text{-}1.021\phi^{-1}) \quad -1.483(1\text{-}0.375\phi^{-1})(1\text{-}0.8\phi^{-1})]$$

$$v_i'(\phi) = [\,-2(1\text{-}1.5\phi^{-1}) \quad -3.267(1\text{-}1.174\phi^{-1})]$$

which is, indeed, polynomial. The actual responses become

$$z_i'(\phi)\ \mathbf{T}_{Ri}(\phi)\ \mathbf{T}_{Pi}(\phi) = [(1\text{-}0.7\phi^{-1})(1\text{-}1.5\phi^{-1}) \quad 2\phi^{-1}(1\text{-}1.5\phi^{-1})]$$

Algorithm II. Since the second row of $[\mathbf{\Omega}_C(\phi) \quad \mathbf{\Omega}_F(\phi)]$ contains an invariant zero factor, select the first response for modification. Thus

$$\mathbf{T}_{Pi}(\phi) = Diag[\,\upsilon_{Pi1}(\phi) \quad 1\,]$$

The cancellation condition (6.129), written for the first column of the $\Omega_F(\phi)$ matrix is

$$\upsilon_{Pi1}(\phi)(1-0.5\phi^{-1})(1-0.6\phi^{-1}) - 20(1-0.5\phi^{-1})(1-0.7\phi^{-1})(1-\phi^{-1})$$

$$= (1-0.7\phi^{-1})(1-1.5\phi^{-1})\beta_i(\phi)$$

Since there are two invariant zeroes, $\upsilon_{Pi1}(\phi)=\alpha_0+\alpha_1\phi^{-1}$. Writing (6.132) for $\phi=\zeta_1=0.7$ and $\phi=\zeta_2=1.5$ yields

$$\alpha_0+\alpha_1/0.7 = 0 \qquad 0.6(\alpha_0+\alpha_1/1.5) - 3.555 = 0$$

leading to the solution

$$\upsilon_{Pi1}(\phi) = 11.11(1-0.7\phi^{-1})$$

With this, the transformation is obtained as

$$w_i'(\phi) = [0.7408(1-0.5\phi^{-1}) \qquad 0.3704(1-0.8\phi^{-1})]$$

$$v_i'(\phi) = [\,-2\phi^{-1} \qquad -2.593\phi^{-1}\,]$$

This is clearly simpler than the transformation yielded by Algorithm I. Now the actual fault responses are

$$z_i'(\phi)\,\mathbf{T}_{Ri}(\phi)\,\mathbf{T}_{Pi}(\phi) = [11.11(1-0.7\phi^{-1}) \qquad 2\phi^{-1}]$$

Clearly, only the first response is affected by the adjustment.

Almost homogeneous specification. Finally, we will show the design for this important special case.

a. Let us design a polynomial residual generator for the example system so that $z_i'(\phi)=[1 \quad 0]$. By (6.144) and (6.146):

$$\upsilon_{Ri1}(\phi) = 1 \qquad \upsilon_{Pi1}(\phi) = (1-0.7\phi^{-1})(1-1.5\phi^{-1})$$

Then from (6.145) and (6.147):

$$w_i'(\phi) = [\,-0.083(1-0.5\phi^{-1})(1-0.6\phi^{-1}) \qquad 0.183(1-0.6\phi^{-1})(1-0.8\phi^{-1})]$$

$$v_i'(\phi) = [\,0 \qquad 0.067\phi^{-1}(1-0.075\phi^{-1})]$$

b. Now let us specify $z_i'(\phi)=[0 \quad 2]$. Then

$$v_{Ri2}(\phi) = \phi^{-1} \qquad v_{Pi2}(\phi) = 1\text{-}1.5\phi^{-1}$$

yielding

$$w_i'(\phi) = [\ 1.666(1\text{-}0.5\phi^{-1})(1\text{-}\phi^{-1}) \qquad -1.666(1\text{-}0.4\phi^{-1})(1\text{-}0.8\phi^{-1})]$$

$$v_i'(\phi) = [\ -\ 2\phi^{-1}(1\text{-}1.5\phi^{-1}) \qquad -3.333\phi^{-1}(1\text{-}1.15\phi^{-1})]$$

6.7. NOTES AND REFERENCES

The basic parity relation formulation arose from earlier work on dynamic balance calculations (Gertler and Almásy, 1973) and was published in (Gertler and Singer, 1985, 1990). Similar ideas can be found in (Mironovskii, 1980; Ben-Haim, 1980; Chow and Willsky, 1984). Some techniques of the naive design with parity relations were first tried out, and partially suggested, in the work of John Shutty (1985) and Amartur Sundar (1985).

The Chow-Willsky scheme is due, of course, to Chow and Willsky. It appeared first in Edward Chow's dissertation (1980) and then published in their celebrated paper (1984). The link to our parity relation techniques was studied by Qiang Luo (1990) and utilized in a simulated distillation column application (Gertler and Luo, 1989; Gertler, Luo and Fang, 1990). Xiaowen Fang (1993) solved the conversion of the scheme into the generic formulation.

The use of the fault system matrix in parity relation implementation was proposed by Ramin Monajemy (1993). It was subsequently utilized in the parity relation design of directional residuals (Gertler and Monajemy, 1993, 1995), in what we refer to here as Algorithm I. Note that the fault system matrix was also employed, independently, by R. Nikoukhah (1994), in the design of what we may call the generic parity relation equivalent of the Kalman filter.

The design of parity relations for explicit response specifications was first outlined in (Gertler and Monajemy, 1993) and then expanded in (Gertler, 1995a). The latter paper contains, for the first time, Algorithm II; the key to this technique was the identity (6.102), found in (Gantmakher, 1959). Algorithm III was also published there, only for homogeneous specifications, where its application is simple and straightforward. These results have been summarized also in (Gertler, 1997).

Note that the Chow-Willsky scheme enjoys overwhelming recognition in the literature. This is well deserved, for historical reasons and because of the merits of the technique. It is somewhat unfortunate, though, that in the minds of many, the Chow-Willsky scheme is equated with the parity relation approach, without realizing that it is, however important, a special case.

7
Design for Structured Residuals

7.0. INTRODUCTION

As it was explained in Chapter 5, designing the residual set with a certain structure is one of the ways of enhancing its fault isolation capabilities. Structured residuals are characterized by selective fault responses; any residual responds only to a specific subset of faults, and to any fault only a specific subset of the residuals responds. In Chapter 5 we gave a geometric interpretation to structured residuals and mentioned that there is also a Boolean vector-matrix representation; in this chapter, we will pursue the latter approach.

While faults need to be detected and isolated, disturbances need to be ignored. This will be achieved by integrating them into the residual structure with the appropriate response.

Recall also from Chapter 5 that structured residuals are tested individually, in parallel, against pre-defined thresholds. The outcome of each test is *fired/not fired*, usually represented by the binary values "*1*" and "*0*":

$$\epsilon_i(t) = \begin{bmatrix} 1 & \text{if } |r_i(t)| \geq \kappa_i \\ 0 & \text{if } |r_i(t)| < \kappa_i \end{bmatrix} \tag{7.1}$$

where $\epsilon_i(t)$ is the binary test result on the i-th residual and κ_i is its threshold. The set or vector of simultaneous test bits

$$\epsilon(t) = \begin{bmatrix} \epsilon_1(t) \\ \cdot \\ \cdot \\ \cdot \\ \epsilon_n(t) \end{bmatrix} \tag{7.2}$$

is the observed *fault signature* or *fault code*.

In this chapter, we will explore the structural properties of structured residual sets, assuming first single faults and then extending the results to multiple faults. This will lead to the specification of structured residual generators. We will show then how such generators can be implemented, using parity equations. The existence criteria for such implementations will also be investigated. An optimization-based design approach which combines structural properties with quantitative performance measures such as fault sensitivity will be described in Chapter 10.

7.1. RESIDUAL STRUCTURES FOR SINGLE FAULT ISOLATION

For faults, it may be reasonable to assume that only one is present in the system at a time. The occurrence of faults, in general, is not very frequent, and in many cases it may be assumed that any fault gets repaired before another one appears. Residual structures for the isolation of single faults are relatively simple and, as far as the number of faults is concerned, unrestricted. In situations when the single fault assumption is not allowable, structures capable of handling multiple faults need to be constructed; these are more complex and/or more restricted than the single fault structures. Disturbances, on the other hand, usually occur simultaneously with one another and with faults. However, their isolation normally is not required, only the residuals need to be decoupled from (made insensitive of) them.

In this section, after defining some basic structural concepts, structures for single fault isolation will be explored. It will also be shown how disturbance decoupling can be incorporated into these schemes.

7.1.1. Structural Definitions

The structure matrix of a residual set. The structure matrix Φ of a residual set expresses the cause-effect relationships between faults and disturbances as inputs and residuals as outputs. Each column of the matrix represents a fault or disturbance and each row a residual. A *"1"* in the intersection means the fault/disturbance does affect the residual while a *"0"* means it does not.

Example 7.1.

Consider a system with three faults p_1, p_2, p_3, two residuals r_1, r_2 and the structure matrix

	p_1	p_2	p_3
r_1	1	1	0
r_2	1	0	1

or

$$\Phi = \begin{bmatrix} 1 & 1 & 0 \\ 1 & 0 & 1 \end{bmatrix}$$

Then p_1 affects both residuals, p_2 affects only the first and p_3 affects only the second; r_1 is affected by the first and second faults while r_2 is affected by the first and the third. ⁝⁝⁝

Ideally, the observed fault signature, in response to a single fault $p_j(t)$, *is the* j*-th column of the structure matrix*, that is,

$$\boxed{\epsilon(t|p_j) = \varphi_{\cdot j}} \tag{7.3}$$

Thus $\varphi_{\cdot j}$ is the *ideal signature* or *code* of fault p_j.

Canonical structure matrices. A structure matrix will be referred to as *row-canonical* if each row contains the same number of "*0*" elements, each in a different configuration. A structure will be referred to as *column-canonical* if each column contains the same number of "*0*" elements, each in a different configuration.

Undetectability in a structure. A fault or disturbance is *undetectable* in a residual structure if its column in the structure matrix contains only "*0*" elements. Note that while undetectability is undesirable for a fault, clearly this is the desirable behavior as far as disturbances are concerned.

Indistinguishability in a structure. Two faults or disturbances are *indistinguishable* in a structure if their respective columns in the structure matrix are identical.

Weakly isolating structure. We will refer to a structure as *weakly isolating* if all columns in the structure matrix are different and nonzero. Obviously, with such structure, all faults are detectable and all *single* faults are mutually distinguishable.

Example 7.2.

The structure shown in Example 7.1 is clearly weakly isolating. ⁝⁝⁝

Unidirectional strongly isolating structure. A structure will be referred to as *unidirectional strongly isolating* if it is weakly isolating and if no column in the structure matrix can be obtained from any other column by turning an arbitrary number of "*1*"s into "*0*"s or by turning an arbitrary number of "*0*"s into "*1*"s. We make the following observations:

1. Unidirectional strong isolation requires that, for each pair of columns, there be a position where one column has a "*1*" and the other a "*0*" and vice versa.
2. A column-canonical structure always satisfies the requirement described in Observation 1.

Bidirectional strongly isolating structure. A structure is *bidirectional strongly isolating of degree 1* if it is weakly isolating and if no column can be obtained from any other column by changing an (arbitrary) single element. Similarly, a structure is bidirectional strongly isolating *of degree k* if no column can be obtained from any other column by changing up to k (arbitrary) elements. The following observations are in order:

1. Bidirectional strong isolation of degree k requires that, for each pair of columns, there be at least $k+1$ positions where the two columns are different.
2. Bidirectional strong isolation of degree *1* is implied in unidirectional strong isolation. Consequently, a column-canonical structure always satisfies the requirement described in Observation 1, at least for degree *1*.

Example 7.3.

Consider the following three structures, all weakly isolating:

$$\Phi_1 = \begin{bmatrix} 1 & 1 & 0 \\ 1 & 0 & 0 \\ 0 & 0 & 1 \\ 0 & 1 & 1 \end{bmatrix} \qquad \Phi_2 = \begin{bmatrix} 1 & 1 & 0 \\ 1 & 1 & 0 \\ 1 & 0 & 1 \\ 0 & 1 & 1 \end{bmatrix} \qquad \Phi_3 = \begin{bmatrix} 1 & 1 & 1 \\ 1 & 1 & 0 \\ 1 & 0 & 0 \\ 1 & 0 & 1 \end{bmatrix}$$

Φ_1 is clearly column-canonical. It is unidirectional strongly isolating, because turning "*0*"s into "*1*"s or "*1*"s into "*0*"s in any column upsets the 2/2 ratio of "*0*"s and "*1*"s and thus can not lead to another column. Also, the structure is bidirectional strongly isolating of degree 1 since changing any element ("*1*" to "*0*" or "*0*" to "*1*') can not lead to any other column.

Φ_2 is not column-canonical. However, it does have "*1*" vs. "*0*" pairs, in either direction, between each pair of columns, thus it is unidirectional strongly isolating. This implies that it is also bidirectional of degree *1*.

Finally, Φ_3 is not column-canonical but it has two different elements between any two columns so it is bidirectional strong isolating of degree *1*. However, it is not unidirectional strong isolating; e.g., the second or third columns can be obtained from the first by turning two of its "*1*"s into "*0*". ⁞⁞⁞

7.1.2. Canonical Structures

Generally, we will construct column-canonical residual structures from a canonical set of rows. Below we will first give a rationale for this strategy and then will introduce some combinatorial properties of canonical structures.

Column-canonical structures are robust. The outcome of a test on a single residual depends on the size of the fault, if any, and on any noise and modeling error affecting the residual. In the absence of faults, the noise and modeling error may cause a test to fire, resulting in a situation known as *false alarm*. If a fault is present, the tests may still not fire, if the size of the fault is small and/or if noise or modeling error act against the effect of the fault in the residual; this situation is known as *missed detection*.

To reduce the frequency of false alarms, which are usually considered the more detrimental of the two situations, the thresholds are normally set high. This, however, increases the frequency of missed detections. In particular, if the size of the fault is small, some of the residuals which are supposed to respond do not trigger their respective test, resulting in what is called *partial firing*. Partial firing is a rather frequent situation which can lead to misclassification of the fault - unless the residual structure is carefully designed. Obviously, to avoid misclassification under partial firing, the residual structure has to be *unidirectional strongly isolating*.

As we pointed out before, column-canonical structures are sufficient (though not necessary) for unidirectional strong isolation. Though they are somewhat more restrictive, their symmetry makes it significantly easier to work with them than with general unidirectional structures.

Row-canonical structures follow naturally. Recall the generic residual generator with $W(\phi)=I$ (primary set):

$$o(t) = y(t) - M(\phi)u(t) = \blacktriangle y(t) - S(\phi)\blacktriangle u(t) \tag{7.4}$$

Each scalar equation contains one of the output sensor faults and, in general, all the input faults. Thus this residual set, in general, has a row-canonical structure, with $\mu - 1$ "*0*" elements in each row (where μ is the number of outputs). There may be additional "*0*" elements, if one or more of the input faults do not affect the concerned residual, but these are incidental.

Now recall from Chapter 6 that, normally, the "maximum" homogeneous specification attainable was almost full homogeneous, or full almost homogeneous. That is, the maximum number of "*0*" elements in a row, attainable in general by transformation, is $\mu - 1$. Thus if we generate residuals so that each one has a different structure and the maximum attainable number of zeroes (without utilizing incidental zeroes), then the resulting set, which may include parts of the primary set, is row-canonical, each row having $\mu - 1$ "0" elements.

Note that the row behavior of a residual structure does not play any role in its fault isolation properties. However, it is easier to construct a column-canonical structure if the rows we are using are also canonical.

Two basic constraints. For simple implementation, the number n of residuals should be kept low and the number τ of "0" elements in each column should be made high. These parameters are subject to two constraints:

1. *There must be enough different patterns.* That is, the number of possible ways how τ zeroes can be positioned in columns of length n should be at least as large as the number ρ of faults:

$$\binom{n}{\tau} \geq \rho \tag{7.5}$$

2. *There must be enough "0" elements.* Each of the n rows contributes $\mu - 1$ "0" elements, or fewer if not all the potential zeroes are utilized; these have to provide the τ zeroes needed for each of the ρ columns:

$$n(\mu - 1) \geq \rho\tau \tag{7.6}$$

With μ and ρ given, the smallest integer solution for n is sought, and then the largest integer for τ is found. Note that the square set $n=\rho$, $\tau=\mu - 1$ is always a solution and in many cases the simplest one; in other cases however it may be possible to find a simpler structure.

Table 7.1 shows the simplest row- and column-canonical sets for various values of μ and ρ. The numbers in the table indicate n/τ and "S" stands for square set.

Notice that the described canonical structure *does not place any limit on the number of faults* that may be handled in the system, no matter what the number of plant outputs is (as long as it exceeds *1*).

Example 7.4.

a. Consider first a system with $\mu=3$, $\rho=5$. The simplest solution is $n=5$, $\tau=2$. A possible structure is

TABLE 7.1. Row- and column-canonical structures (shown is n/τ).

ρ \ μ	2	3	4	5	6	7	8	9
2	s	-	-	-	-	-	-	-
3	s	s	-	-	-	-	-	-
4	s	s	s	-	-	-	-	-
5	s	s	s	s	-	-	-	-
6	s	s	4/2	s	s	-	-	-
7	s	s	s	s	s	s	-	-
8	s	s	s	6/3	s	s	s	-
9	s	s	6/2	s	s	6/4	s	s
10	s	s	s	5/2	6/3	5/3	s	s
11	s	s	s	s	s	s	s	s
12	s	s	8/2	6/2	s	6/3	s	6/4
13	s	s	s	s	s	s	s	s
14	s	s	s	7/2	s	7/3	6/3	7/4
15	s	s	10/2	s	6/2	10/4	s	s

$$\Phi = \begin{bmatrix} 0 & 1 & 1 & 1 & 0 \\ 0 & 0 & 1 & 1 & 1 \\ 1 & 0 & 0 & 1 & 1 \\ 1 & 1 & 0 & 0 & 1 \\ 1 & 1 & 1 & 0 & 0 \end{bmatrix}$$

b. Now consider $\mu=4$, $\rho=6$. The simplest solution turns out to be $n=4$, $\tau=2$. With these, the only possible structure is

$$\Phi = \begin{bmatrix} 0 & 1 & 1 & 0 & 1 & 0 \\ 0 & 0 & 1 & 1 & 0 & 1 \\ 1 & 0 & 0 & 1 & 1 & 0 \\ 1 & 1 & 0 & 0 & 0 & 1 \end{bmatrix}$$

Note that here all the possible column structures have been utilized and only the order of the columns may be varied. ⁝⁝⁝

The combinatorics of row-canonical structures. The number of different canonical row structures is

$$K = \binom{\rho}{\mu - 1} \tag{7.7}$$

The number of different row-canonical sets that may then be composed of these is

$$N = \binom{K}{n} \tag{7.8}$$

which, apart from trivial systems, is a very large number. However, only a fraction of these row-canonical sets is also column-canonical.

Example 7.5.

For the second system in Example 7.4, with $\mu=4$ and $\rho=6$, the number of canonical equation structures is $K=20$ and the number of different row-canonical sets is $N=4845$. However, with $\mu=4$ and $\rho=10$, K grows to 120 and N becomes 8214570. ⁝⁝⁝

7.1.3. Handling Disturbances

For disturbance decoupling, the residual structure needs to be designed so that the disturbances are undetectable. This requires zero columns in the structure matrix. Since in any particular row the number of zeroes normally can not exceed $\mu - 1$, this is also the maximum number of (independently acting) disturbances for which decoupling can be specified, that is

$$\rho_D \leq \mu - 1 \tag{7.9}$$

where ρ_D is the number of (independently acting) decoupled disturbances.

Normally, decoupling from disturbances needs to be combined with the isolation of faults. Thus at least one zero per row has to be reserved for structuring the residuals for fault isolation. That is, (7.9) changes to

$$\rho_D \leq \mu - 2 \tag{7.10}$$

At the same time, from the point of view of residual structuring, the number of available zeroes is reduced by those used for disturbance decoupling, so (7.6) changes to

$$n(\mu - \rho_D - 1) \geq \rho\tau \tag{7.11}$$

This, however, does not place a constraint on the number of faults which can be handled.

Example 7.6.

a. Design a residual structure for $\mu=4$, $\rho_D=2$, $\rho=2$. From (7.11), $n=2$, $\tau=1$, which satisfies (7.5) as well. The resulting structure is

	q_1	q_2	p_1	p_2
r_1	0	0	1	0
r_2	0	0	0	1

b. Now let us design a structure for $\mu=5$, $\rho_D=2$, $\rho=3$. The minimum complexity solution is $n=3$, $\tau=2$, yielding the structure

	q_1	q_2	p_1	p_2	p_3
r_1	0	0	0	0	1
r_2	0	0	0	1	0
r_3	0	0	1	0	0

7.2. RESIDUAL STRUCTURES FOR MULTIPLE FAULT ISOLATION

When more than one fault is present in the system simultaneously then the effect of these faults on the residuals is adding up. Strictly speaking, the combination of the effects is algebraic and, in principle, they may even completely cancel each other. Such complete cancellation is, however, very rare and even if it happens usually it is only temporary. Therefore we may, in general, assume that the combination is logical, that is, ***the ideal signature in response to a combination of faults is the logical OR combination of the component signatures***. That is:

$$r(t|p_i \,\&\, p_j) = \varphi_{\cdot i} \cup \varphi_{\cdot j}$$

$$r(t|p_i \,\&\, p_j \,\&\, p_k) = \varphi_{\cdot i} \cup \varphi_{\cdot j} \cup \varphi_{\cdot k} \quad \text{etc.} \tag{7.12}$$

Column-canonical structures designed for single fault isolation are not in general column-canonical for multiple faults and may not even be weakly isolating. It is possible to construct residual structures with multiple fault isolation in mind, but these are more complex than the single fault structures and/or place stricter restrictions on the number of faults that can be handled. Below we will describe two approaches. The first is simple but restricts the number of faults to that of the system outputs. The second has no such restriction but is rather complex and may restrict the degree of fault multiplicity.

7.2.1. Diagonal Structure

If the number of faults equals that of the outputs, that is, $\rho=\mu$, then the number of canonical row structures from (7.7) is $K=\rho$, with $\rho - 1$ "0" elements in each row. These form a column-canonical structure with $\tau=\rho - 1$. That is, the structure matrix is square $\rho \cdot \rho$, with only one "1" element in each row and column. Since the order of rows (columns) is arbitrary, this structure may be considered *diagonal*.

A diagonal structure is very advantageous for multiple fault isolation. All possible multiple faults have distinct signatures. Further, the signatures for faults of the same degree of multiplicity form column-canonical sets.

The full structure, containing the signatures of all the single and multiple faults, is not unidirectional strongly isolating. Partial firing in a multiple fault situation may produce the signature of one of the component faults, single or composite itself. This, of course, is what is expected.

The only disadvantage of the diagonal structure is that it restricts the number of faults which can be handled to that of the system outputs. Note that this can not be avoided by "decomposing" the system and creating separate diagonal structures for subsets of faults; the faults of the other subsets still affect the residuals even if they are not represented in the structure. Decomposition is only possible if subsets of the faults can be associated with subsets of the outputs - but then the number of outputs in a subset is being reduced as well.

Example 7.7.

Let us consider a diagonal structure for $\rho=\mu=4$ and see all the multiple fault signatures:

	p_1	p_2	p_3	p_4	$p_1\&p_2$	$p_1\&p_3$	$p_1\&p_4$	$p_2\&p_3$	$p_2\&p_4$	$p_3\&p_4$
r_1	1	0	0	0	1	1	1	0	0	0
r_2	0	1	0	0	1	0	0	1	1	0
r_3	0	0	1	0	0	1	0	1	0	1
r_4	0	0	0	1	0	0	1	0	1	1

$p_1\&p_2\&p_3$	$p_1\&p_2\&p_4$	$p_1\&p_3\&p_4$	$p_2\&p_3\&p_4$	$p_1\&p_2\&p_3\&p_4$
1	1	1	0	1
1	1	0	1	1
1	0	1	1	1
0	1	1	1	1

7.2.2. Full Row Canonical Sets

Multiple fault isolation can be achieved even if $\rho > \mu$, by using the full row-canonical structure. The main price to be paid is the significantly greater complexity of such structures. Further, isolation is not possible for all degrees of fault multiplicity.

Consider first the $\mu=3$ case, with arbitrary $\rho>3$. The number of canonical single-fault row structures is

$$K = \binom{\rho}{2} \tag{7.13}$$

each row containing 2 zeroes. Because of the perfect symmetry, this set is also column-canonical; by (7.6), the number of zeroes per column is $\rho - 1$. There are n possible double faults so the structure matrix for double faults will be square. When constructing this matrix, a "0" element will result whenever both component faults have a "0" in the single fault structure. In each row, there is just one such position thus the matrix will be row and column-canonical with a single zero in each row and column. Thus strong isolation of double faults is possible. However, for any triple fault the signature will be all "1", thus the structure does not support even weak isolation of such faults.

Consider now $\mu=4$, with arbitrary $\rho>4$. The number of canonical rows in the single-fault structure is now

$$K = \binom{\rho}{3} \tag{7.14}$$

This is also the number of *triple faults* so the structure matrix for triple faults is square. The single fault matrix has three zero elements per row; the triple fault matrix has one. For double faults, the number of columns is $\binom{\rho}{2}$, with 3 zeroes per row; the structure is also column-canonical. Thus there are column-canonical structures for single, double and triple faults; quadruple faults however can not be isolated.

In general, with $\rho>\mu$ faults, the number of canonical (single fault) row structures is

$$K = \binom{\rho}{\mu - 1} \tag{7.15}$$

This is also the number of columns in the structure for $(\mu\text{-}1)$-tuple faults which thus have a square matrix with one zero per row/column. For a fault multiplicity of κ, $1 \leq \kappa \leq \mu\text{-}1$, row and column-canonical structures are obtained with $\binom{\rho}{\kappa}$ columns and $\binom{\mu - 1}{\kappa}$ zero elements in each row. Thus *with μ outputs in the system, strong isolation (within each degree of multiplicity) is possible for 1, 2, ... $(\mu - 1)$ simultaneous faults.*

Just like in the case of the diagonal structure, the system is not (and can not be) unidirectional strongly isolating between the different degrees of fault multiplicity. Partial firing in a multiple fault situation may well yield the signature of one of the component faults.

Note that the diagonal structure is, in fact, the limit case of the full row-canonical structure. Now $\rho = \mu$ and the full row-canonical set consists of ρ rows, with $\rho - 1$ zeroes in each. The maximum degree of fault multiplicity for which a canonical structure is obtained is $\rho - 1$, with one "0" element per row. However, there is only one possible ρ-tuple fault (with an all "1" signature) so all multiple faults can be isolated.

Example 7.8.

We will demonstrate a full row-canonical structure for $\mu = 4$ and $\rho = 5$. The number of rows is $K = 10$ and the maximum isolable fault multiplicity is 3. There are 10 possible double faults, 10 triple faults and 5 quadruple faults. The full structure matrix is shown in Table 7.2 (the faults are represented by their indices). ⁝⁝⁝

7.3. PARITY EQUATION IMPLEMENTATION OF STRUCTURED RESIDUALS

The implementation of structured residual generators by parity equations is straightforward, using the single residual implementation techniques explored in Chapter 6. Here we will demonstrate how those methods apply to general row-canonical structures and to the diagonal structure. Special attention will be paid to the existence conditions, and the important concepts of *attainability* of a particular row structure and *isolability* of certain faults will be introduced.

7.3.1. Row-by-Row Implementation

Consider a residual structure and define the matrix $S_i(\phi)$ so that it contains those columns of the fault transfer matrix $S(\phi)$ where the i-th row structure has "0" elements. In a row-canonical structure, there are $\mu - 1$ such elements thus $S_i(\phi)$ has μ rows and $\mu - 1$ columns. Note that the $S_i(\phi)$

TABLE 7.2. Full row-canonical set and multiple faults (μ=4, ρ=5).

1	2	3	4	5	12	13	14	15	23	24	25	34	35	45
0	0	0	1	1	0	0	1	1	0	1	1	1	1	1
0	0	1	0	1	0	1	0	1	1	0	1	1	1	1
0	0	1	1	0	0	1	1	0	1	1	0	1	1	1
0	1	0	0	1	1	0	0	1	1	1	1	0	1	1
0	1	0	1	0	1	0	1	0	1	1	1	1	0	1
0	1	1	0	0	1	1	0	0	1	1	1	1	1	0
1	0	0	0	1	1	1	1	1	0	0	1	0	1	1
1	0	0	1	0	1	1	1	1	0	1	0	1	0	1
1	0	1	0	0	1	1	1	1	1	0	0	1	1	0
1	1	0	0	0	1	1	1	1	1	1	1	0	0	0

123	124	125	134	135	145	234	235	245	345	1234	1235	1245	1345	2345
0	1	1	1	1	1	1	1	1	1	1	1	1	1	1
1	0	1	1	1	1	1	1	1	1	1	1	1	1	1
1	1	0	1	1	1	1	1	1	1	1	1	1	1	1
1	1	1	0	1	1	1	1	1	1	1	1	1	1	1
1	1	1	1	0	1	1	1	1	1	1	1	1	1	1
1	1	1	1	1	0	1	1	1	1	1	1	1	1	1
1	1	1	1	1	1	0	1	1	1	1	1	1	1	1
1	1	1	1	1	1	1	0	1	1	1	1	1	1	1
1	1	1	1	1	1	1	1	0	1	1	1	1	1	1
1	1	1	1	1	1	1	1	1	0	1	1	1	1	1

matrix is different for each row in the structure. The i-th residual generator can be specified as

$$\mathbf{w}_i'(\phi)\mathbf{S}_i(\phi) = \mathbf{0}' \tag{7.16a}$$

$$\text{and } \mathbf{w}_i'(\phi)\mathbf{s}_{\cdot k}(\phi) \neq 0 \text{ for all } \mathbf{s}_{\cdot k}(\phi) \text{ outside } \mathbf{S}_i(\phi) \tag{7.16b}$$

Here (7.16a) is clearly an almost full homogeneous specification which can be solved by any of the homogeneous implementation methods discussed in

Chapter 6. However, these solutions do not guarantee that the condition (7.16b) is satisfied as well; the latter needs to be addressed separately. This will be discussed below, in Subsection 7.3.2.

Alternatively, the implementation problem can be formulated as

$$w_i'(\phi)[S_i(\phi) \quad s_{\cdot g}(\phi)] = [0 \ldots 0 \quad z_{ig}(\phi)\upsilon_{ig}(\phi)] \tag{7.17a}$$

$$\text{and} \quad w_i'(\phi)s_{\cdot k}(\phi) \neq 0 \quad \text{for all } s_{\cdot k}(\phi) \text{ outside } [S_i(\phi) \quad s_{\cdot g}(\phi)] \tag{7.17b}$$

Here (7.17a) is a full almost homogeneous specification which can be handled by the techniques introduced in Chapter 6. Again, the condition (7.17b) is not implied in the solution.

The implementation, of course, depends also on the specified computational behavior of the residual generator (ARMA or MA); the different variants have been explored in Chapter 6. As it was pointed out there, the non-homogeneous part of the specification, if any, is not completely guaranteed; the realization, stability and polynomial computational behavior of the residual generator may require the modification of the response via $\upsilon_{ig}(\phi)$.

It is important to see that, of the $\rho - \mu+1$ nonzero responses ("1"s in the structure), only one can in general be specified. This happens explicitly in (7.17a) with $z_{ig}(\phi)$, and implicitly during the solution of (7.16a). The other nonzero responses "float," that is, the (7.16b) and (7.17b) conditions only state that they are nonzero. Of course, once the generator has been designed, the exact responses to those faults can also be computed.

Example 7.9.

Consider first a static system with $\mu=3$, $\rho=4$ and

$$S = \begin{bmatrix} 1 & 3 & -2 & 0 \\ 2 & 4 & 1 & -1 \\ -1 & 0 & 2 & 4 \end{bmatrix}$$

Note that all $3 \cdot 3$ determinants of this system are nonzero. Design a single-output residual generator (one row of a generator) so that its row pattern is

$$\varphi_i' = [0 \quad 0 \quad 1 \quad 1]$$

We will use the almost homogeneous approach, Eqs. (7.17a,b), and include the third response in the specification. Let us choose the response as

$$z_i' = [\,0 \quad 0 \quad 5\,]$$

For this system,

$$[S_i \quad s_{.3}] = \begin{bmatrix} 1 & 3 & -2 \\ 2 & 4 & 1 \\ -1 & 0 & 2 \end{bmatrix}$$

The above matrix has full rank, so (7.17a) yields the transformation as

$$w'_i = z_i' [S_i \quad s_{.3}]^{-1} = [\,-1.333 \quad 1 \quad 0.667\,]$$

We did not (and could not) specify the response to the fourth fault; this can now be computed as

$$z_{i4} = w_i' s_{.4} = 1.667$$

Thus (7.17b) is satisfied and the desired row pattern has been obtained. ⁞⁞⁞

Example 7.10.

Consider now a dynamic system with $\nu=3$, $\mu=3$, $\rho=4$. Assume a fault transfer matrix as shown in Table 7.3. Design a row of a residual generator with the structure

$$\varphi_i' = [\,0 \quad 0 \quad 1 \quad 1\,]$$

Now we will approach this as an almost full homogeneous specification and use

TABLE 7.3

$s_{.1}(\phi)$	$s_{.2}(\phi)$	$s_{.3}(\phi)$	$s_{.4}(\phi)$
$\dfrac{3\phi - 1.6}{(\phi - 0.5)(\phi - 0.6)}$	$\dfrac{3}{\phi - 0.6}$	$\dfrac{1}{\phi - 0.5}$	$\dfrac{1}{\phi - 0.6}$
$\dfrac{2}{\phi - 0.6}$	$\dfrac{4\phi - 3}{(\phi - 0.6)(\phi - 0.8)}$	$\dfrac{1}{\phi - 0.8}$	$\dfrac{1}{\phi - 0.6}$
$\dfrac{1}{\phi - 0.5}$	$\dfrac{1}{\phi - 0.8}$	$\dfrac{2\phi - 1.3}{(\phi - 0.5)(\phi - 0.8)}$	0

the direct ("naive") version of Algorithm III for implementation (c.f. Subsection 6.3.1, Eqs. (6.45)-(6.47)). To find the transformation, we need to use the columns $s_{.1}(\phi)$ and $s_{.2}(\phi)$. Choose the last element of the transformation for arbitrary assignment; thus

$$s_{\mu.}(\phi) = s_{3.}(\phi) = \begin{bmatrix} \dfrac{1}{\phi - 0.5} & \dfrac{1}{\phi - 0.8} \end{bmatrix}$$

$$S^{\mu}(\phi) = S^{3}(\phi) = \begin{bmatrix} \dfrac{3\phi - 1.6}{(\phi - 0.5)(\phi - 0.6)} & \dfrac{3}{\phi - 0.6} \\ \\ \dfrac{2}{\phi - 0.6} & \dfrac{4\phi - 3}{(\phi - 0.6)(\phi - 0.8)} \end{bmatrix}$$

The inverse of $S^3(\phi)$ is found to be

$$[S^3(\phi)]^{-1} = \frac{1}{\phi - 0.667} \begin{bmatrix} 0.667(\phi - 0.75)(\phi - 0.5) & -0.5(\phi - 0.5)(\phi - 0.8) \\ \\ -0.333(\phi - 0.5)(\phi - 0.8) & 0.5(\phi - 0.535)(\phi - 0.8) \end{bmatrix}$$

Here $(\phi - 0.667)$ is the invariant zero polynomial of the subsystem $S^3(\phi)$. Now choose $w_{i3}(\phi) = -(1 - 0.667\phi^{-1})$. Then from (6.47)

$$[w_{i1}(\phi) \quad w_{i2}(\phi)] = -w_{i3}(\phi)s_{3.}(\phi)[S^3(\phi)]^{-1} = [0.333(1 - \phi^{-1}) \quad 0.133\phi^{-1}]$$

The response to the third and fourth faults is not specified in this design. They are determined, implicitly, by the choice of $w_{i3}(\phi)$ and can be computed, with $w_i'(\phi) = [w_{i1}(\phi) \quad w_{i2}(\phi) \quad w_{i3}(\phi)]$, as

$$[z_{i3}(\phi) \quad z_{i4}(\phi)] = w_i'(\phi)[s_{.3}(\phi) \quad s_{.4}(\phi)] = [-1.667\phi^{-1} \quad 0.333\phi^{-1}]$$ ∷∷

7.3.2. Existence Conditions

As shown in Chapter 6, the solvability condition of (7.16a) is

$$Rank\, S_i(\phi) \le \mu - 1 \tag{7.18a}$$

This, of course, is always satisfied if $S_i(\phi)$ contains $\mu - 1$ columns. However, (7.18a) allows more than $\mu - 1$ "0" elements in the row structure, if the accompanying columns in the fault transfer matrix are linearly dependent.

The condition for (7.16b), given (7.16a), is

$$\text{Rank}\,[S_i(\phi) \quad s_{\cdot k}(\phi)] = \text{Rank}\,S_i(\phi)+1 \quad \text{for all } s_{\cdot k} \text{ outside } S_i \qquad (7.18b)$$

In the full almost homogeneous formulation, the solvability condition of (7.17a) is (7.18a), plus the satisfaction of (7.18b) for the selected column $s_{\cdot g}(\phi)$, while (7.17b) requires the satisfaction of (7.18b) for all other columns outside $S_i(\phi)$. Thus the (7.18) conditions completely describe both situations.

If (7.18b) *is not satisfied then the given row structure is not attainable.* That is, it is not possible to design a residual generator for the given system which follows the specified row structure.

Example 7.11.

a. Consider again the system of Example 7.9. Condition (7.18b) is satisfied for both $s_{\cdot 3}$ and $s_{\cdot 4}$, with $S_i = [s_{\cdot 1} \quad s_{\cdot 2}]$ in both cases. Thus the specified structure is attainable.

b. Change the third column to

$$s_{\cdot 3} = s_{\cdot 1} - s_{\cdot 2} = [\,-2 \quad -2 \quad -1\,]'$$

With this,

$$\text{Rank}\,[S_i \quad s_{\cdot 3}] = \text{Rank}\begin{bmatrix} 1 & 3 & -2 \\ 2 & 4 & -2 \\ -1 & 0 & -1 \end{bmatrix} = \text{Rank}\begin{bmatrix} 1 & 3 \\ 2 & 4 \\ -1 & 0 \end{bmatrix} = 2$$

thus (7.18b) is violated for $s_{\cdot 3}$ and the structure is not attainable. Obviously, $[S_i \quad s_{\cdot 3}]$ is now singular so the transformation can not even be computed.

c. Now change the fourth column to $s_{\cdot 4} = [\,-2 \quad -2 \quad -1\,]'$ (and keep the third column as it was originally). Then (7.18b) is violated for $s_{\cdot 4}$ and the structure is not attainable. The transformation now can be computed but the response it yields for the fourth fault is

$$w_i' s_{\cdot 4} = 0$$

which violates the specified structure.

Example 7.12.

a. Consider the system of Example 7.10. Of the columns shown in Table 7.2, any triplet has full rank thus any row structure with two zeroes is attainable.

b. Consider now $S_i(\phi)=[s_{.1}(\phi) \quad s_{.2}(\phi)]$ with a new third column

$$s_{.3}(\phi) = s_{.2}(\phi) - s_{.1}(\phi) = \begin{bmatrix} \dfrac{0.1}{(\phi - 0.5)(\phi - 0.6)} & \dfrac{2\phi - 1.4}{(\phi - 0.6)(\phi - 0.8)} & \dfrac{0.3}{(\phi - 0.5)(\phi - 0.8)} \end{bmatrix}'$$

With this,

$$Rank\,[S_i(\phi) \quad s_{.3}(\phi)] = Rank\,S_i(\phi) = 2$$

Thus the row structure $\varphi_i' = [0 \quad 0 \quad 1 \quad 1]$ is not attainable.

c. Another example of linearly dependent columns is

$$s_{.3}(\phi) = (\phi - 0.5)s_{.1}(\phi) - (\phi - 0.8)s_{.2}(\phi) = \begin{bmatrix} \dfrac{0.8}{(\phi - 0.6)} & \dfrac{-2(\phi - 1)}{(\phi - 0.6)} & 0 \end{bmatrix}'$$

This, of course, also makes the structure $[0 \quad 0 \quad 1 \quad 1]$ unattainable. ⁝⁝⁝

Row loss under non-attainability. Non-attainability of some row structures reduces the set of available canonical rows. This, however, does not necessarily jeopardize the final objective, namely the design of an isolating (strongly isolating, column-canonical) *column structure*.

The number of row structures lost can be traced relatively easily if there is only a single rank defect. Define as the *core of a rank defect* the smallest set of columns in $S(\phi)$ having less than full column rank. Denote by λ, $2 \leq \lambda \leq \mu$, the *size* of the core, that is, the number of columns in it. Any larger set of not more than μ columns which fully includes the core will also exhibit a rank defect. Condition (7.18b) is violated if and only if adding the column $s_{.k}(\phi)$ to the set completes the core. This happens in those row structures which contain one "1" and $\lambda - 1$ "0" elements within the core.

The number of possible patterns within such cores is the number of ways of placing $\lambda - 1$ zeroes into λ positions, which is $\binom{\lambda}{\lambda-1} = \lambda$. For each pattern within the core, there are $\binom{\rho-\lambda}{\mu-\lambda}$ ways of placing the remaining $\mu - \lambda$ zeroes into the remaining $\rho - \lambda$ positions. Thus the total number of row structures lost due to a single rank defect with a core of λ columns is

$$K_\lambda = \lambda \binom{\rho - \lambda}{\mu - \lambda} \tag{7.19}$$

In general, the row loss increases with decreasing core size; see Example 7.14. below.

With multiple rank defects, the situation is more complex. There may be multiple cores and the row losses may partially overlap.

Example 7.13.

Consider a system with $\mu=3$ and $\rho=4$. The full row-canonical set consists of the following six row structures:

	p_1	p_2	p_3	p_4
r_1	0	0	1	1
r_2	0	1	0	1
r_3	0	1	1	0
r_4	1	0	0	1
r_5	1	0	1	0
r_6	1	1	0	0

A column-canonical structure consists of $n=4$ rows with $\tau=2$ zeroes per column. Now assume that there is a rank defect in the $S(\phi)$ matrix, arising from the triplet of columns $[s_{.2} \;\; s_{.3} \;\; s_{.4}]$. This makes the last three row structures unattainable; they are replaced by the row $[1 \;\; 0 \;\; 0 \;\; 0]$. With this, a column-canonical set is no longer possible. However, the remaining set

	p_1	p_2	p_3	p_4
r_1	0	0	1	1
r_2	0	1	0	1
r_3	0	1	1	0
r_{456}	1	0	0	0

is still unidirectional strongly isolating. ⁞⁞⁞

Example 7.14.

Consider now a system with $\mu=4$ and $\rho=6$. The full row-canonical set contains *20* rows (which we are not showing here). For a column-canonical set, $n=4$ rows are needed, with $\tau=2$ zeroes per column. The loss of rows upon a single rank defect, with different core sizes, is

$\lambda=4 \qquad K_\lambda = 4$

$\lambda=3 \qquad K_\lambda = 9$

$\lambda=2 \qquad K_\lambda = 12$

Assume a single rank defect arising from $[s_{\cdot 4} \quad s_{\cdot 5} \quad s_{\cdot 6}]$. Then the remaining 11 rows still allow the formation of a column-canonical structure, namely:

	p_1	p_2	p_3	p_4	p_5	p_6
r_1	0	0	1	0	1	1
r_2	0	1	0	1	0	1
r_3	1	0	0	1	1	0
r_4	1	1	1	0	0	0

Notice that none of the remaining rows contains two zeroes within the core. ⁝⁝⁝

Special case: strictly output faults. The transfer matrix columns accompanying strictly output faults originate from a diagonal matrix and thus have only one nonzero element each, each in a different position. Whether they appear in the $S_i(\phi)$ matrix or in the additional $s_{\cdot k}(\phi)$ column, they cause the rank of $[S_i(\phi) \quad s_{\cdot k}(\phi)]$ depend only on a submatrix of its non-diagonal part.

Example 7.15.

Consider a static system with

$$[S_i \quad s_{\cdot k}] = \begin{bmatrix} 1 & 1 & 0 & 0 \\ 1 & 1 & 0 & 0 \\ 2 & 4 & 1 & 0 \\ 3 & 2 & 0 & 1 \end{bmatrix}$$

Clearly,

$$Rank\,[S_i \quad s_{\cdot k}] = 3$$

The full first two columns have full rank and the defect in their top part becomes manifest only by the addition of the two diagonal columns. Accordingly, the core size is $\lambda=4$. ⁝⁝⁝

Isolability. A most important special case of non-attainability is when the core size is $\lambda=2$. Now two columns of the fault transfer matrix exhibit direct linear dependence, in the polynomial sense, that is

$$\alpha_j(\phi)s_{.j}(\phi) + \alpha_k(\phi)s_{.k}(\phi) = 0 \tag{7.20}$$

where $\alpha_j(\phi)$ and $\alpha_k(\phi)$ are arbitrary nonzero polynomials. If this is the case then no residual can be generated which is decoupled from one of the faults but not from the other. Thus it is not possible to obtain different signatures for the two faults. Then ***the two faults are not isolable.***

Non-isolability is a fundamental deficiency of the system. It can not be remedied by any mathematical manipulation; the only cure is changing the physical system (adding a sensor).

Since the isolation of multiple faults requires the entire set of row-canonical structures, any rank defect (row loss) affects the isolability of multiple faults. In particular, fault combinations involving different sets of $\lambda - 1$ elements from a core, but otherwise identical, can not be distinguished from each other.

Example 7.16.

The two columns

$$[s_{.1}(\phi) \quad s_{.2}(\phi)] = \begin{bmatrix} \dfrac{3\phi - 1.6}{(\phi - 0.5)(\phi - 0.6)} & \dfrac{3\phi - 1.6}{(\phi - 0.6)(\phi - 0.8)} \\ \dfrac{2}{\phi - 0.6} & \dfrac{2\phi - 1}{(\phi - 0.6)(\phi - 0.8)} \\ \dfrac{1}{\phi - 0.5} & \dfrac{1}{\phi - 0.8} \end{bmatrix}$$

are linearly dependent, in that

$$(\phi - 0.5)s_{.1}(\phi) - (\phi - 0.8)s_{.2}(\phi) = 0$$

Subsequently, the corresponding two faults are not isolable.

Example 7.17.

Consider the following system

$$\begin{bmatrix} y_1(t) \\ y_2(t) \end{bmatrix} = \begin{bmatrix} m_{11}(\phi) & m_{12}(\phi) \\ m_{21}(\phi) & 0 \end{bmatrix} \begin{bmatrix} u_1(t) \\ u_2(t) \end{bmatrix}$$

and seek the isolation of the faults of the two output sensors and two input actuators. Then the fault transfer matrix is

$$S(\phi) = \begin{bmatrix} 1 & 0 & m_{11}(\phi) & m_{12}(\phi) \\ 0 & 1 & m_{21}(\phi) & 0 \end{bmatrix}$$

The last column is $m_{12}(\phi)$ times the first one thus the two faults (the first sensor and the second actuator fault) are not isolable. This, of course, is always the case if two (or more) faults affect only a single input-output equation. ⁝⁝⁝

7.3.3. Implementation of the Diagonal Structure

With a diagonal residual structure, the fault transfer matrix $S(\phi)$ is square. Further, each row represents a full almost homogeneous specification, which leads to the implementation equation

$$w_i'(\phi)S(\phi) = [0 \ldots 0 \quad z_{ii}(\phi)\upsilon_{ii}(\phi) \quad 0 \ldots 0] \qquad i=1\ldots\mu \tag{7.21}$$

While this corresponds to (7.17a) in the general row-by-row implementation, now there are no additional conditions of the kind of (7.17b), requiring non-zero fault responses outside the implementation set. Further, (7.21) contains the same $S(\phi)$ matrix for all rows $i=1\ldots\mu$. Therefore it is more effective to combine the row equations as

$$W(\phi)S(\phi) = Z(\phi)\Upsilon(\phi) = Diag[z_{11}(\phi)\upsilon_{11}(\phi) \quad \ldots \quad z_{\mu\mu}(\phi)\upsilon_{\mu\mu}(\phi)] \tag{7.22}$$

This leads to the simple solution

$$\boxed{W(\phi) = Diag[z_{11}(\phi)\upsilon_{11}(\phi) \quad \ldots \quad z_{\mu\mu}(\phi)\upsilon_{\mu\mu}(\phi)]\, S^{-1}(\phi)} \tag{7.23}$$

Clearly, the condition for this solution to exist is

$$\boxed{Rank\, S(\phi) = \mu} \tag{7.24}$$

That is, *any* rank defect of $S(\phi)$ bars a diagonal implementation.

Recall from Chapter 6 that it may not be possible to satisfy arbitrary response specifications $z_{11}(\phi)\ldots z_{\mu\mu}(\phi)$. Realization of the algorithm may require delays in the response while stability or polynomial computational behavior may require the cancellation of the invariant zero polynomial of the fault system, in part or full. These are achieved by the appropriate choice of the response modifiers $\upsilon_{ii}(\phi)$.

Example 7.18.

Consider the system of Example 7.10, without the last fault, so that $S(\phi)=[s_{.1}(\phi) \quad s_{.2}(\phi) \quad s_{.3}(\phi)]$. Find a diagonal implementation. Since the faults are all strictly input, all responses must contain ϕ^{-1}. Further, with three strictly input faults in a third order system, there will be no invariant zeroes. Thus the simplest specification is

$$Z(\phi) = I \qquad \Upsilon(\phi) = \phi^{-1} I$$

The inverse of the fault transfer matrix is obtained as

$$S^{-1}(\phi) = \frac{1}{10}\begin{bmatrix} 7(\phi - 0.643) & -5(\phi - 0.660) & -(\phi - 1.5) \\ -3(\phi - 0.667) & 5(\phi - 0.640) & -\phi \\ -2(\phi - 1) & -0.8 & 6(\phi - 0.667) \end{bmatrix}$$

which is polynomial (no invariant zeroes). The transformation from (7.23) is $W(\phi) = \phi^{-1}S^{-1}(\phi)$, that is,

$$W(\phi) = \begin{bmatrix} 0.7(1 - 0.643\phi^{-1}) & -0.5(1 - 0.660\phi^{-1}) & -0.1(1 - 1.5\phi^{-1}) \\ -0.3(1 - 0.667\phi^{-1}) & 0.5(1 - 0.640\phi^{-1}) & -0.1 \\ -0.2(1 - \phi^{-1}) & -0.08\phi^{-1} & 0.6(1 - 0.667\phi^{-1}) \end{bmatrix}$$

7.4. ALTERNATIVE INTERPRETATIONS OF STRUCTURED PARITY EQUATIONS

So far, we have presented structured parity equations (that is, parity equations generating structured residuals) as a *transformation* applied to the primary residuals

$$r(t) = W(\phi)o(t) = W(\phi)[y(t) - M(\phi)u(t)] \tag{7.25}$$

The primary residuals $o(t)$ have been computed from the *overall* input-output relationship of the monitored plant. In this section, we will show

- how the same transformation can be re-interpreted as a series of variable eliminations, and

- how the overall model equations can be replaced, as the primary set, by models describing *units* of the plant.

7.4.1. Secondary Equations by Elimination

Consider the internal form of the primary residuals

$$o(t) = S(\phi)p(t) \tag{7.26}$$

where $p(t)$ stands for both faults and disturbances and $S(\phi)$ has at least as many columns as rows. Select two rows, both containing the fault $p_j(t)$

$$o_l(t) = s_{lj}(\phi)p_j(t) + s_{l\cdot}^{\,j}(\phi)p^j(t) \tag{7.27a}$$

$$o_k(t) = s_{kj}(\phi)p_j(t) + s_{k\cdot}^{\,j}(\phi)p^j(t) \tag{7.27b}$$

where the superscript indicates that the respective element has been removed from the vector. Now express $p_j(t)$ from the l-th row and substitute it into the k-th

$$p_j(t) = \frac{1}{s_{lj}(\phi)}\left[o_l(t) - s_{l\cdot}^{\,j}(\phi)p^j(t)\right] \tag{7.28a}$$

$$o_k(\phi) = \frac{s_{kj}(\phi)}{s_{lj}(\phi)}o_l(t) + \left[s_{k\cdot}^{\,j}(\phi) - \frac{s_{kj}(\phi)}{s_{lj}(\phi)}s_{l\cdot}^{\,j}(\phi)\right]p^j(\phi) \tag{7.28b}$$

Thus the new residual

$$r_i(\phi) = \frac{s_{kj}(\phi)}{s_{lj}(\phi)}o_l(t) - o_k(\phi) = \left[\frac{s_{kj}(\phi)}{s_{lj}(\phi)}s_{l\cdot}^{\,j}(\phi) - s_{k\cdot}^{\,j}(\phi)\right]p^j(\phi) \tag{7.29}$$

is unaffected by the fault $p_j(\phi)$. Of course

$$r_i^*(\phi) = s_{kj}(\phi)o_l(t) - s_{lj}(\phi)o_k(\phi) = \left[s_{kj}(\phi)\,s_{l\cdot}^{\,j}(\phi) - s_{lj}(\phi)s_{k\cdot}^{\,j}(\phi)\right]p^j(\phi) \tag{7.30}$$

is also a solution.

From a set of μ primary equations, $\mu - 1$ can be used for the elimination of faults, while one needs to be reserved as the receiving equation. This allows the elimination of $\mu - 1$ faults. Let us collect the faults for which zero response has been specified into the vector p_0 and the remaining faults into p_l. Select the last one as the receiving equation and decompose the set of primary equations as

$$\begin{bmatrix} o^{\mu}(t) \\ o_{\mu}(t) \end{bmatrix} = \begin{bmatrix} S_0^{\mu}(\phi) & S_1^{\mu}(\phi) \\ s_{0\mu\cdot}(\phi) & s_{1\mu\cdot}(\phi) \end{bmatrix} \begin{bmatrix} p_0(t) \\ p_1(t) \end{bmatrix} \tag{7.31}$$

Now express $p_0(t)$ from the first $\mu - 1$ equations and substitute it into the μ-th

$$p_0(t) = [S_0^{\mu}(\phi)]^{-1} [o^{\mu}(t) - S_1^{\mu}(\phi) p_1(t)] \tag{7.32a}$$

$$o_{\mu}(t) = s_{0\mu\cdot}(\phi)[S_0^{\mu}(\phi)]^{-1} o^{\mu}(t) + [s_{1\mu\cdot}(\phi) - s_{0\mu\cdot}(\phi)[S_0^{\mu}(\phi)]^{-1} S_1^{\mu}(\phi)\,] \, p_1(\phi) \tag{7.32b}$$

Then the new residual

$$r_i(t) = s_{0\mu\cdot}(\phi)[S_0^{\mu}(\phi)]^{-1} o^{\mu}(t) - o_{\mu}(t)$$

$$= [s_{0\mu\cdot}(\phi)[S_0^{\mu}(\phi)]^{-1} S_1^{\mu}(\phi) - s_{1\mu\cdot}(\phi)] \, p_1(\phi) \tag{7.33}$$

is unaffected by the faults p_0.

Eq. (7.33) implies the transformation

$$w_i'(\phi) = [\, s_{0\mu\cdot}(\phi) \; [S_0^{\mu}(\phi)]^{-1} \quad - 1] \tag{7.34}$$

This is identical with our earlier results for almost full homogeneous specification, with the choice of $w_{i\mu}(\phi) = -1$, see Eq. (6.47).

An important special case arises when the fault vector contains only actuator and sensor faults. If all the output sensor faults are involved then the fault transfer matrix is

$$S(\phi) = [M(\phi) \quad I] \tag{7.35}$$

Now the faults to eliminate are the input related ones (actuator and input sensor faults). While the primary equations contain only one output sensor fault each, the transformed equations will contain more, by as much as the number of the eliminated input faults.

Example 7.19.

Consider the following system:

$$y_1(t) = m_{11}(\phi) u_1(t) + m_{12}(\phi) u_2(t)$$

$$y_2(t) = m_{21}(\phi) u_1(\phi) + m_{22}(\phi) u_2(t)$$

Construct a column-canonical set of residuals for the two actuator faults and two output sensor faults. Four parity equations are needed two of which are the primary equations (we omitted the t and ϕ arguments)

$$o_1 = y_1 - m_{11}u_1 - m_{12}u_2 = \Delta y_1 + m_{11}\Delta u_1 + m_{12}\Delta u_2$$

$$o_2 = y_2 - m_{21}u_1 - m_{22}u_2 = \Delta y_2 + m_{21}\Delta u_1 + m_{22}\Delta u_2$$

The equation unaffected by Δu_1 is obtained by (7.30) as

$$r_3 = m_{21}o_1 - m_{11}o_2 = m_{21}\Delta y_1 - m_{11}\Delta y_2 + (m_{12}m_{21} - m_{11}m_{22})\Delta u_2$$

Similarly, the residual unaffected by Δu_2 is

$$r_4 = m_{22}o_1 - m_{12}o_2 = m_{22}\Delta y_1 - m_{12}\Delta y_2 + (m_{11}m_{22} - m_{12}m_{21})\Delta u_1$$

7.4.2. Plant Unit Models as Primary Equations

Physical systems (plants) usually consist of subsystems or units. While the unit outputs are usually considered outputs of the overall system as well, the unit inputs in many cases include outputs of other units (and thus are not inputs of the overall system). If the models of the individual subsystems are available then the overall system model can be derived from these by substitution and then the parity relations designed in the usual way.

An alternative and, in many cases, more practical approach involves turning the unit model equations directly into residual generators. These unit residuals can then be considered as the "primary" set from which additional residuals can be obtained by transformation. The independence of unit residuals follows from their physical nature, namely that they describe different plant units. They, however, have no standard structure (unlike the input-output primary residuals which contain just one output fault each). Otherwise, the process of implementation is the same as with input-output primary residuals.

Example 7.20.

Consider the system shown in Figure 7.1. The system consists of two units, referred to as A and B. The two unit-models are

$$y_1(t) = m_{A1}(\phi)u_1(t) + m_{A2}(\phi)u_2(t)$$

$$y_2(t) = m_{B1}(\phi)y_1(t) + m_{B2}(\phi)u_2(t)$$

Let us design a column-canonical residual generator for the two actuator faults Δu_1 and Δu_2 and the two sensor faults Δy_1 and Δy_2. The internal form of the unit residuals is (dropping the t and ϕ arguments)

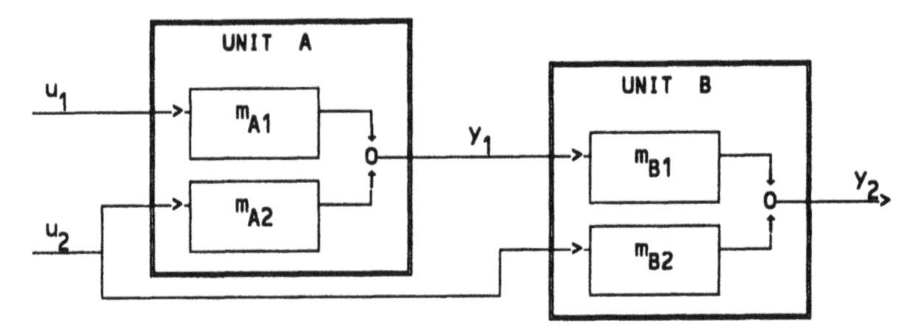

FIGURE 7.1. System with two units.

$$o_A = \blacktriangle y_1 + m_{A1} \blacktriangle u_1 + m_{A2} \blacktriangle u_2$$

$$o_B = \blacktriangle y_2 - m_{B1} \blacktriangle y_1 + m_{B2} \blacktriangle u_2$$

The negative sign in the second equation reflects the fact that $\blacktriangle y_1$ here is an input sensor fault. The structure of the two unit residuals is shown below. Also shown is the structure of the two additional residuals needed for a column-canonical set.

	$\blacktriangle y_1$	$\blacktriangle y_2$	$\blacktriangle u_1$	$\blacktriangle u_2$
$o_A = r_1$	1	0	1	1
$o_B = r_2$	1	1	0	1
r_3	1	1	1	0
r_4	0	1	1	1

The two additional residuals are derived from the unit residuals by elimination (transformation). They are obtained as

$$r_3 = m_{B2} o_A - m_{A2} o_B = (m_{B2} + m_{A2} m_{B1}) \blacktriangle y_1 - m_{A2} \blacktriangle y_2 + m_{A1} m_{B2} \blacktriangle u_1$$

$$r_4 = m_{B1} o_A + o_B = \blacktriangle y_2 + m_{A1} m_{B1} \blacktriangle u_1 + (m_{B2} + m_{A2} m_{B1}) \blacktriangle u_2$$

7.5. NOTES AND REFERENCES

The idea of generating residuals with structural properties may be traced to a number of independent sources (Ben-Haim, 1980; Chow and Willsky, 1984; Gertler and Singer, 1985; Massoumnia, 1986). Strong isolation and canonical structures were first described in (Gertler and Singer, 1985, 1990); see also (Gertler and Luo, 1989). Further studies into structural properties are due to Ziyang Qiu (1994).

The concepts of non-attainable structures and non-isolable faults appeared in (Gertler and Singer, 1985, 1990; Gertler and Luo, 1989). The rank conditions for structured residual design were studied in (Luo, 1990) and later spelled out in (Gertler and Kunwer, 1993, 1995). Row loss under non-attainability is first described here.

8
Design for Directional Residuals

8.0. INTRODUCTION

Directional residuals were introduced in Chapter 5 as an alternative way of enhancing the fault isolation utility of a residual set. In a directional set, the response to each particular (single) fault is confined to a straight line in the space of residuals. This directional property is valid at all times, including the course of any transient in response to fault changes.

Fault isolation in a directional setting amounts to determining which response line an observed residual vector, or a series of such vectors, lies the closest to. Isolation of multiple faults may be attempted by decomposing the observed residual vector along the response directions, provided the latter are independent.

The specification of a directional response consists of a constant directional vector $\boldsymbol{\beta}$ and a scalar transfer function $\gamma(\phi)$, so that

$$\boxed{r(t|p_j) = \boldsymbol{\beta}_j \, \gamma_j(\phi) \, p_j(t)} \tag{8.1}$$

It is essential that the dynamics of the response be the same for each element of the residual vector, otherwise the directional property would not be maintained during fault transients. Note that it is not really necessary to specify the size of each element in the response, it would suffice to prescribe their ratios. Absolute specifications, however, fit better with our implementation approach.

Disturbance decoupling can be incorporated into the directional framework by requiring that the response to the disturbance be zero:

$$r(t|q_j) = 0 \tag{8.2}$$

In this chapter, we will elaborate on the properties of directional specifications, including cases when independent responses can not be attained or no directional behavior is possible at all. We will also show how directional residual generators can be implemented with parity equations.

8.1. DIRECTIONAL SPECIFICATIONS

8.1.1. Directional Specification Without Disturbances

Eq. (8.1) can be re-phrased as

$$z_{F.j}(\phi) = \boldsymbol{\beta}_j \gamma_j(\phi) \tag{8.3}$$

where $z_{F.j}(\phi)$ is the vector specification for the j-th fault response. The complete specification is obtained by combining the ρ_F individual fault responses

$$Z_F(\phi) = [z_{F.1}(\phi) \quad z_{F.2}(\phi) \quad \ldots \quad z_{F.\rho F}(\phi)] \tag{8.4}$$

(8.4) can also be written as

$$\boxed{Z_F(\phi) = \mathbf{B}\,\boldsymbol{\Gamma}(\phi)} \tag{8.5}$$

where

$$\mathbf{B} = [\boldsymbol{\beta}_1 \quad \boldsymbol{\beta}_2 \ldots \boldsymbol{\beta}_{\rho F}]$$

$$\boldsymbol{\Gamma}(\phi) = Diag[\gamma_1(\phi) \quad \gamma_2(\phi) \ldots \gamma_{\rho F}(\phi)] \tag{8.6}$$

Let us write the residual response now with the internal form of the generic residual generator:

$$Z_F(\phi) = W(\phi) S_F(\phi) \tag{8.7}$$

where $S_F(\phi)$ is the fault transfer matrix of the monitored system. Observe that, with μ outputs, ρ_F faults and n transformed residuals, $\mathbf{Z}_F$ is $n\cdot\rho_F$, $\mathbf{W}$ is $n\cdot\mu$ and S_F is $\mu\cdot\rho_F$. We wish to design the transformation so that it is implementable and that *all fault responses have independent directions*.

Since the rows of the transformation are μ-dimensional they can only be designed to satisfy no more than μ independent conditions. Thus implementation requires

$$Rank S_F(\phi) \leq \mu \tag{8.8}$$

For independent response directions, the ρ_F columns of $\mathbf{Z}_F(\phi)$ must be linearly independent, that is,

$$Rank \mathbf{Z}_F(\phi) = \rho_F \tag{8.9}$$

By (8.5), this implies $Rank \mathbf{B}=\rho_F$ since the diagonal $\mathbf{\Gamma}(\phi)$ always has full rank. From (8.7), with (2.127), the condition (8.9) requires

$$\boxed{Rank S_F(\phi) = \rho_F} \tag{8.10}$$

That is, *for independent response directions, the fault transfer matrix must have full column rank*. Now by comparing (8.10) with (8.8),

$$\boxed{\rho_F \leq \mu} \tag{8.11}$$

which follows also directly from (8.10). That is, *the number of faults for which independent* (and, in fact, any arbitrary) *directional responses can be designed is limited to the number of plant outputs*. Finally, (8.9) also implies

$$n \geq \rho_F \tag{8.12}$$

That is, *for independent response directions, the number of transformed residuals can not be less than the number of faults*. Note that, in general, there is no point in having more than ρ_F residual elements.

Of course, independent response directions may also be made orthogonal, a property further enhancing the isolation of faults.

Basis set specification. An important special case of the specification (8.5) arises when $\rho_F = \mu$ and $\mathbf{B} = I$, yielding the diagonal specification

$$Z_F(\phi) = \Gamma(\phi) \tag{8.13}$$

Further, with $\gamma_1(\phi) = \gamma_2(\phi) = \ldots = \gamma_\mu(\phi) = \gamma^x(\phi)$, the specification becomes

$$\boxed{Z_F^x(\phi) = I\,\gamma^x(\phi)} \tag{8.14}$$

This latter system, in which each element of the residual vector reacts to a single fault, and where all the responses have the same dynamics, will be referred to as the *basis set of residuals*.

The basis set is very convenient for fault isolation, both in a directional and in a structured framework. Also, as we will show later, further directional (and, with some qualifications, structured) sets can be derived from it by an additional transformation.

8.1.2. Directional Specification with Disturbances

If the residuals need to be directional for the faults $p_1 \ldots p_{\rho F}$ while being decoupled from the disturbances $q_1 \ldots q_{\rho D}$, then the combined response specification is

$$Z(\phi) = [z_{F \cdot 1}(\phi) \ldots z_{F \cdot \rho F}(\phi) \;\; \underset{1}{0} \ldots \underset{\rho_D}{0}] \tag{8.15}$$

where, in general, $z_{F \cdot j}(\phi)$ is described by (8.3). If basis set behavior is required for the faults then the fault response specifications become

$$z_{\cdot j}(\phi) = [0 \ldots 0 \;\; \underset{j}{1} \;\; 0 \ldots 0]' \, \gamma^x(t) \tag{8.16}$$

The response expressed with the internal form of the generic residual generator is now

$$Z(\phi) = W(\phi)[S_F(\phi) \;\; S_D(\phi)] \tag{8.17}$$

Since the rows of the transformation are μ-dimensional, implementation is possible only if

$$Rank[S_F(\phi) \;\; S_D(\phi)] \leq \mu \tag{8.18}$$

Further, for nonzero and independent fault responses, it is necessary that the columns of $S_F(\phi)$ be independent of the columns of $S_D(\phi)$ and of each other, that is,

$$Rank[S_F(\phi) \quad S_D(\phi)] = RankS_D(\phi) + \rho_F \tag{8.19}$$

(8.18) and (8.19) imply that

$$\boxed{\rho_F \leq \mu - RankS_D(\phi)} \tag{8.20}$$

The generator can be implemented with $n = \rho_F$ residuals, where n may be much smaller than μ. Then the specification (8.15) becomes

$$\boxed{Z(\phi) = [\mathbf{B}\,\Gamma(\phi) \quad 0]} \tag{8.21}$$

where $\mathbf{B}$ and $\Gamma(\phi)$ are $\rho_F \cdot \rho_F$ square matrices and $\Gamma(\phi)$ is diagonal. Similarly, the basis set specification becomes

$$\boxed{Z^x(\phi) = [I\,\gamma^x(\phi) \quad 0]} \tag{8.22}$$

where I is a $\rho_F \cdot \rho_F$ unit matrix.

8.2. PARITY EQUATION IMPLEMENTATION OF DIRECTIONAL RESIDUALS

The parity equation implementation of directional residual generators follows directly from the results of Subsections 8.1.1 and 8.1.2, and the techniques introduced in Chapter 6.

Directional residual generation without disturbance decoupling. Consider first the arbitrary specification (8.4). The generator may be implemented separately for each row. If $\rho_F = \mu$ then the row-specification $z_{Fi\cdot}(\phi)$ is full non-homogeneous. Assume that $S_F(\phi)$ has full rank, then the row-transformation $w_i'(\phi)$ can be obtained as

$$w_i'(\phi) = z_{Fi\cdot}(\phi)\,\mathbf{T}(\phi)\,S_F^{-1}(\phi) \qquad i=1\ldots\mu \tag{8.23}$$

Here we included the response modifier matrix $\mathbf{T}(\phi)=\mathbf{T}_P(\phi)\mathbf{T}_R(\phi)= Diag[\upsilon_1(\phi) \ldots \upsilon_\mu(\phi)]$ (refer to Chapter 6). Recall that $\mathbf{T}_P(\phi)$ serves for making the residual generator stable (or polynomial) while $\mathbf{T}_R(\phi)$ serves to make it causal. The element $\upsilon_j(\phi)$, $j=1\ldots\mu$, serves to modify the response element $z_{Fij}(\phi)$. Under directional design, the $\mathbf{T}(\phi)$ matrix must be the same for all rows $i=1\ldots\mu$, in order to maintain identical response dynamics within each column.

Since the $S_F(\phi)$ matrix is also the same in the computation of each row-transformation, it is advantageous to compute the entire set $w_1'(\phi)\ldots w_\mu'(\phi)$ of transformations in a single step. Thus

$$W(\phi) = Z_F(\phi)\, \mathbf{T}(\phi)\, S_F^{-1}(\phi) \tag{8.24}$$

Here $W(\phi)$ is a $\mu{\cdot}\mu$ matrix, with $w_1'(\phi)\ldots w_\mu'(\phi)$ as its rows. Note that (8.24) can be written, with (8.5), as

$$W(\phi) = \mathbf{B}\,\mathbf{\Gamma}(\phi)\, \mathbf{T}(\phi)\, S_F^{-1}(\phi) \tag{8.25}$$

where both $\mathbf{\Gamma}(\phi)$ and $\mathbf{T}(\phi)$ are diagonal matrices. This suggests that those two matrices might be joined together; the simplest implementation is clearly obtained with $\mathbf{\Gamma}(\phi) = \boldsymbol{I}$.

If the basis set specification is considered then the single-row specifications are full almost homogeneous and in (8.24) $Z_F(\phi)$ is replaced by $Z_F^x(\phi)=\boldsymbol{I}\,\gamma^x(\phi)$ so that the transformation becomes

$$W^x(\phi) = \gamma^x(\phi)\upsilon^x(\phi)S_F^{-1}(\phi) \tag{8.26}$$

where $\upsilon^x(\phi)$ is the common response modifier.

Directional residual generation with disturbance decoupling. Consider the specification (8.15) and assume that $\rho_F+\rho_D=\mu$ and $[S_F(\phi)\quad S_D(\phi)]$ has full rank. Then (8.23) becomes

$$w_i'(\phi) = [z_{Fi\cdot}(\phi)\mathbf{T}(\phi) \quad 0'][S_F(\phi) \quad S_D(\phi)]^{-1} \qquad i=1\ldots\rho_F \tag{8.27}$$

while (8.24) and (8.25) are replaced by

$$W(\phi) = [Z_F(\phi)\mathbf{T}(\phi) \quad 0][S_F(\phi) \quad S_D(\phi)]^{-1} \tag{8.28}$$

$$W^x(\phi) = [I\,\gamma^x(\phi)\upsilon^x(\phi) \quad 0][S_F(\phi) \quad S_D(\phi)]^{-1} \tag{8.29}$$

where now $W(\phi)$ and $W^x(\phi)$ are $\rho_F \cdot \mu$ matrices while $Z_F(\phi)$ and I are $\rho_F \cdot \rho_F$ matrices.

Arbitrary responses from the basis set. Recall that, in general, a residual vector is generated as $r(t) = W(\phi)o(t)$, where $o(t) = y(t) - M(\phi)u(t)$ is the set of primary residuals. Similarly, the basis residuals are computed as

$$r^x(t) = W^x(\phi)o(t) \tag{8.30}$$

Now it is obvious from (8.5), (8.21), (8.24) and (8.26) that a residual vector with arbitrary directions $\mathbf{B}$, but still with common dynamics $\gamma^x(\phi)\upsilon^x(\phi)$, can be obtained from the basis residuals as

$$r(t) = \mathbf{B}\,r^x(t) \tag{8.31}$$

Eq. (8.31) is valid for systems with or without disturbance decoupling.

Implementation considerations. The usual implementation considerations (see Chapter 6) apply to the directional design. In particular:

1. Causality. The response modifier for causality is $\upsilon_{Rj}(\phi) = \phi^{\delta o - \delta j}$, $j = 1 \ldots \rho_F$ (c.f. (6.115)).

2. Algorithm I. If a residual generator is implemented according to (8.24), using Algorithm I, then the response modifiers need to be $\upsilon_{Pj}(\phi) = \pi_U(\phi)/\pi_{Uj}(\phi)$ for stability or $\upsilon_{Pj}(\phi) = \pi(\phi)/\pi_j(\phi)$ for a polynomial algorithm (c.f. (6.120) and (6.127)).

3. Algorithm II. A generator may be implemented by (8.23), using Algorithm II, so that all but one of the response dynamics are frozen as specified. To do this, the modifier is chosen as $\mathbf{T}_{Pi}(\phi) = Diag[1 \ldots 1 \quad \upsilon_{Pig}(\phi) \quad 1 \ldots 1]$ where now $\upsilon_{Pig}(\phi)$ is allowed to be different for each row i. The modifiers are computed so that the unstable zeroes (or all zeroes) are canceled out (see (6.124) and (6.130)). With different modifiers in each row, the g-th fault response loses its directional properties during fault transients. However, all the other responses are directional and of minimum complexity.

4. *Basis set*. The basis set specification is implemented by (8.26), so that the common response modifier $\upsilon^{x}(\phi)$ contains the unstable zeroes (or all zeroes). If an arbitrary specification is then implemented from the basis set, still with common response dynamics, Eq. (8.31), then the stable (polynomial) behavior is sustained. (If, however, the response dynamics are also changed to arbitrary ones, then the stable (polynomial) behavior may be lost.)

5. *Strictly output faults*, that is, faults of the output sensors, are accompanied by columns in the fault transfer matrix which have only one nonzero element each. Such faults do not create any difficulty and do not require special treatment if directional residuals are generated by parity relations. (They do in observer based design, see for example (White and Speyer, 1987; Park et al, 1994b.) The only special circumstance is the fact that, with part of the $S(\phi)$ matrix being diagonal, the rank of the entire matrix hinges on a single minor (c.f. Example 7.15).

Example 8.1.

Consider the system shown in Table 8.1 (which is a slightly modified version of Table 7.2). Notice that the first and second faults are strictly input.

Let us design a directional residual generator for the first three faults so that the responses are

$$Z(\phi) = \begin{bmatrix} 1 & 1 & 1 \\ 1 & 1 & -1 \\ 1 & -2 & 0 \end{bmatrix}$$

TABLE 8.1

$s_{.1}(\phi)$	$s_{.2}(\phi)$	$s_{.3}(\phi)$	$s_{.4}(\phi)$
$\dfrac{3\phi - 1.6}{(\phi - 0.5)(\phi - 0.6)}$	$\dfrac{3}{\phi - 0.6}$	$\dfrac{1}{\phi - 0.5}$	$\dfrac{1}{\phi - 0.6}$
$\dfrac{2}{\phi - 0.6}$	$\dfrac{4\phi - 3}{(\phi - 0.6)(\phi - 0.8)}$	$\dfrac{1}{\phi - 0.8}$	$\dfrac{1}{\phi - 0.6}$
$\dfrac{1}{\phi - 0.5}$	$\dfrac{1}{\phi - 0.8}$	$\dfrac{\phi^2 + 0.7\phi - 0.9}{(\phi - 0.5)(\phi - 0.8)}$	-0.2

Here the three response directions are orthogonal. The inverse of the fault transfer matrix for this system is

$$S^{-1}(\phi) =$$

$$\frac{1}{0.6(\phi+1)}\begin{bmatrix} 0.4(\phi^2+0.5\phi-0.75) & -0.3(\phi^2+0.367\phi-0.7) & -0.1(\phi-1.5) \\ -0.2(\phi^2+0.2\phi-0.6) & 0.3(\phi^2+0.333\phi-0.64) & -0.1\phi \\ -0.2(\phi-1) & -0.08 & 0.6(\phi-0.667) \end{bmatrix}$$

Observe that $\delta_o=1,\ \delta_1=2,\ \delta_2=2,\ \delta_3=1$. Thus for causality, we need

$$\mathbf{T}_R(\phi) = Diag\,[\,\phi^{-1}\quad \phi^{-1}\quad 1\,]$$

The system has one invariant zero (as expected, since $\nu\text{-}\rho_1=3\text{-}2=1$) which is on the unit circle. Thus for stability

$$\mathbf{T}_P(\phi) = \alpha(1+\phi^{-1})I$$

is necessary. With this, and $\alpha=0.6$, the transformation is obtained as

$$W(\phi) =$$

$$\begin{bmatrix} 0.36\phi^{-1}-0.18\phi^{-2} & -0.09\phi^{-1}+0.018\phi^{-2} & 0.6-0.6\phi^{-1}+0.15\phi^{-2} \\ 0.4-0.04\phi^{-1}-0.18\phi^{-2} & 0.07\phi^{-1}+0.018\phi^{-2} & -0.6+0.2\phi^{-1}+0.15\phi^{-2} \\ 0.8+0.28\phi^{-1}-0.54\phi^{-2} & -0.9-0.31\phi^{-1}+0.594\phi^{-2} & 0.1\phi^{-1}+0.15\phi^{-2} \end{bmatrix}$$

∎

Example 8.2.

Let us design a directional residual generator for the same system, so that the first two responses are diagonal and that stabilization does not affect their dynamics. The response specification is

$$Z(\phi) = \begin{bmatrix} 1 & 0 & 1 \\ 0 & 1 & 1 \\ 0 & 0 & 1 \end{bmatrix}$$

The third response specification is basically dummy. For causality, we need $\mathbf{T}_R(\phi)=[\phi^{-1}\quad \phi^{-1}\quad 1]$. The implementation needs to be done row-by-row, with stabilizing modifiers

$$\mathbf{T}_{Pi}(\phi) = [1 \quad 1 \quad \upsilon_{Pi3}(\phi)] \qquad i=1,2,3$$

Here the fixed 1 elements make sure that the first two responses are not modified. The $\upsilon_{Pi3}(\phi)$ elements have to be so chosen that the stability condition is satisfied for the transformation. As it was shown in Chapter 6 (Subsection 6.6.3), the polynomial conditions may be written for any column of $S^{-1}(\phi)$. Using the first column, the conditions are

$$0.4(1+0.5\phi^{-1} - 0.75\phi^{-2}) - 0.2(1 - \phi^{-1})\upsilon_{P13}(\phi) = \alpha_1(\phi)(1+\phi^{-1})$$

$$-0.2(1+0.2\phi^{-1} - 0.6\phi^{-2}) - 0.2(1 - \phi^{-1})\upsilon_{P23}(\phi) = \alpha_2(\phi)(1+\phi^{-1})$$

$$-0.2(1 - \phi^{-1})\upsilon_{P33}(\phi) = \alpha_3(\phi)(1+\phi^{-1})$$

These can be satisfied with

$$\upsilon_{P13}(\phi) = -0.25 \qquad \upsilon_{P23}(\phi) = -0.1 \qquad \upsilon_{P33}(\phi) = 1+\phi^{-1}$$

Since the three dynamic responses in the third column are different, the response to the third fault is *not directional*. The resulting transformation is

$$\mathbf{W}(\phi) = \begin{bmatrix} 0.75 - 0.5\phi^{-1} & -0.5+0.35\phi^{-1} & -0.25+0.25\phi^{-1} \\ -0.3+0.2\phi^{-1} & 0.5 - 0.32\phi^{-1} & -0.1 \\ -0.333+0.333\phi^{-1} & -0.1333\phi^{-1} & 1 - 0.667\phi^{-1} \end{bmatrix}$$

■

Example 8.3.

Assume that in the previous system the $s_{.3}(\phi)$ column belongs to a disturbance. Design a residual generator which is directional (diagonal) for the two faults and is decoupled from the disturbance. We need only two residuals and the response specification is

$$\mathbf{Z}(\phi) = \begin{bmatrix} 1 & 0 & 0 \\ 0 & 1 & 0 \end{bmatrix}$$

For causality and stability, choose

$$\mathbf{T}(\phi) = 0.6\phi^{-1}(1+\phi^{-1})\mathbf{I}$$

This leads to the transformation

$W(\phi) =$

$$\begin{bmatrix} 0.4+0.2\phi^{-1} - 0.3\phi^{-2} & -0.3 - 0.11\phi^{-1}+0.21\phi^{-2} & -0.1\phi^{-1}+0.15\phi^{-2} \\ -0.2 - 0.04\phi^{-1}+0.12\phi^{-2} & 0.3+0.1\phi^{-1} - 0.192\phi^{-2} & -0.1\phi^{-1} \end{bmatrix}$$

8.3. LINEARLY DEPENDENT COLUMNS

As pointed out in (8.10), independent response directions can only be attained if the fault transfer matrix $S_F(\phi)$ has full (column) rank. In the following, we will investigate the consequences and the possible causes of linear dependencies among the columns of $S_F(\phi)$.

Types of linear dependencies. Recall that the set of vectors $\{s_1(\phi), s_2(\phi) \dots s_p(\phi)\}$ is linearly dependent if the relationship

$$\alpha_1(\phi)s_1(\phi) + \alpha_2(\phi)s_2(\phi) + \dots + \alpha_p(\phi)s_p(\phi) = \mathbf{0} \tag{8.32}$$

is satisfied while at least one of the coefficients $\alpha_j(\phi)$ is nonzero. In general, the coefficients are *polynomials*. As a special case, (8.32) includes

$$\alpha_1 s_1(\phi) + \alpha_2 s_2(\phi) + \dots + \alpha_p s_p(\phi) = \mathbf{0} \tag{8.33}$$

where the α_j coefficients are *constant*.

Fault responses under linear dependencies. Assume that there are $p_F > \mu$ faults and a directional residual generator has been designed for μ of the faults, with the independent responses

$$\begin{aligned} \mathbf{Z}_F(\phi)\mathbf{T}(\phi) &= \mathbf{W}(\phi)[s_{F\cdot 1}(\phi) \;\; s_{F\cdot 2}(\phi) \dots s_{F\cdot\mu}(\phi)] \\ &= [\boldsymbol{\beta}_1\gamma_1(\phi)\upsilon_1(\phi) \;\; \boldsymbol{\beta}_2\gamma_2(\phi)\upsilon_2(\phi) \dots \boldsymbol{\beta}_\mu\gamma_\mu(\phi)\upsilon_\mu(\phi)] \end{aligned} \tag{8.34}$$

Now any column $s_{F\cdot j}(\phi)$ outside the design set will linearly depend on the μ columns in the design set. If the relationship is polynomial then the response to the j-th fault is

$$\begin{aligned} &\mathbf{W}(\phi)s_{F\cdot j}(\phi) \\ &= -[\boldsymbol{\beta}_1\alpha_1(\phi)\gamma_1(\phi)\upsilon_1(\phi) + \dots + \boldsymbol{\beta}_\mu\alpha_\mu(\phi)\gamma_\mu(\phi)\upsilon_\mu(\phi)] \,/\, \alpha_j(\phi) \end{aligned} \tag{8.35}$$

This is clearly not directional since the dynamics are, in general, different in each element of the response vector. However, if the linear dependency relationship is of the constant type, Eq. (8.33), and if the *same dynamic response* $\gamma(\phi)\upsilon(\phi)$ has been chosen for all faults in the design set (including the delay for realization, if any), then the response to the j-th fault becomes

$$W(\phi)s_{F\cdot j}(\phi) = -[\alpha_1 \boldsymbol{\beta}_1 + \alpha_2 \boldsymbol{\beta}_2 + \ldots + \alpha_\mu \boldsymbol{\beta}_\mu]\, \gamma(\phi)\upsilon(\phi) / \alpha_j \tag{8.36}$$

which is directional. Obviously, since now there are more than μ responses in the μ-dimensional residual space, the responses are not independent.

Linear dependencies and disturbance decoupling. Assume now that there are ρ_F faults and ρ_D disturbances and $\rho_F+\rho_D>\mu$. A directional generator has been designed for the first $n<\rho_F$ faults, with decoupling from the ρ_D disturbances, so that $n+\rho_D=\mu$. Then any column $s_{F\cdot j}(\phi)$ outside the design set is a linear combination of all columns within the design set, that is,

$$\alpha_1(\phi)s_{F\cdot 1}(\phi) + \ldots + \alpha_n(\phi)s_{F\cdot n}(\phi) + \alpha_{n+1}(\phi)s_{D\cdot 1}(\phi) + \ldots + \alpha_\mu(\phi)s_{D\cdot \rho D}(\phi)$$
$$+\, \alpha_j(\phi)s_{F\cdot j}(\phi) = 0 \tag{8.37}$$

With the disturbance responses being zero, the response to the j-th fault, $W(\phi)s_{F\cdot j}(\phi)$, will be in the n-dimensional space of the designed responses but, in general, will not be directional. However, if the α coefficients in (8.37) are constant and if the first n responses have been designed with the same dynamics $\gamma(\phi)\upsilon(\phi)$, then

$$W(\phi)s_{F\cdot j}(\phi) = -[\alpha_1 \boldsymbol{\beta}_1 + \ldots + \alpha_n \boldsymbol{\beta}_n]\, \gamma(\phi)\upsilon(\phi) / \alpha_j \tag{8.38}$$

which is directional.

Linear dependencies and the E and F matrices. It is easy to show that linear row-dependencies in the combined $[E' \quad F']'$ matrix of the state-space representation result in constant-type dependencies in the $S(\phi)$ matrix. Assume that

$$\alpha_1 e_{\cdot 1} + \alpha_2 e_{\cdot 2} + \ldots + \alpha_p e_{\cdot p} = 0$$
$$\alpha_1 f_{\cdot 1} + \alpha_2 f_{\cdot 2} + \ldots + \alpha_p f_{\cdot p} = 0 \tag{8.39}$$

Now recall that $s_{\cdot j}(\phi) = C(\phi I - A)^{-1} e_{\cdot j} + f_{\cdot j}$. With this, it follows that

$$\alpha_1 s_{\cdot 1}(\phi) + \alpha_2 s_{\cdot 2}(\phi) + \ldots + \alpha_p s_{\cdot p}(\phi) \tag{8.40}$$
$$= C(\phi I - A)^{-1}(\alpha_1 e_{\cdot 1} + \alpha_2 e_{\cdot 2} + \ldots + \alpha_p e_{\cdot p}) + (\alpha_1 f_{\cdot 1} + \alpha_2 f_{\cdot 2} + \ldots + \alpha_p f_{\cdot p}) = 0$$

It can also be shown that if the combined $\boldsymbol{E}$ and $\boldsymbol{F}$ matrices have full column rank, then any linear dependency in the $S(\phi)$ matrix can only be of polynomial type. (The proof is lengthy and is omitted here.) This includes the important case of $\rho > \mu$. Now the $S(\phi)$ matrix is "flat" and its columns are certainly linearly dependent; in general, this is a polynomial dependence (unless the combined $\boldsymbol{E}$ and $\boldsymbol{F}$ matrices have a column-rank defect).

The effect of core-size. We defined, in Subsection 7.3.2, the *core of a rank-defect* as the smallest set of columns exhibiting linear dependence. The fault responses exist in a space of dimension $Rank S_F(\phi)$; if $Rank S_F(\phi) < \rho_F$, then this is a subspace of the ρ_F-dimensional residual space. Any set of responses whose associated columns in $S_F(\phi)$ contain a core are subject to a loss of dimension and, thus, of independence. The smallest set of responses that may suffer such loss is the core; the responses belonging to a core of size λ (with a single rank-defect) are forced into a subspace of dimension λ-1.

Non-isolable faults. If a core-size is $\lambda = 2$ then the two responses are restricted to the *same line*. Thus it is not possible to design a residual generator which would produce distinct response directions for the concerned faults. That is, such faults are *not isolable*. Note that the condition of non-isolability is the same, whether the generator is designed for directional or structured responses.

Example 8.4.

Consider again the system of Table 8.1, with the first three faults and only the first *two* outputs. The three columns of the reduced $S(\phi)$ matrix must be linearly dependent; the relationship is found to be

$$(\phi - 1.5)s_{.1}(\phi) + \phi s_{.2}(\phi) - 6(\phi - 0.667)s_{.3}(\phi) = 0$$

which is of polynomial type. In a (diagonal) state-space realization of this system, $\boldsymbol{F}=0$ and

$$\boldsymbol{E} = \begin{bmatrix} 1 & 0 & 1 \\ 2 & 3 & 0 \\ 0 & 1 & 1 \end{bmatrix}$$

which has full rank. Now let us design a directional residual generator for the first two faults, e.g., with the diagonal response

$$Z(\phi) = \begin{bmatrix} 1 & 0 \\ 0 & 1 \end{bmatrix}$$

The inverse of the reduced $S(\phi)$ matrix was computed in Example 7.10 and was found to have $\delta_o=1$, $\delta_1=\delta_2=2$, with an invariant zero at $\phi=0.667$. This allows the choice of

$$\Upsilon(\phi) = Diag[\phi^{-1} \quad \phi^{-1}]$$

With this, the transformation is

$$W(\phi) = \frac{1}{1-0.667\phi^{-1}} \begin{bmatrix} 0.667-0.834\phi^{-1}+0.25\phi^{-2} & -0.5+0.65\phi^{-1}-0.2\phi^{-2} \\ -0.333+0.433\phi^{-1}-0.133\phi^{-2} & 0.5-0.667\phi^{-1}+0.214\phi^{-2} \end{bmatrix}$$

The response this transformation generates to the third fault is

$$z_{.3}(\phi) = \frac{0.167}{1-0.667\phi^{-1}} \begin{bmatrix} \phi^{-1}-1.5\phi^{-2} \\ \phi^{-1} \end{bmatrix}$$

which is clearly *not directional.* ⁝⁝

Example 8.5.

Consider the full system shown in Table 8.1. The four columns must be linearly dependent and the relationship is found as

$$s_{.1}(\phi) + s_{.2}(\phi) - s_{.3}(\phi) - 5s_{.4}(\phi) = 0$$

which is clearly of constant type. The reason for this can be seen in the E and F matrices of the underlying (diagonal) state-space representation:

$$E = \begin{bmatrix} 1 & 0 & 1 & 0 \\ 2 & 3 & 0 & 1 \\ 0 & 1 & 1 & 0 \end{bmatrix} \qquad F = \begin{bmatrix} 0 & 0 & 0 & 0 \\ 0 & 0 & 0 & 0 \\ 0 & 0 & 1 & -0.2 \end{bmatrix}$$

which exhibit the same linear relationship. Now try to apply the transformation, obtained in Example 8.1 (from a directional specification for the first three faults) to the fourth column $s_{.4}(\phi)$. The resulting response is

$$z_{.4}(\phi) = 0.12\,(1+\phi^{-1}) \begin{bmatrix} -1+2\phi^{-1} \\ 1+2\phi^{-1} \\ -\phi^{-1} \end{bmatrix}$$

This response is not directional, in spite of the constant-type linear dependence, because the transformation was not implemented with three *identical* dynamic responses (two of the response modifiers did contain a one-step delay while the third did not). So let us redesign the residual generator for the first three faults, now with

$$\mathbf{Z}(\phi) = \begin{bmatrix} 1 & 1 & 1 \\ 1 & 1 & -1 \\ 1 & -2 & 0 \end{bmatrix} \qquad \boldsymbol{\Upsilon}(\phi) = 0.6\phi^{-1}(1+\phi^{-1})\mathbf{I}$$

(Here we provided a delay for all three responses though realizability does not necessitate this for the third.) The new transformation is

$$\mathbf{W}(\phi) =$$

$$\begin{bmatrix} 0.2-0.04\phi^{-1}+0.02\phi^{-2} & -0.01\phi^{-1}-0.062\phi^{-2} & 0.4\phi^{-1}-0.25\phi^{-2} \\ 0.2+0.36\phi^{-1}-0.38\phi^{-2} & -0.01\phi^{-1}+0.098\phi^{-2} & -0.8\phi^{-1}+0.55\phi^{-2} \\ 0.8+0.28\phi^{-1}-0.54\phi^{-2} & -0.9-0.31\phi^{-1}+0.594\phi^{-2} & 0.1\phi^{-1}+0.15\phi^{-2} \end{bmatrix}$$

Applying this transformation to the fourth fault yields the directional response

$$z_{.4}(\phi) = 0.12\,\phi^{-1}(1+\phi^{-1}) \begin{bmatrix} 1 \\ 3 \\ -1 \end{bmatrix}$$

8.4. NOTES AND REFERENCES

The idea of directional responses originated from Beard (1971) and Jones (1973). Important contributions to directional residual generation in the framework of diagnostic observers include Massoumnia (1986), White and Speyer (1987), Park and Rizzoni (1994) and Edelmayer et al (1997).

Directional residual generation by dynamic parity relations was first suggested in (Gertler, 1991). The technique was then developed in (Monajemy, 1993) and in (Gertler and Monajemy, 1993 and 1995). The considerations about linearly dependent columns are shown here for the first time.

9
Residual Generation for Parametric Faults

9.0. INTRODUCTION

Parametric (or multiplicative) faults, as it was explained in Chapter 5, affect the *parameters* of the monitored system, causing a *discrepancy* between the system model $M(\phi)$ and the real system $M^{\circ}(\phi)$. In the absence of any other fault, disturbance and noise, the plant output is

$$y(t) = M^{\circ}(\phi)u(t) = M(\phi)u(t) + \blacktriangle M_F(\phi)u(t) \tag{9.1}$$

where $\blacktriangle M_F(\phi)$ is the model discrepancy attributed to parametric faults. It was also pointed out in Chapter 5 that modeling errors cause similar discrepancies and, in fact, the distinction between parametric faults and modeling errors is mainly historical (in that faults occur during plant operation while model errors usually are there from the beginning) and, to some extent, arbitrary. As we mentioned there, modeling errors may be considered as *parametric disturbances*. In the presence of such disturbances, (9.1) expands to

$$y(t) = M(\phi)u(t) + \blacktriangle M_F(\phi)u(t) + \blacktriangle M_D(\phi)u(t) \tag{9.2}$$

where $\blacktriangle M_D(\phi)$ is the model discrepancy attributed to modeling errors.

Similarly to our treatment of additive faults and disturbances, we will want to detect and isolate the parametric faults while ignoring the parametric disturbances. This calls for the generation of residuals which exhibit structured or directional behavior with respect to the faults and are decoupled from the disturbances.

A special property of parametric faults and disturbances, as visible from Eq. (9.2), is that their coefficients are *variables*. This is what distinguishes them from their additive counterparts which have time-invariant (though usually dynamic) coefficients. The time-varying coefficients result in the need for the *on-line design* of the residual generator. However, the decoupling problem to be solved at each sample is *static* thus keeping the computational complexity of the algorithms manageable. While there are other possible representations of parametric faults and disturbances, workable design techniques can be mainly based on the *underlying parameter* representation. In this it is assumed that all model discrepancies root in the discrepancies of a limited set of (physically meaningful) parameters.

The number of uncertain (underlying) parameters may be rather high, especially when model errors are concerned. This may prohibit *exact decoupling* from such disturbances and faults. To alleviate this problem, two *approximate decoupling* techniques will also be introduced in Chapter 10. One of these methods utilizes the *singular value decomposition* of the momentary fault/disturbance entry matrix. The other implies the minimization of the momentary residual norm under equality constraints. Though these need to be applied on-line (when dealing with parametric disturbances), the decoupling problem, just like above, is static at each sample.

9.1. REPRESENTATION OF PARAMETRIC FAULTS

The equations (9.1) and (9.2) imply a direct representation of parametric faults and modeling errors. A similar representation can be applied in the framework of state-space models. In Chapter 5 (Subsection 5.1.2), an alternative representation was introduced which relies on the concept of underlying parameters. These representation techniques will be further explored below, with the generation of residuals in mind. It will be shown that, in this context, the direct representation approaches, though simple, usually have limited utility.

Recall that for the design of a (generic) residual generator, the primary residuals have to be expressed as

$$[\textit{primary residuals}] = [\textit{coefficient matrix}] \; x \; [\textit{fault vector}] \tag{9.3}$$

A transformation is then applied to the primary residuals so that the desired fault response (including decoupling from faults/residuals) is achieved, that is

$$[transformation] \times [coefficient\ matrix] = [specified\ response] \tag{9.4}$$

Recall also that the transformation may have undesirable side-effects if there are linear dependencies in the coefficient matrix.

9.1.1. Direct Representation

Discrepancies in the transfer function parameters. Assume first that the parametric faults and model errors are represented directly as discrepancies in the transfer function parameters, $\blacktriangle M(\phi) = [\blacktriangle M_F(\phi) \quad \blacktriangle M_D(\phi)]$. Then, from (9.2), the internal form of the primary residuals is

$$o(t) = \blacktriangle M(\phi) u(t) \tag{9.5}$$

This formulation, however, is not suitable for residual generation. To allow the generation of residuals with specified responses, the equation needs to be made consistent with (9.3), that is, it needs to be written as

$$o(t) = I\, v(t) \qquad where\ \ v_k(t) = \blacktriangle m_{k.}(\phi) u(t) \tag{9.6}$$

This implies that residual design (including decoupling) is not possible for an individual element $\blacktriangle m_{kj}(\phi)$ of the discrepancy matrix. The smallest cluster of discrepancies for which such design is possible is a row of the $\blacktriangle M(\phi)$ matrix. Therefore, decoupling from modeling errors in a row makes any parametric fault appearing in the same row undetectable. In the extreme but not unlikely situation when modeling errors affect all rows, decoupling from these leads to complete insensitivity to faults.

Further, the coefficient matrix of the clustered discrepancies is the unit matrix. Decoupling the i-th residual from the k-th cluster requires

$$w'_i(\phi)[0 \ldots 0 \ \underset{k}{1} \ 0 \ldots 0]' = 0 \quad that\ is,\ \ w_{ik}(\phi) = 0 \tag{9.7}$$

If there are also additive faults $p(t)$ in the system then the same transformation acts on these as well. The response to the additive faults is

$$z'_i(\phi) = w'_i(\phi) S_F(\phi) \tag{9.8}$$

Now if a column of $S_F(\phi)$ is the k-th unit-matrix column (which is the case for the k-th output sensor fault) then the effect of this fault is wiped out completely. Further, if the i-th residual is decoupled from several rows of the model discrepancy matrix, and a column of $S_F(\phi)$ has nonzero elements only in those rows, then the effect of the respective fault is also completely eliminated.

Discrepancies in the state-space parameters. Now assume that the parametric discrepancies are associated with the matrices of the system's state-space description (from which they propagate to the transfer function matrix). Then

$$x(t+1) = (A+\blacktriangle A)x(t) + (B+\blacktriangle B)u(t)$$

$$y(t) = (C+\blacktriangle C)x(t) + (D+\blacktriangle D)u(t) \tag{9.9}$$

where A, B, C and D are the known model matrices and $\blacktriangle A$, $\blacktriangle B$, $\blacktriangle C$ and $\blacktriangle D$ are the discrepancies. This can be written as

$$x(t+1) = Ax(t) + Bu(t) + v_x(t)$$

$$y(t) = Cx(t) + Du(t) + v_y(t) \tag{9.10}$$

where

$$v_x(t) = \blacktriangle Ax(t) + \blacktriangle Bu(t) \qquad v_y(t) = \blacktriangle Cx(t) + \blacktriangle Du(t) \tag{9.11}$$

are the vectors of clustered discrepancies, the clustering taking place along the rows of the state and output equations. (9.10) leads to the following input-output representation

$$y(t) = [C(\phi I\text{-}A)^{-1}B+D]u(t) + C(\phi I\text{-}A)^{-1}v_x(t) + v_y(t) \tag{9.12}$$

yielding the primary residuals

$$o(t) = C(\phi I\text{-}A)^{-1}v_x(t) + I\,v_y(t) \tag{9.13}$$

Again, residual design for individual parameter-fault responses is not possible; the response can only be specified for row-clusters of faults and model errors. Decoupling from an element of $v_y(t)$ has the same consequences as the ones described in (9.7) and (9.8). To decouple from the k-th element of $v_x(t)$, in turn, it is required that

$$w'_i(\phi)[C(\phi I\text{-}A)^{-1}]_{\cdot k} = 0 \tag{9.14}$$

Now for a strictly input actuator or sensor fault $p_j(t)$,

$$s_{F\cdot j}(\phi) = C(\phi I\text{-}A)^{-1}b_{\cdot j} \tag{9.15}$$

Thus (9.14) neutralizes the k-th row of $b_{\cdot j}$ in the j-th fault response $w'_i(\phi)s_{F\cdot j}(\phi)$. If a residual is decoupled from several rows of $v_x(t)$ and the $b_{\cdot j}$

column contains nonzero elements only in those rows, then the response to the j-th fault is completely eliminated.

Based on the above considerations, we believe that the direct representation of parametric faults and modeling errors, in the framework of residual generation, *should in general be avoided*. The exceptions are those cases where the structure of parametric faults and, especially, model errors is simple enough so that the side-effects of decoupling do not make other faults undetectable.

9.1.2. Underlying Parameter Representation

The underlying parameter representation was introduced in Chapter 5. Here it will be first recollected the way it was presented there, using the ARMA form of the input-output relationship, then a variation based on the MA form will be introduced. In general, the MA form is much simpler than the ARMA form. As it will be seen, with the underlying parameter representation decoupling from parametric faults and model errors does not, in general, cause unwanted decoupling from other faults. However, the residual generator becomes *time-varying*.

Input-output relation in ARMA form. Let us express the transfer function matrix as $\mathbf{M}(\phi, \boldsymbol{\theta})$ where $\boldsymbol{\theta}=[\theta_1 \ldots \theta_\upsilon]'$ is a set of underlying parameters. The underlying parameters are subject to discrepancies (faults and uncertainties/errors) so that

$$\boldsymbol{\theta}^\circ = \boldsymbol{\theta}^\# + \blacktriangle\boldsymbol{\theta} \tag{9.16}$$

where $\blacktriangle\boldsymbol{\theta}=[\blacktriangle\theta_1 \ldots \blacktriangle\theta_\upsilon]'$ is the discrepancy of the parameter-set while $\boldsymbol{\theta}^\#$ is the value-set yielding the actual model $\mathbf{M}(\phi)$ and $\boldsymbol{\theta}^\circ$ is that for the true plant $\mathbf{M}^\circ(\phi)$. Then $\blacktriangle\mathbf{M}(\phi)$ can be approximated as

$$\blacktriangle\mathbf{M}(\phi) = \sum_{k=1}^{\upsilon} \mathbf{M}_{\theta k}(\phi) \blacktriangle\theta_k \tag{9.17}$$

where

$$\mathbf{M}_{\theta k}(\phi) = \partial\mathbf{M}(\phi, \boldsymbol{\theta}) / \partial\theta_k \Big|_{\boldsymbol{\theta} = \boldsymbol{\theta}^\#} \qquad k=1\ldots\upsilon \tag{9.18}$$

Recall that, in general, the transfer function parameters are nonlinear functions of the underlying parameters. Therefore the accuracy of the approximation (9.17) deteriorates as the deviation from the model values grows.

With the above approximation, the internal form of the generic residual generator can be written as

$$o(t) = \blacktriangle M(\phi)u(t) = [\sum_{k=1}^{\upsilon} M_{\theta k}(\phi)\blacktriangle \theta_k]\, u(t) \tag{9.19}$$

By rearranging the sum, we obtain

$$o(t) = \sum_{k=1}^{\upsilon} [M_{\theta k}(\phi)\, u(t)] \blacktriangle \theta_k = N(t) \blacktriangle \boldsymbol{\theta} \tag{9.20}$$

where

$$N(t) = [n_{\cdot 1}(t) \dots n_{\cdot \upsilon}(t)] \qquad n_{\cdot k}(t) = M_{\theta k}(\phi)u(t), \quad k=1\dots\upsilon \tag{9.21}$$

If the underlying parameter vector can be decomposed as $\boldsymbol{\theta}=[\ \boldsymbol{\theta}'_F \quad \boldsymbol{\theta}'_D]'$, where $\boldsymbol{\theta}_F$ contains the underlying parameters subject to faults and $\boldsymbol{\theta}_D$ contains those subject to modeling errors then, with the appropriate decomposition $N(t)=[\ N_F(t) \quad N_D(t)]$, (9.20) can be further written as

$$o(t) = N_F(t) \blacktriangle \boldsymbol{\theta}_F + N_D(t) \blacktriangle \boldsymbol{\theta}_D \tag{9.22}$$

Input-output relation in MA form. Recall the moving average form of the input-output relationship

$$H(\phi)y(t) = G(\phi)u(t) \tag{9.23}$$

where $H(\phi)$ and $G(\phi)$ are polynomial matrices. With parametric faults and model errors this becomes

$$[H(\phi)+\blacktriangle H(\phi)]y(t) = [G(\phi)+\blacktriangle G(\phi)]u(t) \tag{9.24}$$

leading to the primary residuals

$$o^*(t) = H(\phi)y(t) - G(\phi)u(t) = \blacktriangle G(\phi)u(t) - \blacktriangle H(\phi)y(t) \tag{9.25}$$

The matrices G and H depend on the underlying parameters $\boldsymbol{\theta}$ thus their discrepancies may be expressed with the $\blacktriangle\boldsymbol{\theta}$ underlying parameter discrepancies as

$$\blacktriangle G(\phi) = \sum_{k=1}^{\upsilon} G_{\theta k}(\phi)\blacktriangle \theta_k \qquad \blacktriangle H(\phi) = \sum_{k=1}^{\upsilon} H_{\theta k}(\phi)\blacktriangle \theta_k \tag{9.26}$$

where

$$G_{\theta_k}(\phi) = \frac{\partial G(\phi,\ \boldsymbol{\theta})}{\partial \theta_k}\Bigg|_{\boldsymbol{\theta} = \boldsymbol{\theta}^\#} \qquad H_{\theta_k}(\phi) = \frac{\partial H(\phi,\ \boldsymbol{\theta})}{\partial \theta_k}\Bigg|_{\boldsymbol{\theta} = \boldsymbol{\theta}^\#}$$

$$k=1\ldots\upsilon \tag{9.27}$$

With these, the primary residuals can be written as

$$\boldsymbol{o}^*(t) = [\sum_{k=1}^{\upsilon} \boldsymbol{G}_{\theta_k}(\phi)\blacktriangle\theta_k]\ \boldsymbol{u}(t) - [\sum_{k=1}^{\upsilon} \boldsymbol{H}_{\theta_k}(\phi)\blacktriangle\theta_k]\ \boldsymbol{y}(t) \tag{9.28}$$

Rearranging this latest equation leads to

$$\boldsymbol{o}^*(t) = \sum_{k=1}^{\upsilon} [\boldsymbol{G}_{\theta_k}(\phi)\boldsymbol{u}(t) - \boldsymbol{H}_{\theta_k}(\phi)\boldsymbol{y}(t)]\ \blacktriangle\theta_k = \boldsymbol{\Xi}(t)\blacktriangle\boldsymbol{\theta} \tag{9.29}$$

where

$$\boldsymbol{\Xi}(t) = [\boldsymbol{\xi}_{.1}(t)\ \ldots\ \boldsymbol{\xi}_{.\upsilon}(t)]$$

$$\boldsymbol{\xi}_{.k}(t) = \boldsymbol{G}_{\theta_k}(\phi)\boldsymbol{u}(t) - \boldsymbol{H}_{\theta_k}(\phi)\boldsymbol{y}(t) \qquad k=1\ldots\upsilon \tag{9.30}$$

Finally, if the underlying parameter vector is decomposed as $\boldsymbol{\theta}=[\ \boldsymbol{\theta}_F'\ \ \boldsymbol{\theta}_D'\ ]'$ then, with the appropriate decomposition of $\boldsymbol{\Xi}(t)$,

$$\boldsymbol{o}^*(t) = \boldsymbol{\Xi}_F(t)\ \blacktriangle\boldsymbol{\theta}_F + \boldsymbol{\Xi}_D(t)\ \blacktriangle\boldsymbol{\theta}_D \tag{9.31}$$

Relationship between $N(t)$ and $\boldsymbol{\Xi}(t)$. Let us write the transfer function as $\boldsymbol{M}(\phi)=\bar{\boldsymbol{G}}(\phi)/\bar{h}(\phi)$. Then, for this representation and with (9.18) and (9.21)

$$\boldsymbol{n}_{.k}(t) = \frac{[\partial\bar{\boldsymbol{G}}(\phi,\ \boldsymbol{\theta})\ /\ \partial\theta_k]\ \bar{h}(\phi) - [\partial\bar{h}(\phi,\ \boldsymbol{\theta})\ /\ \partial\theta_k]\ \bar{\boldsymbol{G}}(\phi)}{[\bar{h}(\phi)]^2}\ \boldsymbol{u}(t) \tag{9.32}$$

On the other hand, with $\bar{\boldsymbol{H}}(\phi)=\boldsymbol{I}\bar{h}(\phi)$ and using (9.27) and (9.30),

$$\bar{\boldsymbol{\xi}}_k(t) = [\partial\bar{\boldsymbol{G}}(\phi,\ \boldsymbol{\theta})\ /\ \partial\theta_k]\ \boldsymbol{u}(t) - [\partial\bar{h}(\phi,\ \boldsymbol{\theta})\ /\ \partial\theta_k]\ \boldsymbol{y}(t) \tag{9.33}$$

The output is related to the input as $\boldsymbol{y}(t)=\boldsymbol{M}^\circ(\phi)\boldsymbol{u}(t)$, however, the true plant $\boldsymbol{M}^\circ(\phi)$ is generally not known. The input-output relationship can, though, be approximated with the model $M(\phi)$ as $M(\phi)u(t)$. With this, (9.33) becomes

$$\xi_k(t) \approx \left[[\partial \bar{G}(\phi, \theta) / \partial \theta_k] - [\partial \bar{h}(\phi, \theta) / \partial \theta_k] \, \bar{G}(\phi)/\bar{h}(\phi) \right] u(t) \quad (9.34)$$

Thus

$$\boxed{\bar{\xi}_k(t) \approx \bar{h}(\phi)\, \bar{n}_{\cdot k}(t)} \qquad k=1 \ldots \upsilon \quad (9.35)$$

Example 9.1.

Consider again the system of Example 5.3. There we derived the partial derivatives of the ARMA form (the full transfer function) with respect to the two poles α_1 and α_2 as underlying parameters. Let us now derive the partial derivatives from the MA form. The $G(\phi)$ and $H(\phi)$ matrices are

$$G(\phi) = \begin{bmatrix} \beta_{11}\phi^{-1} & \beta_{12}\phi^{-1} \\ \beta_{11}\beta_{21}\phi^{-2} & \beta_{12}\beta_{21}\phi^{-2}+\beta_{22}\phi^{-1}(1+\alpha_1\phi^{-1}) \end{bmatrix}$$

$$H(\phi) = Diag[(1+\alpha_1\phi^{-1}) \quad (1+\alpha_1\phi^{-1})(1+\alpha_2\phi^{-1})]$$

The partial derivatives are obtained as

$$G_{\alpha 1}(\phi) = \begin{bmatrix} 0 & 0 \\ 0 & \beta_{22}\phi^{-2} \end{bmatrix} \qquad G_{\alpha 2}(\phi) = 0$$

$$H_{\alpha 1}(\phi) = Diag[\phi^{-1} \quad \phi^{-1}(1+\alpha_2\phi^{-1})]$$

$$H_{\alpha 2}(\phi) = Diag[0 \quad \phi^{-1}(1+\alpha_1\phi^{-1})]$$

Finally, the $\Xi(t)$ matrix is obtained, from (9.30), as $[\xi_{\cdot 1}(t) \quad \xi_{\cdot 2}(t)]=$

$$\begin{bmatrix} -\phi^{-1}y_1(t) & 0 \\ \beta_{22}\phi^{-2}u_2(t) - \phi^{-1}(1+\alpha_2\phi^{-1})y_2(t) & -\phi^{-1}(1+\alpha_1\phi^{-1})y_2(t) \end{bmatrix}$$

Comparing this to the original result in Example 5.3 verifies that the MA form is, indeed, much simpler than the ARMA form. ⁙

Notice that the primary residuals, as expressed in terms of the parametric faults and disturbances in (9.20)-(9.22), and in (9.29)-(9.31), are consistent with the requirement formulated in (9.3). This implies that they are suitable for residual generation. The coefficient matrices $N(t)$ and $\Xi(t)$ do not have any special structure (e.g., are not the unit matrix) which would increase the likelihood of unwanted decoupling from other faults. However, the coefficient matrices are time-varying (though, for a given value of time, non-dynamic).

9.2. DESIGN FOR PARAMETRIC FAULTS AND MODEL ERRORS

The design of residual generators for parametric faults and model errors follows the pattern of design for additive faults and disturbances. There are differences, though, which follow from the *time-varying* nature of the problem. The most substantial consequence of time-varying coefficient matrices is that the implementation of the residual generator must be done *on-line*, anew for each sample. However, the problem to be solved at each sample is *momentarily static*, that is, it involves manipulations with *constant* (non-polynomial) matrices.

In this section, the specification and implementation issues of time-varying, momentarily static residual generators will be summarized. The underlying parameter representation will be used. The equations will be written for the ARMA format; they apply to the MA format with only minor modifications.

9.2.1. Response Specification

For a momentarily static residual generator, the response specifications are *constant*. The fault response may be structured or directional. Since modeling errors are considered as parametric disturbances, decoupling from them is included in the specification, whether structured or directional, as zero disturbance responses.

A *structured response* is basically specified by its structure matrix. In this respect, all results of Sections 7.1 and 7.2 apply without modification. To completely define the implementation, one of the nonzero responses in each row may be fully specified (as discussed in Section 7.3); such response specification is now a constant.

A *directional response* specification follows the patterns of Section 8.1, except the response vectors are constant (they contain no dynamic factor). Disturbance decoupling is included as zero columns.

Just like in the case of additive fault/disturbance response specifications, a single row of the specification may be

- full or almost full (or less than almost full)
- homogeneous or non-homogeneous.

Also, the diagonal specification set is of special interest.

9.2.2. Implementation

Row-by-row implementation. The implementation of a single line specification is governed by the equation

$$z_i' = w_i'(t)\, N(t) \tag{9.36}$$

Here z_i' is the (constant) specification for the i-th row, $w_i'(t)$ is the (time-varying) transformation and $N(t)$ is the (time-varying) entry matrix associated with the faults and disturbances in the specification.

If the specification is full non-homogeneous, that is, $z_i' = [z_{i1} \ \ldots \ z_{i\mu}]$ and at least one of the elements is nonzero, and if $N(t)$ has full rank then the transformation can be obtained as

$$\boxed{w_i'(t) = z_i' \, [N(t)]^{-1}} \tag{9.37}$$

If the specification is homogeneous, that is, $z'_i = 0'$, then it must be (at most) almost full, that is $z_i' = [z_{i1} \ \ldots \ z_{i,\mu-1}]$. The solution may be obtained by one of the following algorithms (c.f. Subsection 6.3.1):

a. Add a non-homogeneous scalar specification $z_{i\mu}$ with the respective column $n_{\cdot\mu}(t)$, so that the augmented entry matrix $N^A(t) = [\,N(t) \quad n_{\cdot\mu}(t)]$ has full rank. Then the problem is reduced to the non-homogeneous specification and can be solved by (9.37).

b. Choose one element of $w_i'(t)$ arbitrarily. For example, choose $w_{i\mu}(t) = c$. Then (9.36) can be rearranged as (c.f. (6.45))

$$[w_{i1}(t) \ \ldots \ w_{i,\mu-1}(t)]\; N^{\mu}(t) = -\,c\, n_{\mu\cdot}(t) \tag{9.38}$$

where $N^{\mu}(t)$ contains the first μ-1 rows of the $N(t)$ matrix. $N^{\mu}(t)$ is a square matrix; if it is invertible, a solution may be obtained as

$$\boxed{[w_{i1}(t) \ \ldots \ w_{i,\mu-1}(t)] = -\,c\, n_{\mu\cdot}(t)[N^{\mu}(t)]^{-1}} \tag{9.39}$$

Since the residual generator is now momentarily static (unlike in the case of additive fault residual generators), *its realizability is not an issue, neither is its stability or polynomial computational behavior.* This simplifies the implementation quite significantly.

Matrix implementation. If each row of the specification concerns the same set of faults and disturbances then the $N(t)$ matrix is also the same for all rows. This is the case with directional and diagonal specifications. Now the implementation equation is (c.f. (8.7))

$$\mathbf{Z} = \boldsymbol{W}(t)\, \boldsymbol{N}(t) \tag{9.40}$$

where $\mathbf{Z}$ is the specification matrix and $\boldsymbol{W}(t)$ is the transforming matrix. If $N(t)$ is invertible, the latter is obtained as (c.f. (8.24))

$$\boxed{\boldsymbol{W}(t) = \mathbf{Z}\,[\boldsymbol{N}(t)]^{-1}} \tag{9.41}$$

If there are no disturbances then the $\mathbf{Z}$ matrix is square $\mu \cdot \mu$; if, further, the specification is diagonal then $\mathbf{Z}=Diag[z_{11} \ \dots \ z_{\mu\mu}]$. If there are disturbances, so that $N(t)=[\,N_F(t) \quad N_D(t)]$, then $\mathbf{Z}=[\mathbf{Z}_F \quad \mathbf{Z}_D]$; here $\mathbf{Z}_F$ is square $(\mu-\rho_D)\cdot(\mu-\rho_D)$ (where ρ_D is the number of independent disturbances), and may further be diagonal, while $\mathbf{Z}_D=\mathbf{0}$ (c.f. (8.28)).

9.2.3. Existence Conditions

The existence conditions, expressed in terms of the rank properties of the $N(t)$ matrix, follow the pattern of the conditions for residual generators with additive faults and disturbances (there expressed in terms of the rank properties of the $S(\phi)$ matrix). A significant difference, though, follows from the time-varying nature of $N(t)$. A rank defect of this matrix may be

- temporary, if it arises from a particular input sequence $\boldsymbol{u}(t)$, or
- permanent, if it arises from the properties of the $\boldsymbol{M}_{\theta k}(\phi)$ derivative matrices.

With a temporary rank defect, it may not be possible to satisfy the response specification in the particular sample. With a permanent rank defect, it may not be possible at all.

In this subsection, the existence conditions in terms of the $N(t)$ matrix will be reviewed. The situations leading to permanent rank defect will then be explored in the next subsection.

Conditions for single residual implementation. Consider the implementation equation (9.36). Arbitrary response specifications can be satisfied only if

$$\rho \leq \mu \quad and \quad Rank\, N(t) = \rho \tag{9.42}$$

where ρ is the number of faults and disturbances. However, solutions may be obtained even if (9.42) is not satisfied, provided the specification is consistent with the rank defects of $N(t)$. In particular, faults and disturbances for which zero response has been specified may have linearly dependent columns in $N(t)$.

Conditions for structured residual generation. Consider a single residual specification which is part of a structured specification set. Recall (from Chapter 7) that the structure specification for the row includes μ-1 zeroes and an arbitrary number of ones. However, the explicit specification z_i' contains only the zero responses and one of the nonzero responses; the remaining non-zero responses float (are specified only to the extent that they are nonzero). Thus the existence condition for the implementation is (c.f. (7.18)):

$$Rank\, N_i(t) \leq \mu - 1 \tag{9.43a}$$

$$Rank\, [N_i(t) \quad \mathbf{n}_{\cdot k}(t)] = Rank\, N_i(t) + 1 \quad for\ all\ \mathbf{n}_{\cdot k}\ outside\ N_i \tag{9.43b}$$

where $N_i(t)$ contains all those columns of $N(t)$ which are associated with a zero specification in z_i'.

Conditions for directional residual generation. The conditions for directional response generation, without disturbance decoupling, are (c.f. (8.10) and (8.11))

$$\rho_F \leq \mu \quad and \quad Rank\, N_F(t) = \rho_F \tag{9.44a}$$

With disturbance decoupling, the conditions become (c.f. (8.10) and (8.20))

$$\rho_F \leq \mu - Rank\, N_D(t) \quad and \quad Rank\, N_F(t) = \rho_F \tag{9.44b}$$

The above conditions apply also to diagonal response specifications.

9.2.4. Sources of Permanent Rank Defects

As mentioned earlier, permanent rank defects of $N(t)$ arise from the special behavior of the partial derivative matrices $M_{\theta k}(\phi)$. In this subsection, the possible sources and propagation mechanism of such rank defects will be explored. First, however, some linear dependency concepts will be recalled and clarified.

Types of linear dependencies. Everything we are going to say here about sets of polynomial vectors applies also to sets of rational vectors, which can be seen by expanding the latter to common denominator. Recall that a set of polynomial vectors $[m_1(\phi) \;\; m_2(\phi) \ldots m_\upsilon(\phi)]$ is linearly dependent if the relationship

$$\beta_1(\phi)m_1(\phi) + \beta_2(\phi)m_2(\phi) + \ldots + \beta_\upsilon(\phi)m_\upsilon(\phi) = 0 \qquad (9.45)$$

is satisfied with at least one nonzero $\beta_j(\phi)$ coefficient. In general, the $\beta_1(\phi)$, $\beta_2(\phi) \ldots \beta_\upsilon(\phi)$ coefficients are polynomials. A special case of linearly dependent polynomial vectors is when the coefficients are $\beta_1, \beta_2 \ldots \beta_\upsilon$ constants. We are now adding a third type of linear dependence, which we will refer to as *dependence in steady state*, when the steady state equivalents of the polynomial vectors are linearly related:

$$\beta_1[m_1(\phi)]_{\phi=1} + \beta_2[m_2(\phi)]_{\phi=1} + \ldots + \beta_\upsilon[m_\upsilon(\phi)]_{\phi=1} = 0 \qquad (9.46)$$

Clearly (9.45), with constant or polynomial coefficients, implies (9.46), but a set of polynomial vectors may be dependent in steady state even if (9.45) is not satisfied.

When non-polynomial vectors, such as $n_{\cdot k}(t)$ are generated as a result of manipulations with polynomial matrices, they may inherit polynomial relationships. However, a relationship

$$\beta_1(\phi)n_{\cdot 1}(t)+\beta_2(\phi)n_{\cdot 2}(t) + \ldots + \beta_\upsilon(\phi)n_{\cdot \upsilon}(t) = 0 \qquad (9.47)$$

does *not* mean that the momentary constant vectors $n_{\cdot 1}(t)$, $n_{\cdot 2}(t) \ldots n_{\cdot \upsilon}(t)$ are linearly dependent. What (9.47) represents is

$$\beta_1^o n_{\cdot 1}(t)+\beta_1^1 n_{\cdot 1}(t-1)+\beta_1^2 n_{\cdot 1}(t-2) + \ldots + \beta_2^o n_{\cdot 2}(t)+\beta_2^1 n_{\cdot 2}(t-1)+\beta_2^2 n_{\cdot 2}(t-2)$$
$$+ \ldots = 0 \qquad (9.48)$$

which says nothing about the relationship between the momentary values $n_{\cdot 1}(t)$, $n_{\cdot 2}(t) \ldots n_{\cdot \upsilon}(t)$. Of course, (9.47) with *constant* coefficients does represent a linear relationship among the momentary values. Also, (9.47) in general implies that $n_{\cdot 1}(t)$, $n_{\cdot 2}(t) \ldots n_{\cdot \upsilon}(t)$ are linearly dependent in steady state.

Example 9.2.

Consider pairwise the following polynomial vectors:

$$m_1(\phi) = [(1 - 0.5\phi^{-1})(1 - 0.6\phi^{-1}) \qquad (1 - 0.8\phi^{-1})(1 - 0.6\phi^{-1})]'$$

$$m_2(\phi) = [3(1 - 0.5\phi^{-1})(1 - 0.6\phi^{-1}) \qquad 3(1 - 0.8\phi^{-1})(1 - 0.6\phi^{-1})]'$$

$$m_3(\phi) = [(1 - 0.5\phi^{-1})(1 - 0.2\phi^{-1}) \qquad (1 - 0.8\phi^{-1})(1 - 0.2\phi^{-1})]'$$

$$m_4(\phi) = [(1 - 0.4\phi^{-1})(1 - 0.75\phi^{-1}) \qquad (1 - 0.8\phi^{-1})(1 - 0.7\phi^{-1})]'$$

* Obviously, $m_1(\phi)$ and $m_2(\phi)$ are linearly dependent with constant coefficients since $m_2(\phi)=3m_1(\phi)$.
* There is polynomial linear dependence between $m_3(\phi)$ and $m_1(\phi)$ (and also $m_2(\phi)$) since $(1 - 0.2\phi^{-1})m_1(\phi)=(1 - 0.6\phi^{-1})m_3(\phi)$.
* The latter are also linearly dependent in steady state since $m_1(1)=[0.2 \quad 0.08]'$ and $m_3(1)=[0.4 \quad 0.16]'=2m_1(1)$.
* $m_4(\phi)$ is linearly related to each of the other vectors (only) in steady state since $m_4(1)=[0.15 \quad 0.06]'=0.75m_1(1)$. ⁝⁝⁝

Similar parameters. We will refer to the parameters θ_1 and θ_2 as *similar* if they affect the *entire* $M(\phi, \boldsymbol{\theta})$ matrix in a linearly related way, that is, if the full partial derivative matrices are linearly related

$$\beta_1(\phi)M_{\theta 1}(\phi) + \beta_2(\phi)M_{\theta 2}(\phi) = 0 \tag{9.49}$$

For a set of parameters $\theta_1, \theta_2 \ldots \theta_\upsilon$, (9.49) expands as

$$\beta_1(\phi)M_{\theta 1}(\phi) + \beta_2(\phi)M_{\theta 2}(\phi) + \ldots + \beta_\upsilon(\phi)M_{\theta \upsilon}(\phi) = 0 \tag{9.50}$$

Since $n_{\cdot k}(t)=M_{\theta k}(\phi)u(t)$, (9.50) propagates directly to the columns of the $N(t)$ matrix, yielding

$$\beta_1(\phi)n_{\cdot 1}(t) + \beta_2(\phi)n_{\cdot 2}(t) + \ldots + \beta_\upsilon(\phi)n_{\cdot \upsilon}(t) = 0 \tag{9.51}$$

Clearly, if the β coefficients are constant, (9.51) represents a linear relationship among the columns of $N(t)$. If they are polynomial, the linear relationship is not present in general, only in steady state.

A linear relationship among the $n_{\cdot 1}(t), n_{\cdot 2}(t) \ldots n_{\cdot \upsilon}(t)$ columns makes some of the parity equation structures unattainable (for structured residuals) and it blocks the generation of independent response directions (for directional residuals). If the linear relationship concerns a pair of parameters, as in (9.49), then the respective faults/disturbances are not isolable.

Example 9.3.

Consider a system with one input and two outputs, and assume that there are three underlying parameters θ_1, θ_2 and θ_3, affecting the transfer functions the following way:

$$M(\phi) = \left[\frac{\theta_1}{1-(\theta_2+2\theta_3)\phi^{-1}} \quad \frac{\theta_1\phi^{-1}}{[1-(\theta_2+2\theta_3)\phi^{-1}](1-0.8\phi^{-1})} \right]'$$

The three partial derivative matrices are obtained as

$$M_{\theta 1}(\phi) = \left[\frac{1}{1-(\theta_2+2\theta_3)\phi^{-1}} \quad \frac{\phi^{-1}}{[1-(\theta_2+2\theta_3)\phi^{-1}](1-0.8\phi^{-1})} \right]'$$

$$M_{\theta 2}(\phi) = \left[\frac{\theta_1\phi^{-1}}{[1-(\theta_2+2\theta_3)\phi^{-1}]^2} \quad \frac{\theta_1\phi^{-2}}{[1-(\theta_2+2\theta_3)\phi^{-1}]^2(1-0.8\phi^{-1})} \right]'$$

$$M_{\theta 3}(\phi) = \left[\frac{2\theta_1\phi^{-1}}{[1-(\theta_2+2\theta_3)\phi^{-1}]^2} \quad \frac{2\theta_1\phi^{-2}}{[1-(\theta_2+2\theta_3)\phi^{-1}]^2(1-0.8\phi^{-1})} \right]'$$

For θ_2 and θ_3 there is a constant type relationship $M_{\theta 3}(\phi)=2M_{\theta 2}(\phi)$ thus faults of these two parameters are not isolable. This, of course, may be expected from the way they appear in the transfer functions. $M_{\theta 1}(\phi)$ is polynomially related to the other partial derivative matrices, in that

$$M_{\theta 2}(\phi) = \theta_1\phi^{-1}/[1-(\theta_2+2\theta_3)\phi^{-1}] \cdot M_{\theta 1}(\phi)$$

Thus faults of θ_1 may in general be isolated from the other faults. In steady state, however, isolation is not possible. ⁞⁞⁞

Linearly dependent rows. Another important situation arises when the rows of the $M(\phi)$ matrix are linearly dependent:

$$\alpha_1(\phi)m_{1.}(\phi) + \alpha_2(\phi)m_{2.}(\phi) + \dots + \alpha_\mu(\phi)m_{\mu.}(\phi) = 0' \tag{9.52}$$

If the coefficients $\alpha_1(\phi)$, $\alpha_2(\phi)$ ... $\alpha_\mu(\phi)$ are not affected by the underlying parameters then (9.52) applies to all the partial derivative matrices, with the *same* coefficients:

$$\alpha_1(\phi)[M_{\theta k}(\phi)]_{1.} + \alpha_2(\phi)[M_{\theta k}(\phi)]_{2.} + \ldots + \alpha_\mu(\phi)[M_{\theta k}(\phi)]_{\mu.} = 0' \qquad k=1\ldots\upsilon \tag{9.53}$$

Since any element of the $N(t)$ matrix is computed as

$$n_{ik}(t) = [M_{\theta k}(\phi)]_{i.}\, u(t) \tag{9.54}$$

the row relationship (9.51) is transferred to $N(t)$ so that its rows follow

$$\alpha_1(\phi)n_{1.}(t) + \alpha_2(\phi)n_{2.}(t) + \ldots + \alpha_\mu(\phi)n_{\mu.}(t) = 0' \tag{9.55}$$

If the $\alpha_1, \alpha_2 \ldots \alpha_\mu$ coefficients are constant, this is a general linear relationship among the rows. With polynomial coefficients, the rows become linearly dependent only in steady state, if the input signal allows such state, or due to some special temporal properties of the input. In any of these cases, when $Rank M(\phi) \le \upsilon$, the row relations imply a column relation of the same type, and may result in the usual restrictions on residual generation (unattainable row structures and dependent directional responses).

Example 9.4.

Consider again a system with a single input and two outputs and two uncertain parameters θ_1 and θ_2. Let the transfer functions be

$$M(\phi) = \begin{bmatrix} \dfrac{1-\theta_1\phi^{-1}}{1-\theta_2\phi^{-1}} & c\dfrac{1-\theta_1\phi^{-1}}{1-\theta_2\phi^{-1}} \end{bmatrix}'$$

The second row is a constant times the first: obviously the two outputs are not independent. The partial derivatives are obtained as

$$M_{\theta 1}(\phi) = \begin{bmatrix} \dfrac{-\phi^{-1}}{1-\theta_2\phi^{-1}} & c\dfrac{-\phi^{-1}}{1-\theta_2\phi^{-1}} \end{bmatrix}'$$

$$M_{\theta 2}(\phi) = \begin{bmatrix} \dfrac{\phi^{-1}}{(1-\theta_2\phi^{-1})^2} & c\dfrac{\phi^{-1}}{(1-\theta_2\phi^{-1})^2} \end{bmatrix}'$$

Now define

$$v_1(t) = \frac{\phi^{-1}}{1-\theta_2\phi^{-1}}u(t) \qquad v_2(t) = \frac{\phi^{-1}}{(1-\theta_2\phi^{-1})^2}u(t)$$

Then the fault-entry matrix can be written as

$$N(t) = \begin{bmatrix} -v_1(t) & v_2(t) \\ -cv_1(t) & cv_2(t) \end{bmatrix}$$

This matrix is clearly singular for any input $u(t)$. ⁙

9.2.5. Fault Isolation with Output Filtering

The above results, concerning linearly dependent rows of the transfer function matrix $M(\phi)$, have a consequence which is practically very important. Since polynomial relations among the rows of the transfer function matrix, in general, do not affect the rank of the $N(t)$ matrix, it follows that an arbitrary number of parametric faults can be isolated from even a single consistency relation, by simply filtering its residual. This possibility, though, is subject to the following conditions:

(i) there is no constant-type similarity among the underlying parameters in the original consistency relation(s);

(ii) there is no constant-type linear relationship among the filters;

(iii) the input is sufficiently rich.

Note that condition (ii) is satisfied even with such a simple arrangement as a series of delays. However, the input richness (persistent excitation) requirement increases with the number of parameters and may be difficult to satisfy in the course of normal plant operation, only during transients. More about this may be found in Chapter 13.

Example 9.5.

This example will illustrate the isolation of two parametric faults from a single-input single-output system, with the help of a second equation obtained simply by applying a one-step delay to the output. Let the transfer function be

$$M(\phi) = \beta / (1+\alpha\phi^{-1})$$

Let β and α be the parameters of interest, with nominal values $\beta=1$ and $\alpha= -0.5$. The second row is simply $\phi^{-1}M(\phi)$. The partial derivatives are

$$M_\beta(\phi) = \begin{bmatrix} 1/(1-0.5\phi^{-1}) \\ \phi^{-1}/(1-0.5\phi^{-1}) \end{bmatrix} \qquad M_\alpha(\phi) = \begin{bmatrix} -\phi^{-1}(1-\phi^{-1}+0.25\phi^{-2}) \\ -\phi^{-2}(1-\phi^{-1}+0.25\phi^{-2}) \end{bmatrix}$$

and the $N(t)$ matrix is

$$N(t) = [M_\beta(\phi) \quad M_\alpha(\phi)]\, u(t)$$

Assume that the input is unit step; with zero initial conditions, the determinant of the $N(t)$ matrix is obtained, for different values of t, as

t	0	1	2	5	8
Det $N(t)$	0	1	1.25	0.5039	0.1075

As it can be seen, the condition of the matrix deteriorates as the system approaches steady state (in steady state, the determinant is zero). The full $N(t)$ matrix, e.g., for $t=5$, and its inverse are

$$N(5) = \begin{bmatrix} 1.96875 & -3.5625 \\ 1.9375 & -3.25 \end{bmatrix} \qquad N^{-1}(5) = \frac{1}{0.5039} \begin{bmatrix} -3.25 & 3.5625 \\ -1.9375 & 1.96875 \end{bmatrix}$$

Let us design the residual generator with diagonal response. Assume first that $\Delta\beta=0.1$. With this, and for $t=5$, we obtain

$$o(5) = \begin{bmatrix} 0.19688 \\ 0.19375 \end{bmatrix} \qquad r(5) = N^{-1}(5)\, o(5) = \begin{bmatrix} 0.1 \\ 0 \end{bmatrix}$$

as expected. Now assume that $\Delta\alpha = -0.05$. This leads to

$$o(5) = \begin{bmatrix} 0.1919 \\ 0.1729 \end{bmatrix} \qquad r(5) = N^{-1}(5)\, o(5) = \begin{bmatrix} -0.0153 \\ 0.0623 \end{bmatrix}$$

The expected residual would be $[0 \quad 0.05]'$. The discrepancy is due to the fact that $M(\phi)$ is a nonlinear function of α and the relationship has been approximated by the derivative at the nominal value $\alpha = -0.5$. Recomputing the derivative at $\alpha = -0.525$ yields $[-3.8357 \quad -3.4558]'$ for the second column of $N(5)$ which leads exactly to the expected residual. (This, however, only serves to support the claim concerning the source of discrepancy, and can not be used in practice since the value and direction of the fault is not known.)

A link to identification. When the fault detection and isolation algorithm is designed with unit diagonal response to parametric faults then the residuals are, in fact, estimates of the parameter changes.This technique thus must be related to the identification of the same parametric faults. While the identification approach will be discussed in some detail in Chapter 13, some preliminary consideration to this relationship will be given in the following paragraphs.

The link between the parity equation based isolation of parametric faults and their traditional identification is most obvious when (i) the parameters of interest (the "underlying" parameters) are those of the transfer function, (ii) the plant is single-output and time-shifted variants of the original input-output relation are used as additional consistency relations, and (iii) the parity equations are written in MA form.

Let us consider the simplest, single-input single-output case. Write the original residual as

$$o^*(t) = h(\phi)y(t) - g(\phi)u(t) = \Delta g(\phi)u(t) - \Delta h(\phi)y(t) \tag{9.56}$$

$$= [1 + h^1\phi^{-1} + \ldots + h^{\nu}\phi^{-\nu}]\, y(t) - [g^o + g^1\phi^{-1} + \ldots + g^{\tau}\phi^{-\tau}]\, u(t)$$

$$= [\Delta g^o + \Delta g^1\phi^{-1} + \ldots + \Delta g^{\tau}\phi^{-\tau}]\, u(t) - [\Delta h^1\phi^{-1} + \ldots + \Delta h^{\nu}\phi^{-\nu}]\, y(t)$$

With the additional equations defined as

$$\mathbf{o}^*(t) = [1 \quad \phi^{-1} \quad \ldots \quad \phi^{-\nu-\tau}]'\, o^*(t) \tag{9.57}$$

the input and output matrices become

$$\mathbf{g}(\phi) = [1 \quad \phi^{-1} \quad \ldots \quad \phi^{-\nu-\tau}]'\, g(\phi)$$

$$\mathbf{h}(\phi) = [1 \quad \phi^{-1} \quad \ldots \quad \phi^{-\nu-\tau}]'\, h(\phi) \tag{9.58}$$

Thus the partial derivatives with respect to g^k, $k=0\ldots\tau$, and h^k, $k=1\ldots\nu$, are

$$\mathbf{g}_{gk}(\phi) = [1 \quad \phi^{-1} \quad \ldots \quad \phi^{-\nu-\tau}]'\, \phi^{-k} \qquad \mathbf{g}_{hk}(\phi) = \mathbf{0}$$

$$\mathbf{h}_{hk}(\phi) = [1 \quad \phi^{-1} \quad \ldots \quad \phi^{-\nu-\tau}]'\, \phi^{-k} \qquad \mathbf{h}_{gk}(\phi) = \mathbf{0} \tag{9.59}$$

leading to the time-varying fault entry matrix

$$\Xi(t) = \begin{bmatrix} u(t) & \ldots & u(t-\tau) & -y(t-1) & \ldots & -y(t-\nu) \\ u(t-1) & \ldots & u(t-\tau-1) & -y(t-2) & \ldots & -y(t-\nu-1) \\ \vdots & & & & & \\ u(t-\nu-\tau) & \ldots & u(t-\nu-2\tau) & -y(t-\nu-\tau-1) & \ldots & -y(t-2\nu-\tau) \end{bmatrix} \tag{9.60}$$

This is basically the same data matrix encountered when estimating $\boldsymbol{\theta}$ with the usual least squares identification technique. The only difference is that, under least squares identification, the data matrix would contain more than $\nu+\tau+1$ rows, which are minimally necessary to estimate the same number of parameters.

For diagonal response, the residual is obtained as

$$r(t) = [\Xi(t)]^{-1} o^*(t) \tag{9.61}$$

On the other hand, as it will be shown in Chapter 13, replacing the outputs $y(t)$ in the usual least-squares algorithm with the residuals $o^*(t)$ results in estimating the parameter changes $\Delta\boldsymbol{\theta}$ instead of the parameters $\boldsymbol{\theta}$. That is

$$\Delta\hat{\boldsymbol{\theta}} = [\Xi'(t)\,\Xi(t)]^{-1}\,\Xi'(t)\,o^*(t) \tag{9.62}$$

Thus $r(t)$ is the estimate of $\Delta\boldsymbol{\theta}$ using the minimum possible data-set, which is the limiting case of the general least-squares algorithm.

The data matrix in (9.60) also reveals that the sufficiently rich input signal requirement of parametric fault isolation corresponds to the usual persistent excitation condition of least squares identification. For $\nu+\tau+1$ parameters, the input has to be persistently exciting of order $\nu+\tau+1$ in either case. However, while this is to be achieved over a long data-set in identification, it applies to the minimum data-set in the case of parity relation based isolation.

Let us finally investigate the situation when

$$m^\circ(\phi) = \frac{g^\circ(\phi)}{h^\circ(\phi)} \qquad\qquad m(\phi) = \frac{g(\phi)}{h(\phi)} = \frac{g^\circ(\phi)(1-\beta\phi^{-1})}{h^\circ(\phi)(1-\alpha\phi^{-1})} \tag{9.63}$$

that is, when the model contains an excess pole-zero pair relative to the true system. In identification, this situation is referred to as over-parametrization, and is known to lead to non-identifiability. We will explore the properties of the fault-entry matrix in this case. The partial derivatives with respect to the β and α parameters are

$$\begin{aligned} g_\beta(\phi) &= -\phi^{-1} g^\circ(\phi) \qquad\qquad & g_\alpha(\phi) &= 0 \\ h_\beta(\phi) &= 0 & h_\alpha(\phi) &= -\phi^{-1} h^\circ(\phi) \end{aligned} \tag{9.64}$$

Then, for each shifted output $i=1\ldots\nu+\tau$,

$$\xi_{i\beta}(t) = -\phi^{-i-1} g^{\circ}(\phi)u(t) \qquad \xi_{i\alpha}(t) = \phi^{-i-1} h^{\circ}(\phi)y(t) \tag{9.65}$$

Since $y(t)=m^{\circ}(\phi)u(t)$, this implies

$$\xi_{i\alpha}(t) = \phi^{-i-i} g^{\circ}(\phi)u(t) = -\xi_{i\beta}(t) \qquad \textit{for } i=1...\nu+\tau \tag{9.66}$$

That is, the redundant pole and zero are similar parameters and therefore faults associated with them are not isolable. Thus over-parametrization has been seen as a special case of similar parameters.

9.3. MIXED ADDITIVE AND MULTIPLICATIVE FAULTS

If some of the faults and disturbances are additive while others are multiplicative then, in general, the convenient linear transformation machinery breaks down. The reason is that, in this case, one needs to work formally with products of several time-functions and shift-operator polynomials, but the latter are meaningful only if the time-functions they are applied to are linear. In particular, with $u(t)$ and $v(t)$ two arbitrary time-functions and $w(\phi)$ a shift-operator polynomial, the expression $w(\phi)u(t)v(t)$ is undefined because, in general, $[w(\phi)u(t)]v(t) \neq [w(\phi)v(t)]u(t) \neq w(\phi)[u(t)v(t)]$.

This problem can be circumvented if it is possible to decompose the design into an off-line stage, involving manipulations only with shift-operator polynomials, and an on-line stage, involving only time-functions. This possibility exists when the response specification for the additive faults and disturbances is homogeneous (all zero). Under such specification, intermediate residuals can be generated in the off-line stage which are decoupled from the additive faults/disturbances. These serve then as the basis for the on-line implementation stage, where the multiplicative fault and disturbance response specifications are satisfied (while the resulting transformations of the intermediate residuals remain decoupled from the additive faults/disturbances).

Note that homogeneous responses to additive faults and disturbances represent some important practical situations. If the responses are homogeneous also on the multiplicative side, we are dealing with disturbance decoupling combined with structural design for all faults. If the multiplicative fault responses are non-homogeneous then directional design for these, combined with decoupling from additive disturbances, may be the case.

Assume first that, for a particular scalar residual, the multiplicative part of the response specification is also homogeneous. Then, assuming that no output filtering is applied, the maximum number of faults/disturbances that can be handled is

$$\rho_A + \rho_M = \mu - 1 \tag{9.67}$$

where ρ_A is the number of additive faults and disturbances and ρ_M is that of the multiplicative ones. Decoupling (on-line) from the ρ_M multiplicative faults and disturbances requires ρ_M+1 independent intermediate residuals. To generate an intermediate residual decoupled from ρ_A additive faults and disturbances, in turn, requires ρ_A+1 primary residuals. For the polynomial independence of the intermediate residuals, each new set has to contain at least one primary residual not occurring in the other sets. From μ primary residuals,

$$\mu - \rho_A = \rho_M+1 \tag{9.68}$$

such sets may be selected, exactly the number needed for the on-line stage.

With non-homogeneous specification on the multiplicative side, and no output filtering, the maximum number of faults/disturbances becomes

$$\rho_A+\rho_M = \mu \tag{9.69}$$

The (on-line) implementation of the non-homogeneous response for ρ_M multiplicative faults and disturbances requires ρ_M independent intermediate residuals. From μ primary residuals, exactly $\mu - \rho_A = \rho_M$ such residuals may be obtained.

If we allow polynomial dependences at the level of intermediate residuals, which is equivalent to output filtering at this level, then the limits (9.67) and (9.69) may be exceeded. Further, if the primary set already has some special structure, this may be utilized in the implementation.

Example 9.6.

Consider a system with a single input and three outputs, subject to an actuator fault and two parametric faults. Generate a single residual with the response specification $z'=[0 \quad 0.5 \quad 0.8]$ (where the zero response belongs to the actuator fault).

The system is described by

$$y_i(t) = m_i(\phi, \theta_1, \theta_2)u(t) \qquad i=1,2,3$$

The three primary residuals are

$$o_i(t) = y_i(t) - m_i(\phi)u(t) = m_i(\phi)\blacktriangle u(t) + n_{i1}(t)\blacktriangle\theta_1 + n_{i2}(t)\blacktriangle\theta_2 \qquad i=1,2,3$$

where

$$n_{ij}(t)=m_{i\theta_j}(\phi)u(t) \qquad i=1,2,3; \quad j=1,2$$

First generate two intermediate residuals (off-line), one from $\{o_1(t), o_2(t)\}$, the other from $\{o_2(t), o_3(t)\}$, so that they are both decoupled from $\Delta u(t)$:

$$r_{o1}(t) = m_2(\phi)o_1(t) - m_1(\phi)o_2(t) = n_{o11}(t)\Delta\theta_1 + n_{o12}(t)\Delta\theta_2$$

$$r_{o2}(t) = m_3(\phi)o_2(t) - m_2(\phi)o_3(t) = n_{o21}(t)\Delta\theta_1 + n_{o22}(t)\Delta\theta_2$$

where

$$n_{o11}(t) = [m_2(\phi)m_{1\theta 1}(\phi) - m_1(\phi)m_{2\theta 1}(\phi)]\, u(t), \quad \text{etc.}$$

Second, generate the final residual (on-line) as

$$r(t) = w_1(t)r_{o1}(t) + w_2(t)r_{o2}(t)$$

so that the specification $r(t|\Delta\theta_1)=0.8\Delta\theta_1$, $r(t|\Delta\theta_2)=0.5\Delta\theta_2$ is met, requiring

$$[w_1(t) \quad w_2(t)] = [0.8 \quad 0.5] \begin{bmatrix} n_{o11}(t) & n_{o12}(t) \\ n_{o21}(t) & n_{o22}(t) \end{bmatrix}^{-1}$$

∎

Example 9.7.

Consider a system with $\mu=3$ outputs, $\rho_A=2$ additive faults and $\rho_M=3$ multiplicative faults. Assume that all primary residuals (o_1, o_2, o_3) depend on all faults. Design a structured residual generator for the five faults.

A possible column canonical structure is

	p_1	p_2	$\Delta\theta_1$	$\Delta\theta_2$	$\Delta\theta_3$
r_1	0	0	1	1	1
r_2	0	1	1	1	0
r_3	1	1	1	0	0
r_4	1	1	0	0	1
r_5	1	0	0	1	1

Four intermediate residuals are necessary, each one derived off-line from two of the primary residuals:

	p_1	p_2	$\blacktriangle\theta_1$	$\blacktriangle\theta_2$	$\blacktriangle\theta_3$	
r_{o1}	0	1	1	1	1	(from o_1 and o_2)
r_{o2}	0	1	1	1	1	(from o_2 and o_3)
r_{o3}	1	0	1	1	1	(from o_1 and o_2)
r_{o4}	1	0	1	1	1	(from o_2 and o_3)

Of the five final residuals,

r_1 is obtained directly from the three primary residuals off-line

r_2 is obtained from the intermediates r_{o1} and r_{o2} on-line

r_5 is obtained from the intermediates r_{o3} and r_{o4} on-line

r_3 and r_4 are obtained directly from the three primary residuals on-line.

9.4. NOTES AND REFERENCES

The idea of applying parity relations to parametric faults was first outlined in (Gertler et al, 1985). The methodology was then explored by Moid Kunwer (1992) and also in (Gertler, 1992), and the results were summarized in (Gertler and Kunwer, 1993 and 1995). The difficulties with the popular additive representation, described here in Subsection 9.1.1, were pointed out in (Gertler, 1994). Some considerations about linear independence and persistent excitation in relation to parity equations are given here for the first time.

The link between parity relations designed for parametric faults and fault detection by parameter estimation will be explored in more detail in Chapter 13. We will comment on various contributions in Section 13.5.

The application of parity relations for the detection of parametric faults has been studied extensively by the group of Rolf Isermann, especially by Thomas Höfling (Höfling, 1993; Höfling and Pfeufer, 1994; Höfling and Isermann, 1995). In their approach, the parity relations are written for the continuous-time model. This is advantageous because the plant parameters appear in the continuous-time model in a more direct way than in the discrete-time model. However, computing higher order derivatives from sampled observations is very noise sensitive and requires special algorithms. An approach to this problem, the "integral filter," will be described in Section 13.4.

An important related activity has been undertaken by Michele Basseville, Albert Benveniste and their coworkers (Basseville et al, 1987; Benveniste et al, 1987; Zhang et al, 1994). Their primary interest lies in the detection (and isolation) of small parametric faults based on extended sets of observations. The essence of their approach is to test if the recent set is consistent with a reference model. Test statistics are generated from what we would call parity relation residuals, computed with the reference model. The tests rely on the assumption that the changes are small and thus their "local approach" applies.

10
Robustness in Residual Generation

10.0. INTRODUCTION

Model based fault detection and isolation algorithms rely on the mathematical model of the monitored plant and, naturally, it is of critical importance how they are affected by inaccuracies and uncertainties of the plant model. Loosely speaking, an algorithm is called robust if it is able to operate in spite of model(ing) errors.

Unfortunately, robustness in fault detection and isolation is at least as elusive as it is important. There is no algorithm which is robust under arbitrary model error conditions. To design an algorithm for robustness, some rather detailed information is necessary about the nature of errors and uncertainties, and such information is seldom available. But even if it is, what can be achieved is rather limited and generally falls into one of the following categories:

1. Perfect decoupling of the residuals from uncertainties in a limited number of plant or model parameters;
2. Approximate decoupling of the residuals from uncertainties in a greater number of parameters or a set of variants of the complete model;
3. Selection of the set of model equations, from a larger number of possible equations, which are (almost) best for some selected model error robustness or fault sensitivity measure.

Several algorithms published in the literature under the claim of robustness actually decouple the residuals only from additive disturbances. These do not satisfy our definition of model error insensitivity and therefore we consider them only "quasi robust." Some of these algorithms will be discussed here, though, because their approach is similar to the techniques used to achieve model error robustness.

The algorithms which approach model error robustness by approximate decoupling rely on a variety of optimization techniques. The most important of these are

a. Rank reduced approximation by singular value decomposition;

b. Constrained least squares optimization;

c. H-infinity optimization.

The selection of the best set of model equations also utilizes an optimization procedure; this is a discrete search procedure under structural constraint. Optimization based algorithms are intellectually appealing but their utility may be limited. This is because, in general, optimal design is the *best* that can be achieved for the *given system* with the *chosen specification* - but there is no guarantee that the resulting optimal performance will be really *satisfactory*.

We will start this chapter with the recollection of the mechanism and conditions of perfect decoupling, as applied to additive disturbances and parametric uncertainties, combined with various fault isolation strategies. Then the technique of rank reduction by singular value decomposition will be introduced and applied to approximate decoupling from additive disturbances and parametric uncertainties. This will be followed by a least squares based technique, for approximate decoupling from model parameter uncertainties. Multiple model variants and their approximation by singular value decomposition will be discussed next. H-infinity concepts will be introduced briefly. Finally, a discrete search algorithm will be described which finds the locally or globally best parity equation set, by some measure of robustness or sensitivity, under structured fault isolation.

10.1. PERFECT DECOUPLING FROM DISTURBANCES

We will start by summarizing the mechanism and conditions of perfectly decoupling the residuals from disturbances. Because of their formal similarity, additive disturbances and model parameter uncertainties (multiplicative disturbances) will be handled in parallel. Various fault isolation schemes will be considered, including fault detection, structured isolation and directional isolation, each combined with the decoupling from the disturbances.

10.1.1. Additive Disturbances

We will consider first the case of both the faults and disturbances being additive. The residual is described by the following equation

$$r(t) = W(\phi)[y(t) - M(\phi)u(t)] = W(\phi)[S_D(\phi)q(t) + S_F(t)p(t)] \tag{10.1}$$

where, as in the earlier chapters, $q(t)$ denotes the vector of additive disturbances and $p(t)$ the vector of additive faults. It will be assumed that neither of these vectors is empty. Further, the number of primary residuals (and that of the plant outputs) is μ.

Specifications for perfect decoupling. Perfect decoupling from the disturbance vector requires

$$r(t|q) = W(\phi)S_D(\phi)q(t) = 0 \tag{10.2}$$

that is, the response specification $Z_D(\phi)$ is

$$Z_D(\phi) = W(\phi)S_D(\phi) = 0 \tag{10.3}$$

a. Fault detection. If the objective is only fault detection, with disturbance decoupling, then (10.3) can be implemented as a homogeneous specification. Alternatively, it can be made a non-homogeneous specification by including the response to one of the faults, as

$$Z(\phi) = W(\phi)[S_D(\phi) \quad s_{F\cdot k}(\phi)] = [0 \quad z_{F\cdot k}(\phi)] \tag{10.4}$$

b. Structured residuals. If disturbance decoupling is combined with structured isolation of faults then the decoupling condition (10.3) becomes part of the structure specification. In the structure matrix Φ, the disturbance part is

$$\Phi_D = 0 \tag{10.5}$$

In the design specification for the i-th residual, the disturbance response appears as

$$z_i'(\phi) = w_i'(\phi)[S_D(\phi) \quad S_{Fi}(\phi)] = [0 \quad 0] \tag{10.6}$$

where $S_{Fi}(\phi)$ contains the columns of $S_F(\phi)$ which belong to the i-th *zero response set* (those faults the response to which in $r_i(t)$ is specified as zero). (10.6) is a homogeneous specification; it may be expanded into a non-homogeneous one by including a fault outside the zero response set, as

$$z_i'(\phi) = w_i'(\phi)[S_D(\phi) \quad S_{Fi}(\phi) \quad s_{F\cdot k}(\phi)] = [0 \quad 0 \quad z_{Fik}(\phi)] \tag{10.7}$$

c. *Directional residuals.* If disturbance decoupling is combined with directional design for the faults then the disturbance responses are simply

$$r(t|q_j) = 0 \tag{10.8}$$

so that the combined specification is

$$Z(\phi) = W(\phi)[S_D(\phi) \quad S_F(t)] = [0 \quad Z_F(\phi)] \tag{10.9}$$

where $Z_F(\phi)$ is the specified fault response.

Existence criteria for perfect disturbance decoupling. We will recall (from earlier chapters) the existence criteria of disturbance decoupling under the various fault isolation strategies. After defining the *number of independent residuals*, we will derive a common underlying criterion, which we will call *the fundamental limitation of disturbance decoupling*.

a. *Fault detection.* All faults must be detectable from the residuals decoupled from the disturbances, that is,

$$W(\phi)s_{F\cdot k}(\phi) \neq 0 \qquad \textit{for all } k \tag{10.10}$$

This requires that

$$Rank\,[S_D(\phi) \quad s_{F\cdot k}(\phi)] = Rank\,S_D(\phi) + 1 \qquad \textit{for all } k \tag{10.11}$$

b. *Structured residuals.* Now all faults outside the i-th zero response set must be detectable from the i-th residual, that is (10.10) must be satisfied for *all k outside the i-th zero response set*. This requires that (c.f. (7.18b))

$$Rank\,[S_D(\phi) \quad S_{Fi}(\phi) \quad s_{F\cdot k}(\phi)] = Rank\,[S_D(\phi) \quad S_{Fi}(\phi)] + 1$$
$$\textit{for all } k \textit{ outside the i-th zero response set} \tag{10.12}$$

c. *Directional residuals.* Now the residuals have to simultaneously provide decoupling from the disturbances and the specified directional response to the faults. This requires (c.f. (8.19))

$$Rank\,[S_D(\phi) \quad S_F(\phi)] = Rank\,S_D(\phi) + \rho_F \tag{10.13}$$

The fundamental limitation of disturbance decoupling. Define the number of *independent primary residuals* as

$$m = Rank\,[S_D(\phi) \quad S_F(\phi)] \leq \mu \tag{10.14}$$

where μ is the total number of primary residuals. In general, $m=\mu$, but exceptionally $m<\mu$, if there are fewer than μ disturbances and faults, or if the matrix $[S_D(\phi) \quad S_F(\phi)]$ has rank defect.

a. Fault detection. Clearly,

$$Rank\,[S_D(\phi) \quad s_{F\cdot k}(\phi)] \leq m \tag{10.15}$$

Thus from (10.11)

$$Rank\,S_D(\phi) = Rank\,[S_D(\phi) \quad s_{F\cdot k}(\phi)] - 1 \tag{10.16}$$

which implies

$$\boxed{Rank\,S_D(\phi) \leq m - 1 \leq \mu - 1} \tag{10.17}$$

b. Structured residuals. Applying (10.14) to (10.12) yields

$$Rank\,[S_D(\phi) \quad S_{Fi}(\phi)] \leq m - 1 \tag{10.18}$$

This, with at least one fault in the zero response set, implies

$$\boxed{Rank\,S_D(\phi) \leq m - 2 \leq \mu - 2} \tag{10.19}$$

c. Directional residuals. From (10.13), with (10.14)

$$Rank\,S_D(\phi) \leq m - \rho_F \tag{10.20}$$

This, with the assumption that there are at least two faults, again implies (10.19).

We may conclude that perfect disturbance decoupling is subject to the fundamental limitation expressed by (10.17), if only fault detection is concerned, and by (10.19) if isolation is also involved. Note that (10.19) only provides a necessary condition; more stringent sufficient conditions may arise from the isolation strategy, as expressed by (10.12) and (10.13). It is a matter of design consideration how the complete design freedom, represented by the m independent primary residuals, is divided between disturbance decoupling and fault isolation - but (10.17) or (10.19) may never be violated.

10.1.2. Parametric Disturbances (Model Uncertainty)

Now we turn our attention to the case when both the faults and disturbances are related to the system model, and are represented by discrepancies between the assumed and true values of the underlying (system) parameters $\boldsymbol{\theta}$. The residuals are now governed by the equation

$$r(t) = W(t)[y(t) - M(\phi)u(t)] = N_D(t)\Delta\boldsymbol{\theta}_D + N_F(t)\Delta\boldsymbol{\theta}_F \tag{10.21}$$

where $\Delta\boldsymbol{\theta}_D$ are the uncertainties of the underlying parameters while $\Delta\boldsymbol{\theta}_F$ are faults of those parameters. The coefficient matrices $N_D(t)$ and $N_F(t)$ are defined in Section 9.1, equation (9.21); recall that these are time varying number matrices. The specifications and conditions for perfect decoupling basically follow those of the additive case, with the matrices $S_D(\phi)$ and $S_F(\phi)$ replaced by $N_D(t)$ and $N_F(t)$, respectively. We will summarize the results below, then comment on an extension, involving filtered or simply delayed residuals.

Specifications for perfect decoupling. These will follow the pattern of equations (10.3), (10.4), (10.6), (10.7) and (10.9). The decoupling specification proper is

$$Z_D = W(t)N_D(t) = 0 \tag{10.22}$$

Combined with a nonzero fault response in a detection scheme it becomes

$$Z = W(t)[N_D(t) \quad n_{F\cdot k}(t)] = [0 \quad z_{F\cdot k}] \tag{10.23}$$

For a single residual in a structured scheme, a homogeneous specification is

$$z_i' = w_i'(t)[N_D(t) \quad N_{Fi}(t)] = [0 \quad 0] \tag{10.24}$$

where $N_{Fi}(t)$ is the coefficient matrix of the i-th zero response fault set. An almost homogeneous specification for the same case is

$$z_i' = w_i'(t)[N_D(t) \quad N_{Fi}(t) \quad n_{F\cdot k}(t)] = [0 \quad 0 \quad z_{Fik}] \tag{10.25}$$

Finally, for directional residuals

$$Z = W(t)[N_D(t) \quad N_F(t)] = [0 \quad Z_F] \tag{10.26}$$

Recall that the residual generator $W(t)$ or $w_i'(t)$ is also time varying and numerical.

Existence criteria. Here we will follow the pattern of equations (10.11), (10.12), (10.13), (10.14), (10.17) and (10.19). The criterion for detection is

$$Rank\,[N_D(t) \quad n_{F\cdot k}(t)] = Rank\,N_D(t) + 1 \quad \textit{for all k} \tag{10.27}$$

For structured isolation

$$Rank\,[N_D(t) \quad N_{Fi}(t) \quad n_{F\cdot k}(t)] = Rank\,[N_D(t) \quad N_{Fi}(t)] + 1$$
$$\textit{for all k outside the i-th zero response set} \tag{10.28}$$

and for directional isolation

$$Rank\,[N_D(t) \quad N_F(t)] = Rank\,N_D(t) + \rho_F \tag{10.29}$$

(10.27) implies the fundamental limitation of disturbance decoupling as

$$Rank\,N_D(t) \leq m - 1 \leq \mu - 1 \tag{10.30}$$

while (10.28), with at least one fault in the zero response set, and (10.29) with two faults, lead to

$$Rank\,N_D(t) \leq m - 2 \leq \mu - 2 \tag{10.31}$$

where now

$$m = Rank\,[N_D(t) \quad N_F(t)] \leq \mu \tag{10.32}$$

Within and in addition to the above limits, the rank considerations of Chapter 5 apply, especially those concerning permanent rank-defects described in Subsection 9.2.4.

Rank expansion by multiple samples. As it was pointed out in Subsection 9.2.5, the rank-limit on $N(t)$ may be expanded by adding filtered variants of the primary residuals. The $n_{i\cdot}(t)$ rows arising from the filtered residuals are independent of the original rows provided certain conditions are met. Filtering may be as simple as a series of time-delays and the resulting rank expansion may be quite significant; by using τ samples, the rank-limit in (10.32) changes from μ to $\tau\cdot\mu$. The most important of the conditions is that the input signal must be persistently exciting, at least of order τ, over a sliding window of τ samples. This requirement, though, may be difficult to satisfy in the course of normal plant operation, in which case rank expansion may only be possible during plant transients.

10.2. APPROXIMATE DECOUPLING WITH SINGULAR VALUE DECOMPOSITION

As shown above, the number of disturbances (or disturbances and faults), from which the residuals can be perfectly decoupled, is restricted. Since in practice the number of actual disturbances may be quite high, such a restriction may affect the utility of the diagnostic algorithm. To alleviate this problem, perfect decoupling may be replaced with approximate decoupling. Now the number of disturbances is not restricted anymore but, of course, the quality of decoupling deteriorates as the number of disturbances is increased.

In this section, an approximate decoupling approach will be introduced. This approach involves a rank reduced approximation of the disturbance entry matrix, using the technique of singular value decomposition. The resulting matrix is the best approximation of the original one in a least squares sense.

Approximate decoupling by rank reduced approximation provides significant flexibility, allowing for the implementation of a variety of fault isolation strategies. These will also be discussed in this section.

10.2.1. Singular Value Decomposition

The singular value decomposition technique is the extension, to general rectangular matrices, of the spectral decomposition procedure of square matrices (see Subsection 2.5.2). Consider a rectangular matrix $\mathbf{\Pi}$ with k rows and l columns, and assume that $l \geq k$ (the matrix is "flat"). The same results will apply, with slight modifications, also to "tall" matrices $(l \leq k)$. Notice that $\mathbf{\Pi}\,\mathbf{\Pi}'$ and $\mathbf{\Pi}'\,\mathbf{\Pi}$ are both square symmetrical matrices, having the following properties:

a. Both $\mathbf{\Pi}\,\mathbf{\Pi}'$ and $\mathbf{\Pi}'\,\mathbf{\Pi}$ have real non-negative eigenvalues. Further, $\mathbf{\Pi}'\,\mathbf{\Pi}$ has the same set of eigenvalues as $\mathbf{\Pi}\,\mathbf{\Pi}'$, plus l-k zero eigenvalues.

b. The eigenvectors $\boldsymbol{\gamma}_1, \boldsymbol{\gamma}_2 \dots$ of $\mathbf{\Pi}\,\mathbf{\Pi}'$ form an orthogonal set. Therefore, after normalization,

$$\mathbf{\Gamma}'\,\mathbf{\Gamma} = \mathbf{I} \qquad \text{where} \qquad \mathbf{\Gamma} = [\boldsymbol{\gamma}_1 \ \dots \ \boldsymbol{\gamma}_k] \tag{10.33}$$

This implies that the right-side and left-side eigenvectors are the same. Similarly, the eigenvectors $\boldsymbol{\psi}_1, \boldsymbol{\psi}_2, \dots$ of $\mathbf{\Pi}'\,\mathbf{\Pi}$ form an orthogonal set.

Now the singular value decomposition of $\mathbf{\Pi}$ is defined as

$$\boxed{\mathbf{\Pi} = \mathbf{\Gamma}\,S\,\mathbf{\Psi}'} \tag{10.34}$$

Here

$$S = Diag[\sigma_1 \quad \sigma_2 \quad \dots \quad \sigma_k] \tag{10.35}$$

and $\sigma_1, \sigma_2, \dots, \sigma_k$ are the non-negative square-roots of the eigenvalues of $\Pi\Pi'$, referred to as the *singular values* of Π. The singular values in S are arranged in the order of decreasing magnitude. If $Rank\Pi < k$ then the last k - $Rank\Pi$ singular values are zero.

Γ is a $k \cdot k$ matrix the columns γ_j of which are the orthonormal eigenvectors of $\Pi\Pi'$; they are arranged in the order of the singular values they belong to.

Ψ is an $l \cdot k$ matrix the columns ψ_j of which are orthonormal eigenvectors of $\Pi'\Pi$ belonging to the common eigenvalues, and arranged in their order. For the nonzero singular values, Ψ is computed as

$$\Psi = \Pi' \Gamma S^{-1} \tag{10.36}$$

(For the zero eigenvalues the ψ_j vectors can be computed in an indirect way, but they are irrelevant for the decomposition.)

Finally, Eq. (10.34) may be written in spectral form as

$$\Pi = \sum_{j=1}^{k} \sigma_j \gamma_j \psi_j' \tag{10.37}$$

To see the validity of the above decomposition, consider the following:

(i) $\sigma_1^2 \dots \sigma_k^2$ and $\gamma_1 \dots \gamma_k$ are indeed the eigenvalues and eigenvectors of $\Pi\Pi'$ since, by (10.34),

$$\Pi\Pi' = (\Gamma S \Psi')(\Psi S \Gamma') = \Gamma \Lambda \Gamma' \tag{10.38}$$

where $\Psi'\Psi = I$ and $\Lambda = S^2$.

(ii) (10.36) yields a set of orthonormal vectors since

$$\psi_j' \psi_g = \gamma_j' \Pi\Pi' \gamma_g / \sigma_j \sigma_g = \gamma_j' \left[\sum_{i=1}^{k} \lambda_i \gamma_i \gamma_i' \right] \gamma_g / \sigma_j \sigma_g = \delta_{jg} \lambda_j \gamma_j' \gamma_g / \sigma_j \sigma_g \tag{10.39}$$

where we utilized the spectral form of $\Pi\Pi'$ and the orthonormality of the $\gamma_1 \dots \gamma_k$ set.

(iii) $\sigma_1^2 \dots \sigma_k^2$ and $\psi_1 \dots \psi_k$ are indeed eigenvalues and eigenvectors of $\Pi' \Pi$ since

$$\Pi' \Pi = (\Psi S \Gamma')(\Gamma S \Psi') = \Psi \Lambda \Psi' \tag{10.40}$$

with $\Gamma' \Gamma = I$ and $\Lambda = S^2$ (the additional zero eigenvalues do not appear in this expression but they do not contribute to the expansion either).

(iv) (10.34), with (10.36), indeed returns Π since

$$\Gamma S \Psi' = \Gamma S S^{-1} \Gamma' \Pi = \Pi \tag{10.41}$$

Rank reduction by SVD. Consider the spectral decomposition of the matrix Π, (10.37). The essence of rank reduction is that, by omitting some of the last terms from the sum, an approximation

$$\boxed{\Pi_* = \sum_{j=1}^{k^*} \sigma_j \, \gamma_j \, \psi_j'} \tag{10.42}$$

is obtained so that

$$Rank\, \Pi_* = k^* < k \tag{10.43}$$

and the error introduced by the approximation is as little as possible since the terms with the smallest σ_j values have been left out. As shown in the references, the approximation (10.42) is optimal also in the sense that it minimizes the norm

$$L = \sum_{i=1}^{k} \sum_{j=1}^{l} [\pi_{*ij} - \pi_{ij}]^2 \tag{10.44}$$

(10.44) has an important implication for the implementation of rank reduction. As it is seen, (10.42) minimizes the sum of the error squares between Π and Π_*, with no respect to the size of the elements. In many cases it makes more sense to require that the errors be proportional to the vector (column) sizes. Therefore it may be advisable to normalize the columns of Π before singular value decomposition, by writing it as

$$\Pi = \Pi_\upsilon \, Diag[\upsilon_1 \quad \upsilon_2 \dots] \tag{10.45}$$

where the columns of Π_υ are normalized (have unit length) and the coefficients $\upsilon_1, \upsilon_2 \ldots$ are the lengths of the respective columns in the original Π matrix.

An additional advantageous property of this approach follows from the orthogonality of the γ_j vectors (Lou et al, 1986). With this and Eq. (10.42)

$$\gamma_j' \Pi_* = 0 \qquad j=k^*+1 \ldots k \tag{10.46}$$

That is, the vectors $\gamma_{k^*+1} \cdots \gamma_k$ (the ones not used in the rank reduced approximation) provide perfect decoupling transformations from the columns of Π_*. Of course, decoupling by these vectors from the columns of the original Π matrix will be approximate but it is optimal. Define the index describing the goodness of such decoupling

$$L_W = \sum_i \sum_j (w_{i\cdot}\, \pi_{\cdot j})^2 \tag{10.47}$$

(where W is an arbitrary approximate decoupling matrix). It has been shown (Lou et al, 1986) that this index is minimized by the choice

$$W = [\gamma_{k^*+1} \cdots \gamma_k]' \tag{10.48}$$

and the value of the index under the optimal choice is

$$L_{W,\,opt} = \sum_{j=k^*+1}^{k} \sigma_j^2 \tag{10.49}$$

That is, the quality of decoupling by the eigenvectors γ_j is measured by the square of their accompanying singular values σ_j.

10.2.2. Rank Reduced Approximation for Additive Disturbances

We wish to find a rank reduced approximation for the disturbance transfer matrix $S_D(\phi)$. To avoid the difficulty of applying SVD to a transfer function matrix, we need to convert the system (10.1) into state-space form

$$x(t+1) = Ax(t) + Bu(t) + E_D\, q(t) + E_F\, p(t)$$

$$y(t) = Cx(t) + Du(t) + F_D\, q(t) + F_F\, p(t) \tag{10.50}$$

With this,

$$S_D(\phi) = C(\phi I - A)^{-1} E_D + F_D \tag{10.51}$$

Now recall (c.f. (6.81)) that

$$Rank\, S_D(\phi) = Rank\, \Gamma_D^+(\phi) - \nu \tag{10.52}$$

where ν is the system order and $\Gamma_D^+(\phi)$ is the disturbance system matrix

$$\Gamma_D^+(\phi) = \begin{bmatrix} \phi I - A & -E_D \\ C & F_D \end{bmatrix} \tag{10.53}$$

Thus the rank reduction can be performed on the $[-E_D' \quad F_D']'$ matrix to which SVD is easy to apply.

Define

$$\Pi_D = [-E_D' \quad F_D']' \tag{10.54}$$

This is usually a tall matrix with $\nu+\mu$ rows and $\rho_D < \nu+\mu$ columns. It will be decomposed as

$$\Pi_D = \sum_{j=1}^{\rho_D} \sigma_j \gamma_j \psi_j' \tag{10.55}$$

where still γ_j are the eigenvectors of $\Pi_D \Pi_D'$ and ψ_j are those of $\Pi_D' \Pi_D$. This is then approximated as

$$\Pi_{D*} = \sum_{j=1}^{m*} \sigma_j \gamma_j \psi_j' \tag{10.56}$$

where $m*$ is the desired rank of the approximation. This matrix still has ρ_D columns; for the purpose of the residual generator design, it is finally replaced by

$$\Pi_D^* = m* \textit{ independent columns of } \Pi_{D*} \tag{10.57}$$

The submatrices E_D^* and F_D^* of this Π_D^* matrix can then be placed in the system matrix $\Gamma_D^+(\phi)$, or the rank-reduced disturbance transfer matrix $S_D^*(\phi)$ can be computed with them, using (10.51). Any residual generator designed for "perfect" decoupling from Π_D^* will provide perfect decoupling also from the approximation Π_{D*} and approximate decoupling from the original Π_D.

Design strategies with rank reduced approximation. Recall that, in many respects, faults and disturbances can be handled the same or in a similar way. Therefore we will treat them here together.

Soft and hard response specifications. With the rank reduced approximation technique, there are four ways how the residual response to a particular fault/disturbance can be specified:

a. *Soft zero.* The response is specified to stay close to zero but perfect decoupling is not guaranteed. Such soft zero is specified for a fault/disturbance by including its entry column in the approximated matrix.

b. *Hard zero.* The transformation is designed for perfect decoupling. Faults/disturbances with hard zero specification are not included in the approximated entry matrix but a separate decoupling condition is applied to them.

c. *Hard one.* The transformation is designed for a nonzero response with a specified value. Separate nonzero response conditions are applied to these faults/disturbances.

d. *Soft one.* The response is not zero but its value is not specified (it "floats"). These are the faults which are not explicitly included in the design of the transformation (for the particular residual) but still need to be detectable.

Decoupling from a disturbance may be implemented with a soft or a hard *zero*. The same applies to decoupling from a fault for structured isolation. The faults from which a structured residual is not decoupled may be handled with a soft or a hard *one*. Directional residual generation clearly requires hard *ones* (and occasionally hard zeroes). Of course, each hard response (whether *zero* or *one*) uses up one degree of design freedom thus hard responses should be specified with caution.

As we mentioned earlier, there is no guarantee the results of the optimal implementation will be really *satisfactory*. The problems may be eased by including the most important performance specifications among the hard constraints. Also, it may be helpful to adapt the specifications to the natural inclinations of the system, for example by choosing an isolation structure which requires decoupling from faults having the smallest gains.

Disturbance decoupling only. Recall that $m = Rank S(\phi) \leq \mu$; this can be computed from (10.52) by replacing S_D and Γ_D^+ with $S = [S_D \quad S_F]$ and Γ^+. Assume that the system contains $\rho_D \geq m$ disturbances with $Rank S_D(\phi) = m$, and we aim for disturbance decoupling with no residual enhancement for the faults. Then the disturbances can be handled with soft *zeroes*; a hard *one* response is specified for one of the faults while the other fault responses float.

A rank reduced equivalent Π_D^* is found for the disturbance entry matrix Π_D, with $m^* = m - 1$. The rank-reduced transfer matrix $S_D^*(\phi)$ is computed from this, which is then supplemented with the entry column of the fault selected for hard one response, $s_{F\cdot g}(\phi)$, leading to the implementation equation (for the i-th residual)

$$w_i'(\phi)[S_D^*(\phi) \quad s_{F\cdot g}(\phi)] = [0\ldots0 \quad z_{ig}(\phi)] \tag{10.58}$$

where $z_{ig}(\phi)$ is the hard response specification. The transformation may be obtained the usual way, by inverting the combined entry matrix, directly or by means of the system matrix. The existence criterion (10.11) applies, with $S_D^*(\phi)$ substituted for $S_D(\phi)$.

It should be emphasized that the satisfaction of (10.11) alone says little about the size of the responses and there is no guarantee that a particular *soft one* response will exceed the *soft zero* responses.

Disturbance decoupling with structured isolation. Assume now that, in addition to decoupling from ρ_D disturbances, the residuals have to be structured for a set of faults. This implies decoupling also from a fault subset which is different for each residual. The disturbances and the faults in the particular decoupling subset may be handled with soft *zero*es. A hard *one* response is specified to one of the faults outside the subset while the remaining fault responses float.

Let us collect the entry columns of the faults in the i-th zero response subset into Π_{Fi}. Now a rank reduced equivalent Π_{DFi}^* needs to be found for Π_D and Π_{Fi} together, with $m^*=m-1$. A rank-reduced combined transfer matrix $S_{DFi}^*(\phi)$ is then computed from this, which is used in place of $S_D^*(\phi)$ in Eq. (10.58), so that

$$w_i'(\phi)[S_{DFi}^*(\phi) \quad s_{F\cdot g}(\phi)] = [0\ldots0 \quad z_{ig}(\phi)] \tag{10.59}$$

where, again, $s_{F\cdot g}(\phi)$ and $z_{ig}(\phi)$ are the entry vector and specified response for the fault selected for hard one response. The solution may be obtained by the inversion of the combined entry matrix. The existence condition (10.12) applies, with $[S_D(\phi) \quad S_{Fi}(\phi)]$ replaced by $S_{DFi}^*(\phi)$.

Disturbance decoupling with directional isolation. Now assume that, in addition to decoupling from ρ_D disturbances, the residuals have to be directional for $\rho_F < m$ faults. All fault responses are specified as hard *one*s in this case. The disturbance responses are specified as soft *zero*es, and the rank of the disturbance entry matrix is reduced to $m - \rho_F$.

Accordingly, an equivalent $\Pi_D^*(\phi)$ is found, with $m^*=m-\rho_F$ from which $S_D^*(\phi)$ is computed. This latter appears in the implementation equation, together with the fault entry matrix $S_F(\phi)$ (both matrices are invariant over the residual set):

$$w_i'(\phi)[S_D^*(\phi) \quad S_F(\phi)] = [0 \ldots 0 \quad z_{i,m^*+1}(\phi) \ldots z_{i,m}(\phi)] \tag{10.60}$$

Here $z_{ij}(\phi)$, $j=m^*+1 \ldots m$ are the fault responses representing the directional specification. The transformation can be obtained by the inversion of the combined entry matrix. The condition for (10.60) to have a solution is (10.13), with $S_D^*(\phi)$ replacing $S_D(\phi)$.

Example 10.1.

In this example, we will demonstrate how rank reduced approximation can be applied under the various design strategies. We will also show that, although the technique always works, the resulting residual generator may be of limited utility, especially when the rank reduction is significant.

We will consider a third order system with three outputs, three disturbances and two faults. The transfer matrices are shown in Table 10.1. The equivalent state space representation is

TABLE 10.1

$$S_F(\phi) = \frac{\begin{bmatrix} 5\phi^{-1}(1-.8\phi^{-1})(1-.54\phi^{-1}) & -\phi^{-1}(1-.5\phi^{-1})(1-.8\phi^{-1}) \\ 2\phi^{-1}(1-.5\phi^{-1})(1-.9\phi^{-1}) & \phi^{-1}(1-.5\phi^{-1})(1-.4\phi^{-1}) \\ \phi^{-1}(1-.6\phi^{-1})(1-1.1\phi^{-1}) & 2\phi^{-1}(1-.5\phi^{-1})(1-.6\phi^{-1}) \end{bmatrix}}{(1-.5\phi^{-1})(1-.6\phi^{-1})(1-.8\phi^{-1})}$$

$$S_D(\phi) = \frac{\begin{bmatrix} 3\phi^{-1}(1-.8\phi^{-1})(1-.533\phi^{-1}) & 2\phi^{-1}(1-.6\phi^{-1})(1-.8\phi^{-1}) & 4\phi^{-1}(1-.8\phi^{-1})(1-.525\phi^{-1}) \\ 3\phi^{-1}(1-.5\phi^{-1})(1-.733\phi^{-1}) & 3\phi^{-1}(1-.5\phi^{-1})(1-.6\phi^{-1}) & 3\phi^{-1}(1-.5\phi^{-1})(1-.8\phi^{-1}) \\ 2\phi^{-1}(1-.6\phi^{-1})(1-.65\phi^{-1}) & 5\phi^{-1}(1-.6\phi^{-1})(1-.62\phi^{-1}) & \phi^{-1}(1-.6\phi^{-1})(1-.8\phi^{-1}) \end{bmatrix}}{(1-.5\phi^{-1})(1-.6\phi^{-1})(1-.8\phi^{-1})}$$

$$A = \begin{bmatrix} 0.5 & 0 & 0 \\ 0 & 0.6 & 0 \\ 0 & 0 & 0.8 \end{bmatrix} \qquad C = \begin{bmatrix} 1 & 1 & 0 \\ 0 & 1 & 1 \\ 1 & 0 & 1 \end{bmatrix}$$

$$E_F = \begin{bmatrix} 2 & 0 \\ 3 & -1 \\ -1 & 2 \end{bmatrix} \qquad E_D = \begin{bmatrix} 1 & 2 & 1 \\ 2 & 0 & 3 \\ 1 & 3 & 0 \end{bmatrix} \qquad F_F = 0 \quad F_D = 0$$

a. First we will design a single residual generator for disturbance decoupling and fault detection. Accordingly, the following assignments are made:

disturbance 1: soft zero

disturbance 2: soft zero

disturbance 3: soft zero

fault 1: soft one (floating)

fault 2: hard one

Thus a rank 2 approximation is needed for the three disturbances. The design equation is

$$w_i'(\phi)[S_D^*(\phi) \quad s_{F\cdot 2}(\phi)] = [0 \quad 0 \quad 1]$$

The singular values of the E_D matrix are

$$Diag[\sigma_1 \quad \sigma_2 \quad \sigma_3] = Diag[4.3649 \quad 3.1464 \quad 0.2184]$$

The approximation will be obtained by discarding the last singular value, yielding

$$E_{D*} = \begin{bmatrix} 1.1457 & 1.9583 & 0.9059 \\ 1.9583 & 0.0120 & 3.0269 \\ 0.9059 & 3.0269 & 0.0608 \end{bmatrix} \qquad m^*=2$$

Observe the close resemblance to the original matrix. Of this approximation, the first two columns will be used, together with the second column of E_F, to form the design matrix

$$[E_D^* \quad e_{F\cdot 2}] = \begin{bmatrix} 1.1457 & 1.9583 & 0 \\ 1.9583 & 0.0120 & -1 \\ 0.9059 & 3.0269 & 2 \end{bmatrix}$$

The actual design is performed using the system matrix with the above as $\boldsymbol{E}$. For causality, a response modifier $[0 \quad 0 \quad \phi^{-1}]$ is needed. The generator is obtained as

$$w_i'(\phi) = [-0.6761+0.4202\phi^{-1} \quad 0.9609 - 0.5910\phi^{-1} \quad -0.3185+0.0771\phi^{-1}]$$

Finally, the actual responses are obtained as

$$w_i'(\phi)\, s_{D1}(\phi) = 0.2174\, \phi^{-1}$$

$$w_i'(\phi)\, s_{D2}(\phi) = -0.0620\, \phi^{-1}$$

$$w_i'(\phi)\, s_{D3}(\phi) = -0.1402\, \phi^{-1}$$

$$w_i'(\phi)\, s_{F1}(\phi) = -1.7772\, \phi^{-1}$$

$$w_i'(\phi)\, s_{F2}(\phi) = 1.0000\, \phi^{-1}$$

The design may be considered successful; the soft zero responses are relatively small, the floating one response is of decent size and the hard one response, of course, is exact. The success is partly due to the fact that the discarded singular value is small compared to the others.

b. Now a single residual generator will be designed for decoupling from the disturbances and from the first fault, and for a nonzero response to the second fault. The fault responses may be part of a structured scheme. The assignments are:

disturbance 1: soft zero

disturbance 2: soft zero

disturbance 3: soft zero

fault 1: soft zero

fault 2: hard one

Now a rank 2 approximation is needed for the three disturbances and the first fault. The design equation is

$$w_i'(\phi)[S^*_{DFi}(\phi) \quad s_{F\cdot 2}(\phi)] = [0 \quad 0 \quad 1]$$

The matrix to approximate is

$$[E_D \quad e_{F\cdot 1}] = \begin{bmatrix} 1 & 2 & 1 & 2 \\ 2 & 0 & 3 & 3 \\ 1 & 3 & 0 & -1 \end{bmatrix}$$

which has the singular values

$$Diag[\sigma_1 \quad \sigma_2 \quad \sigma_3] = Diag[5.3573 \quad 3.6322 \quad 1.0519]$$

The approximation is obtained by discarding the last singular value, yielding

$$E_{DFi^*} = \begin{bmatrix} 1.4120 & 1.8051 & 1.4071 & 1.4441 \\ 1.7720 & 0.1079 & 2.7747 & 3.3076 \\ 0.7687 & 3.1094 & -0.2285 & -0.6880 \end{bmatrix} \qquad m^* = 2$$

Notice that the accuracy of the approximation has deteriorated relative to Case a. The first two columns of E_{DFi^*} will be used, together with the second column of E_F, to form the design matrix

$$[E^*_{DFi} \quad e_{F\cdot 2}] = \begin{bmatrix} 1.4120 & 1.8051 & 0 \\ 1.7720 & 0.1079 & -1 \\ 0.7687 & 3.1094 & 2 \end{bmatrix}$$

The generator is obtained as

$$w_i'(\phi) = [-0.8853 + 0.5419\phi^{-1} \quad 1.8572 - 1.1251\phi^{-1} \quad -0.8712 + 0.3363\phi^{-1}]$$

Finally, the actual responses are computed as

$$w_i'(\phi)\, s_{D1}(\phi) = 1.1733\, \phi^{-1}$$
$$w_i'(\phi)\, s_{D2}(\phi) = -0.5550\, \phi^{-1}$$
$$w_i'(\phi)\, s_{D3}(\phi) = 1.1592\, \phi^{-1}$$
$$w_i'(\phi)\, s_{F1}(\phi) = -1.5833\, \phi^{-1}$$
$$w_i'(\phi)\, s_{F1}(\phi) = 1.0000\, \phi^{-1}$$

Though the design is technically perfect, the results are disappointing. The hard response is exact but all the soft zeroes are quite large, in fact most of them larger than the "one" response. This is because now the discarded singular value is relatively large.

c. Now in addition to disturbance decoupling, the single residual will be designed for exact responses to the two faults (perhaps as part of a directional scheme). The assignments are:

disturbance 1: soft zero

disturbance 2: soft zero

disturbance 3: soft zero

fault 1: hard zero

fault 2: hard one

This calls for a rank *1* approximation for the three disturbances. The design equation is

$$w_i'(\phi)[S_D^*(\phi) \quad s_{F\cdot 1}(\phi) \quad s_{F\cdot 2}(\phi)] = [0 \quad 0 \quad 1]$$

Of the singular values of the E_D matrix, only the first one can be kept, yielding the rank *1* approximation

$$E_{D*} = \begin{bmatrix} 1.2745 & 1.4872 & 1.3141 \\ 1.4872 & 1.7354 & 1.5335 \\ 1.3141 & 1.5335 & 1.3550 \end{bmatrix} \qquad m^*=1$$

Notice that this shows very little resemblance of the original matrix. The first column of E_{D*} will be combined with the full E_F, to form the design matrix

$$[E_D^* \quad E_F] = \begin{bmatrix} 1.2745 & 2 & 0 \\ 1.4872 & 3 & -1 \\ 1.3141 & -1 & 2 \end{bmatrix}$$

The generator is obtained as

$$w_i'(\phi) = [0.5389 - 0.2387\phi^{-1} \quad -2.3092+1.3009\phi^{-1} \quad 1.9240 - 0.9928\phi^{-1}]$$

Finally, the actual responses are computed as

$$w_i'(\phi)\, s_{D1}(\phi) = -\, 1.4629\, \phi^{-1}$$

$$w_i'(\phi)\, s_{D2}(\phi) = 3.7707\, \phi^{-1}$$

$$w_i'(\phi)\, s_{D3}(\phi) = -\, 2.8480\, \phi^{-1}$$

$$w_i'(\phi)\, s_{F1}(\phi) = < 10^{-4}$$

$$w_i'(\phi)\, s_{F1}(\phi) = 1.0000\, \phi^{-1}$$

Again, the hard responses are accurate but disturbance decoupling is practically useless, though there are no technical errors in the design. This is due to the fact that now three independent vectors have been approximated with one, and one of the two discarded singular values is quite large.

10.2.3. Rank Reduced Approximation for Parametric Disturbances

When dealing with approximate decoupling from parametric disturbances, singular value decomposition can be applied directly to the coefficient matrices $N_D(t)$ and $N_F(t)$. This is a significant conceptual simplification relative to the additive case. However, the residual generator is time varying, therefore its design, including the rank reduced approximation, needs to be performed on-line. This, of course, increases the computational complexity of the algorithm.

Now it is the disturbance coefficient matrix $N_D(t)$ (or its equivalent containing part of the fault coefficient matrix $N_F(t)$) which is expanded as

$$N_D(t) = \sum_{j=1}^{m} \sigma_j(t)\, \gamma_j(t)\, \psi_j'(t) \tag{10.61}$$

(where we assumed that $N_D(t)$ is a flat or square matrix), then approximated as

$$N_{D*}(t) = \sum_{j=1}^{m^*} \sigma_j(t)\, \gamma_j(t)\, \psi_j'(t) \tag{10.62}$$

and finally replaced with

$$N_D^*(t) = m^*\ \textit{independent columns of}\ N_{D*}(t) \tag{10.63}$$

Note that now we may utilize in the design the orthogonality property (c.f. (10.46))

$$\gamma_j'(t)\, N_D^*(t) = 0 \qquad j = m^*+1 \ldots m \tag{10.64}$$

Design strategies. The design strategies discussed in the previous subsection apply to parametric disturbances and faults as well. We will revisit them here, especially since the orthogonality property (10.64) introduces a new element in the design.

Disturbance decoupling only. To decouple the residual from $\rho_D \geq m$ disturbances, with no residual enhancement for the faults, the disturbances can be handled as soft *zero*es while a hard *one* response is specified for one of the faults (the other fault responses float). A rank reduced equivalent $N_D^*(t)$ is found for the disturbance entry matrix $N_D(t)$, with $m^*=m-1$. This is supplemented with the entry column of the fault selected for hard one response, $n_{F \cdot g}(t)$, leading to the implementation equation (for the i-th residual)

$$w_i'(t)[N_D^*(t) \quad n_{F \cdot g}(t)] = [0 \ldots 0 \quad z_{ig}] \tag{10.65}$$

where z_{ig} is the hard response specification. The transformation may be obtained the usual way, by inverting the combined entry matrix. Alternatively, the orthogonality property (10.64) may be utilized so that

$$w_i'(t) = \alpha_i(t)\, \boldsymbol{\gamma}_m'(t) \tag{10.66}$$

Here $\boldsymbol{\gamma}_m'(t)$ is the m-th eigenvector of $N_D(t)N_D'(t)$ (which is the same for all i) and the coefficient $\alpha_i(t)$ is computed from the fault response specification as

$$\alpha_i(t)\, \boldsymbol{\gamma}_m'(t) n_{F \cdot g}(t) = z_{ig} \tag{10.67}$$

The condition for (10.65) to have a solution, and for the solution to provide detection for all other faults, is given in (10.27), with N_D replaced by N_D^*.

Disturbance decoupling with structured isolation. Now each residual is decoupled from the same ρ_D disturbances and from a subset of faults which is different for each residual. The disturbances, and the faults in the particular decoupling subset, will be handled with soft *zero*es while a hard *one* response is specified to one of the faults outside the subset (the remaining fault responses float). Denote as $N_{Fi}(t)$ the coefficient matrix of the i-th decoupling subset. Then a rank reduced equivalent $N_{DFi}^*(t)=[N_D(t) \quad N_{Fi}(t)]^*$ needs to be found, with $m^*=m-1$. This is used in place of $N_D^*(t)$ in Eq. (10.65)

$$w_i'(t)[N_{DFi}^*(t) \quad n_{F \cdot g}(t)] = [0 \ldots 0 \quad z_{ig}] \tag{10.68}$$

where, again, $n_{F \cdot g}(t)$ and z_{ig} are the entry vector and specified response for the fault selected for hard one response. The solution may be obtained by the inversion of the combined entry matrix, or via the orthogonality property; in the latter case, (10.66) and (10.67) become

$$w_i'(t) = \alpha_i(t)\boldsymbol{\gamma}_{im}'(t) \tag{10.69}$$

$$\alpha_i(t)\boldsymbol{\gamma}_{im}'(t)\boldsymbol{n}_{F\cdot g}(t) = z_{ig} \tag{10.70}$$

Here $\boldsymbol{\gamma}_{im}'(t)$ is the m-th eigenvector of $N_{DFi}(t)N_{DFi}'(t)$ which is now different for every i. The existence condition (10.28) applies, with $[N_D \quad N_{Fi}]$ replaced by $[N_D \quad N_{Fi}]^*$.

Disturbance decoupling with directional isolation. Now the residual set is decoupled from ρ_D disturbances and is directional for $\rho_F < m$ faults. All fault responses are specified as hard *ones* (or zeroes) in this case. The disturbance responses are specified as soft *zero*es, and the rank of the disturbance entry matrix is reduced to $m - \rho_F$. Accordingly, an equivalent $N_D^*(t)$ is found for $N_D(t)$, with $m^* = m - \rho_F$. This appears in the implementation equation, together with the fault entry matrix $N_F(t)$ (both matrices are invariant over the residual set):

$$w_i'(t)[N_D^*(t) \quad N_F(t)] = [0\ldots0 \quad z_{i,m^*+1}\ldots z_{im}] \tag{10.71}$$

Here z_{ij}, $j = m^*+1 \ldots m$ are the fault responses representing the directional specification. The transformation can be obtained by the inversion of the combined entry matrix or by orthogonality. In the latter case, $w_i'(t)$ is the linear combination of the ρ_F eigenvectors eliminated from the spectral form:

$$w_i'(t) = \sum_{j=m^*+1}^{m} \alpha_{i,j}(t)\,\boldsymbol{\gamma}_j'(t) \tag{10.72}$$

Note that the set of eigenvectors $\boldsymbol{\gamma}_j'(t)$, $j = m^*+1 \ldots m$, is the same for all i. The coefficients $\alpha_{i,j}(t)$ are computed from the hard *one* specifications as

$$[\alpha_{i,m^*+1}(t)\ldots\alpha_{i,m}(t)]\,\boldsymbol{\Gamma}^*(t)\,N_F(t) = [z_{i,m^*+1}\ldots z_{im}] \tag{10.73}$$

where

$$\boldsymbol{\Gamma}^*(t) = \begin{bmatrix} \boldsymbol{\gamma}_{m^*+1}'(t) \\ \cdot \\ \cdot \\ \boldsymbol{\gamma}_m'(t) \end{bmatrix} \tag{10.74}$$

The condition for (10.71) to have a solution is given in (10.29), with N_D replaced by N_D^*. If (10.29) holds, that is, $Rank N_D^*(t) = m - \rho_F$, then the $\rho_F \cdot \rho_F$

matrix $\Gamma^*(t)N_F(t)$ has full rank. To see this, recall that with $[N_D^*(t) \quad N_F(t)]$ non-singular, $Rank\Gamma^*(t)[N_D^*(t) \quad N_F(t)] = Rank\Gamma^*(t) = \rho_F$. But $\Gamma^*(t)[N_D^*(t) \quad N_F(t)] = [\Gamma^*(t)N_D^*(t) \quad \Gamma^*(t)N_F(t)]$ and $\Gamma^*(t)N_D^*(t) = \mathbf{0}$, so it follows that $Rank\Gamma^*(t)N_F(t) = \rho_F$. Thus (10.73) can be solved by inversion.

Example 10.2.

The following example will illustrate the mechanics of the rank reduced approximation for parametric disturbances and faults. Consider a system with three outputs, four parametric disturbances and two parametric faults. Implement a momentary residual generator which (approximately) decouples from the disturbances and provides unit response gain for one of the faults. Note that, with three outputs, exact decoupling would be possible only from two disturbances.

Assume that the momentary disturbance and fault entry matrices are

$$N_D(t) = \begin{bmatrix} 1 & 3 & -3 & 2 \\ 2 & -2 & 1 & 2 \\ -1 & 1 & 1 & -2 \end{bmatrix} \qquad N_F(t) = \begin{bmatrix} 0 & 4 \\ 2 & -2 \\ 6 & 4 \end{bmatrix}$$

A singular value decomposition of $N_D(t)$ is needed for the approximation. The eigenvalues of $N_D(t)\,N_D'(t)$ are found as *24.540, 17.608, 0.85165*, with the normalized eigenvectors

$$\Gamma(t) = [\gamma_1(t) \quad \gamma_2(t) \quad \gamma_3(t)] = \begin{bmatrix} 0.9666 & 0.0325 & 0.2544 \\ -0.1112 & -0.8412 & 0.5292 \\ -0.2310 & 0.5398 & 0.8094 \end{bmatrix}$$

The three singular values are

$$\sigma_1(t) = 4.9538 \qquad \sigma_2(t) = 4.1962 \qquad \sigma_3(t) = 0.9228$$

The first three normalized eigenvectors of $N_D'(t)\,N_D(t)$ are obtained as

$$[\psi_1(t) \quad \psi_2(t) \quad \psi_3(t)] = N_D'(t)\Gamma(t)Diag[1/\sigma_1(t) \quad 1/\sigma_2(t) \quad 1/\sigma_3(t)]$$

$$= \begin{bmatrix} 0.1969 & -0.5218 & 0.5457 \\ 0.5836 & 0.5528 & 0.5572 \\ -0.6544 & -0.0951 & 0.6232 \\ 0.4386 & -0.6427 & -0.0555 \end{bmatrix}$$

Note that $N'_D(t)\, N_D(t)$ has an additional eigenvector, which belongs to its zero eigenvalue, but this is irrelevant for the rank reduced approximation.

The *rank 2* approximation of $N_D(t)$ is obtained as

$$N_{D*}(t) = \sigma_1(t)\gamma_1(t)\psi'_1(t)+\sigma_2(t)\gamma_2(t)\psi'_2(t)$$

$$= \begin{bmatrix} 0.8708 & 2.8704 & -3.1473 & 2.0127 \\ 1.7332 & -2.2727 & 0.6963 & 2.0267 \\ -1.4075 & 0.5844 & 0.5337 & -1.9575 \end{bmatrix}$$

By inspection, this is reasonably close to the original matrix, apparently because the neglected singular value is relatively small.

The transformation is obtained as $w'_i(t)=\alpha_i(t)\gamma'_3(t)$ and $\alpha_i(t)$ is computed from the specified fault response. Choosing the first fault for a hard one response this is

$$\alpha_i(t)[0.2544 \quad 0.5292 \quad 0.8094]\,[0 \quad 2 \quad 6]' = 1$$

yielding $\alpha_i(t)=0.16906$. Thus

$$w'_i(t) = [0.0430 \quad 0.0895 \quad 0.1368]$$

Now let us check the responses. The disturbance responses are

$$w'_i(t)N_D(t) = [0.0851 \quad 0.0869 \quad 0.0973 \quad -0.0087]$$

and the fault responses are

$$w'_i(t)N_F(t) = [1 \quad 0.5405]$$

This provides reasonably good detection for both faults, again because the ignored singular value was relatively small.

We wish to stress here that the results of the above example, or of any other example, are not indicative of the general performance of the algorithm. The actual behavior is time-varying and depends on the size and direction of the columns in the disturbance and fault entry matrices. The actual performance may be better or worse than what is seen in the example. If the detection or isolation performance turns out to be very poor, it may be necessary or advisable to revise the diagnostic strategy.

10.3. APPROXIMATE DECOUPLING WITH CONSTRAINED LEAST SQUARES

The approach introduced in this section achieves optimal approximate disturbance decoupling, combined with structured or directional fault isolation, via the minimization of a quadratic performance index under linear equality constraints. Each component of the performance index implements a *soft zero*, representing (approximate) decoupling from a disturbance or (approximate) structuring for a fault. Each element of the set of constraints implements a *hard one*, representing a fault response, or a *hard zero*, representing exact decoupling.

Just like in the case of the singular value decomposition approach, the number of soft zeroes in the performance index is not limited; however, the number of constraints is. Intuitively, the constrained least squares approach is more appealing than the singular value decomposition approach since the former relies on a performance index expressed directly in terms of the residuals. Also, constrained least squares appears as the more manageable approach as far as computational complexity is concerned.

The constrained least squares technique will be seen as straightforward for parametric faults and disturbances, where each time sample is characterized by "static" (numerical) coefficient matrices. However, we have not been able to apply it to additive faults and disturbances in a dynamic system.

10.3.1. The Constrained Least Squares Formulation

In this introduction of the constrained least squares formulation, it will be assumed that all elements of the performance index are disturbance responses while all elements in the constraint-set are fault responses. This picture will then be refined in the next subsection.

Consider the multiplicative disturbances $\blacktriangle\theta_{Dj}$, $j=1\dots\rho_D$, and the multiplicative faults $\blacktriangle\theta_{Fj}$, $j=1\dots\rho_F$. Define $\blacktriangle\theta^{\circ}_{Dj}$, $j=1\dots\rho_D$, as the "nominal sizes" of the disturbances $\blacktriangle\theta_{Dj}$. Now seek the transformation $w_i'(t)$ so that the quadratic performance index

$$J_i(t) = \sum_{j=1}^{\rho_D} [r_i(t|\blacktriangle\theta^{\circ}_{Dj})]^2 \tag{10.75}$$

is minimized while the following equality constraints are satisfied

$$r_i(t|\blacktriangle\theta_{Fj}) = z_{ij}\blacktriangle\theta_{Fj} \qquad j=1\dots\rho_F \tag{10.76}$$

The individual responses are obtained as

$$r_i(t|\blacktriangle\theta^\circ_{Dj}) = w_i'(t)n_{Dj}(t)\blacktriangle\theta^\circ_{Dj} \tag{10.77}$$

$$r_i(t|\blacktriangle\theta_{Fj}) = w_i'(t)n_{Fj}(t)\blacktriangle\theta_{Fj} \tag{10.78}$$

Now define

$$r'_{iD}(t) = [r_i(t|\blacktriangle\theta^\circ_{D1}) \ldots r_i(t|\blacktriangle\theta^\circ_{D,\rho D})] \tag{10.79}$$

which is obtained as

$$r'_{iD}(t) = w_i'(t)\, N_D(t)\, \Theta^\circ \tag{10.80}$$

where $\Theta^\circ = Diag[\blacktriangle\theta^\circ_{D1} \ldots \blacktriangle\theta^\circ_{D,\rho D}]$. With this, the performance index (10.75) is expressed as

$$J_i(t) = r'_{iD}(t)\, r_{iD}(t) = w_i'(t)\, Q(t)\, w_i(t) \tag{10.81}$$

where

$$Q(t) = N_D(t)\, (\Theta^\circ)^2\, N'_D(t) \tag{10.82}$$

As it can be seen, the nominal disturbance sizes act as weighting factors for the different disturbance responses in the performance index. The set of constraints (10.76) is described as

$$w_i'(t)\, N_F(t) = [z_{i1} \ldots z_{i,\rho F}] = z_i' \tag{10.83}$$

The solution will be sought using the conventional Lagrange multiplier technique. The constrained performance index is

$$J_i^c(t) = w_i'(t)\, Q(t)\, w_i(t) + [w_i'(t)\, N_F(t) - z_i']\, \beta_i(t) \tag{10.84}$$

where $\beta_i(t)$ is a ρ_F vector of Lagrange multipliers. The derivatives with respect to the transformation and the multipliers are (see the Appendix in Chapter 4 for definitions)

$$\partial J_i^c(t)/\,\partial w_i(t) = 2\, w_i'(t)\, Q(t) + \beta_i'(t)\, N_F'(t) = 0' \tag{10.85}$$

$$\partial J_i^c(t)/\,\partial \beta_i(t) = w_i'(t)\, N_F(t) - z_i' = 0' \tag{10.86}$$

That is

$$[w_i'(t) \quad \beta_i'(t)] \begin{bmatrix} 2Q(t) & N_F(t) \\ N_F'(t) & 0 \end{bmatrix} = [\,0' \quad z_i'\,] \tag{10.87}$$

Define $R(t)$ as

$$\begin{bmatrix} 2Q(t) & N_F(t) \\ N_F'(t) & 0 \end{bmatrix}^{-1} = \left[\begin{array}{c|c} \cdot & \cdot \\ \hline R(t) & \cdot \end{array}\right] \begin{matrix} \mu \\ \rho_F \end{matrix} \qquad \begin{matrix} \mu & \rho_F \end{matrix} \tag{10.88}$$

Then the transformation is obtained as

$$w_i'(t) = z_i'\, R(t) \tag{10.89}$$

The solution is subject to the following conditions:

a. No solution exists if $RankN_F(t) < \rho_F$. This situation arises when
   (i) $RankN_F(t) = \mu < \rho_F$ (too many constraints) or
   (ii) $RankN_F(t) < \rho_F \leq \mu$ (linearly dependent constraints).

b. If $RankN_F(t) = \rho_F = \mu$ then the constraints completely determine the solution (there is no minimization for disturbance decoupling).

c. If $RankN_D(t) + RankN_F(t) = \mu$ then the solution yields the perfectly decoupling transformation.

d. If $RankN_D(t) + RankN_F(t) < \mu$ then the solution is not unique (too few conditions).

e. If $RankN_D(t) = \mu$ (more than μ-1 independent disturbances) then $Q(t)$ has full rank and (provided $RankN_F(t) = \rho_F$) $R(t)$ can be computed as

$$R(t) = [N_F'(t)\, Q^{-1}(t)\, N_F(t)]^{-1}\, N_F'(t)\, Q^{-1}(t) \tag{10.90}$$

(c.f. (6.86)). If $Q(t)$ is not invertible then (10.90) cannot be computed but (10.87) still has a solution.

10.3.2. Design Strategies with Constrained Least Squares

Soft and hard response specifications. *Soft zeroes* are implemented as components in the performance index; they represent (approximate) decoupling from disturbances and (approximate) structuring for certain faults. *Hard ones*, representing specified fault responses, are implemented as constraints. It is also possible to implement as constraints *hard zeroes* if exact decoupling from some faults or disturbances is necessary. Finally, responses not included in either the performance index or the constraints float (behave as *soft ones*).

Note that the conditions of the constrained least squares solution do not restrict the number of disturbances but the quality of decoupling certainly suffers if this number is too high. The number of faults (constraints) is restricted to μ and, obviously, the quality of decoupling deteriorates seriously as this limit is approached. Further, there is no guarantee in general (for any combination of fault/disturbance sizes) that a soft zero response will be smaller than a soft one.

Disturbance decoupling only. All the disturbance responses are included in the performance index. It is necessary to specify the response for at least one of the faults, otherwise the implementation would involve unconstrained minimization for the disturbance responses which yields the trivial zero transformation. With a single constraint, $\rho_F = 1$ and $N_F(t)$ contains a single column $n_{F\cdot g}(t)$. Thus (10.87) and (10.90) change to

$$[w_i'(t) \quad \beta_i'(t)] \begin{bmatrix} 2Q(t) & n_{F\cdot g}(t) \\ n'_{F\cdot g}(t) & 0 \end{bmatrix} = [\,0' \quad z_i'] \tag{10.91}$$

$$R(t) = [n'_{F\cdot g}(t)\, Q^{-1}(t)\, n_{F\cdot g}(t)]^{-1}\, n'_{F\cdot g}(t)\, Q^{-1}(t) \tag{10.92}$$

Notice that $R(t)$ is now a row vector.

Detection of the faults (other than the one used in the constraint) is possible if the transformation is not orthogonal to any of the fault entries, that is, if

$$w_i'(t) n_{F\cdot j}(t) \neq 0 \qquad j = 1,2\ldots \tag{10.93}$$

Unlike in the case of perfect decoupling or rank reduced implementation, we do not know of any direct rank condition on the entry matrices that could be checked beforehand. Note also that satisfaction of (10.93) does not guarantee sufficient fault sensitivity compared to the disturbance responses.

Disturbance decoupling with structured fault isolation. Now the decoupling subset of faults, which is different for each residual, joins the disturbances in the performance index. Thus (10.82) changes to

$$Q_i(t) = [N_D(t) \quad N_{Fi}(t)]\, (\Theta_i^\circ)^2\, [N_D(t) \quad N_{Fi}(t)]' \tag{10.94}$$

where $N_{Fi}(t)$ is the entry matrix of the faults from which the i-th residual is to be decoupled and Θ_i° contains the nominal sizes not only for the disturbances but also for faults in the i-th decoupling subset. Again, a *hard one* fault response for each residual needs to be specified, outside the respective decoupling subset, to avoid trivial transformation. Thus $\rho_F = 1$ and Eqs. (10.91) and (10.92) hold, but $Q(t)$ and $R(t)$ are replaced with $Q_i(t)$ and $R_i(t)$. Also, the detection condition (10.93) needs to be satisfied for all faults outside the decoupling subset.

Disturbance decoupling with directional fault isolation. Now all disturbances are included in the performance index and all faults are in the constraint set. The algorithm developed in Subsection 10.3.1 applies without any modification.

Example 10.3.

Let us revisit the system introduced in Example 10.2 and implement the residual generator now using the constrained least-squares approach. Assume that the nominal size for all disturbances is 1, thus $\Theta°=I$. As in the previous example, let us fix the response to the first fault at 1.

From (10.82),

$$Q(t) = \begin{bmatrix} 23 & -3 & -5 \\ -3 & 13 & -7 \\ -5 & -7 & 7 \end{bmatrix}$$

Then from (10.92),

$$R(t) = [0.0413 \quad 0.0873 \quad 0.1376]$$

Finally, from (10.89), with $z_{i1} = 1$, the transformation is obtained as $w_i'(t) = R(t)$. Notice how remarkably close this transformation is to the one obtained in Example 10.2 by rank reduced approximation.

Checking the responses shows

$$w_i'(t)N_D(t) = [0.0784 \quad 0.0868 \quad 0.1010 \quad -0.0179]$$

$$w_i'(t)N_F(t) = [1 \quad 0.5408]$$

These, again, are close to the rank reduced responses. Note that the least squares performance index for the constrained least squares implementation is 0.023558 while for the rank reduced implementation it is 0.0244341. ⁝⁝⁝

The above example demonstrates that, computationally, the constrained least squares technique is less demanding than the rank reduced technique. While the latter approach requires finding the eigenvalues and eigenvectors for a $p_D \cdot p_D$ matrix, constrained least squares normally involves two matrix inversions, of size p_D and p_F.

The reader should be warned again that the results of the example are not indicative of the performance of the algorithm in general; the actual performance may vary significantly by time and from one system to another.

10.4. ROBUSTNESS TO MULTIPLE MODELS

Often the uncertain model can be characterized by a finite set of known model variants. This is the case when a nonlinear system operates in a number of operating points, and a different linearized model applies in each one. Using transfer function notation, this situation is described as

$$[M(\phi),\ S_F(\phi),\ S_D(\phi)]_l \qquad l=1 \ldots \lambda \tag{10.95}$$

where l refers to the model variant, or with state variables as

$$[A,\ B,\ C,\ D,\ E_F,\ E_D,\ F_F,\ F_D]_l \qquad l=1 \ldots \lambda \tag{10.96}$$

If the operating point can be identified then the algorithm should be switched on-line to the appropriate model. This, of course, provides perfect robustness. If however the operating point is varying but unknown then the algorithm needs to be designed so that it functions in an optimal way over the entire set of models.

Two issues will be addressed here:

1. How to design the residual generator for additive faults and disturbances, in the presence of multiple models;
2. How to deal with the model error created by the presence of multiple models.

We will discuss these issues in connection with the Chow-Willsky scheme, for which a partial solution of the first problem was originally proposed (Lou et al, 1986). Consider the input-output-state equation of the scheme

$$Y(t) = J\,x(t-s) + K\,U(t) + L_F P(t) + L_D Q(t) \tag{10.97}$$

where the meaning and structure of the vectors $Y(t)$, $U(t)$, $P(t)$ and $Q(t)$, and matrices J, K, L_F and L_D are defined in Subsection 6.1.3. The residuals are computed as

$$r(t) = W\,[Y(t) - K\,U(t)] \tag{10.98}$$

which implies

$$r(t) = W\,[\,J\,x(t-s) + L_F P(t) + L_D Q(t)] \tag{10.99}$$

With multiple model variants, the system is characterized as

$$[\,J,\,K,\,L_F,\,L_D\,]_l \qquad l=1\,\dots\,\lambda \tag{10.100}$$

10.4.1. Design for Additive Faults and Disturbances

We will address two sub-problems here:

a. How to decouple from additive disturbances (and the state) in the presence of multiple models;

b. How to design for specified nonzero fault responses in this situation.

Note that, with the Chow-Willsky scheme, decoupling from the state vector must be part of the design, no matter what the design strategy is. Also, nonzero response is part of the specification if any fault isolation scheme is involved, and may be present even if the objective is only the detection of the faults. Therefore the two steps of the design (*a* and *b* above) must be integrated at some point.

a. Design for decoupling. Assume first that the design is aimed at detection only. Then the residual must be decoupled from the state $x(t-s)$ and from the disturbances $Q(t)$. Thus with a single model, the decoupling equation is

$$W[\,J \quad L_D\,] = 0 \tag{10.101}$$

With multiple models, this becomes

$$W[\,J_1,\,L_{D1}\,\dots\,J_\lambda,\,L_{D\lambda}\,] = W\Pi = 0 \tag{10.102}$$

where Π is defined implicitly in the equation. If Π has full rank m, which is usually the case, (10.102) cannot be satisfied without eliminating the fault responses as well. Thus Π needs to be replaced by a rank reduced approximation Π_* or by the reduced size equivalent of the latter Π^*. The decoupling equation then becomes

$$W\Pi^* = 0 \tag{10.103}$$

Finally, in accordance with (10.46)-(10.49), the optimal set of decoupling vectors is

$$W = \Gamma^* = [\gamma_{m^*+1}\,\cdots\,\gamma_m]' \tag{10.104}$$

where γ_j, $j=m^*+1\,\dots\,m$, are the singular vectors of Π not used in Π^*.

The actual degree of rank reduction, $m - m^*$, depends on the additional requirements the design has to satisfy, such as the detection only of faults or their isolation in a structured or directional scheme. It may be reasonable to reduce the combined model Π^* to the size of a single model $[\,J_l,\,L_{Dl}\,]$. Then the division of the available design freedom between disturbance decoupling and fault isolation follows the same pattern as if the design was done with a single model.

b. Nonzero response specification. The singular value decomposition approach to multiple models is difficult to extend to the case when part of the response specification is nonzero. This is because the rank reduced multiple model coefficient matrix, combined with identical response specifications for the model variants, would lead to a contradictory set of equations. Instead, a least squares approach will be followed here for the responses with nonzero specification. This will then be combined with the decoupling part of the specification - in one of two ways shown below.

Least squares formulation for nonzero responses. We will consider the generation of a single residual $r_i(t)$. First the decoupling requirements will be ignored and it will be assumed that the residual only has to satisfy a nonhomogeneous specification

$$r_i(t) = w_i' \, L_F \, p(t) = z_{Fi}' \, p(t) \tag{10.105}$$

Assume that L_F is square and invertible, so that with a single model the transformation could be obtained as

$$w_i' = z_{Fi}' \, L_F^{-1} \tag{10.106}$$

With multiple models $L_{F1} \dots L_{F\lambda}$, and identical response specifications under the various models, (10.106) cannot be solved. Instead, a least squares index can be formulated as

$$N = \sum_{l=1}^{\lambda} (w_i' L_{Fl} - z_{Fi}')(L_{Fl}' w_i - z_{Fi}) \tag{10.107}$$

This index is similar to (10.44), but it is applied to the multimodel case and extended to nonzero specification. It is the squared error of the response gains, summed over the faults (implicitly) and the model variants (explicitly). The optimal transformation w_i' is then obtained by the minimization of N.

The partial derivative of the index, with respect to w_i' is

$$\frac{\partial}{\partial w_i'} \sum_{l=1}^{\lambda} [w_i' L_{Fl} L_{Fl}' w_i - 2 z_{Fi}' L_{Fl}' w_i + z_{Fi}' z_{Fi}] = 2 \sum_{l=1}^{\lambda} [w_i' L_{Fl} L_{Fl}' - z_{Fi}' L_{Fl}'] \tag{10.108}$$

Setting this equal to zero and solving for w_i' yields

$$w_i' = z_{Fi}' \sum_{l=1}^{\lambda} L_{Fl}' \left[\sum_{l=1}^{\lambda} L_{Fl} L_{Fl}' \right]^{-1} \tag{10.109}$$

Combination with decoupling. The above least squares solution for the non-zero specifications must be combined with decoupling from the state and the disturbances. This can be done

- by expanding the least squares formulation to include the zero responses for decoupling; or
- by constructing w_i' as the linear combination of the singular vectors which satisfy the decoupling requirement, and computing their weights from the least squares formulation.

In the first approach, the matrices L_{Fl}, $l=1 \dots \lambda$, are replaced by $L_l = [J_l, L_{Dl}, L_{Fl}]$ and the specification z_{Fi}' is replaced by $[0 \dots 0 \;\; z_{Fi}']$. Thus (10.109) becomes

$$w_i' = [0 \dots 0 \;\; z_{Fi}'] \sum_{l=1}^{\lambda} L_l' \, [\sum_{l=1}^{\lambda} L_l L_l']^{-1} \tag{10.110}$$

With the second approach, singular value decomposition is performed for the $[J_1, L_{D1} \dots J_\lambda, L_{D\lambda}]$ matrix and Γ^* is found (c.f. (10.104)). The transformation is then sought as

$$w_i' = [\alpha_{i,m^*+1} \cdots \alpha_{i,m}] \, \Gamma^* \tag{10.111}$$

Substituting this into (10.107) yields the least squares solution for the $\alpha_{i,k}$ coefficients as

$$[\alpha_{i,m^*+1} \cdots \alpha_{i,m}] = z_{Fi}' \sum_{l=1}^{\lambda} L_{Fl}' \Gamma^{*\prime} \, [\sum_{l=1}^{\lambda} \Gamma^* L_{Fl} L_{Fl}' \Gamma^{*\prime}]^{-1} \tag{10.112}$$

Note that the design of the residual generator for decoupling and specified nonzero response, discussed in paragraphs *a.* and *b.* above, can also be done using algebraic average models for J, L_D and L_F (instead of singular value decomposition and/or least squares optimization). The design computations are simpler with the average models, however the resulting residual generator is suboptimal (for the least squares indices).

10.4.2. Model Errors Caused by Multiple Models

With multiple models, (10.97) changes to

$$\boldsymbol{Y}(t|l) = \boldsymbol{J}_l\boldsymbol{x}(t - s) + \boldsymbol{K}_l\boldsymbol{U}(t) + \boldsymbol{L}_{Fl}\boldsymbol{P}(t) + \boldsymbol{L}_{Dl}\boldsymbol{Q}(t) \tag{10.113}$$

However, since it is not known which model is valid, the residual in (10.98) needs to be computed with some equivalent $\bar{\boldsymbol{K}}$ matrix, as

$$\boldsymbol{r}(t) = \boldsymbol{W}[\boldsymbol{Y}(t) - \bar{\boldsymbol{K}}\boldsymbol{U}(t)] \tag{10.114}$$

Thus the internal form of the residual, (10.99), becomes

$$\boldsymbol{r}(t|l) = \boldsymbol{W}[\boldsymbol{J}_l\boldsymbol{x}(t - s) + (\boldsymbol{K}_l - \bar{\boldsymbol{K}})\boldsymbol{U}(t) + \boldsymbol{L}_{Fl}\boldsymbol{P}(t) + \boldsymbol{L}_{Dl}\boldsymbol{Q}(t)] \tag{10.115}$$

The second term on the right-hand side is the effect on the residual of the model error. The objective of the design now is to reduce this effect as much as possible. This will be done in two steps:

a. First $\bar{\boldsymbol{K}}$ will be chosen to minimize the combined size of the model errors over the model variants;

b. Then $\boldsymbol{W}$ will be selected to provide (approximate) decoupling from the remaining effect.

The two step procedure is justified by the fact that, as it will be seen, the optimum of $\bar{\boldsymbol{K}}$ is independent of the choice of $\boldsymbol{W}$. Note that step *a* is necessary since the decoupling in step *b* is usually approximate. The design of $\boldsymbol{W}$ needs then to incorporate also decoupling from $\boldsymbol{x}(t - s)$ and whatever additional disturbance and fault responses may be specified.

a. Minimizing the model error effect. The desire to minimize the model error effect in (10.115), over the model variants $l = 1 \ldots \lambda$, leads to the following least squares index:

$$M(t) = \sum_{l=1}^{\lambda} [\boldsymbol{W}\boldsymbol{K}_l\boldsymbol{U}(t) - \boldsymbol{W}\bar{\boldsymbol{K}}\boldsymbol{U}(t)]' [\boldsymbol{W}\boldsymbol{K}_l\boldsymbol{U}(t) - \boldsymbol{W}\bar{\boldsymbol{K}}\boldsymbol{U}(t)] \tag{10.116}$$

The derivative of this index with respect to $\bar{\boldsymbol{K}}$ is

$$\frac{\partial}{\partial\bar{\boldsymbol{K}}} M(t) = \sum_{l=1}^{\lambda} 2\boldsymbol{W}'\boldsymbol{W}(\boldsymbol{K}_l - \bar{\boldsymbol{K}})\boldsymbol{U}(t)\boldsymbol{U}'(t) = 2\boldsymbol{W}'\boldsymbol{W} \sum_{l=1}^{\lambda} (\boldsymbol{K}_l - \bar{\boldsymbol{K}})\boldsymbol{U}(t)\boldsymbol{U}'(t) \tag{10.117}$$

Setting this equal to zero for all $\boldsymbol{U}(t)$ yields the optimal equivalent model as

$$\bar{K} = \frac{1}{\lambda} \sum_{l=1}^{\lambda} K_l \tag{10.118}$$

which is simply the algebraic average of the model variants.

b. Decoupling from the remaining model errors. Now the matrix to decouple from is

$$\Pi = [(K_1 - \bar{K}) \dots (K_\lambda - \bar{K})] \tag{10.119}$$

With this, the results of Subsection 10.4.1 apply. Usually, rank reduction needs to be performed on Π. Then Γ^*, containing the singular vectors not used in the reduced representation, can be utilized as the basis for the transformation in (10.112). Alternatively, a combined least squares design may be performed as in (10.110). The J_l matrices may be included in Π or made part of L_l in (10.110), with zero response specification.

10.5. H-INFINITY DESIGN OF RESIDUAL GENERATORS

During the past fifteen years, much attention has been paid in the control community to robust controller design, using the so called H-infinity (H_∞) approach. Several attempts have also been made at applying this technique to fault detection and isolation, in order to achieve robustness. In this section, we will outline the main ideas of residual generator design by the H_∞ methodology, as applied to additive faults and disturbances. We will also comment on some of the limitations of this approach. Because of these and the complex mathematics involved, we will not venture into a full-blown description of the design procedure.

H-infinity fundamentals. Below we will summarize some fundamental concepts and results of the H_∞ methodology.

Consider a vector time function $v(t)$ and its z-transform $V(z)$. Observe that $V(e^{j\vartheta})$, $0 \leq \vartheta < 2\pi$, is the z-transform along the unit circle in the z domain. Now define the following norms:

$$|v(t)| = \left[\lim_{N \to \infty} \frac{1}{N} \sum_{\tau=0}^{N-1} v'(t-\tau) v(t-\tau) \right]^{1/2} \tag{10.120}$$

$$\| V(z) \| = \left[\frac{1}{2\pi} \int_0^{2\pi} V'(e^{j\vartheta})\, V(e^{j\vartheta})\, d\vartheta \right]^{1/2} \tag{10.121}$$

According to Parseval's theorem, the two norms are equal, that is,

$$|v(t)| = \| V(z) \| \qquad (10.122)$$

Consider now a rational, proper and stable transfer function matrix $F(z)$. Denote by $\sigma(F)$ the singular values of the matrix $F(e^{j\vartheta})$; clearly, these are functions of ϑ. Further, denote by $\underline{\sigma}(F)$ and $\bar{\sigma}(F)$ the smallest and largest of the singular values at any given ϑ. Finally, consider

$\inf_{\vartheta} \underline{\sigma}(F)$ the smallest value over ϑ of the smallest σ

$\sup_{\vartheta} \bar{\sigma}(F)$ the largest value over ϑ of the largest σ.

Now assume that the signal $v(t)$ is applied to the system described by $F(z)$, so that its output is $F(\phi)v(t)$. Then the following fundamental relationship holds:

$$\inf_{\vartheta} \underline{\sigma}(F) \| V(z) \| \leq \| F(z)V(z) \| \leq \sup_{\vartheta} \bar{\sigma}(F) \| V(z) \| \qquad (10.123)$$

With (10.122), this becomes

$$\boxed{\inf_{\vartheta} \underline{\sigma}(F) |v(t)| \leq |F(\phi)v(t)| \leq \sup_{\vartheta} \bar{\sigma}(F) |v(t)|} \qquad (10.124)$$

Application to additive faults and disturbances. Consider our usual setup with additive faults and disturbances, described by (10.1), which we repeat here:

$$r(t) = W(\phi)[S_F(\phi)p(t) + S_D(\phi)q(t)] \qquad (10.125)$$

The detection test will be designed for the norm $|r(t)|$ of the residual. In a worst-case scenario, we need to consider the smallest fault response versus the largest disturbance response. For this, let us assign *norm bounds* for the faults and disturbances, as

$$|q(\phi)| \leq q_o \qquad |p(t)| \geq p_o \qquad (10.126)$$

The two bounds have different meanings, namely

- q_o is the upper bound which the norm of the disturbances does not exceed;

- p_o. is a lower bound of the fault norm under which it is acceptable if the fault is not detected.

Applying the left-hand side of (10.124) to the fault response of the residual generator, and taking (10.126) into account, one obtains the following *lower bound*:

$$|r(t|p)| = |W(\phi)S_F(\phi)p(t)| \geq \inf_{\vartheta} \underline{\sigma}(WS_F)|p(t)| \geq \inf_{\vartheta} \underline{\sigma}(WS_F)p_o \qquad (10.127)$$

Similarly, applying the right-hand side of (10.124) to the disturbance response yields, with (10.126), the following *upper bound*:

$$|r(t|q)| = |W(\phi)S_D(\phi)q(t)| \leq \sup_{\vartheta} \bar{\sigma}(WS_D)|q(t)| \leq \sup_{\vartheta} \bar{\sigma}(WS_D)q_o \qquad (10.128)$$

It is reasonable to aim for a design in which the ratio of the smallest fault response versus the largest disturbance response is as high as possible. With p_o and q_o given, this leads to the performance index

$$\boxed{J = \frac{\inf_{\vartheta} \underline{\sigma}(WS_F)}{\sup_{\vartheta} \bar{\sigma}(WS_D)}} \qquad (10.129)$$

A transformation $W_{opt}(\phi)$ is then sought which maximizes this index.

One may assess the performance of the generator with the optimal design and the fault and disturbance bounds. The threshold for the residual norm $|r(t)|$ can be set at the upper bound of the disturbance response, that is,

$$\kappa = \sup_{\vartheta} \bar{\sigma}(W_{opt}S_D)q_o \qquad (10.130)$$

For worst-case design, the smallest possible fault response, acting together with the largest possible disturbance in the opposite direction, has to exceed the threshold, that is,

$$\inf_{\vartheta} \underline{\sigma}(W_{opt}S_F)p_o - \sup_{\vartheta} \bar{\sigma}(W_{opt}S_D)q_o > \kappa \qquad (10.131)$$

This, together with (10.130) and (10.129) implies that, for guaranteed worst-case operation,

$$\boxed{J_{opt} > 2\frac{q_o}{p_o}} \tag{10.132}$$

is required, where J_{opt} is the performance index under the optimal transformation $W_{opt}(\phi)$. Unfortunately, there is no guarantee whatsoever that the optimal design satisfies (10.132). If the latter is not met with the given p_o and q_o then the specification (10.126) may need to be relaxed.

The actual design, that is, the computation of the optimal transformation $W_{opt}(\phi)$, followed by the setting of the threshold and the evaluation of the performance, can be executed in the framework of so called *inner-outer factorization*. For this, the reader is referred to books by Vidyasagar (1985) and Francis (1987). Computer packages for inner-outer factorization have also been developed (Qiu, 1994).

Let us mention finally several features which may limit the utility of the H_∞ design approach in fault detection and isolation, especially in an on-line setting.

1. The test is designed in terms of long-term residual norms (averages) instead of momentary residual behavior;
2. The test is designed for the entire residual vector with no means to take into account and access individual components;
3. Because of the chain of bounds (inequalities) involved, the design is extremely conservative;
4. As with any optimization-based method, there is no guarantee that the optimal solution will be practically useful;
5. The design computations are rather complex.

10.6. ROBUSTNESS VIA COMBINATORIAL OPTIMIZATION

The method described in this section is different from the ones discussed so far in this chapter, in that it achieves optimal robustness by searching over a discrete set of possible parity relations, rather than by adjusting residual generator parameters. This approach can be applied in connection with structured parity relations. It relies on the fact that, apart from very small systems, there is a combinatorial multitude of possible equation structures of which only a fraction is needed to form the diagnostic set.

The selection of the optimal set will be based on some scalar measure of robustness or sensitivity for each possible equation structure. These measures, introduced in Subsection 5.2.2, will be briefly reviewed. This will be followed by a recapitulation of the combinatorics of parity relation structures, from Subsection 7.1.2. The similarity of parity relations designed for homogeneous or almost homogeneous specifications, discussed in Chapter 6, will then be reviewed. Finally, two algorithms will be described for the optimization proper, both looking for the column canonical structure resulting in the best performance in terms of the selected measure. One algorithm uses global search over the possible structures; this does (eventually) find the true optimum solution but may, because of the combinatorial nature of the problem, be prohibitive computationally. The other algorithm utilizes a local search procedure; this is computationally manageable but may lead to a local optimum.

10.6.1. Performance Measures for Parity Relations

Here we summarize the performance measures introduced in Subsection 5.2.2.

a. *Triggering limit* for the j-th fault in the i-th parity relation (in steady state):

$$\eta_{ij} = \kappa_i / [\mathbf{w}_i'(\phi)\mathbf{s}_{F\cdot j}(\phi)]_{\phi=1} \tag{10.133}$$

where κ_i is the threshold for the i-th residual. Note that a large triggering limit signifies low sensitivity.

b. *Sensitivity ratio* for the j-th fault in the i-th equation, in steady state:

$$\zeta_{ij} = \frac{p_j^\circ\,[\mathbf{w}_i'(\phi)\mathbf{s}_{F\cdot j}(\phi)]_{\phi=1}}{\kappa_i} \tag{10.134}$$

where p_j° is the nominal size of the j-th fault. A large sensitivity ratio signifies high sensitivity.

c. *Sensitivity condition* for the i-th equation:

$$\xi_i = \max_j \zeta_{ij} / \min_j \zeta_{ij} \tag{10.135}$$

A large sensitivity condition indicates that the equation is poorly balanced among the various faults.

d. *Limit model error* for the j-th underlying parameter in the i-th equation:

$$\vartheta_{ij} = \kappa_i / \mathbf{w}'_i(\phi)\mathbf{n}_{\cdot j}(t) \tag{10.136}$$

Note that $\mathbf{n}_{\cdot j}(t)$ depends on the operating point but may be considered constant for a particular operating point. A large value of the limit error indicates that the residual is robust with respect to uncertainties of the concerned parameter.

10.6.2. Structure and Performance

Equation structures and sets. Consider a system with μ outputs (all assumed independent) and ρ faults. As it was shown in Chapter 6, structured parity relations normally cannot have more than $\mu - 1$ zeroes (there may be more if the fault transfer matrix $S_F(\phi)$ has special properties). The full row-canonical set of parity relations (see Subsection 7.1.2) contains all the different row patterns, with $\mu - 1$ zeroes each. There are

$$K = \binom{\rho}{\mu - 1} \tag{10.137}$$

different row structures in this set.

Usually, column canonical sets are sought for structured fault isolation. The number of parity relations n needed for such a set is governed by the inequalities (c.f. (7.5) and (7.6))

$$\binom{n}{\tau} \geq \rho \qquad\qquad n(\tau - 1) \geq \rho\tau \tag{10.138}$$

where τ is the number of zeroes in each column of the scheme. For any n, the number of ways how n different row structures can be selected from the full set of K parity relations is

$$N = \binom{K}{n} \tag{10.139}$$

which may be a very large number. However, only a fraction of these n-tuples of rows form column canonical structures.

Similar solutions. As it was seen in Chapter 6, there are two ways to design parity relations with $\mu - 1$ zeroes:

a. By specifying only the zero responses (almost full homogeneous specification). Then the solution involves choosing a scalar transfer function arbitrarily and the transformation $w_i'(\phi)$ is scaled by this transfer function.

b. By specifying one nonzero response in addition to the zero responses (full almost homogeneous specification). Then again the transformation is determined by the position of the zeroes, while the position of the nonzero response and its specified transfer function only act as a scaling factor.

Thus given the pattern of the zeroes in an equation, all transformations are similar, that is, if $\tilde{w}_i'(\phi)$ is a solution then all other solutions are given as

$$w_i'(\phi) = \alpha(\phi)\, \tilde{w}_i'(\phi) \tag{10.140}$$

where $\alpha(\phi)$ is a polynomial or a rational function.

The similarity of all transformations for a given zero pattern has important implications as far as the performance measures are concerned. In particular:

(i) All sensitivity ratios ζ_{ij}, $j=1 \ldots \rho$, depend on $\alpha(\phi)$ the same way, therefore the sensitivity condition ξ_i is invariant to $\alpha(\phi)$.

(ii) The steady state triggering limits for the faults, η_{ij}, and the limit model errors for the parameters, ϑ_{ij}, vary together if $\alpha(\phi)$ changes. Thus the effect of changing $\alpha(\phi)$ is the same as that of changing the threshold; the robustness of the equation is not affected.

Thus we conclude that the fundamental sensitivity and robustness properties of any parity relation in a row canonical set are completely determined by the structure of its zeroes.

10.6.3. Finding the Optimal Set

Here we will describe the global and local search procedures to find the column canonical subset which is optimal for a selected performance measure.

Preparing the equation set. Before the actual optimization, the equation set needs to be prepared. This involves the following activities:

1. List all the equations in the row canonical set.
2. Establish the threshold κ_i for each equation of the set. In lack of other criteria, the threshold may be selected so that $min_i\ \zeta_{ij} = 1$.
3. Compute for each equation the performance criterion for which the optimization will be done. This may be the sensitivity condition ξ_i (which does not depend on the threshold) or one of the partial robustness measures ϑ_{ij} (which do depend on the threshold chosen in step 2).
4. Arrange the equations in the order of the selected measure, so that the best equation is on the top.
5. Eliminate from the set those equations which perform very poorly (for example, for which $\xi_i > 10$).

Minimax criterion. Observe that a set of parity relations performs as good, for the selected measure, as the poorest performer in the set. Hence we will seek the column canonical structure which optimizes the worst performance in the set.

Global optimization. We start from the top of the ordered list of equations and work our way downward until we find the first column canonical subset of n equations. That is:

- Start with the first n equations. If they form a column canonical subset, this is the solution.
- Otherwise, consider the first $n+1$ equations. Of these, an n-tuple can be

selected $n+1$ ways (of which one has already been tried). If any of these is column canonical, this is the solution.

- Otherwise, consider the first $n+2$ equations. Of these, an n-tuple can be selected $(n+2)\cdot(n+1)/2$ ways (of which $n+1$ have already been tried). If any of these subsets is column canonical, this is the solution.

This procedure is continued until the first column canonical subset is found. Obviously, the number of combinations increases fast as we move down the list. Since the list is ordered by a performance measure and not by any structural rule, it is unpredictable how far we need to go before we hit the first column canonical subset. Though checking for the structure involves only manipulations with relatively small binary matrices, the combinatorial number of possibilities may easily become computationally prohibitive.

Example 10.4.

a. With $\mu=4$ independent outputs and $\rho=10$ faults, the number of row canonical equations is $K=120$. To form a column canonical subset from row canonical equations in this system requires $n=10$ equations. Global search over the first 25% of the equation list then involves checking $3\cdot10^7$ structures.

b. With $\mu=6$ and $\rho=14$, the equation list contains 2002 entries. With $n=14$ equations in a column canonical subset, to check over the top 2% of the list involves $2.37\cdot10^{10}$ structures while doing the same over the top 10% of the list involves $1.18\cdot10^{21}$ structures. ▩

Example 10.5.

The following example has been developed from a model of an automobile engine (Kamei et al, 1987). The system has two inputs (throttle position command, fuel rate command) and four outputs (torque, air/fuel ratio, intake manifold pressure, throttle position). The model order is eight. Additive faults on the two actuators and four output sensors are considered.

Now $\mu=4$ and $\rho=6$. There are $6\cdot5\cdot4/(1\cdot2\cdot3)=20$ canonical row structures. Further, the minimum number of equations to form a column canonical structure is $n=4$, with $\tau=2$ zeroes per column. Table 10.2. shows the 20 row structures, ordered by their sensitivity condition values ξ_i, which are also shown. The last four structures have very high sensitivity condition and should not appear in any final structure.

A global search was performed on the set yielding the following column canonical subset, optimal for the sensitivity condition:

#2	1.77	1	1	0	1	0	0
#5	2.09	1	0	1	0	1	0
#6	3.00	0	0	0	1	1	1
#11	4.49	0	1	1	0	0	1

TABLE 10.2. Full row-canonical set.

	ξ_i	$\blacktriangle u_1$	$\blacktriangle u_2$	$\blacktriangle y_1$	$\blacktriangle y_2$	$\blacktriangle y_3$	$\blacktriangle y_4$
#1	1.28	0	1	1	1	0	0
#2	1.77	1	1	0	1	0	0
#3	1.79	0	1	0	1	0	1
#4	1.85	0	0	1	0	1	1
#5	2.09	1	0	1	0	1	0
#6	3.00	0	0	0	1	1	1
#7	3.06	1	0	0	1	1	0
#8	3.12	1	0	1	1	0	0
#9	3.37	0	1	0	1	1	0
#10	3.51	0	0	1	1	0	1
#11	4.49	0	1	1	0	0	1
#12	4.52	1	1	1	0	0	0
#13	5.36	0	1	0	0	1	1
#14	5.41	1	1	0	0	1	0
#15	5.51	0	0	1	1	1	0
#16	8.32	0	1	1	0	1	0
#17	28.6	1	0	0	0	1	1
#18	34.2	1	0	1	0	0	1
#19	67.5	1	0	0	1	0	1
#20	155	1	1	0	0	0	1

Note that, starting from the top of the set of 20 equations, it was necessary to cover more than half of the list to find the first column canonical subset. ▩

Local search procedure. The local search procedure starts from an arbitrary column canonical subset and uses a series of systematic row replacements to improve its performance while maintaining column canonicality.

Structural distance. The procedure utilizes the concept of structural distance between row canonical equations. Observe that interchanging a 0 and a 1 in an equation structure with $\mu - 1$ zeroes results in an equation structure with the same number of zeroes. The minimum number of such 0-1 interchanges needed to bring one structure into another is called the *structural distance* between the two equations. (For example, the structural distance between 111100 and

111010 is one while between 111100 and 001111 it is two.) The distance can be conveniently obtained by halving the number of 1's in the exclusive OR of the two Boolean patterns. The maximum possible distance in a system with μ outputs and ρ faults is $min(\mu - 1, \rho - \mu + 1)$. For a set of equations, a distance matrix may be set up (at least conceptually), showing the structural distances between all possible pairs.

Primary row replacement. Given any column canonical initial structure, the equation with the poorest performance is selected for replacement. In the basic form of the algorithm, only equations with distance 1 from the original equation will be considered as substitutes; there are $(\mu - 1)\cdot(\rho - \mu + 1)$ such potential substitutes for any equation. The search starts with the distance 1 substitute which has the best performance. The candidate substitute equation disqualifies if it is already a member of the structure or if its performance is worse than that of the equation to be replaced.

Complementary row replacement. Replacing a single equation would upset the column canonical structure of the subset; a complementary replacement is necessary to restore it. There may be several equations in the subset which, if appropriately replaced, restore at least the number of ones/zeroes in each column. For each of these, there is only one replacement equation. A candidate for complementary substitute disqualifies if it is already in the structured subset or if its performance is worse than that of the equation subject to the primary replacement. Also, the resulting structure may exhibit repeated columns and thus be useless. If no suitable complementary replacement is found, the primary replacement has to be modified, moving to the substitute equation with the next best performance.

Once the equation with the poorest performance has been successfully replaced, the procedure is repeated for the worst equation in the resulting structured set, and so on. The search stops if no distance 1 replacement (with suitable complementary replacement) can be found for the worst equation in the latest structured set.

Since the search is limited to distance 1 replacements, it may easily stop in a local optimum. Convergence toward a global optimum can be improved by allowing distance 2, 3, etc. replacements, preferably once all the distance 1 replacements have been exhausted. (In fact, after a higher distance replacement, a series of distance 1 replacements may again be possible.) This, however, requires a significantly more complex algorithm, especially the complementary replacements may become quite complicated.

Initial structure. The column canonical initial subset needed to start the search may be generated according to some arbitrary symmetrical pattern. If it turns out to contain equations which are unattainable for the given system, other symmetrical patterns may be attempted.

Example 10.6.

We will demonstrate the local search procedure using the system of Example 10.5.

Initial column canonical subset. An arbitrary column canonical subset is

#12	1 1 1	0 0 0
#13	0 1 0	0 1 1
#15	0 0 1	1 1 0
#19	1 0 0	1 0 1

I. First primary replacement. We will first seek a distance 1 replacement for #19, starting at the top of the equation list.

0.	#19	1 0 0	1 0 1	
	#1	0 1 1	1 0 0	
	EXOR	1 1 1	0 0 1	distance 2 (not used)
1.	#19	1 0 0	1 0 1	
	#2	1 1 0	1 0 0	
	EXOR	0 1 0	0 0 1	distance 1

With this replacement, the temporary subset is shown below. It has an excess 1 in the second column and an excess 0 in the sixth column.

#12	1 1 1	0 0 0	→	1 0 1	0 0 1	#18
#13	0 1 0	0 1 1				
#15	0 0 1	1 1 0				
#2	1 1 0	1 0 0				
	↑	↑				

This can be repaired by replacing #12; the necessary replacement turns out to be #18 which, though better than #19, is still in the unacceptable range ($\xi_{18} = 34.2$). Therefore another distance 1 replacement is sought for #19.

2. The next distance 1 replacement is #3. With this, #13 has to be changed to restore the column canonical subset, but the replacement for #13 would be #17 which is still in the unacceptable region.

3. The next distance 1 replacement for #19 is #6. With this, the temporary subset is

#12	1 1 1 0 0 0	
#13	0 1 0 0 1 1 → 1 1 0 0 0 1	#20
#15	0 0 1 1 1 0 → 1 0 1 1 0 0	#8
#6	0 0 0 1 1 1	
	↑ ↑	

Now there is an excess 0 in the first column and an excess 1 the fifth. Candidates for complementary replacement are #13, introducing #20 which is unacceptable, and #15 introducing #8. With the latter, the structure is

#12	1 1 1 0 0 0
#13	0 1 0 0 1 1
#8	1 0 1 1 0 0
#6	0 0 0 1 1 1
	↑ ↑ ↑ ↑

Though the number of 1's and 0's is correct in this scheme, there are repeated columns so it is not column canonical.

4. The next distance 1 replacement for #19 is #7. This yields the temporary subset

#12	1 1 1 0 0 0	
#13	0 1 0 0 1 1	
#15	0 0 1 1 1 0 → 0 0 1 1 0 1	#10
#7	1 0 0 1 1 0	
	↑ ↑	

To restore the fifth and sixth columns, #15 needs to be changed. The replacement is #10, yielding the new subset

#12	1 1 1 0 0 0
#13	0 1 0 0 1 1
#10	0 0 1 1 0 1
#7	1 0 0 1 1 0

This is column canonical. So the replacements have been

primary: #19 → #7 complementary: #15 → #10.

II. Second primary replacement. Now we seek a distance 1 replacement for the poorest equation in the subset obtained in Step I above, which is #13. The first such equation in the list is #3. With this

```
#12    1 1 1   0 0 0
#3     0 1 0   1 0 1
#10    0 0 1   1 0 1  →  0 0 1   0 1 1   #4
#7     1 0 0   1 1 0
               ↑ ↑
```

To restore the columns, #10 is replaced by #4. This yields

```
#12    1 1 1   0 0 0
#3     0 1 0   1 0 1
#4     0 0 1   0 1 1
#7     1 0 0   1 1 0
```

This is column canonical. Thus the replacements are

primary: #13 → #3 complementary: #10 → #4.

III. Third primary replacement. Next we attempt to replace #12.

1. #1 is a distance 1 replacement. To restore the columns, #3 needs to be replaced by #20 which is not acceptable.

2. The next distance 1 replacement is #2. Now the columns can be restored by #7 → #5 or by #3 → #11. Either complementary replacement leads to subsets with repeated columns which, of course, cannot be used.

3. #12 → #5 (distance 1) requires one of the complementary replacements #7 → #2 or #4 → #11, both resulting in structures with repeated columns.

4. #12 → #8 (distance 1) primary replacement comes with complementary #7 → #14, but #14 is worse than #12 originally replaced.

5. #12 → #11 (distance 1) again comes with complementary replacements which lead to repeated columns.

Thus #12 cannot be replaced. That is, the structure obtained in Step II cannot be further improved in a local search procedure restricted to distance 1 replacements. Notice that this subset of equations is slightly worse than the one obtained by global search (the worst equation is #12 instead of #11).

10.7. NOTES AND REFERENCES

The idea of and need for robustness was brought up in very clear form in the celebrated paper by Chow and Willsky (1984). Robustness aspects are covered in various surveys by Paul Frank (e.g., 1991, 1992). The field has been surveyed recently, with an explicit focus on robustness, by Ron Patton (1994, 1995). Note that the literature of fault detection and diagnosis is full of papers claiming robustness, though most of the reported methods do not go beyond decoupling from additive disturbances.

The conditions for perfect decoupling were treated in various papers by Paul Frank; they were spelled out in the form shown here in (Gertler, 1992; Gertler and Kunwer, 1995).

The singular value decomposition (SVD) technique for the rank reduced approximation of a rectangular matrix was developed by Eckart and Young (1936). It was introduced into fault diagnosis by Lou et al (1986), who used it to approximate multiple model variants for robustness under varying operating conditions. Patton and Chen (1991, 1993) employed this method recently to approximate the disturbance entry matrix in the state space representation, for the robust design of a diagnostic observer. The application of SVD to parity relations designed for parametric faults and disturbances was explored in (Kunwer, 1992) and reported in (Gertler and Kunwer, 1993, 1995). Application to parity relations for additive faults and disturbances is being shown here for the first time.

The constrained least squares approach to parity relations for multiplicative faults and disturbances was originally proposed by the author. It was subsequently investigated in (Kunwer, 1992) and reported in (Gertler and Kunwer, 1993, 1995).

Viswanadham and Minto (1988) were the first to apply the H_∞ approach to residual generation in fault detection and diagnosis in the presence of model errors. Ding and Frank, in a series of articles (e.g., 1989, 1991), adopted H_∞ concepts and techniques to deal with additive faults and disturbances. This direction was further explored and refined in (Qiu, 1994), and some new results reported in (Qiu and Gertler, 1993, 1994).

The combinatorial optimization technique described in Section 10.6 was proposed by the author. It was developed in (Luo, 1990) and published in (Gertler and Luo, 1989).

An alternative approach to the robustness problem rests on the fact that the uncertainty caused by model errors depends on the operating conditions. It may be possible to model this dependence, theoretically or empirically, and vary the test thresholds accordingly. Such *adaptive thresholding* techniques were proposed by Emami-Naeini et al (1988) and by Horak and coworkers (Horak and Goblirsch, 1986; Horak, 1988), and further explored by Frank (1993).

11
Statistical Testing of Residuals

11.0. INTRODUCTION

As it was pointed out in Chapter 5, the monitored plant is usually subjected to random noise. As unknown inputs, these noises affect the residuals and interfere with the detection and isolation of faults. In general, this situation requires a decision process which involves testing the residuals against thresholds or uncertainty regions. In many practical situations, only limited information is available concerning the noise and therefore the thresholds have to be chosen empirically.

If the statistical properties of the noise, together with the way it affects the plant outputs, are known (or can be reasonably approximated or assumed) then the fault detection and isolation problem can be formulated in the framework of statistical decision making. Usually it is reasonable to assume that the residuals are the sums of two components, one caused by the noise (which is random, with zero mean) while the other by the faults (which is deterministic but unknown). Thus the residuals may be considered as random variables (vectors) whose mean is determined by the faults. The fault detection problem is then posed as testing for the zero mean hypothesis while the isolation problem becomes a decision among a set of alternative hypotheses.

This avenue will be explored in the present chapter. First the propagation of noise to the residuals will be described. Then detection and isolation schemes will be introduced which rely on the statistical properties of the noise-induced part of the residuals. The isolation schemes will also utilize the geometric properties of the residuals, arising from the isolation enhancement techniques introduced in the previous chapters (namely the structured and directional residual designs). The resulting methodology is a variant of the generalized likelihood ratio approach, with the maximum likelihood estimation performed under constraints. Other applications of the generalized likelihood ratio concept to fault detection and diagnosis have been discussed by Basseville and Nikiforov in their recent book (1993).

Obviously, the performance of the statistical decision procedure is improved if it is based on a series of observations, rather than on a single one. However, this also results in a significant increase of the computational complexity of the decision process. This will also be explored, together with the possibility of reducing complexity, by condensing the series of observations into empirical averages.

An *on-line approach* to data processing will be assumed throughout this chapter. That is, a series of observations will be meant as one taken through a window having constant length and sliding with real time. An alternative would be working with cumulative data-sets; this approach has been explored in great detail in the book by Basseville and Nikiforov (1993).

If the residual time-series is uncorrelated, the statistical tests are relatively easy to implement. The Kalman filter, and the equivalent input-output residual generators (Nikoukhah, 1994) produce uncorrelated residuals. This, however, consumes the available design freedom and isolation enhancement is not possible. With parity relations and observers, the design freedom is used for isolation enhancement but the residuals are correlated in time. This results in increased complexity of the tests, a fact which will be examined below. Time correlatedness may be removed by recursive "whitening filters," a technique also outlined in this chapter. Such filters, however, interfere with the geometric properties of the residuals and therefore can only be applied in schemes where the elements of the residual vector are tested individually.

11.1. NOISE PROPERTIES AND PROPAGATION

11.1.1. Basic Assumptions

Let us recall the basic plant formulation from Chapter 5, recaptured here in Figure 11.1. The variables are as follows:

$u(t)$ is the vector of controlled and measured inputs

$p(t)$ is the vector of additive faults

$q(t)$ is the vector of additive disturbances

$v(t)$ is the vector of noise

$y(t)$ is the vector of outputs.

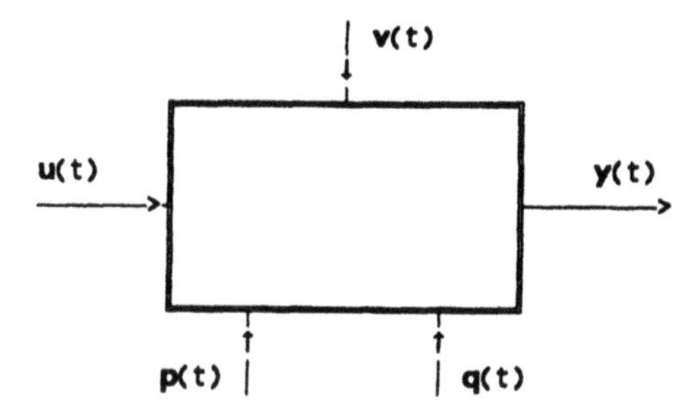

FIGURE 11.1. Plant variables.

The input-output relationship is described as (c.f. Eq. (5.12))

$$y(t) = M(\phi)u(t) + S_F(\phi)p(t) + S_D(\phi)q(t) + S_N(\phi)v(t) \tag{11.1}$$

For the sake of simplicity, the disturbance input $q(t)$ will not be shown as a separate entity in the rest of this chapter but will be considered as included in the fault vector $p(t)=[p_1(t) \ldots p_l(t)]'$.

The noise $v(t)$ is assumed to be random, with the following properties:

a. $E\{v(t)\} = 0$ (zero mean)

b. $\Phi_{vv}(\tau) = E\{v(t)v'(t-\tau)\} = 0, \quad \tau \neq 0$ (whiteness)

c. $\Phi_{vv}(0) = E\{v(t)v'(t)\} = I$ (uncorrelatedness)

These assumptions are not restrictive since

a. any nonzero noise-mean can be represented as a disturbance input;
b. any time-correlated noise may be represented by a white source noise acting through a transfer function matrix, with the latter included in $S_N(\phi)$;
c. any deviation from unit zero-shift covariance may be represented by a constant multiplier matrix included in $S_N(\phi)$.

In addition, it will also be assumed in general that $v(t)$ is normally distributed. This assumption is somewhat restrictive, though it may be justified by the central limit theorem, which states that the sum of a large number of random variables (random effects) can always be characterized approximately with the normal distribution (see Subsection 3.5.2).

Define the residual vector as $r(t)=[r_1(t) \ldots r_n(t)]'$. This is generated from the observations $u(t)$ and $y(t)$ as

$$r(t) = W(\phi)\,[y(t) - M(\phi)u(t)] \tag{11.2}$$

where $W(\phi)$ is the residual generator matrix. The internal form of the residuals is obtained, by substituting (11.1), as

$$r(t) = r_F(t) + r_N(t) = W(\phi)S_F(\phi)p(t) + W(\phi)S_N(\phi)v(t) \tag{11.3}$$

where $r_F(t)$ is the fault-induced part of the residual while $r_N(t)$ is its noise-induced part. The fault-induced part is deterministic, though unknown and time varying. The noise-induced part is random with zero mean. This we will interpret as *the residual having a time-varying mean, contributed entirely by the faults and represented by* $r_F(t)$, so that

$$\boxed{\mu_r(t) = r_F(t)} \tag{11.4}$$

Further, we will consider *the covariance of the residual originating entirely from the noise and represented by* $r_N(t)$, that is

$$\boxed{\Psi_{rr}(\tau) = \Psi_{rNrN}(\tau) = \Phi_{rNrN}(\tau) = \mathrm{E}[r_N(t)r_N'(t-\tau)]} \tag{11.5}$$

These interpretations are crucial to the approach discussed in this chapter.

11.1.2. The Noise Transfer Function

In Chapter 6, the residual generator was designed for a prescribed fault response $\mathbf{Z}(\phi)$, so that

$$r_F(t) = \mathbf{Z}(\phi)\mathbf{T}(\phi)p(t) \tag{11.6}$$

where $\mathbf{T}(\phi)$ is a response modifier matrix, which may be necessary for the causality and stability (or polynomial behavior) of the generator. Thus, with (11.3),

$$W(\phi) = \mathbf{Z}(\phi)\mathbf{T}(\phi)S_F^{-1}(\phi) \tag{11.7}$$

where it has been assumed that $S_F(\phi)$ is square and invertible. With this, the noise induced part of the residual is

$$r_N(t) = \mathbf{Z}(\phi)\mathbf{T}(\phi)S_F^{-1}(\phi)S_N(\phi)v(t) \tag{11.8}$$

System matrix representation. Let us represent (11.1), without the disturbance inputs, in the state space:

$$x(t+1) = Ax(t) + Bu(t) + Ep(t) + Kv(t)$$

$$y(t) = Cx(t) + Du(t) + Fp(t) + Lv(t) \tag{11.9}$$

Recall, from Chapter 6, the fault system matrix and its inverse

$$\Gamma^+(\phi) = \begin{bmatrix} \phi I - A & -E \\ C & F \end{bmatrix} \tag{11.10}$$

$$[\Gamma^+(\phi)]^{-1} = \frac{\Omega^+(\phi)}{\pi^+(\phi)} = \frac{1}{\pi^+(\phi)} \begin{bmatrix} \Omega_A^+(\phi) & \Omega_E^+(\phi) \\ \Omega_C^+(\phi) & \Omega_F^+(\phi) \end{bmatrix} \tag{11.11}$$

As it was shown in Chapter 6, the following relationships hold:

$$[S_F(\phi)]^{-1} = [C(\phi I - A)^{-1}E + F]^{-1} = \Omega_F^+(\phi) / \pi^+(\phi) \tag{11.12}$$

$$[S_F(\phi)]^{-1} S_N(\phi) = [C(\phi I - A)^{-1}E + F]^{-1} [C(\phi I - A)^{-1}K + L]$$

$$= [- \Omega_C^+(\phi)K + \Omega_F^+(\phi)L] / \pi^+(\phi) \tag{11.13}$$

With this, (11.7) and (11.8) become

$$W(\phi) = Z(\phi)\mathrm{T}(\phi)\Omega_F^+(\phi) / \pi^+(\phi) \tag{11.14}$$

$$r_N(t) = Z(\phi)\mathrm{T}(\phi) \frac{1}{\pi^+(\phi)} [- \Omega_C^+(\phi)K + \Omega_F^+(\phi)L] \, v(t) \tag{11.15}$$

Finally, decomposing the modifier matrix as $\mathrm{T}(\phi) = \mathrm{T}_P(\phi)\mathrm{T}_R(\phi)$, where T_P is the stabilizer matrix and T_R is the causality (realizability) matrix, (11.14) and (11.15) can be written as

$$W(\phi) = Z(\phi)\mathrm{T}_P(\phi)\Omega_F(\phi) / \pi(\phi) \tag{11.16}$$

$$r_N(t) = Z(\phi)\mathrm{T}_P(\phi) \frac{1}{\pi(\phi)} [- \Omega_C(\phi)K + \Omega_F(\phi)L] \, v(t) \tag{11.17}$$

Here $\Omega_F(\phi)$, $\Omega_C(\phi)$ and $\pi(\phi)$ are realizable (matrix) polynomials.

Equations (11.16) and (11.17) reveal that by choosing $\mathrm{T}_R(\phi)$ so that the residual generator $\mathbf{W}(\phi)$ is realizable, the noise induced part of the residual is, of course, also causal. Further, if $\mathbf{T}_P(\phi)$ is so chosen that the computational algorithm is made stable (or polynomial), by canceling the unstable factors of $\pi(\phi)$ (or the entire $\pi(\phi)$ polynomial), then the noise induced part of the residual also becomes stable (or polynomial).

Example 11.1.

We will consider a third order system with two outputs, two fault inputs and two noise inputs. The noise is white and normal, with zero mean and unit covariance. The fault-to-output and noise-to-output transfer functions are:

$$S_F(\phi) = \begin{bmatrix} \dfrac{10(\phi - 0.7)(\phi - 0.4)}{(\phi - 0.6)(\phi - 0.5)} & \dfrac{2.2}{\phi - 0.5} \\ \dfrac{10(\phi - 1)(\phi - 0.7)}{(\phi - 0.8)(\phi - 0.6)} & \dfrac{1}{\phi - 0.8} \end{bmatrix}$$

$$S_N(\phi) = \begin{bmatrix} \dfrac{4(\phi - 0.525)}{(\phi - 0.6)(\phi - 0.5)} & \dfrac{3(\phi - 0.567)}{(\phi - 0.6)(\phi - 0.5)} \\ \dfrac{3}{\phi - 0.6} & \dfrac{3(\phi - 0.667)}{(\phi - 0.8)(\phi - 0.6)} \end{bmatrix}$$

The diagonal state-space representation of the system is

$$A = \begin{bmatrix} 0.5 & 0 & 0 \\ 0 & 0.6 & 0 \\ 0 & 0 & 0.8 \end{bmatrix} \qquad E = \begin{bmatrix} 2 & 2.2 \\ -2 & 0 \\ -1 & 1 \end{bmatrix} \qquad K = \begin{bmatrix} 1 & 2 \\ 3 & 1 \\ 0 & 2 \end{bmatrix}$$

$$C = \begin{bmatrix} 1 & 1 & 0 \\ 0 & 1 & 1 \end{bmatrix} \qquad F = \begin{bmatrix} 10 & 0 \\ 10 & 0 \end{bmatrix} \qquad L = \begin{bmatrix} 0 & 0 \\ 0 & 0 \end{bmatrix}$$

We will design a residual generator for the original specification

$$\mathbf{Z}(\phi) = \begin{bmatrix} 1 & -1 \\ 1 & 1 \end{bmatrix}$$

This is directional response with directions $\boldsymbol{\beta}_1 = [1 \quad 1]'$, $\boldsymbol{\beta}_2 = [-1 \quad 1]'$ and $\gamma_1(\phi) = \gamma_2(\phi) = 1$ original response dynamics. The determinant of the fault system matrix, and the parts of interest of its adjoint, are obtained as

$$\pi^+(\phi) = -12(\phi - 1.5)(\phi - 0.7)$$

$$\boldsymbol{\Omega}_C^+(\phi) = \begin{bmatrix} -(\phi - 0.6) & 1.2(\phi - 1.05) & 2.2(\phi - 0.6) \\ 10(\phi - 1)(\phi - 0.7) & -3(\phi - 0.7) & -10(\phi - 0.7)(\phi - 0.4) \end{bmatrix}$$

$$\boldsymbol{\Omega}_F^+(\phi) = \begin{bmatrix} (\phi - 0.6)(\phi - 0.5) & -2.2(\phi - 0.8)(\phi - 0.6) \\ -10(\phi - 1)(\phi - 0.7)(\phi - 0.5) & 10(\phi - 0.8)(\phi - 0.7)(\phi - 0.4) \end{bmatrix}$$

The polynomial degrees are $\delta_o = 2$, $\delta_1 = 2$, $\delta_2 = 3$, which calls for a one step delay in the second fault response. Further, the invariant zero at 1.5 is outside the unit circle and therefore needs to be canceled. Thus the response modifier is

$$\mathbf{T}(\phi) = \begin{bmatrix} (1 - 1.5\phi^{-1}) & 0 \\ 0 & \phi^{-1}(1 - 1.5\phi^{-1}) \end{bmatrix}$$

The generator transfer function may be obtained by (11.16); we omit the numerical results here. The noise-to-residual transfer function, computed by (11.17), is

$$\frac{1}{1 - 0.7\phi^{-1}} [\mathbf{Q}^1\phi^{-1} + \mathbf{Q}^2\phi^{-2} + \mathbf{Q}^3\phi^{-3}]$$

with

$$\mathbf{Q}^1 = \begin{bmatrix} -0.6167 & 0.3000 \\ 1.0500 & 0.3000 \end{bmatrix} \quad \mathbf{Q}^2 = \begin{bmatrix} 1.9017 & 1.0250 \\ -2.4317 & -1.4750 \end{bmatrix} \quad \mathbf{Q}^3 = \begin{bmatrix} -1.1083 & -0.8750 \\ 1.1083 & 0.8750 \end{bmatrix}$$

Notice that in the plant transfer function the gain of the first fault is about ten times larger than that of the second fault. Yet the residual generator has been designed for identical size responses to the two faults. This will have important consequences as we further develop this example in the later course of the chapter.

11.1.3. Computing the Covariance of the Residuals

In this subsection, we will outline an algorithm for the computation of the autocovariance matrix of a residual vector sequence. The algorithm will utilize the assumption that the driving noise sequence is uncorrelated in time.

Let us rewrite (11.17) as

$$r_N(t) = \frac{Q(\phi)}{s(\phi)}\, \nu(t) \tag{11.18}$$

where the polynomials

$$Q(\phi) = Q^0 + Q^1\phi^{-1} + \ldots + Q^\sigma\phi^{-\sigma}$$

$$s(\phi) = 1 + s^1\phi^{-1} + \ldots + s^\sigma\phi^{-\sigma} \tag{11.19}$$

are computed from (11.17). Define the noise-induced part of the residual time series

$$R_N(t) = [r'_N(t) \;\; r'_N(t-1) \;\; \ldots \;\; r'_N(t-m)]' \tag{11.20}$$

We wish to compute the autocovariance matrix

$$\Psi_R = \mathrm{E}\{R_N(t)\, R'_N(t)\} = \begin{bmatrix} \Psi_{rr}(0) & \Psi_{rr}(1) & \ldots & \Psi_{rr}(m) \\ \Psi'_{rr}(1) & \Psi_{rr}(0) & \ldots & \Psi_{rr}(m-1) \\ \cdot & & & \\ \cdot & & & \\ \cdot & & & \\ \Psi'_{rr}(m) & \Psi'_{rr}(m-1) & \ldots & \Psi_{rr}(0) \end{bmatrix} \tag{11.21}$$

where

$$\Psi_{rr}(\tau) = \mathrm{E}\{r_N(t)\, r'_N(t-\tau)\} \tag{11.22}$$

Computing the cross-covariances. First we will compute the cross-covariances

$$\Psi_{r\nu}(\tau) = \mathrm{E}\{r_N(t)\, \nu'(t-\tau)\} \qquad \tau = 0 \ldots \sigma \tag{11.23}$$

With (3.165) and taking into account that $\Psi_{vv}(\tau)=0,\ \tau\neq 0$,

$$\Psi_{rv}(\tau) = Q^{\tau} I - s^1 \Psi_{rv}(\tau-1) - \ldots - s^{\sigma} \Psi_{rv}(\tau-\sigma) \tag{11.24}$$

$\Psi_{rv}(\tau)=0$ for $\tau<0$, since the plant plus residual generator system $Q(\phi)/s(\phi)$ is causal, so its outputs do not depend on future inputs, and the input sequence is uncorrelated. Thus the cross-covariances may be computed recursively as

$$\Psi_{rv}(0) = Q^0$$

$$\Psi_{rv}(1) = Q^1 - s^1 \Psi_{rv}(0)$$

$$\vdots$$

$$\Psi_{rv}(\sigma) = Q^{\sigma} - s^1 \Psi_{rv}(\sigma-1) - \ldots - s^{\sigma} \Psi_{rv}(0) \tag{11.25}$$

Computing the residual autocovariances. Now let us write the residual autocovariances $\Psi_{rr}(\tau)$, for $\tau=0\ldots-\sigma$. By (3.169):

$$\Psi_{rr}(0) = \Psi_{rv}(0)\,[Q^0]' + \ldots + \Psi_{rv}(\sigma)\,[Q^{\sigma}]' - s^1 \Psi_{rr}(1) - \ldots - s^{\sigma} \Psi_{rr}(\sigma)$$

$$\Psi_{rr}(-1) = \Psi_{rv}(0)\,[Q^1]' + \ldots + \Psi_{rv}(\sigma-1)\,[Q^{\sigma}]' - s^1 \Psi_{rr}(0) - \ldots - s^{\sigma} \Psi_{rr}(\sigma-1)$$

$$\vdots$$

$$\Psi_{rr}(-\sigma) = \Psi_{rv}(0)[Q^{\sigma}]' - s^1 \Psi_{rr}(-\sigma+1) - \ldots - s^{\sigma} \Psi_{rr}(0) \tag{11.26}$$

Utilizing the fact that $\Psi_{rr}(-\tau)=\Psi'_{rr}(\tau)$ and rearranging the equations yields

$$\Psi_{rr}(0) + s^1 \Psi_{rr}(1) + \ldots + s^{\sigma} \Psi_{rr}(\sigma) = \Psi_{rv}(0)[Q^0]' + \ldots + \Psi_{rv}(\sigma)[Q^{\sigma}]'$$

$$\Psi'_{rr}(1) + s^1 \Psi_{rr}(0) + \ldots + s^{\sigma} \Psi_{rr}(\sigma-1) = \Psi_{rv}(0)[Q^1]' + \ldots + \Psi_{rv}(\sigma-1)[Q^{\sigma}]'$$

$$\vdots$$

$$\Psi'_{rr}(\sigma) + s^1 \Psi'_{rr}(\sigma-1) + \ldots + s^{\sigma} \Psi_{rr}(0) = \Psi_{rv}(0)[Q^{\sigma}]' \tag{11.27}$$

Let us add now the transpose of σ of the equations, for example

$$\Psi_{rr}(1) + s^1 \Psi_{rr}(0) + \ldots + s^{\sigma} \Psi'_{rr}(\sigma-1) = Q^1 \Psi'_{rv}(0) + \ldots + Q^{\sigma} \Psi'_{rv}(\sigma-1)$$

$$\vdots$$

$$\Psi_{rr}(\sigma) + s^1 \Psi_{rr}(\sigma-1) + \ldots + s^{\sigma} \Psi_{rr}(0) = Q^{\sigma} \Psi'_{rv}(0) \tag{11.28}$$

Then (11.27) and (11.28) can be merged into the scheme

$$
\left[\begin{array}{c|ccc|ccc}
I & s^1 I & s^2 I \ldots & s^\sigma I & 0 & 0 \ldots & 0 \\
s^1 I & s^2 I & s^3 I \ldots & 0 & I & 0 \ldots & 0 \\
\vdots & & & & & & \\
s^{\sigma-1} I & s^\sigma I & 0 \ldots & 0 & s^{\sigma-2} I & s^{\sigma-3} I \ldots & 0 \\
s^\sigma I & 0 & 0 \ldots & 0 & s^{\sigma-1} I & s^{\sigma-2} I \ldots & I \\
\hline
s^1 I & I & 0 \ldots & 0 & s^2 I & s^3 I \ldots & 0 \\
\vdots & & & & & & \\
s^\sigma I & s^{\sigma-1} I & s^{\sigma-2} I \ldots & I & 0 & 0 \ldots & 0
\end{array}\right]
\left[\begin{array}{c}
\Psi_{rr}(0) \\
\hdashline
\Psi_{rr}(1) \\
\vdots \\
\Psi_{rr}(\sigma) \\
\hdashline
\Psi'_{rr}(1) \\
\vdots \\
\Psi'_{rr}(\sigma)
\end{array}\right]
=
\left[\begin{array}{c}
Q_0 \\
Q_1 \\
\vdots \\
Q_\sigma \\
\hdashline
Q'_1 \\
\vdots \\
Q'_\sigma
\end{array}\right]
\tag{11.29}
$$

where $Q_0 \ldots Q_\sigma$ are the right-hand sides of the equations (11.27) and where I and 0 are $n \cdot n$ matrices. The scheme can be solved for $\Psi_{rr}(0) \ldots \Psi_{rr}(\sigma)$ by the inversion of the coefficient matrix.

An alternative to the scheme (11.29) would be to expand (11.27) into scalar equations and solve for the elements of the $\Psi_{rr}(0) \ldots \Psi_{rr}(\sigma)$ matrices. With n residuals, the number of scalar equations and unknowns is $\sigma n^2 + n(n+1)/2$. In comparison, the size of the coefficient matrix in (11.29) is $n(2\sigma+1)$ (in either direction) and

$$
n(2\sigma+1) < \sigma n^2 + \frac{n(n+1)}{2} \qquad \text{if} \qquad \frac{4\sigma+1}{2\sigma+1} < n \tag{11.30}
$$

Since $(4\sigma+1)/(2\sigma+1) < 2$, the scheme is always more efficient than the scalar calculation as long as $n > 1$.

Once $\Psi_{rr}(\tau)$ has been computed for $\tau = 0 \ldots -\sigma$, additional autocovariance matrices may be obtained recursively, from (3.168), as

$$
\begin{aligned}
&\Psi_{rr}(\tau) = - s^1 \Psi_{rr}(\tau+1) - \ldots - s^\sigma \Psi_{rr}(\tau+\sigma) \\
&\Psi_{rr}(-\tau) = \Psi'_{rr}(\tau), \qquad\qquad \tau = -\sigma-1, -\sigma-2, \ldots
\end{aligned}
\tag{11.31}
$$

Polynomial systems. If the noise-to-residual transfer function is polynomial, that is, $s(\phi)=1$ in (11.18), then the computation of the residual autocovariance becomes significantly simpler. Now the cross-covariances in (11.25) are

$$\Psi_{rv}(\tau) = \begin{cases} Q^{\tau} & \text{for } 0\leq\tau\leq\sigma \\ 0 & \text{otherwise} \end{cases} \tag{11.32}$$

and the autocovariances (11.26) become

$$\Psi_{rr}(\tau) = \begin{cases} \sum_{j=0}^{\sigma-\tau} Q^{j}\,[Q^{\tau+j}]' & \text{for } 0\leq\tau\leq\sigma \\ 0 & \text{otherwise} \end{cases} \tag{11.33}$$

Example 11.2.

We will compute the covariances for the residual generator designed in Example 11.1. Refer to Equations (11.18) and (11.19) and notice that here $\sigma=3$, $Q^0=0$ and $s^1=-0.7$ while $s^2=s^3=0$.

By (11.25), the cross-covariances are obtained as

$$\Psi_{rv}(0) = 0$$

$$\Psi_{rv}(1) = Q^1$$

$$\Psi_{rv}(2) = Q^2 + 0.7\,\Psi_{rv}(1)$$

$$\Psi_{rv}(3) = Q^3 + 0.7\,\Psi_{rv}(2)$$

yielding

$$\Psi_{rv}(1) = \begin{bmatrix} -0.6167 & 0.3000 \\ 1.0500 & 0.3000 \end{bmatrix} \quad \Psi_{rv}(2) = \begin{bmatrix} 1.4700 & 1.2350 \\ -1.6967 & -1.2650 \end{bmatrix} \quad \Psi_{rv}(3) = \begin{bmatrix} -0.0793 & -0.0105 \\ -0.0794 & -0.0105 \end{bmatrix}$$

The right-hand sides of Equations (11.27) are now

$$Q_o = \Psi_{rv}(1)\,[Q^1]' + \Psi_{rv}(2)\,[Q^2]' + \Psi_{rv}(3)\,[Q^3]'$$

$$Q_1 = \Psi_{rv}(1)\,[Q^2]' + \Psi_{rv}(2)\,[Q^3]'$$

$$Q_2 = \Psi_{rv}(1)\,[Q^3]'$$

$$Q_3 = 0$$

which yield

$$Q_0 = \begin{bmatrix} 4.6288 & -6.0509 \\ -4.9836 & 7.0871 \end{bmatrix} \qquad Q_1 = \begin{bmatrix} -3.5751 & 3.7670 \\ 5.2916 & -5.9831 \end{bmatrix} \qquad Q_2 = \begin{bmatrix} 0.4210 & -0.4210 \\ -1.4262 & 1.4262 \end{bmatrix}$$

Finally, the coefficient matrix in (11.29) is

$$\left[\begin{array}{c|ccc|ccc} I & -0.7I & 0 & 0 & 0 & 0 & 0 \\ -0.7I & 0 & 0 & 0 & I & 0 & 0 \\ 0 & 0 & 0 & 0 & -0.7I & I & 0 \\ 0 & 0 & 0 & 0 & 0 & -0.7I & I \\ \hdashline -0.7I & I & 0 & 0 & 0 & 0 & 0 \\ 0 & -0.7I & I & 0 & 0 & 0 & 0 \\ 0 & 0 & -0.7I & I & 0 & 0 & 0 \end{array}\right]$$

where

$$I = \begin{bmatrix} 1 & 0 \\ 0 & 1 \end{bmatrix} \qquad 0 = \begin{bmatrix} 0 & 0 \\ 0 & 0 \end{bmatrix}$$

With these, (11.29) yields the autocovariances

$$\Psi_{rr}(0) = \begin{bmatrix} 4.1690 & -4.6014 \\ -4.6014 & 5.6841 \end{bmatrix} \qquad \Psi_{rr}(1) = \begin{bmatrix} -0.6568 & 2.0706 \\ 0.5460 & -2.0042 \end{bmatrix}$$

$$\Psi_{rr}(2) = \begin{bmatrix} -0.0388 & 0.0232 \\ -0.0388 & 0.0232 \end{bmatrix} \qquad \Psi_{rr}(3) = \begin{bmatrix} -0.0271 & 0.0163 \\ -0.0271 & 0.0163 \end{bmatrix}$$

Notice in $\Psi_{rr}(0)$ that the correlation coefficient between the two elements of the residual, $r_1(t)$ and $r_2(t)$, is $-4.6014/\sqrt{(4.1690 \cdot 5.6841)} = -0.9452$. That is, the two elements are strongly negatively correlated. This means that most of the noise response is centered around the $[-1 \quad 1]$ direction in the residual space, which is the second fault response direction. This is because the residual generator, which has been designed for a relatively large response to the second fault, in spite of its small gain in the plant (see Example 11.1), strongly amplifies everything in this direction.

11.2. DETECTION TESTING

Testing the residuals for the purpose of fault detection usually involves testing for the zero mean hypothesis. Zero residual mean represents no fault while the alternative hypothesis, namely that the mean of the residual is nonzero, represents the presence of fault:

$$\begin{bmatrix} H^\circ : \mu_r = 0 \quad \textit{(no fault)} \\ H^1 : \mu_r \neq 0 \quad \textit{(fault)} \end{bmatrix} \tag{11.34}$$

Note that the alternative hypothesis is composite - in general, the faults are continuous-valued therefore the fault-induced residuals are so as well.

Various test statistics may be constructed, using a single observation of the residual, a series of observations or the empirical average of the latter. These involve various degrees of computational complexity. Tests for these statistics can be designed so that their false alarm probabilities are identical. This does not imply, however, that they yield the same decision if applied to a particular observation. A comparison of the various strategies is possible on the basis of their test power, that is, provided they have the same false alarm rate, on the basis of their probability of missed detection. The latter, obviously, depends on the size of the residual mean and is affected also by the distribution of the noise-induced part of the residual sequence.

We will explore the testing of scalar residuals then move to residual vectors. First, however, the χ^2 distribution will be revisited.

χ^2 testing. In many cases, when the underlying noise distribution is normal or can be approximated as such, the detection testing of the residuals may be implemented as a χ^2 test. Consider a vector $r=[r_1 \dots r_n]'$, whose elements

are jointly normally distributed with zero mean and Ψ_r covariance. By the transformation

$$r^* = \Psi_r^{-1/2} r \tag{11.35}$$

the resulting r^* vector has independent elements, each with unit variance, since

$$E\{r^* r^{*\prime}\} = E\{\Psi_r^{-1/2} r r' \Psi_r^{-1/2}\} = \Psi_r^{-1/2} E\{r r'\} \Psi_r^{-1/2}$$

$$= \Psi_r^{-1/2} \Psi_r \Psi_r^{-1/2} = I \tag{11.36}$$

Now recall that for such a vector $r^* = [r_1^* \dots r_n^*]'$, the statistic $w_n = r^{*\prime} r^*$ follows the central χ^2 distribution, with n degrees of freedom. Thus the statistic

$$w_n = r^{*\prime} r^* = r' \Psi_r^{-1} r \tag{11.37}$$

may be tested accordingly.

It is worth recalling that for the original distribution, and in general with nonzero mean, the equation of the constant density contour

$$(r - \mu_r)' \Psi_r^{-1} (r - \mu_r) = c \tag{11.38}$$

describes a hyperellipsoid in the space of the r variable. The transformation (11.35) converts this into

$$(r^* - \mu_{r*})' (r^* - \mu_{r*}) = c \tag{11.39}$$

which describes a hypersphere in the space of r^*, with $\mu_{r*} = \Psi_r^{-1/2} \mu_r$.

11.2.1. Testing Scalar Residuals

Here we will discuss the detection testing of

- a scalar residual $r(t)$
- a scalar time series $R(t) = [r(t), \dots r(t-m)]'$
- the window average of the scalar series $\bar{r}(t, m) = \frac{1}{m+1} \sum_{j=0}^{m} r(t-j)$

The tests will be designed for a selected false alarm rate α. The three approaches may then be compared on the basis of their respective missed detection rates $(1-\beta)$. Normal distribution and zero nominal mean will be assumed throughout. The simplified notation $f(r)$ or $f[r(t)]$ will be used for the density function $f_r(\rho, 0)$ or $f_r[r(t), 0]$.

Single observation of scalar residual. The simplest possible detection strategy is to test each observation of the scalar residual individually and declare a fault whenever the test fires. The test is thus designed on the basis of the density function for a single observation of a scalar residual which is

$$f(r) = \frac{1}{\sqrt{(2\pi)}\sigma_r} \exp\left[-\frac{r^2}{2\sigma_r^2}\right] \tag{11.40}$$

where σ_r^2 is the variance of the residual. The test threshold κ is so chosen that

$$\int_{-\kappa}^{\kappa} f(r)\, dr = 1 - \alpha \tag{11.41}$$

Finally, the test is performed as

$$\text{if } |r(t)| \begin{cases} < \kappa & \text{then no fault} \\ \geq \kappa & \text{then fault} \end{cases} \tag{11.42}$$

A reasonable alternative strategy is to test each observation individually but declare a fault based on the test results over a sliding window, if (i) *any* test fired within the window or (ii) if *all* tests fired within the window. The computational simplicity of single observation testing is thus retained but the thresholds have to be designed on the basis of the distribution of the sequence. This is

$$f(R) = \frac{1}{(2\pi)^{(m+1)/2}\, |\Psi_R|^{1/2}} \exp\left[-1/2\, R' \Psi_R^{-1} R\right] \tag{11.43}$$

where Ψ_R is the correlation (covariance) matrix of R. For sub-strategy (i) (fault declared if any test is firing), the thresholds are chosen so that

$$\int_{r(t)=-\kappa}^{\kappa} \cdots \int_{r(t-m)=-\kappa}^{\kappa} f(R)\, dr(t-m) \ldots dr(t) = 1 - \alpha \tag{11.44}$$

The test is then performed as

$$\begin{cases} \text{if } |r(t-\tau)| < \kappa \text{ for all } \tau = 0 \ldots m \text{ then no fault} \\ \text{otherwise fault} \end{cases} \tag{11.45}$$

For sub-strategy (ii) (fault declared if all tests are firing), the selection of the threshold is more involved.

Scalar time sequence. Now a single test is applied to a series of $m+1$ consecutive scalar observations. Their joint distribution is described by (11.43). The acceptance region is the inside of a constant density contour, chosen according to the desired false alarm rate. This hyperelliptical region can be converted into a hypersphere (see (11.35)-(11.39)) and the test performed on the statistic

$$w_{m+1}(t) = R'(t)\,\Psi_R^{-1}\,R(t) \tag{11.46}$$

as a χ^2 test, namely

$$\text{if } w_{m+1}(t) \begin{cases} < \kappa = \chi^2_{m+1,\alpha} & \text{then no fault} \\ \geq \kappa = \chi^2_{m+1,\alpha} & \text{then fault} \end{cases} \tag{11.47}$$

Here $\kappa = \chi^2_{m+1,\alpha}$ is the threshold obtained from the χ^2 distribution of $m+1$ degrees of freedom, for a tail of α. This approach is computationally somewhat demanding since the transformation (11.46) needs to be performed on-line at every time sample.

Sliding window average. A less demanding procedure ensues if the average of the (scalar) residual is computed over the sliding window and then this is tested like a single observation. The density function of the average is

$$f(\bar{r}) = \frac{1}{\sqrt{(2\pi)}\sigma_{\bar{r}}} \exp\left[-\frac{\bar{r}^2}{2\sigma_{\bar{r}}^2}\right] \tag{11.48}$$

where the variance of the average is computed (off-line) as

$$\sigma_{\bar{r}}^2 = \frac{1}{m+1}\sigma_r^2 + \frac{2}{(m+1)^2}\sum_{\tau=1}^{m}(m+1-\tau)\,\psi_{rr}(\tau) \tag{11.49}$$

(c.f. (3.126)). The threshold is then chosen so that

$$\int_{-\kappa}^{\kappa} f(\bar{r})\,d\bar{r} = 1 - \alpha \tag{11.50}$$

and the test is performed as

$$\text{if } |\bar{r}(t)| \begin{cases} < \kappa & \text{then no fault} \\ \geq \kappa & \text{then fault} \end{cases} \tag{11.51}$$

A comparison of the detection strategies. Figure 11.2 shows a possible joint density function (constant density contours) for two consecutive observations of

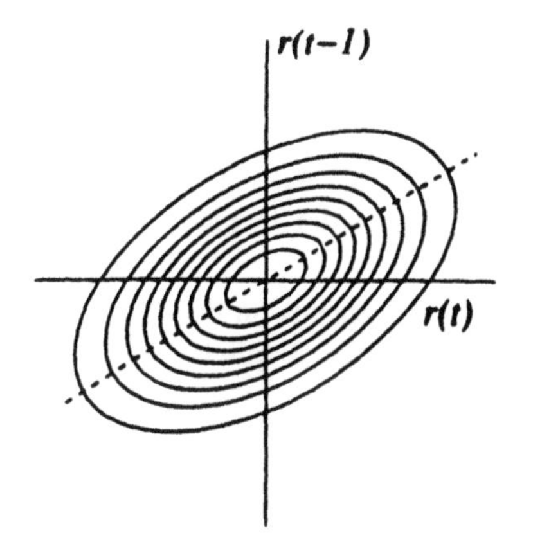

FIGURE 11.2. Joint density.

FIGURE 11.3.a. Single observation test.

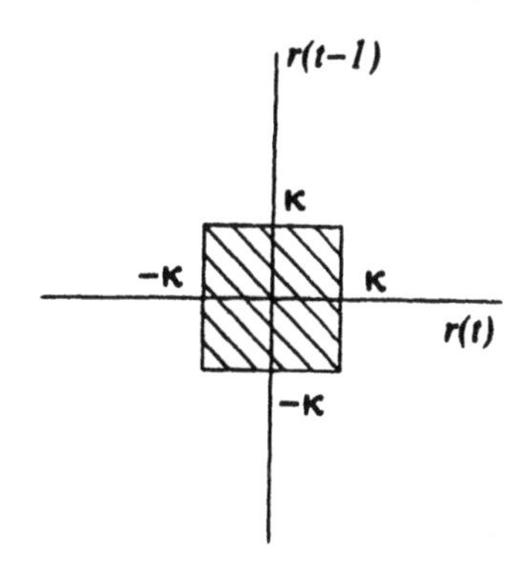

FIGURE 11.3.b. Single obs. test: fault if any test fires.

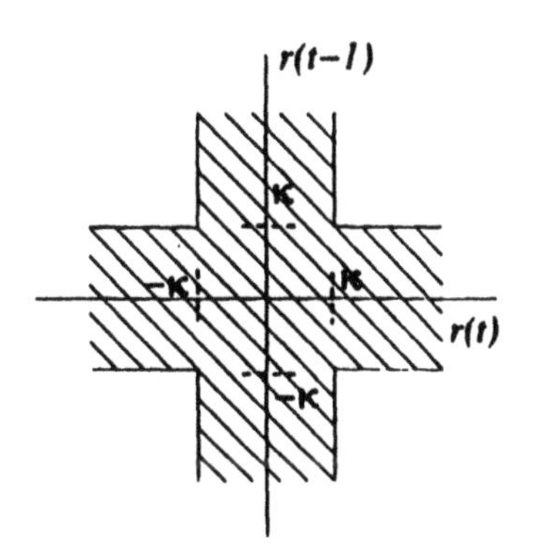

FIGURE 11.3.c. Single obs. test: fault if all tests fire.

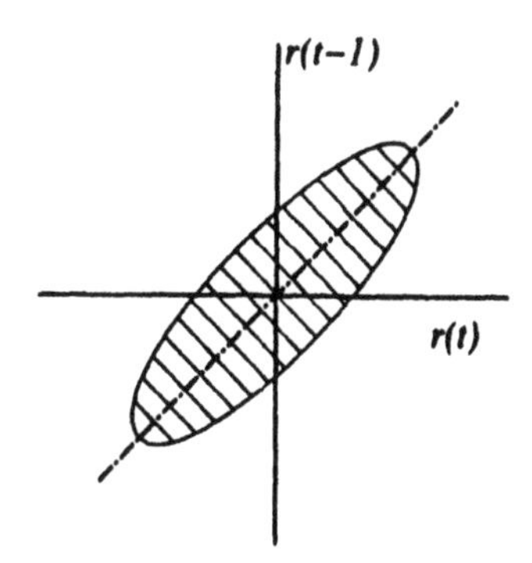

FIGURE 11.3.d. Sliding window test.

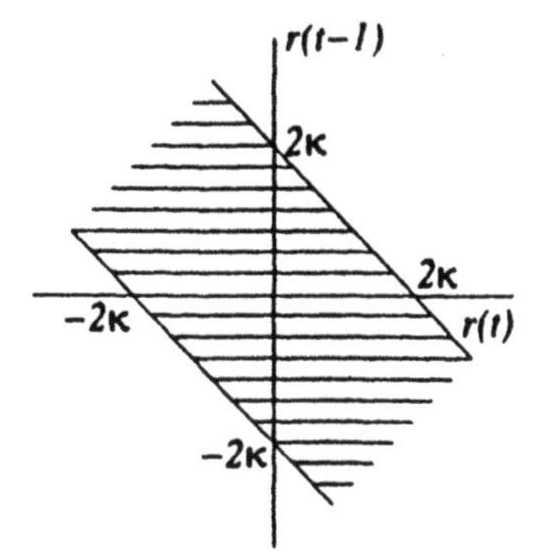

FIGURE 11.3.e. Window average test.

the residual ($m=1$). The contours are slanted ellipses because the observations are correlated. Figures 11.3 (*a* through *e*) show the acceptance/rejection regions for the various detection strategies discussed above. The shape of the various regions is determined by the strategy selected and, with the exception of the constant contour strategy, it is only their size (the thresholds) that depend on the distribution and the chosen false alarm rate.

Missed detections. The probability of missed detection depends on the size and time evolution of the fault-induced part of the residual. For simplicity, assume that a step function fault $p(t)=c\epsilon(t)$ is acting on the system and that the residual generator has been designed with a constant response to this fault. Then the fault, once occurred (perhaps after some initial delay d), contributes a constant mean μ_r to the residual. Thus the joint density function upon this fault, $m+d$ or more samples after its arrival, is

$$f(R|p) = \frac{1}{(2\pi)^{(m+1)/2}\,|\Psi_R|^{1/2}} \exp\left[-1/2\,(R - e\mu_r)'\,\Psi_R^{-1}\,(R - e\mu_r)\right] \tag{11.52}$$

where $e=[1_0 \;\; 1_1 \;\; \ldots \;\; 1_m]'$. The missed detection rate is then the volume of this density function over the region of acceptance Γ_a:

$$\int_{\Gamma_a} f(R|p)\,dR = 1 - \beta \tag{11.53}$$

A possible density function, with a constant fault present, is shown for two observations in Figure 11.4.

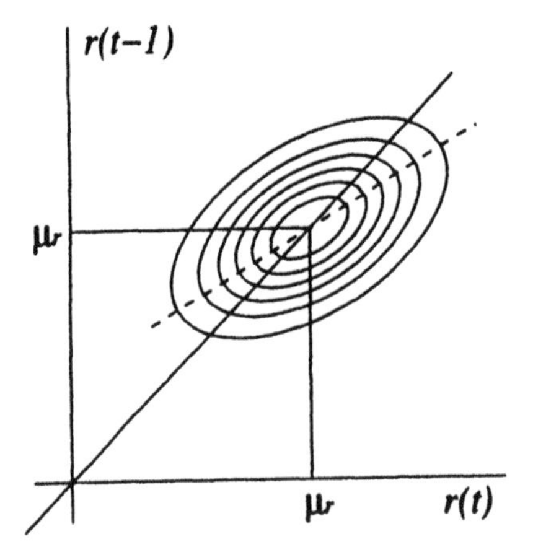

FIGURE 11.4. Density function upon constant fault.

Example 11.3. (Weihua Li)

We will demonstrate the various detection strategies described above using two simple example systems.

1. Assume that a scalar residual is

$$r(t) = 4\,p(t) + [0.6\,/\,(1 - 0.8\phi^{-1})]\,v(t) \qquad \text{(System 1)}$$

where $p(t) = \epsilon(t)$ is a unit-step fault and $v(t)$ is a normally distributed white noise with zero mean and unit variance. Assume that "sequences" consisting of just two consecutive observations, $R(t) = [r(t) \quad r(t-1)]'$ are tested. The covariance matrix of the sequence is obtained as

$$\Psi_R = \begin{bmatrix} 1 & 0.8 \\ 0.8 & 1 \end{bmatrix}$$

Choose a false alarm rate $\alpha = 0.01$. First the test thresholds are computed for the five strategies, namely

a. Testing only $r(t)$;

b. Testing $r(t)$ and $r(t-1)$, declaring fault if either test fired;

c. Testing $r(t)$ and $r(t-1)$, declaring fault if both tests fired;

d. Testing $R'(t)\,\Psi_R^{-1}\,R(t)$;

e. Testing $[r(t) + r(t-1)]/2$.

The thresholds were found by numerically integrating the density function over the various acceptance regions and searching for the threshold value which yields α (except for Case *d.* which was handled by using χ^2 distribution). The thresholds are listed in Table 11.1. In all cases, the thresholds apply to the residual or its mean, except Case *d.* where it applies to the χ^2 statistic. Subsequently, the missed detection rates $(1-\beta)$ were computed for the various regions, again by numerical integration, assuming a unit-step fault input.

TABLE 11.1. Thresholds and missed detection rates - System 1.

strategy	threshold for α=0.01	$1-\beta$ for $p(t)=\epsilon(t)$
a	2.54	0.0728
b	2.74	0.0594
c	2.21	0.0570
d	9.2103 (χ^2)	0.0930
e	2.42	0.0476

TABLE 11.2. Thresholds and missed detection rates - System 2.

strategy	threshold for α=0.01	$1-\beta$ for $p(t)=\epsilon(t)$
a	2.54	0.0728
b	2.74	$2.17 \cdot 10^{-6}$
c	2.21	0.0743
d	9.2108 (χ^2)	$1.68 \cdot 10^{-22}$
e	0.81	$3.39 \cdot 10^{-24}$

Note that one would expect the best performance from Strategy *d*. Notice, however, that this distribution is stretched in the +45° direction, due to its high positive correlation between $r(t)$ and $r(t-1)$. Therefore a mean shift, appearing in the same direction, is missed with a relatively high probability in Case *d*.

2. Now the residual is generated as

$$r(t) = 4\,p(t) + [0.6\,/\,(1 + 0.8\phi^{-1})]\;v(t) \qquad \text{(System 2)}$$

yielding the covariance matrix

$$\Psi_R = \begin{bmatrix} 1 & -0.8 \\ -0.8 & 1 \end{bmatrix}$$

The thresholds and missed detection rates are computed the same way as for System 1. The results are summarized in Table 11.2. Note that now the distribution is elongated in the -45° direction while the mean-shift still appears in the +45° direction; this leads to superior performance in Cases *d* and *e*.

11.2.2. Testing Vector Residuals

Here we will discuss the detection testing of

- a vector residual $r(t)=[r_1(t)\ \dots\ r_n(t)]'$
- a vector time series $R(t)=[r'(t),\ \dots\ r'(t-m)]'$
- the window average of the vector series $\bar{r}(t, m) = \frac{1}{m+1} \sum_{j=0}^{m} r(t-j)$

The treatment in many respects follows the scalar case discussed in the previous subsection.

Single observation of vector residual. The joint density function of the vector residual is

$$f(r) = \frac{1}{(2\pi)^{n/2}\,|\Psi_r|^{1/2}}\, exp\,[\,-1/2\, r'\,\Psi_r^{-1}\, r] \tag{11.54}$$

where Ψ_r is the covariance matrix of r. Clearly, the statistic

$$w_n(t) = r'(t)\,\Psi_r^{-1}\, r(t) \tag{11.55}$$

obeys χ^2 distribution with n degrees of freedom and can be tested as

$$if\; w_n(t) \begin{cases} < \chi^2_{n,\alpha} & then\ no\ fault \\ \geq \chi^2_{n,\alpha} & then\ fault \end{cases} \tag{11.56}$$

Vector residual sequence. The joint density function of the vector sequence is

$$f(R) = \frac{1}{(2\pi)^{n(m+1)/2}\,|\Psi_R|^{1/2}}\, exp\,[\,-1/2\, R'\,\Psi_R^{-1}\, R\,] \tag{11.57}$$

Here the autocovariance matrix Ψ_R is as given in (11.21). Now the statistic

$$w_{n(m+1)}(t) = R'(t)\,\Psi_R^{-1}\, R(t) \tag{11.58}$$

obeys χ^2 distribution with $n(m+1)$ degrees of freedom and may be tested against the threshold $\chi^2_{n(m+1),\alpha}$. Notice that this test strategy requires the on-line multiplication with an $[n(m+1)]\cdot[n(m+1)]$ constant matrix at each observation.

Window average of vector residual. The density function for the window average is

$$f(\bar{r}) = \frac{1}{(2\pi)^{n/2}\,|\Psi_{\bar{r}}|^{1/2}}\, exp\,[\,-1/2\, \bar{r}'\,\Psi_{\bar{r}}^{-1}\, \bar{r}\,] \tag{11.59}$$

where (c.f. (3.131))

$$\Psi_{\bar{r}} = \frac{1}{m+1}\,\Psi_{rr}(0) + \frac{1}{(m+1)^2}\sum_{\tau=1}^{m} (m+1-\tau)\,[\Psi_{rr}(\tau) + \Psi'_{rr}(\tau)] \tag{11.60}$$

The statistic

$$w_n(t) = \bar{r}'(t)\, \Psi_f^{-1}\, \bar{r}(t) \tag{11.61}$$

obeys χ^2 distribution with n degrees of freedom and can be tested against the threshold $\chi^2_{n,\alpha}$.

11.2.3. Fault Sensitivity

While the power of the test is the theoretically correct measure to compare various test strategies, the probability of missed detection is, in general, quite difficult to compute. An easier, and practically not less meaningful, alternative is to characterize the quality of a testing strategy by the size of the fault(s) which, with only the fault-induced part of the residual present, take the test statistic exactly to the threshold. This value of the fault will be called the *triggering limit*. Clearly, this limit depends on the threshold size, which is determined by the selected false alarm rate and by the noise and its propagation to the residual, and on the gain of the particular fault-to-residual transfer. If the fault is equal to the triggering limit then the center of the residual distribution is moved onto the boundary of the acceptance region. Since the acceptance region is convex, the probability of missed detection in this case is never more than 50%; the actual value depends on the distribution and on the position of the mean in the residual space.

Transient response of the generator. The residual generator is a dynamic system so, in response to any fault change, a transient takes place. The fault response is characterized by the design relationship (11.6)

$$r_F(t) = \mathbf{Z}(\phi)\, \mathbf{T}(\phi)\, \mathbf{p}(t) \tag{11.62}$$

If the generator is designed with particular geometric properties (directional or structured residuals) then the fault induced part of the residual, in response to any single fault, is confined to a specific subspace at all times, including the fault transients.

In the context of dynamic response, the residual series $\mathbf{R}(t)$ can be represented as

$$\mathbf{R}(t) = \mathbf{J}(\phi)\mathbf{r}(t) \tag{11.63}$$

where

$$\mathbf{J}(\phi) = [\mathbf{I} \quad \phi^{-1}\mathbf{I} \ldots \phi^{-m}\mathbf{I}]' \tag{11.64}$$

Further, the window average can be written as

$$\bar{r}(t) = \frac{1}{m+1}\,[1 + \phi^{-1} + \ldots + \phi^{-m}]\mathbf{r}(t) = \frac{1}{m+1}\,\frac{1 - \phi^{-m-1}}{1 - \phi^{-1}}\,\mathbf{r}(t) \tag{11.65}$$

Here the polynomial expression corresponds to the direct (additive) computation of the average while the rational transfer function describes its recursive computation.

Steady state triggering limit. Because of the dynamic nature of the fault response, the definition of the triggering limit requires further clarification. One possibility is to define it as the constant value of a single fault which brings the test statistic to its threshold in steady state (in the absence of noise). A less well defined alternative relates the limit to the maximum of the transient response. We will pursue the steady state definition.

Denote the steady state triggering limit for the j-th fault as η_j. The steady state response to the j-th fault of this size is

$$r_F(t|p_j=\eta_j) = [\mathbf{Z}(\phi)\mathbf{T}(\phi)]_{\phi=1}[0 \ldots 0 \;\; \eta_j \;\; 0 \ldots 0]' = \mathbf{v}_j \eta_j \tag{11.66}$$

Upon this residual, the test statistic is equal to its threshold. For single observation testing this means (see (11.55) and (11.56))

$$\eta_j \mathbf{v}_j' \mathbf{\Psi}_r^{-1} \mathbf{v}_j \eta_j = \chi^2_{n,\alpha} \tag{11.67}$$

From this, η_j can be computed. Since the fault induced residual is constant under the steady state assumption, $\mathbf{R}(t|p_j=\eta_j)=[\eta_j \mathbf{v}_j' \;\; \ldots \;\; \eta_j \mathbf{v}_j']'$ and $\bar{\mathbf{r}}(t|p_j=\eta_j)=\mathbf{v}_j\eta_j$; the respective values for the triggering limit can be computed by substituting these into (11.58) and (11.61).

Detection delay. Due to the dynamic nature of the fault response, threshold crossing upon the occurrence of a fault is usually not immediate, even if the fault is large enough to eventually trigger the test. The delay between the onset of the fault and the time the test statistic crosses its threshold is called the detection delay. Due to the presence of noise, this delay is also a random variable. Practically, it can be characterized by the time it takes for the fault induced part of the test statistic, with no noise present, to cross the test threshold. This can be obtained, for any particular fault function $p(t)$ and using the transfer functions of the residual generator, analytically or by simulation. Just like the triggering limit, the delay also depends on the size of the thresholds which, of course, are determined by the noise and its propagation to the residuals.

Example 11.4.

We will find the steady state triggering limits for the residual generator which was designed in Example 11.1. and for which the covariances were computed in Example 11.2. The limits for the first and the second fault will be compared. Also, they will be computed for the three test strategies, namely single observation, series of observations and window average. To keep the example simple, a window of three observations ($m=2$) will be assumed.

The comparison will be performed for two values of the false alarm rate, $\alpha=0.01$ and $\alpha=0.001$. For single observation and for the window average, the degrees of freedom is 2, yielding the thresholds (from a χ^2 distribution table):

$$\chi^2_{2,\,0.01} = 9.210 \qquad \chi^2_{2,\,0.001} = 13.815$$

For the observation series, the degrees of freedom is 6 and

$$\chi^2_{6,\,0.01} = 16.812 \qquad \chi^2_{6,\,0.001} = 22.457$$

To obtain the steady state gain for the faults, consider first the modifier matrix $\mathbf{T}$

$$[\mathbf{T}(\phi)]_{\phi=1} = \begin{bmatrix} 1 - 1.5\phi^{-1} & 0 \\ 0 & \phi^{-1}(1 - 1.5\phi^{-1}) \end{bmatrix}_{\phi=1} = \begin{bmatrix} -0.5 & 0 \\ 0 & -0.5 \end{bmatrix}$$

Thus the gains are

$$\boldsymbol{v}_1 = -0.5\,[1 \quad 1]' \qquad \boldsymbol{v}_2 = -0.5\,[-1 \quad 1]'$$

a. For a single observation, $\boldsymbol{\Psi}_r = \boldsymbol{\Psi}_{rr}(0)$ which was computed in Example 11.2. With this,

$$\boldsymbol{v}_1' \boldsymbol{\Psi}_r^{-1} \boldsymbol{v}_1 = 1.8874 \qquad \boldsymbol{v}_2' \boldsymbol{\Psi}_r^{-1} \boldsymbol{v}_2 = 0.0644$$

Thus the steady state triggering limits for $\alpha=0.01$ are

$$\eta_{1,\,0.01} = \sqrt{\frac{9.210}{1.8874}} = 2.2090 \qquad \eta_{2,\,0.01} = \sqrt{\frac{9.210}{0.0644}} = 11.959$$

Similarly, for $\alpha=0.001$,

$$\eta_{1,\,0.001} = 3.3135 \qquad \eta_{2,\,0.001} = 17.936$$

b. For a series of (three) observations,

TABLE 11.3. Steady state triggering limits.

test	$\eta_{1,0.01}$	$\eta_{1,0.001}$	$\eta_{2,0.01}$	$\eta_{2,0.001}$
single	2.209	3.314	11.96	17.94
series	1.373	1.834	5.640	7.534
average	1.318	1.976	6.049	9.074

$$\Psi_R = \begin{bmatrix} \Psi_{rr}(0) & \Psi_{rr}(1) & \Psi_{rr}(2) \\ \Psi'_{rr}(1) & \Psi_{rr}(0) & \Psi_{rr}(1) \\ \Psi'_{rr}(2) & \Psi'_{rr}(1) & \Psi_{rr}(0) \end{bmatrix}$$

The numerical values can be found in Example 11.2. With those,

$$[\mathbf{v}'_1 \quad \mathbf{v}'_1 \quad \mathbf{v}'_1]\ \Psi_R^{-1} [\mathbf{v}'_1 \quad \mathbf{v}'_1 \quad \mathbf{v}'_1]' = 8.9148$$

$$[\mathbf{v}'_2 \quad \mathbf{v}'_2 \quad \mathbf{v}'_2]\ \Psi_R^{-1} [\mathbf{v}'_2 \quad \mathbf{v}'_2 \quad \mathbf{v}'_2]' = 0.5286$$

The triggering limits obtained with these are shown in Table 11.3.

c. For the window average, by (11.60),

$$\Psi_r = \frac{1}{3}\Psi_{rr}(0) + \frac{2}{9}[\Psi_{rr}(1) + \Psi'_{rr}(1)] + \frac{1}{9}[\Psi_{rr}(2) + \Psi'_{rr}(2)]$$

$$= \begin{bmatrix} 1.0891 & -0.9541 \\ -0.9541 & 1.0091 \end{bmatrix}$$

This yields

$$\mathbf{v}'_1 \Psi_r^{-1} \mathbf{v}_1 = 5.3050 \qquad\qquad \mathbf{v}'_2 \Psi_r^{-1} \mathbf{v}_2 = 0.2517$$

Again, the resulting triggering limits are shown in Table 11.3.

We wish to make the following observations here:

- Clearly, the triggering limits increase if the test is designed for lower false alarm rate.
- The fault sensitivity clearly increases as we move from a single observation to a series or its average (recall that this is just a series of three observations; with a longer window the improvement would be more pronounced).
- Testing the full series yields slightly better sensitivity than working with the

window average, in most cases (one would expect this to happen in all cases - but recall that the triggering limit is not an exact measure of the test power).

- Observe the significant difference between the limits for the two faults. This is because the distribution, in this particular system, is strongly stretched along the second response direction, for reasons explained in Example 11.2.

11.2.4. Residual Filtering

It is usually advantageous to filter the residuals before testing. The filter may serve one of two purposes:

1. Low-pass filtering, to decrease the variance of the residual without affecting its steady-state gain;
2. Whitening filtering, to eliminate the correlation in the residual sequence.

We will discuss low-pass filtering here while whitening filtering will be treated in Section 11.4.

The purpose of low-pass residual filtering is to improve the signal-to-noise ratio in the residual. In general, the noise is in the higher frequency range relative to the faults. This property carries over, though somewhat modified by the plant and the residual generator dynamics, to the noise and fault induced parts of the residual. Thus low-pass filtering usually does improve the signal-to-noise ratio in the residuals, while increasing the detection delay.

The low-pass filter is applied individually to each element of the residual vector (Figure 11.5). The filters, of course, affect the dynamic behavior of the residuals. For directional residuals, the individual filters must be identical, otherwise they interfere with the geometric properties of the residual vector. For structured residuals this is not necessary, since individual filtering does not affect the residual structure.

Denote the filter transfer function, which is usually a rational function, as $k(\phi)$. Now the noise induced part of the i-th residual element after filtering is obtained, with (11.18), as

$$r^{\#}_{Ni}(t) = k(\phi)\, r_{Ni}(t) = k(\phi)\, [\mathbf{q}_{i.}(\phi)/s(\phi)]\, \mathbf{v}(t) \tag{11.68}$$

Here $[\mathbf{q}_{i.}(\phi)/s(\phi)]$ is the noise-to-residual transfer function. In all covariance computations, the expression $k(\phi)[\mathbf{q}_{i.}(\phi)/s(\phi)]$, $i=1...n$, replaces the rows of (11.18). The fault induced part is obtained, with (11.6), as

$$r^{\#}_{Fi}(t) = k(\phi)\, r_{Fi}(t) = k(\phi)\, \mathbf{z}_{i.}(\phi)\, \mathbf{T}(\phi)\, \mathbf{p}(t) \tag{11.69}$$

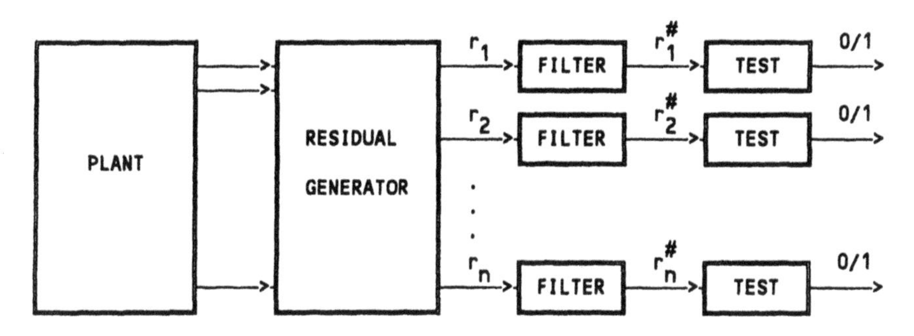

FIGURE 11.5. Residual filtering.

where $z_{i.}(\phi)\ \mathbf{T}(\phi)$ is the fault-to-residual transfer function. The transient fault response of the filtered residual is then computed from this expression.

Due to its low-pass property, the filter reduces the variance of the noise induced part $r_N(t)$ and thus allows the reduction of the threshold. The filter is so designed that its steady state gain is one, so it does not change the fault induced part $r_F(t)$ in steady state. It does, however, affect the transient fault response, increasing the rise-time and reducing the overshoots. Longer rise-time increases the detection delay and reduced overshoots reduce the fault sensitivity - but these are both usually compensated for by the reduction of the threshold.

The filter may be designed using some of the standard techniques such as the Butterworth design. Alternatively, poles close to the unit circle may be chosen arbitrarily and the numerator, which may be a constant, set so that the steady state gain is one. So the simplest first and second order filters with arbitrary pole selection would be

$$k(\phi) = \frac{1-\alpha}{1-\alpha\phi^{-1}} \qquad\qquad k(\phi) = \frac{(1-\alpha_1)(1-\alpha_2)}{(1-\alpha_1\phi^{-1})(1-\alpha_2\phi^{-1})} \tag{11.70}$$

Instead of applying separate filters to the residuals, filtering may be part of the original design of the residual generator. That is, the generator can be designed in such a way that it contains poles close to the unit circle. It should be noted here that polynomial residual generators, which are advantageous from other points of view, usually behave as high-pass filters and thus amplify the noise.

Example 11.5.

We will apply identical low-pass filters to the two residuals of the generator designed in Example 11.1. The filter will be first order with unit steady state gain and a pole at 0.95, that is,

$$k(\phi) = \frac{0.05}{1 - 0.95\phi^{-1}}$$

With this, the transfer function from the faults to the filtered residuals is

$$\frac{0.05(1 - 1.5\phi^{-1})}{1 - 0.95\phi^{-1}} \begin{bmatrix} 1 & -1 \\ 1 & 1 \end{bmatrix} \begin{bmatrix} 1 & 0 \\ 0 & \phi^{-1} \end{bmatrix}$$

Note that identical filtering of the two residual elements does not interfere with the directional property of the residual vector. The transfer function from the noise to the filtered residuals becomes

$$\frac{0.05}{(1 - 0.7\phi^{-1})(1 - 0.95\phi^{-1})} [\mathbf{Q}^1 \phi^{-1} + \mathbf{Q}^2 \phi^{-2} + \mathbf{Q}^3 \phi^{-3}]$$

Here the matrices $\mathbf{Q}^1, \mathbf{Q}^2, \mathbf{Q}^3$ are the ones computed in Example 11.1. Following the procedure of Example 11.2., the autocovariances of the filtered residuals can be computed, yielding

$$\Psi_{r\#r\#}(0) = \begin{bmatrix} 0.0695 & -0.0553 \\ -0.0553 & 0.0513 \end{bmatrix} \qquad \Psi_{r\#r\#}(1) = \begin{bmatrix} 0.0641 & -0.0472 \\ -0.0515 & 0.0439 \end{bmatrix}$$

$$\Psi_{r\#r\#}(2) = \begin{bmatrix} 0.0607 & -0.0447 \\ -0.0492 & 0.0419 \end{bmatrix} \qquad \Psi_{r\#r\#}(3) = \begin{bmatrix} 0.0574 & -0.0423 \\ -0.0469 & 0.0399 \end{bmatrix}$$

Steady state triggering limit. Now we will compute the steady state triggering limits, for single observation testing of the filtered residuals, following the logic of Example 11.4. With $\Psi_{r\#} = \Psi_{r\#r\#}(0)$,

$$[1 \quad 1]\, \Psi_{r\#}^{-1}\, [1 \quad 1]' = 456.23 \qquad [-1 \quad 1]\, \Psi_{r\#}^{-1}\, [-1 \quad 1]' = 20.11$$

The steady state gain for the faults is not affected by this filter so, with the gain 0.5 and the threshold $\chi^2_{2,\,0.01} = 9.21$ (for $\alpha = 0.01$), the triggering limits are obtained as

$$\eta_{1,\,0.01} = \sqrt{\frac{9.21}{0.25 \cdot 456.23}} = 0.284 \qquad \eta_{2,\,0.01} = \sqrt{\frac{9.21}{0.25 \cdot 20.11}} = 1.353$$

These compare with the limits 2.209 and 11.96, respectively, in the unfiltered case (c.f. Example 11.4.).

Detection delay. Without filtering, the step response to the first fault has no delay while that to the second fault has a one step delay. With filtering, the step responses are exponential, which is a source of detection delay. We will compute this delay for the case when the faults are step functions with magnitudes equal to the respective triggering limits without filtering.

The dynamic triggering limits can be obtained by solving for ξ_j, $j=1,2$,

$$\xi_1 [1 \;\; 1] \Psi_{r\#}^{-1} [1 \;\; 1]' \xi_1 = \chi^2_{2,\,0.01}$$

$$\xi_2 [-1 \;\; 1] \Psi_{r\#}^{-1} [-1 \;\; 1]' \xi_2 = \chi^2_{2,\,0.01}$$

This yields $\xi_1 = 0.1421$ and $\xi_2 = 0.6768$. Now the $\xi_j(t)$, $j=1,2$ functions will be the responses of $0.05(1 - 1.5\phi^{-1})/(1 - 0.95\phi^{-1})$, the dynamic part of the fault transfer functions, to step inputs of magnitude 2.209 and 11.96, respectively. These are obtained as

time	$\xi_1(t)$	$\xi_2(t)$
0	0.1145	0.6199
1	0.0497	0.2555
2	- 0.0080	- 0.0433
3	- 0.0628	- 0.3400
4	- 0.1149	- 0.6220
5	- 0.1643	- 0.8895

Clearly, the triggering limits are exceeded, for either fault, at the fifth sample, so the detection delay introduced by the filter (for this particular fault size and shape) is five samples. Note that the residual generator is non-minimum phase which fact is partially responsible for this delay.

11.3. ISOLATION TESTING

In isolation testing, we have to decide among a set of alternative hypotheses

$$H^j, \; j=1,2\ldots : \; \textit{the j-th fault is present} \tag{11.71}$$

The residual generators are so designed that their fault-induced response is enhanced for isolation. The response to a particular single fault is confined, in the residual space,

- to a straight line (directional residuals), or
- to a subspace characterized by the absence of certain coordinate directions (structured residuals), or

- to a single coordinate direction (diagonal residuals).

The purpose of the test is to determine which hypothesis the observations support the most. This will be posed in the likelihood ratio framework as: under which geometric assumption are the actual observations the most likely.

Note that the geometric approach outlined above is the only possibility if directional residuals are concerned. With structured residuals, there is a choice, according to two possible interpretations of the residual properties:

(i) Structured residuals may be interpreted in a logical framework, with a Boolean incidence matrix, where "1"s and "0"s represent if a residual element is or is not affected by a particular fault. In this framework, each residual element is tested separately against an individual threshold. The test outcome is presented as a logical "1" or "0" which becomes an element of the fault code. This is then compared to the columns of the incidence matrix for the isolation decision.

(ii) Alternatively, structured residuals may be interpreted in a geometric framework, with subspaces of the residual space corresponding to the various faults. The test is then of geometric nature, determining the one among these subspaces to which the observations are the closest.

The logical approach can be implemented with the scalar residual testing techniques described in the previous section. This approach is relatively simple but it does not utilize the full information (namely the covariances) about the residuals. It may be the method of choice when simplicity is desired or when the covariances are not available. The geometric approach, which will be discussed in this section, is more complex but it relies on the complete information.

The isolation tests will be designed for the following kinds of observations:

- a single vector observation $\boldsymbol{r}(t)$
- vector time series $\boldsymbol{R}(t)$
- window average of vector time series $\bar{\boldsymbol{r}}(t)$.

Recall that fault isolation is not possible on the basis of a single scalar residual.

The generalized likelihood ratio (GLR) technique. We will apply the GLR technique, suitably modified, with the residual mean $\mu_r(t)$ as the parameter of interest. Start from the joint density function

$$f\,[\boldsymbol{x}(t),\, \boldsymbol{\mu}_r(t)] = K\, exp\, \{\, -1/2\, [\boldsymbol{x}(t) - \boldsymbol{\mu}_x(t)]'\, \boldsymbol{\Psi}_x^{-1}\, [\boldsymbol{x}(t) - \boldsymbol{\mu}_x(t)]\} \tag{11.72}$$

Here $\boldsymbol{x}$ stands for the observation used ($\boldsymbol{r}$, $\boldsymbol{R}$ or $\bar{\boldsymbol{r}}$) and $\boldsymbol{\mu}_x(t)$ is related to $\boldsymbol{\mu}_r(t)$ in a way which depends on the particular kind of observations. Define the simplified log-likelihood function as

$$logL[\boldsymbol{x}(t),\, \boldsymbol{\mu}_r(t)] = -1/2\, [\boldsymbol{x}(t) - \boldsymbol{\mu}_x(t)]'\, \boldsymbol{\Psi}_x^{-1}\, [\boldsymbol{x}(t) - \boldsymbol{\mu}_x(t)] \tag{11.73}$$

Here we omitted *logK* because it cancels out of the likelihood ratios anyway.

The GLR procedure consists of two steps:

1. First the Maximum Likelihood (ML) estimates of the residual mean are computed, from the observations, under the various hypotheses:

$$\hat{\mu}_{rj}(t) = \arg\max_{\mu_r(t)} logL[x(t), \mu_r(t) \mid H^j] \qquad j=1...l \tag{11.74}$$

The hypotheses H^j impose geometric constraints on the estimates.

2. Then conditional likelihood functions are computed, with the observations and the various conditional estimates

$$logL_j(t) = logL[x(t), \hat{\mu}_{rj}(t)] \qquad j=1...l \tag{11.75}$$

According to the basic idea of likelihood ratio testing, the most likely fault is selected as the one which yields the highest log-likelihood value. However, this simple procedure, which produces a single suspect, does not take into account the uncertainty surrounding the decision in the presence of noise. To improve the robustness of the isolation decision, the simple ratio testing will be supplemented with a set of plausibility checks, also performed on the log-likelihood functions.

11.3.1. Geometric Constraints

Directional residuals. Under directional design, the generator response to the j-th fault $p_j(t)$ is specified as

$$r_F(t|p_j) = z_{.j}(\phi)p_j(t) = \beta_j \gamma_j(\phi)p_j(t) \qquad j=1 \ldots l \tag{11.76}$$

where β_j is the j-th response direction vector and $\gamma_j(\phi)$ is the j-th (scalar) response dynamic. The fault $p_j(t)$ may be any time-function but, with the scalar transfer function $\gamma_j(\phi)$, the response always stays on (or moves along) the straight line defined by β_j. Thus the mean of the residual, though time varying, is always confined to this line as well:

$$\mu_r(t|p_j) = \beta_j c_j(t) \qquad j=1 \ldots l \tag{11.77}$$

where $c_j(t)$ is an undefined time-function. (11.77) is the geometric constraint for the ML estimation under directional design, that is,

$$H^j : \mu_r(t) = \beta_j c_j(t) \qquad j=1 \ldots l \tag{11.78}$$

Structured residuals. Structured residuals are primarily specified with their structure, that is the residual elements which do not respond to a particular fault. This is expressed as zeroes in the response specification:

$$r(t|p_j) = z_{.j}(\phi)p_j(t) \quad with\ z_{ij}(\phi)=0\ for\ i=i_1,\ i_2,...;\ j=1...l \tag{11.79}$$

Accordingly, the respective elements in the residual mean $\mu_r=[\mu_{r1}\ ,,,\ \mu_{rm}]'$ will also be zero. For the purpose of the ML estimation, this structural property needs to be expressed as an equality constraint. This can be done by a set of template matrices T_j, $j=1...l$, as

$$T_j\, z_{.j}(\phi) = 0 \qquad T_j\, \mu_r(t|p_j) = 0 \qquad j=1...l \tag{11.80}$$

Each template matrix has n columns and as many rows as the number of zero elements in $z_{.j}(\phi)$; its rows are $[0\ ...\ 0\ \ 1\ \ 0\ ...\ 0]$, the single 1 element assigning a zero position in $z_{.j}(\phi)$.

Expressed with a template matrix, the j-th geometric constraint under structured design is

$$H^j:\ T_j\, \mu_r(t) = 0 \qquad j=1...l \tag{11.81}$$

Example 11.6.

Consider a structured residual generator with the structure matrix

$$\begin{bmatrix} 0 & 1 & 1 & 0 \\ 0 & 0 & 1 & 1 \\ 1 & 0 & 0 & 1 \\ 1 & 1 & 0 & 0 \end{bmatrix}$$

(Recall that in the structure matrix, the rows correspond to elements of the residual vector and the columns to faults. A "0" in an intersection represents zero response while a "1" represents any nonzero response.) The template matrices characterizing the four column structures are

$$T_1 = \begin{bmatrix} 1 & 0 & 0 & 0 \\ 0 & 1 & 0 & 0 \end{bmatrix} \qquad T_2 = \begin{bmatrix} 0 & 1 & 0 & 0 \\ 0 & 0 & 1 & 0 \end{bmatrix}$$

$$T_3 = \begin{bmatrix} 0 & 0 & 1 & 0 \\ 0 & 0 & 0 & 1 \end{bmatrix} \qquad T_4 = \begin{bmatrix} 1 & 0 & 0 & 0 \\ 0 & 0 & 0 & 1 \end{bmatrix}$$

❖

Diagonal residuals. Diagonal residuals are a common special case of the directional and structured designs. Now $n=l$ and each residual element responds to one and only one fault. In the directional framework, this means

$$\boldsymbol{\beta}_j = [0 \dots 0 \underset{j}{1} 0 \dots 0]' \qquad j=1 \dots l \qquad (11.82)$$

In the structural framework, the structure matrix is diagonal, which translates into

$$T_j = \begin{bmatrix} 1 & 0 \dots & \dots 0 \\ \cdot & & \\ \cdot & & \\ \cdot & & \\ 0 \dots 1 & 0 & 0 \dots 0 \\ 0 \dots 0 & 0 & 1 \dots 0 \\ \cdot & & \\ \cdot & & \\ \cdot & & \\ 0 \dots & & \dots 0 \; 1 \end{bmatrix} \begin{matrix} 1 \\ \\ \\ \\ j-1 \\ j+1 \\ \\ \\ \\ n \end{matrix} \qquad j=1 \dots l \qquad (11.83)$$

11.3.2. Conditional Likelihoods from a Single Observation

Now the decision is made on the basis of a single observation $r(t)$. Thus in (11.73) $x(t)=r(t)$ and $\mu_x(t)=\mu_r(t)$. The simplified log-likelihood function becomes

$$logL[r(t), \mu_r(t)] = -1/2\, [r(t) - \mu_r(t)]' \, \Psi_r^{-1} \, [r(t) - \mu_r(t)] \qquad (11.84)$$

The solution depends on the nature of the geometric constraint.

Directional residuals. The constraint (11.78) may be substituted directly into (11.84), so that

$$logL[r(t), \mu_r(t) \mid H^j] = -1/2\, [r(t) - \boldsymbol{\beta}_j c_j(t)]' \, \Psi_r^{-1} \, [r(t) - \boldsymbol{\beta}_j c_j(t)] \qquad (11.85)$$

An estimate of $c_j(t)$ can be obtained from

$$\partial \, logL[r(t), \mu_r(t) \mid H^j] \, / \, \partial \, c_j(t) = \boldsymbol{\beta}_j' \, \Psi_r^{-1} \, [r(t) - \boldsymbol{\beta}_j c_j(t)] = 0 \qquad (11.86)$$

as

$$\hat{c}_j(t) = \boldsymbol{\beta}_j' \, \Psi_r^{-1} \, r(t) \, / \, (\boldsymbol{\beta}_j' \, \Psi_r^{-1} \boldsymbol{\beta}_j) \qquad j=1 \dots l \qquad (11.87)$$

Substituting this into (11.78) yields the conditional estimate of the mean as

$$\hat{\mu}_{rj}(t) = \Pi_{rj}^{D} \, r(t) \qquad j=1 \dots l \qquad (11.88)$$

where

$$\Pi_{rj}^{D} = \beta_j \beta_j' \Psi_r^{-1} / (\beta_j' \Psi_r^{-1} \beta_j) \qquad j=1\ldots l \tag{11.89}$$

Finally, substituting this back into (11.84) yields the conditional log-likelihood function (11.75) as

$$\boxed{logL_j(t) = -1/2\, r'(t)\, \Psi_r^{-1}\, [I - \Pi_{rj}^{D}]\, r(t)} \qquad j=1\ldots l \tag{11.90}$$

Here we utilized that

$$(\Pi_{rj}^{D})' \Psi_r^{-1} \Pi_{rj}^{D} = \frac{\Psi_r^{-1} \beta_j \beta_j' \Psi_r^{-1} \beta_j \beta_j' \Psi_r^{-1}}{\beta_j' \Psi_r^{-1} \beta_j \beta_j' \Psi_r^{-1} \beta_j} = \frac{\Psi_r^{-1} \beta_j \beta_j' \Psi_r^{-1}}{\beta_j' \Psi_r^{-1} \beta_j} = \Psi_r^{-1} \Pi_{rj}^{D} \tag{11.91}$$

Structured residuals. The geometric constraint (11.81) will be incorporated into the log-likelihood function with a Lagrange multiplier $\zeta(t)$:

$$logL[r(t), \mu_r(t) \mid H^j] = -1/2\, [r(t) - \mu_r(t)]'\, \Psi_r^{-1}\, [r(t) - \mu_r(t)] + \zeta'(t) T_j\, \mu_r(t) \tag{11.92}$$

The partial derivatives are (see Appendix 4A for definitions)

$$\partial\, logL[r(t), \mu_r(t) \mid H^j] / \partial\, \mu_r(t) = \Psi_r^{-1} [r(t) - \mu_r(t)] + T_j' \zeta(t) = 0$$

$$\partial\, logL[r(t), \mu_r(t) \mid H^j] / \partial\, \zeta(t) = T_j\, \mu_r(t) = 0 \tag{11.93}$$

yielding the solution

$$\hat{\mu}_{rj}(t) = \Pi_{rj}^{S}\, r(t) \qquad j=1\ldots l \tag{11.94}$$

where

$$\Pi_{rj}^{S} = I - \Psi_r T_j' (T_j \Psi_r T_j')^{-1} T_j \qquad j=1\ldots l \tag{11.95}$$

With this, the conditional log-likelihood function is

$$\boxed{logL_j(t) = -1/2\, r'(t)\, \Psi_r^{-1}\, [I - \Pi_{rj}^{S}]\, r(t)} \qquad j=1\ldots l \tag{11.96}$$

Here we utilized that

$$(\Pi_{rj}^{S})' \Psi_r^{-1} \Pi_{rj}^{S} = \Psi_r^{-1} \Pi_{rj}^{S} \tag{11.97}$$

Note finally that

$$\Psi_r^{-1} [I - \Pi_{rj}^{S}] = T_j' (T_j \Psi_r T_j')^{-1} T_j \tag{11.98}$$

Diagonal residuals. The conditional log-likelihood function can be obtained from (11.90) or (11.96), β_j or T_j substituted according to (11.82) or (11.83). The directional interpretation (11.89)-(11.82) leads to a particularly simple formulation.

11.3.3. Conditional Likelihoods from a Series of Observations

The test is now applied to a series $R(t)$ so that $x(t)=R(t)$ in (11.73). Further, $\mu_x(t)=\mu_R(t)$ but the relationship of this mean to $\mu_r(t)$ requires some consideration. Strictly speaking, $\mu_R(t)=[\mu_r'(t) \ \ \mu_r'(t-1) \dots \mu_r'(t-m)]'$ and the residual mean μ_r may vary over the time window. To avoid the difficulty this would pose in the log-likelihood function, a simplifying assumption will be made namely that the residual mean is constant over each window. That is,

$$\mu_R(t) = [\mu_r'(t) \ \ \mu_r'(t) \dots \mu_r'(t)]' = J \mu_r(t) \tag{11.99}$$

where

$$J = [I_n \ \ I_n \dots I_n]' \atop \quad 0 \quad 1 \qquad m \tag{11.100}$$

With this, the log-likelihood function (11.73) becomes

$$logL[R(t), \mu_r(t)] = -1/2 \, [R(t) - J\mu_r(t)]' \Psi_R^{-1} [R(t) - J\mu_r(t)] \tag{11.101}$$

When $\mu_r(t)$ is estimated from this approximate log-likelihood function, the time-varying residual mean is effectively approximated by its sliding window average. This is also time varying but follows the residual mean with some delay and smoothes out some of its possible variations.

Directional residuals. Substituting (11.78) into (11.101) yields

$$logL[R(t), \mu_r(t) \mid H^j] = -1/2 \, [R(t) - J\beta_j c_j(t)]' \Psi_R^{-1} [R(t) - J\beta_j c_j(t)] \tag{11.102}$$

The partial derivative with respect to $c_j(t)$ is

$$\partial \, logL[R(t), \mu_r(t) \mid H^j] / \partial \, c_j(t) = \beta_j' J' \Psi_R^{-1} [R(t) - J\beta_j c_j(t)] = 0 \tag{11.103}$$

leading to

$$\hat{c}_j(t) = \beta_j' J' \Psi_R^{-1} R(t) / (\beta_j' J' \Psi_R^{-1} J \beta_j) \qquad j=1...l \tag{11.104}$$

and

$$\hat{\mu}_{rj}(t) = \Pi_{Rj}^D R(t) \qquad j=1...l \tag{11.105}$$

where

$$\Pi_{Rj}^D = \beta_j \beta_j' J' \Psi_R^{-1} / (\beta_j' J' \Psi_R^{-1} J \beta_j) \qquad j=1...l \tag{11.106}$$

Finally, the conditional log-likelihood function (11.75) is obtained as

$$\boxed{logL_j(t) = -1/2 \, R'(t) \Psi_R^{-1} [I - J\Pi_{Rj}^D] R(t)} \qquad j=1...l \tag{11.107}$$

Structured residuals. Adding the constraint (11.81) to (11.101) with the Lagrange multiplier $\zeta(t)$:

$$logL[R(t), \mu_r(t) \mid H^j] = -1/2 \, [R(t) - J\mu_r(t)]' \Psi_R^{-1} [R(t) - J\mu_r(t)] + \zeta'(t) T_j \mu_r(t) \tag{11.108}$$

The partial derivatives are

$$\partial \, logL[R(t), \mu_r(t) \mid H^j] / \partial \, \mu_r(t) = J' \Psi_R^{-1} [R(t) - J\mu_r(t)] + T_j' \zeta(t) = 0$$

$$\partial \, logL[R(t), \mu_r(t) \mid H^j] / \partial \, \zeta(t) = T_j \mu_r(t) = 0 \tag{11.109}$$

leading to

$$\hat{\mu}_{rj}(t) = \Pi_{Rj}^S R(t) \qquad j=1...l \tag{11.110}$$

where

$$\Pi_{Rj}^S = [\Gamma - \Gamma T_j' (T_j \Gamma T_j')^{-1} T_j \Gamma] J' \Psi_R^{-1}$$

$$\Gamma = (J' \Psi_R^{-1} J)^{-1} \qquad j=1...l \tag{11.111}$$

With this, the conditional log-likelihood function is

$$logL_j(t) = -1/2\, R'(t)\, \Psi_R^{-1} [I - J\Pi_{Rj}^S]\, R(t) \qquad j=1...l \tag{11.112}$$

11.3.4. Conditional Likelihoods from the Window Average

Now $x(t)=\bar{r}(t)$ and $\Psi_x = \Psi_r$, as given in (11.60). Thus (11.73) becomes

$$logL[\bar{r}(t), \mu_r(t)] = -1/2\, [\bar{r}(t) - \mu_r(t)]'\, \Psi_r^{-1} [\bar{r}(t) - \mu_r(t)] \tag{11.113}$$

Here $\mu_r(t)$ has been directly substituted for $\mu_r(t)$; this implies that the residual mean at time t is being approximated by the mean of the average over the window ending at t. Formally, the log-likelihood function (11.113) resembles closely that for a single observation, (11.84). Thus the results of subsection 11.3.2 can and will be directly adopted to the window average case.

Directional residuals.

$$\hat{\mu}_{rj}(t) = \Pi_{rj}^D\, \bar{r}(t) \qquad j=1...l \tag{11.114}$$

$$\Pi_{rj}^D = \beta_j \beta_j' \Psi_r^{-1} / (\beta_j' \Psi_r^{-1} \beta_j) \qquad j=1...l \tag{11.115}$$

$$logL_j(t) = -1/2\, \bar{r}'(t)\, \Psi_r^{-1}\, [I - \Pi_{rj}^D]\, \bar{r}(t) \qquad j=1...l \tag{11.116}$$

Structured residuals.

$$\hat{\mu}_{rj}(t) = \Pi_{rj}^S\, \bar{r}(t) \qquad j=1...l \tag{11.117}$$

$$\Pi_{rj}^S = I - \Psi_r T_j' (T_j \Psi_r T_j')^{-1} T_j \qquad j=1...l \tag{11.118}$$

$$logL_j(t) = -1/2\, \bar{r}'(t)\, \Psi_r^{-1}\, [I - \Pi_{rj}^S]\, \bar{r}(t) \qquad j=1...l \tag{11.119}$$

$$\Psi_r^{-1}\, [I - \Pi_{rj}^S] = T_j' (T_j \Psi_r T_j')^{-1} T_j \tag{11.120}$$

Example 11.7.

For the directional residual generator designed in Example 11.1. and analyzed in Example 11.2. and 11.4., the Π_{xj}^D matrices under the three testing strategies are:

$$\Pi^D_{r1} = \begin{bmatrix} 0.5399 & 0.4604 \\ 0.5399 & 0.4604 \end{bmatrix} \qquad \Pi^D_{r2} = \begin{bmatrix} 1.6649 & 0.6649 \\ -1.6649 & -0.6649 \end{bmatrix}$$

$$\Pi^D_{R1} = \begin{bmatrix} -0.0201 & 0.0568 & 0.1772 & 0.2389 & 0.2881 & 0.2577 \\ -0.0201 & 0.0568 & 0.1772 & 0.2389 & 0.2881 & 0.2577 \end{bmatrix}$$

$$\Pi^D_{R2} = \begin{bmatrix} 0.7602 & 0.2981 & -0.2680 & -0.7308 & -0.9056 & -0.9806 \\ -0.7602 & -0.2981 & 0.2680 & 0.7308 & 0.9056 & 0.9806 \end{bmatrix}$$

$$\Pi^D_{F1} = \begin{bmatrix} 0.4897 & 0.5097 \\ 0.4897 & 0.5097 \end{bmatrix} \qquad \Pi^D_{F2} = \begin{bmatrix} 0.2895 & -0.7105 \\ -0.2895 & 0.7105 \end{bmatrix}$$

11.3.5. Rank-Defects of the Covariance Matrix

All the test statistics used in detection and isolation involve the inverse of some form of the residual covariance matrix. Obviously, if the covariance matrix is singular then the statistics cannot be computed. We will explore the rank properties of the various covariance matrices (single observation, series of observations, window average) in this subsection. As it will be shown, the requirement of full rank places a restriction on the number of residuals and on the possible polynomial relations among them, and poses a persistent excitation condition on the primary residuals. It will also be seen that tests based on a series of observations of structured residual-sets are particularly prone to rank defect.

Single observation. Write the covariance matrix for a single observation of a vector residual as

$$\Psi_r = \lim_{K\to\infty} \frac{1}{K} \mathbf{\Lambda}'_r(t, K)\, \mathbf{\Lambda}_r(t, K) \tag{11.121}$$

where $\mathbf{\Lambda}'_r(t, K)$ is the regression matrix

$$\mathbf{\Lambda}'_r(t, K) = \begin{bmatrix} r_1(t) & r_1(t-1) & \dots & r_1(t-K+1) \\ \cdot & & & \\ \cdot & & & \\ \cdot & & & \\ r_n(t) & r_n(t-1) & \dots & r_n(t-K+1) \end{bmatrix} \tag{11.122}$$

Clearly,

$$Rank\, \Psi_r = \lim_{K \to \infty} Rank\, \Lambda_r(t, K) \tag{11.123}$$

Consider the first column of $\Lambda'_r(t, K)$. If there is a direct (non-polynomial) linear relationship among the elements $r_1(t) \dots r_n(t)$ then the same relationship applies to all of the other columns, that is, it exists among the full rows, leading to a rank defect. Recall that $r(t)$ is computed from the primary residuals, and assume first that the transformation $W(\phi)$ is polynomial with degree σ. Then

$$r(t) = W(\phi)o(t) = W^0 o(t) + W^1 o(t-1) + \dots + W^\sigma o(t-\sigma) \tag{11.124}$$

Thus each element of $r(t)$ is a linear combination of the $\mu \cdot (\sigma+1)$ scalar primary residual observations $o_1(t), \dots o_\mu(t); \dots \dots o_1(t-\sigma), \dots o_\mu(t-\sigma)$, where μ is the number of independent plant outputs. Now for the n elements of $r(t)$ to be independent, they must be composed of at least n independent observations. This implies the following two requirements:

(i) $n \leq \mu(\sigma+1)$ (11.125)

(ii) of the $\mu \cdot (\sigma+1)$ primary residual observations, at least n must be independent.

Noting that the μ elements of $o(t)$ at any sample are independent, (ii) boils down to a persistent excitation type condition on the time series $o_i(t)$, $i=1 \dots \mu$. This translates into similar persistent excitation requirements for the noise $v(t)$ which drives the primary residuals.

If $W(\phi)$ is rational then it is equivalent to a polynomial transformation with $\sigma \to \infty$. Then Condition (i) is removed but (ii) still stands.

The following is an alternative way to derive the above conditions. Write

$$r(t) = W\, O(t) \tag{11.126}$$

where

$$W = [W^0 \quad W^1 \dots W^\sigma] \tag{11.127}$$

$$O(t) = [o'(t) \quad o'(t-1) \dots o'(t-\sigma)]' \tag{11.128}$$

Then

$$\Lambda'_r(t, K) = W\, \Omega'(t, K) \tag{11.129}$$

where

$$\Omega'(t, K) = [O(t) \quad O(t-1) \quad \ldots \quad O(t-K+1)] \tag{11.130}$$

Now

$$Rank\, \Lambda'_r(t, K) \leq min[Rank\, W,\ Rank\, \Omega'(t, K)] \tag{11.131}$$

With this, Condition (i) corresponds to $Rank W=n$ while (ii) means $Rank\Omega(t, K) \geq n$. (11.131) also reveals that (i) and (ii) are necessary conditions but they do not guarantee that $\Lambda'_r(t, K)$ has full rank.

Series of observations. Write the covariance matrix as

$$\Psi_R = \lim_{K\to\infty} \frac{1}{K} \Lambda'_R(t, K)\, \Lambda_R(t, K) \tag{11.132}$$

where

$$\Lambda'_R(t, K) = \begin{bmatrix} \Lambda'_r(t, K) \\ \cdot \\ \cdot \\ \cdot \\ \Lambda'_r(t-m, K) \end{bmatrix} \tag{11.133}$$

Now the first column of $\Lambda'_R(t, K)$ contains $n\cdot(m+1)$ elements and relies on $(m+\sigma+1)$ observations of the vector $o(t)$, amounting to a total of $\mu\cdot(m+\sigma+1)$ scalar observations. Thus, in this case, Conditions (i) and (ii) are

$$\text{(i)} \quad n \leq \mu \frac{m+\sigma+1}{m+1} \tag{11.134}$$

(ii) at least $n\cdot(m+1)$ of the $\mu\cdot(m+\sigma+1)$ primary residual observations have to be independent.

Note that (11.134) is stricter than (11.125) whenever $m>0$.

There is a third reason, not encountered in the single sample case, why $\Lambda'_R(t, K)$ may lose rank. Assume that there is a polynomial linear relationship among the elements of the single residual sample:

$$\alpha_1(\phi)r_1(t) + \ldots + \alpha_n(\phi)r_n(t) = [\alpha^0+\alpha^1\phi^{-1}+\ldots+\alpha^k\phi^{-k}]'r(t) = 0 \tag{11.135}$$

Note that if there is a linear relationship, it can always be put in polynomial from even if $W(\phi)$ is rational. If there exists such a relationship, it makes a group of rows in $\Lambda'_R(t, K)$ linearly dependent, provided the entire range of the relationship, $k+1$, is within the data window $m+1$. Of course, one does not intend the residuals to be linearly dependent, but if $n>\mu$, which is usually the case with structured design, then polynomial linear dependence follows. This leads to the third criterion for the full rank of $\mathbf{\Lambda}'_R(t, K)$:

(iii) if $n>\mu$ then $m<k$ must be assured.

Window average. Write the covariance matrix as

$$\Psi_r = \lim_{K\to\infty} \frac{1}{K} \mathbf{\Lambda}'_r(t, K)\, \mathbf{\Lambda}_r(t, K) \tag{11.136}$$

where

$$\mathbf{\Lambda}'_r(t, K) = \frac{1}{m+1} \sum_{\tau=0}^{m} \mathbf{\Lambda}'_r(t-\tau, K) \tag{11.137}$$

Now there are n elements in the first column which rely on $\mu\cdot(m+\sigma+1)$ scalar observations of the primary residuals. Thus the necessary conditions for full rank are

(i) $n \le \mu(m+\sigma+1)$ (11.138)

(ii) of the $\mu\cdot(m+\sigma+1)$ observations, at least n must be independent.

Of the three approaches, clearly the window average poses the mildest conditions for the existence of the test statistics. The conditions are the most difficult to satisfy if a series of observations is used, especially when the residuals are designed as a structured set.

11.3.6. Test Implementation

The purpose of the isolation test is to decide which of the possible faults is present. Usually, it is activated once the detection test has indicated the presence of *some* fault. The isolation test is different from the detection test in two important aspects:

- the test is symmetrical and therefore the concepts of false alarm and missed detection do not apply;
- the choice is usually between a multitude of possible faults, in contrast to the simple detection choice of fault/no fault.

Straight comparison of likelihoods. The basic likelihood ratio concept calls for the straight comparison of conditional likelihoods obtained under the various hypotheses. The comparison may be performed pairwise, considering two hypotheses at a time, or globally, considering all. Using the conditional log-likelihood functions, the pairwise decision rule is

$$if\ logL_i(t) > logL_j(t)\ \ then\ \ H^i\ \ is\ accepted\ over\ \ H^j \tag{11.139}$$

The probability of two log-likelihood functions being equal is zero but numerically this may still happen due to limited resolution; in this case the outcome may be *"undecided."* The global decision rule is

$$\boxed{if\ \ logL_i(t) = max\,[logL_j(t),\ j=1...l\,]\ \ then\ \ H^i\ \ is\ accepted} \tag{11.140}$$

Note that $[r(t) - \mu_r(t)]'\,\Phi_r^{-1}\,[r(t) - \mu_r(t)]$, and its equivalents for $\bar{r}(t)$ and $R(t)$, are non-negative numbers which are basically distance measures. A conditional estimate $\hat{\mu}_{ri}(t)$ is more likely than another one, $\hat{\mu}_{rj}(t)$, if it is "closer" to the observations. These distance measures, however, appear with negative sign in the conditional log-likelihood functions, resulting in the somewhat counter-intuitive decision rules.

Apart from the rare case of equal likelihoods, the above decision rules always declare a clear single suspect. In reality, however, the presence of noise may make the isolation outcome uncertain, especially if the effect of the fault on the residuals is small compared to that of the noise. To improve the robustness of the procedure, the straight comparison of conditional likelihood values needs to be replaced or supplemented with a test of plausibility. With this, the test will return an explicit statement of ambiguity when the situation does not allow a sufficiently reliable clear isolation decision.

Plausibility test for pairwise comparison. Let us address first the pairwise comparison of likelihood values. Consider the conditional log-likelihood ratio

$$W_{ij}(t) = logL_i(t) - logL_j(t) \tag{11.141}$$

Observe that this is a scalar statistic. From (11.73) and (11.75), with x representing r, R and $\bar{r}$, this can be written as

$$\begin{aligned} W_{ij}(t) & \qquad\qquad\qquad\qquad\qquad\qquad\qquad\qquad\qquad\qquad\qquad\qquad (11.142) \\ &= -\tfrac{1}{2}[x(t) - \hat{\mu}_{xi}(t)]'\,\Phi_x^{-1}\,[x(t) - \hat{\mu}_{xi}(t)] + \tfrac{1}{2}[x(t) - \hat{\mu}_{xj}(t)]'\,\Phi_x^{-1}\,[x(t) - \hat{\mu}_{xj}(t)] \\ &= [\hat{\mu}_{xi}(t) - \hat{\mu}_{xj}(t)]'\,\Phi_x^{-1}\,x(t) - \tfrac{1}{2}\hat{\mu}_{xi}'(t)\,\Phi_x^{-1}\,\hat{\mu}_{xi} + \tfrac{1}{2}\hat{\mu}_{xj}'(t)\,\Phi_x^{-1}\,\hat{\mu}_{xj}(t) \end{aligned}$$

where $\hat{\mu}_x = \hat{\mu}_r$ if the computations are based on a single observation or window average while $\hat{\mu}_x = J\hat{\mu}_r$ if a full window is used.

When evaluating the conditional likelihood ratio, we will consider the

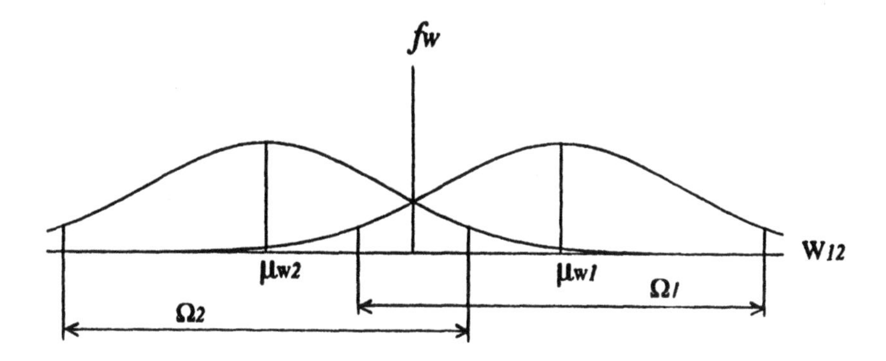

FIGURE 11.6.a. Overlapping plausibility regions.

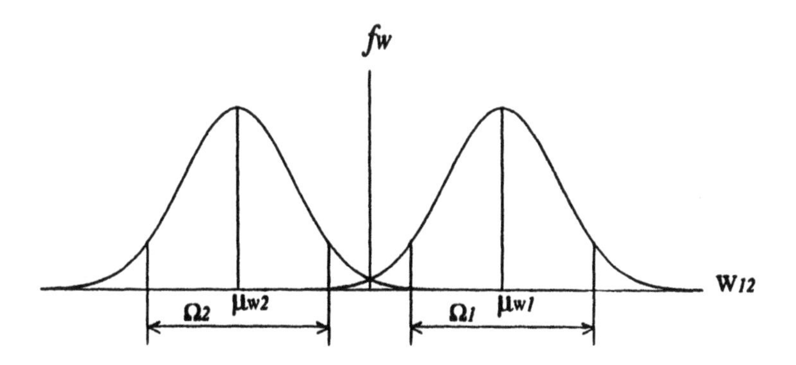

FIGURE 11.6.b. Non-overlapping plausibility regions.

conditional mean estimates as known deterministic quantities (which, of course, is an approximation). Then the only random variable on the right-hand side of (11.142) is $x(t)$. Thus if $x(t)$ is normally distributed, $W_{ij}(t)$ is also a normally distributed random variable. Its mean depends on which hypothesis is assumed valid; under H^i, $\mu_x(t)=\hat{\mu}_{xi}(t)$ and

$$\mu_{Wi}(t) = \mathrm{E}\{W_{ij}(t|H^i)\} = \tfrac{1}{2}[\hat{\mu}_{xi}(t) - \hat{\mu}_{xj}(t)]'\,\Phi_x^{-1}\,[\hat{\mu}_{xi}(t) - \hat{\mu}_{xj}(t)] \quad (11.143)$$

while under H^j, $\mu_x(t)=\hat{\mu}_{xj}(t)$ and

$$\mu_{Wj}(t) = \mathrm{E}\{W_{ij}(t|H^j)\} = -\,\mu_{Wi}(t) \quad (11.144)$$

The variance of $W_{ij}(t)$ in either case is obtained as

$$\sigma_W^2(t) = [\hat{\mu}_{xi}(t) - \hat{\mu}_{xj}(t)]'\,\Phi_x^{-1}\,[\hat{\mu}_{xi}(t) - \hat{\mu}_{xj}(t)] \quad (11.145)$$

The plausibility test is then conducted as follows. We choose a test size α; with this, regions of plausibility Ω_i and Ω_j are defined for H^i and H^j, in the W_{ij} domain, according to the distributions $N[\mu_{Wg}(t), \sigma_W(t)]$, $g=i, j$. The two plausibility regions may overlap (Fig. 11.6). The decision rule for each plausibility test is

$$\text{if } |W_{ij}(t) - \mu_{Wg}(t)| \leq \sigma_W(t)\, \kappa_\alpha \; (g=i, j) \text{ then } H^g \text{ is plausible} \tag{11.146}$$

where κ_α is the threshold of the standard normal distribution for the test size α. The overall decision may have the following outcomes:

$$\left.\begin{array}{l} \text{both } H^i \text{ and } H^j \text{ are plausible} \\ \text{only } H^i \text{ is plausible} \\ \text{only } H^j \text{ is plausible} \\ \text{neither } H^i \text{ nor } H^j \text{ is plausible} \end{array}\right] \tag{11.147}$$

Note that this test is easy to perform since $W_{ij}(t)$ is a scalar variable. However, if there is a large number of potential faults, pairwise comparisons may not be practical.

Plausibility test for global comparison. Now consider the following statistics:

$$w_j(t) = -2 \log L_j(t) = [x(t) - \hat{\mu}_{xj}(t)]' \, \Phi_x^{-1} \, [x(t) - \hat{\mu}_{xj}(t)], \quad j=1,2,\ldots \tag{11.148}$$

Under the H^j hypothesis, $\mu_x(t)=\hat{\mu}_{xj}(t)$. Thus, if $x(t)$ is normal, $w_j(t)$ follows χ^2 distribution, with degrees of freedom ν to be determined below. For a selected test size α, the test is performed against the threshold $\chi^2_{\nu,\alpha}$ as

$$\text{if } w_j(t) \leq \chi^2_{\nu,\alpha} \; (j=1,2,\ldots) \text{ then } H^j \text{ is plausible} \tag{11.149}$$

Plausibility regions Ω_j can be assigned in the original x domain; these may be overlapping (Fig. 11.7). The possible outcomes of the full set of plausibility tests are

$$\left.\begin{array}{l} \text{more than one of the hypotheses is plausible} \\ \text{only one of the hypotheses is plausible} \\ \text{none of the hypotheses is plausible} \end{array}\right] \tag{11.150}$$

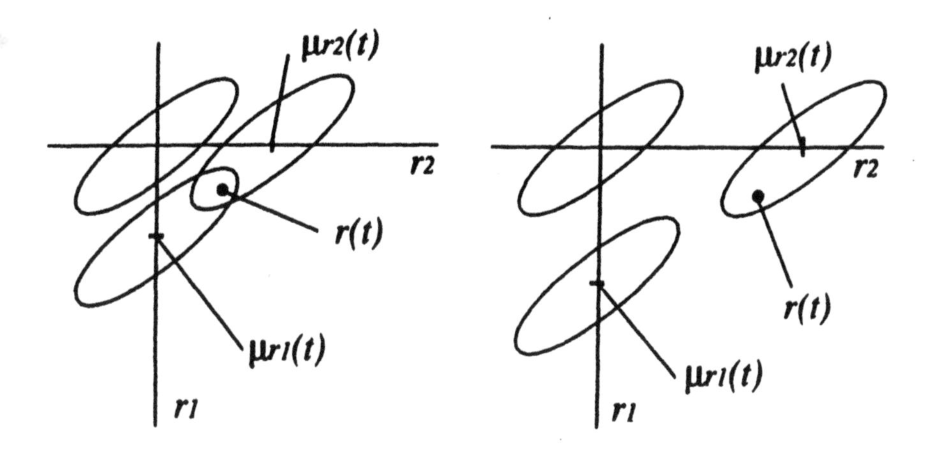

FIGURE 11.7.a. Overlapping regions. **FIGURE 11.7.b.** Non-overlapping regions.

It remains to determine ν, the degrees of freedom in the χ^2 distribution. Recall that ν is the number of independent elements in the transformed variable vector $\Phi_x^{-1/2}[x(t) - \mu_x(t)]$ (c.f. Eq. (11.35), there with $\mu_x(t)=0$). The number of elements in x is n if $x=r$ or $x=\tilde{r}$ and $n\cdot(m+1)$ if $x=R$. However, the degrees of freedom are reduced by linear constraints applied to the observations. In our case it is the computation of the conditional means that imposes such constraints. If the residuals are subject to directional design, only one scalar is computed under each hypothesis, so the number of constraints is 1. With structured design, the number of elements computed in $\hat{\mu}_j$ is $n - \tau_j$, where τ_j is the number of rows in the template matrix T_j (each row setting an element of $\hat{\mu}_j$ to identically zero).

One may, of course, draw the same conclusion from the equations (11.88), (11.94), (11.105), (11.110), (11.114), (11.117), which are the actual constraints imposed on the observations by computing the estimates. In this context, the above statements on the number of constraints are equivalent to the rank properties

$$Rank\ \Pi_{xj}^{D} = 1 \quad for\ all\ j \tag{11.151a}$$

$$Rank\ \Pi_{xj}^{S} = n - \tau_j \tag{11.151b}$$

In summary, the degrees of freedom in the various situations are:

$\nu =$	directional	structured
$x=r,\ x=\tilde{r}$	$n-1$	τ_j
$x=R$	$m \cdot n+n-1$	$m \cdot n+\tau_j$

(11.152)

It should be noted that the above techniques of plausibility testing, just as well as the direct comparison of conditional likelihoods, ignores the fact that the conditional estimates of the mean are also random variables. This decomposition of the GLR procedure into separate estimation and testing stages keeps the analysis technically manageable. The uncertainty present in the conditional estimates will be considered in the next subsection. However, to take this into account in the test, the latter should be based on the conditional likelihood statistic expressed entirely in terms of the observations (as in (11.90), etc.), which would involve more complicated distributions.

Example 11.8.

Consider again the residual generator designed, for two directional responses, in Example 11.1. Recall that the response directions were $\beta_1=[1 \quad 1]'$ and $\beta_2=[1 \quad -1]'$. The covariance matrices were computed in Example 11.2. and 11.4., while the transformations Π^D_{xj} were obtained in Example 11.7.

Case a. Assume that an isolation decision is made from a single observation of the residual which is

$$r(t) = [\, 1 \quad 0.5 \,]'$$

The conditional mean estimates are obtained from (11.88) as

$$\hat{\mu}_{r1}(t) = [0.7701 \quad 0.7701]' \qquad \hat{\mu}_{r2}(t) = [1.9974 \quad -1.9974]'$$

The direct comparison of the conditional likelihoods, by (11.141) and (11.142), yields

$$W_{12}(t) = 1.7203$$

which seems to clearly point at the first fault. Let us now perform the pairwise plausibility analysis. From (11.143)-(11.145),

$$\mu_{W1}(t) = 1.8289 \qquad \mu_{W2}(t) = -1.8289 \qquad \sigma_W(t) = 1.9125$$

Choosing the test size $\alpha=0.05$, one obtains

$$\kappa_{0.05} = 1.96 \qquad \kappa_{0.05}\, \mu_W(t) = 3.7485$$

Thus the plausibility regions are:

$$\Omega_1 : -1.9196 \leq W_{12}(t) \leq 5.5774 \qquad \Omega_2 : -5.5774 \leq W_{12}(t) \leq 1.9196$$

The two regions obviously overlap and, with the observed $W_{12}(t)=1.7203$, either fault hypothesis is plausible. Changing the test size to $\alpha=0.1$ would yield

$$\kappa_{0.1} = 1.645 \qquad \kappa_{0.1}\,\mu_W(t) = 3.1461$$

$$\Omega_1 : -1.3172 \le W_{12}(t) \le 4.9750 \qquad \Omega_2 : -4.9750 \le W_{12}(t) \le 1.3172$$

where the regions still overlap but only the first fault is plausible.

Similar conclusions can be drawn from the global analysis. By (11.148)

$$w_1(t) = 0.0131 \qquad w_2(t) = 3.4600$$

The degrees of freedom is 1. With $\alpha=0.05$, the threshold is $\chi^2_{1,\,0.05}=3.841$ and both fault hypotheses are plausible. However, with $\alpha=0.1$, the threshold becomes $\chi^2_{1,\,0.1} = 2.706$ which leaves only the first hypothesis plausible.

Case b. We will repeat the above analysis for the case when, in the same system, the decision is based on the average of three observations. Assume that

$$\bar{r}(t) = [1 \quad 0.5]'$$

With the appropriate covariance and transformation matrices,

$$\hat{\mu}_{r1}(t) = [0.7445 \quad 0.7445]' \qquad \hat{\mu}_{r2}(t) = [-0.0658 \quad 0.0658]'$$

Now

$$W_{12}(t) = 5.8900$$

which strongly favors the first fault. The pairwise plausibility analysis yields

$$\mu_{W1}(t) = 5.8663 \qquad \mu_{W2}(t) = -5.8663 \qquad \sigma_W(t) = 3.4253$$

Choosing again the test size $\alpha=0.05$, one obtains $\kappa_{0.05}\,\mu_W(t) = 6.7136$ and

$$\Omega_1 : -0.8473 \le W_{12}(t) \le 12.5799 \qquad \Omega_2 : -12.5799 \le W_{12}(t) \le 0.8473$$

The two regions still overlap but now only the first fault is plausible. The global analysis yields

$$w_1(t) = 0.0624 \qquad w_2(t) = 11.8424$$

of which only the first one is inside the threshold $\chi^2_{1,\,0.05}=3.841$. Clearly, *using a window average instead of a single observation* (even if it is just a window of three observations) *improves the reliability of the test.*

11.3.7. Properties of the Conditional Estimates

We will compute here the expectation and the variance of the conditional mean estimate for the three cases (single observation, series of observations and window average). Recall that the residual mean and its conditional estimates are functions of time. Thus expectation in this context is not meant over a time series but, for a given instant of time, over various (noise induced) realizations of the residual.

In general,

$$\hat{\mu}_{rj}(t) = \Pi_{xj}\, x(t) \tag{11.153}$$

where $x(t)$ stands for $r(t)$, $\bar{r}(t)$ and $R(t)$ and Π_{xj} stands for Π^D_{xj} and Π^S_{xj}. Then

$$\mathrm{E}\{\hat{\mu}_{rj}(t)\} = \Pi_{xj}\, \mathrm{E}\{x(t)\} \tag{11.154}$$

and

$$Var\{\hat{\mu}_{rj}(t)\} = \mathrm{E}\{[\hat{\mu}_{rj}(t) - \mathrm{E}\{\hat{\mu}_{rj}(t)\}]\, [\hat{\mu}_{rj}(t) - \mathrm{E}\{\hat{\mu}_{rj}(t)\}]'\}$$

$$= \Pi_{xj}\, \mathrm{E}\{[x(t) - \mathrm{E}\{x(t)\}]\, [x(t) - \mathrm{E}\{x(t)\}]'\}\, \Pi'_{xj} = \Pi_{xj}\, \Psi_x\, \Pi'_{xj} \tag{11.155}$$

Single observation. Now $x(t)=r(t)$ and

$$\mathrm{E}\{r(t)\} = \mu_r(t) \tag{11.156}$$

Further, with (11.89) and (11.95),

$$\Pi^D_{rj}\, \Psi_r\, (\Pi^D_{rj})' = \beta_j \beta_j' / (\beta_j'\, \Psi_r^{-1} \beta_j) \tag{11.157}$$

$$\Pi^S_{rj}\, \Psi_r\, (\Pi^S_{rj})' = [I - \Psi_r\, T_j'\, (T_j'\, \Psi_r\, T_j)^{-1}\, T_j]\, \Psi_r \tag{11.158}$$

Series of observations. Now $x(t)=R(t)$ and

$$\mathrm{E}\{R(t)\} = [\mu_r'(t) \;\; \mu_r'(t-1) \ldots \mu_r'(t-m)]' = \mathbf{M}(t) \tag{11.159}$$

where $\mathbf{M}(t)$ is as defined here. Also, with (11.106) and (11.111),

$$\Pi^D_{Rj}\, \Psi_R\, (\Pi^D_{Rj})' = \beta_j \beta_j' / (\beta_j'\, J'\, \Psi_R^{-1} J \beta_j) \tag{11.160}$$

$$\Pi^S_{Rj}\, \Psi_R\, (\Pi^S_{Rj})' = [I - \Gamma\, T_j'\, (T_j'\, \Gamma\, T_j)^{-1}\, T_j]\, \Gamma \tag{11.161}$$

where $\Gamma=(J'\, \Psi_R^{-1} J)^{-1}$, as defined in (11.111).

Window average. Now $x(t)=\bar{r}(t)$. Observe that

$$\bar{R}(t) = J'\, R(t) \,/\, (m+1) \tag{11.162}$$

Thus

$$E\{\bar{R}(t)\} = J'\, \mathbf{M}(t) \,/\, (m+1) \tag{11.163}$$

Further, with (11.115) and (11.118),

$$\Pi^{D}_{\bar{r}j}\, \Psi_{\bar{r}}\, (\Pi^{D}_{\bar{r}j})' = \beta_j\, \beta_j' \,/\, (\beta_j'\, \Psi_{\bar{r}}^{-1}\, \beta_j\,) \tag{11.164}$$

$$\Pi^{S}_{\bar{r}j}\, \Psi_{\bar{r}}\, (\Pi^{S}_{\bar{r}j})' = [I - \Psi_{\bar{r}}\, T_j'\, (T_j'\, \Psi_{\bar{r}}\, T_j)^{-1}\, T_j]\, \Psi_{\bar{r}} \tag{11.165}$$

Observe that, in general,

$$\Pi_{Rj} \neq \Pi_{\bar{r}j}\, J' \,/\, (m+1) \tag{11.166}$$

Thus the conditional estimate $\hat{\mu}_{rj}(t)$ computed from the window average $\bar{R}(t)$ is different from that computed from the full series $R(t)$, and so are its expectation and variance. This is because, for a correlated series, the window average is not a sufficient statistic for the estimation of the mean.

However, if the residual series is uncorrelated then $\Psi_{rr}(\tau) = 0$ for $\tau \neq 0$. Thus, from (11.60),

$$\Psi_{\bar{r}} = \Psi_{rr}(0) \,/\, (m+1) = \Psi_r \,/\, (m+1) \tag{11.167}$$

Also,

$$\Psi_R^{-1} = (Diag[\Psi_r \;\; \Psi_r \ldots \Psi_r])^{-1} = Diag[\Psi_r^{-1} \;\; \Psi_r^{-1} \;\ldots\; \Psi_r^{-1}] \tag{11.168}$$

and

$$\Gamma = [J'\, \Psi_R^{-1}\, J]^{-1} = [(m+1)\Psi_r^{-1}]^{-1} = \Psi_r \,/\, (m+1) \tag{11.169}$$

With these, and (11.89), (11.95), (11.106), (11.111), (11.115) and (11.118),

$$\Pi_{\bar{r}j} = \Pi_{rj} \qquad and \qquad \Pi_{Rj} = \Pi_{rj}\, J' \,/\, (m+1) \tag{11.170}$$

where again Π_{xj} stands for both Π^{D}_{xj} and Π^{S}_{xj}. Further, with (11.157), (11.158), (11.160), (11.161), (11.164) and (11.165)

$$\Pi_{Rj}\, \Psi_R\, (\Pi_{Rj})' = \Pi_{\bar{r}j}\, \Psi_{\bar{r}}\, (\Pi_{\bar{r}j})' = \Pi_{rj}\, \Psi_r\, (\Pi_{rj})' / (m+1) \tag{11.171}$$

Thus we can conclude that, if the residual series is uncorrelated, then

- working with the full series $R(t)$ or with its window average $\bar{R}(t)$ lead to identical estimates $\hat{\mu}_{rj}(t)$ (and thus to the same expectation and same variance);

- the variance of the estimate obtained with the full series or its window mean is $1/(m+1)$ times the variance obtained with a single observation.

In the general case, when the residual series is correlated, the above simple conclusions do not apply. However, two quite obvious rules of thumb may be suggested:

1. By using a series of observations instead of a single one, the variance of the estimates is expected to decrease (usually) significantly;
2. By representing the full series with its window average, the variance of the estimates is expected to increase somewhat.

Considering also the computational complexity of the various approaches, working with the window average seems to be the most reasonable choice.

Example 11.9.

For the directional residual generator designed in Example 11.1. and analyzed in Example 11.2., the variances of the conditional estimate under the three test strategies are:

$$\Pi_{r1}^{D} \Psi_{r} (\Pi_{r1}^{D})' = 0.1325 \begin{bmatrix} 1 & 1 \\ 1 & 1 \end{bmatrix} \qquad \Pi_{r2}^{D} \Psi_{r} (\Pi_{r2}^{D})' = 3.8815 \begin{bmatrix} 1 & -1 \\ -1 & 1 \end{bmatrix}$$

$$\Pi_{R1}^{D} \Psi_{R} (\Pi_{R1}^{D})' = 0.0280 \begin{bmatrix} 1 & 1 \\ 1 & 1 \end{bmatrix} \qquad \Pi_{R2}^{D} \Psi_{R} (\Pi_{R2}^{D})' = 0.4729 \begin{bmatrix} 1 & -1 \\ -1 & 1 \end{bmatrix}$$

$$\Pi_{f1}^{D} \Psi_{f} (\Pi_{f1}^{D})' = 0.0471 \begin{bmatrix} 1 & 1 \\ 1 & 1 \end{bmatrix} \qquad \Pi_{f2}^{D} \Psi_{f} (\Pi_{f2}^{D})' = 0.9932 \begin{bmatrix} 1 & -1 \\ -1 & 1 \end{bmatrix}$$

- Clearly, the variances decrease significantly when the estimates are based on a series rather than a single observation (recall that the example assumes a very short series, with a longer one the improvement would be more pronounced).
- The estimates which use the window average rather than the full series have somewhat increased variances.
- The variances in the second fault response direction are much higher than in the first, reflecting the fact that, in this particular system, the distribution is strongly elongated in the second response direction.

11.4. WHITENING FILTER

For single (scalar) elements of the residual, a filter can be so designed that, in the presence of no faults, the resulting filtered sequence is uncorrelated. Such filter is usually referred to as the *whitening filter*. Uncorrelated sequences are then easier to test than correlated ones.

Note that individual whitening filters cannot be applied to directional residuals because they spoil the geometric properties of the residual vector. In case of structured residuals, individual filters do not interfere with the structure but they provide whitening only for the scalar time sequences and do not remove the correlation among the different elements of the residual vector. It is possible to design a matrix filter which whitens the complete residual vector, this procedure however is quite complex and, to our understanding, does not allow for the preservation of the geometric residual properties. Therefore, here we will restrict ourselves to the scalar whitening problem.

In the following, we will outline the procedure to design the whitening filter. Equation (11.68) describes the filtered noise-induced residual. Let us denote now $q_{i.}(\phi)$ as $q'(\phi)$ where

$$q'(\phi) = [q_1(\phi) \; \dots \; q_l(\phi)] = [q^0 + q^1\phi^{-1} + \dots + q^\sigma \phi^{-\sigma}]' \qquad (11.172)$$

Now recall (3.169) and apply it to this case (omitting the subscript i)

$$\psi_{r\#r\#}(\tau) = k(\phi)\, \frac{q'(\phi)}{s(\phi)}\, \Psi_{vv}(\tau)\, \frac{q(\phi^{-1})}{s(\phi^{-1})}\, k(\phi^{-1}) \qquad (11.173)$$

For the white input noise, $\Psi_{vv}(\tau)=\delta(\tau)I$ (unit matrix for $\tau=0$, zero otherwise). For the filtered residual to be white (with unit variance), $\psi_{r\#r\#}(\tau)=\delta(\tau)$ is required. Thus the condition for whitening is

$$k(\phi)\, \frac{q'(\phi)}{s(\phi)}\, \frac{q(\phi^{-1})}{s(\phi^{-1})}\, k(\phi^{-1}) = 1 \qquad (11.174)$$

The scalar product of the polynomial vectors $q'(\phi)$ and $q(\phi^{-1})$ can be replaced with the product of the scalar polynomials $\gamma(\phi)$ and $\gamma(\phi^{-1})$, where

$$\gamma(\phi) = \gamma^0 + \gamma^1\phi^{-1} + \dots + \gamma^\sigma\phi^{-\sigma} \qquad (11.175)$$

so that

$$q'(\phi)q(\phi^{-1}) = \gamma(\phi)\gamma(\phi^{-1}) \qquad (11.176)$$

Equation (11.176), expanded by the powers of ϕ, provides $\sigma+1$ conditions for the same number of parameters in $\gamma(\phi)$ (the conditions for ϕ^{-j} and ϕ^{j} are identical). The conditions are nonlinear in the parameters but have an almost triangular structure which allows for a relatively simple solution. This replacement operation is a special case of the procedure known as *spectral factorization*.

Obviously, (11.174) can be satisfied if

$$k(\phi)\,k(\phi^{-1}) = \frac{s(\phi)}{\gamma(\phi)}\;\frac{s(\phi^{-1})}{\gamma(\phi^{-1})} \tag{11.177}$$

However, the filter $k(\phi)$ must be stable (all poles inside the unit circle); it is also desirable to make it minimum phase (all zeroes inside the unit circle). This will need further consideration. Assume first that no root of $\gamma(\phi)$ or $s(\phi)$ is on the unit circle. Expand $\gamma(\phi)/s(\phi)$ as

$$\frac{\gamma(\phi)}{s(\phi)} = \frac{c\,\phi^{-d}\,\gamma_S(\phi)\,\gamma_U(\phi)}{s_S(\phi)\,s_U(\phi)} \tag{11.178}$$

where $\gamma_S(\phi)$ and $s_S(\phi)$ have all their roots inside the unit circle while $\gamma_U(\phi)$ and $s_U(\phi)$ have all their roots outside and none of them has roots in the origin. Now observe that

$$(1 - \alpha\,\phi^{-1})\Big|_{\phi \leftarrow \phi^{-1}} = (1 - \alpha\,\phi) = -\,\alpha\,\phi\,(1 - \frac{1}{\alpha}\phi^{-1}) \tag{11.179}$$

that is, an unstable root in $k(\phi)$ is a stable root in $k(\phi^{-1})$, and vice versa. Thus (11.177) can be satisfied, and the filter made stable and minimum phase, if all roots of $\gamma(\phi)$ and $s(\phi)$, which are inside the unit circle, are assigned to $k(\phi)$ while the roots outside the unit circle are assigned to $k(\phi^{-1})$. Thus

$$\boxed{k(\phi) = \frac{s_S(\phi)\,s_U(\phi^{-1})}{c\,\gamma_S(\phi)\,\gamma_U(\phi^{-1})}\,\phi^{\rho}} \tag{11.180}$$

where

$$\rho = Deg\,\gamma_U(\phi) - Deg\,s_U(\phi) \tag{11.181}$$

and ϕ^{ρ} is needed to restore $k(\phi)$ to realizable form.

Note that the residual generator is normally designed to be stable so $s_U(\phi)=1$. If further it is minimum phase, that is $\gamma_U(\phi)=1$, then

$$k(\phi) = \frac{s(\phi)}{\gamma(\phi)} \tag{11.182}$$

If the residual generator has poles or zeroes on the unit circle then the filter will contain these as zeroes or poles, respectively. That is, if the generator is integrating, then the filter will be differentiating and vice versa.

11.5. NOTES AND REFERENCES

A comprehensive treatment of statistical change detection is given in the book *Detection of Abrupt Changes* by M. Basseville and I. Nikiforov (1993). They address the change detection problem in a broad sense, including both the pure signal processing and the model-based framework. Their approach is fundamentally off-line, utilizing mostly cumulative testing procedures, and the independence of the time series is usually assumed. The material of this chapter is complementary to their work, in that we deal with a purely model-based setup where the correlatedness of the time series is taken into account explicitly, and our test strategies are fundamentally on-line.

The statistical detection of changes in signal and system properties was previously discussed in several articles by M. Basseville and coworkers, see e.g., (Basseville et al, 1987; Basseville, 1988). Further developments of the methodology, with emphasis on small changes, were reported in (Zhang et al, 1994).

The various detection tests described in Section 11.2. are straightforward applications of the standard procedures for testing the zero-mean hypothesis, see for example (Stuart and Ord, 1991). Most of the considerations about fault sensitivity, detection delay and residual filtering, found in this section, were presented in (Gertler and Luo, 1989, and Gertler et al, 1990). Some of these techniques have been successfully applied to automobile engine diagnosis (Gertler et al, 1995).

The generalized likelihood ratio philosophy and technique are described in much detail in (Basseville and Nikiforov, 1993). The method was applied to Kalman filter residuals, which are uncorrelated, already in the seventies (Willsky, 1976; Willsky and Jones, 1976; Chien and Adams, 1976; Friedland, 1979, 1981; Willsky, 1986). The algorithm discussed here in Section 11.3., which takes into account the correlatedness of the residual sequence and the geometric constraints arising from isolation enhancement, was first reported in (Gertler and Yin, 1996). Most of the ideas were tested by Adam Strassel (1996). Some extensions and refinements, such as plausibility testing and the existence of the inverse covariance matrix, appear here for the first time.

The dynamic whitening filter (spectral factorization) is classical material in advanced textbooks, see e.g., (Anderson and Moore, 1979). An application to fault detection and isolation, with diagonal residual set, has been reported recently by Peng and coworkers (1997).

12
Model Identification for the Diagnosis of Additive Faults

12.0. INTRODUCTION

The model-based detection and isolation of faults, of course, requires a mathematical model of the monitored plant. In many cases, such models are not available on the outset of the diagnostic project, but, before anything else can be done, need to be established. In fact, model building turns out to be the most time- and effort-consuming part of most practical diagnostic projects.

While theoretical knowledge of the plant may be utilized in model building, one usually needs to resort to empirical modeling as well. In practice, two basic modeling approaches are commonly used:

1. Build a theoretical (first principles) model from the understanding of the physics (mechanics, chemistry, etc.) of the plant and then identify (some of) the model parameters from experimental data.
2. Consider the plant as a "black box" (or "gray box," utilizing some understanding of its physics) and determine its structure and parameters entirely from experimental data.

Clearly, the two approaches are not completely distinct as both contain, to various degrees, theoretical as well as empirical components.

Since theoretical modeling is process specific, here we will concentrate on the more general empirical approach. The fundamental methodology is model identification by least squares and related methods, reviewed in Chapter 4. In this chapter, we will describe some of the special features and problems, encountered when identification is used to establish the plant model for the detection and isolation of additive faults. Most of the specialities follow from the fact that, in the context of additive faults, our interest lies not with the true parameters of the plant but with the best consistency relations among the observed variables. The use of identification methods in the context of parametric faults will be the subject of the next chapter.

The first section of this chapter is concerned with the effect of parameter bias in the identified model. As it was shown in Chapter 4, such bias is always present if the parameters of an ARMA model have been identified with the basic least squares method from noisy measurements. It will be seen here that the residuals of polynomial parity relations are identical with the prediction errors in parameter estimation. Since the parameter estimates are obtained by optimizing an LS index on the prediction errors, these same estimates (with their bias) are also optimal in the LS sense for the parity relation residuals, as long as the signal and noise properties in the monitoring phase follow those encountered during identification. It will also be shown that parameter bias does not interfere, beyond the above prediction error, with the residual responses to sensor and actuator faults but does so in response to plant faults.

The second section deals with the problem of identifying the model from linearly dependent data. In practice, the identification data is often collected under closed loop control or other circumstances which establish linear relationships among the variables. As shown in Chapter 4, the plant parameters are not identifiable in this situation. It is still possible, however, to determine consistency relations which are valid as long as the relationship among the concerned variables remains unchanged. In these models, the gains of the linearly related variables may shift among each other arbitrarily. This may even be utilized to increase the residual sensitivity with respect to selected faults. It will be shown, however, that gain-shifting may interfere with the isolation of certain actuator and sensor faults.

The third section describes the powerful and practically important technique of direct identification of structured parity relations. Each parity relation structure corresponds to an underlying model; some of these belong to a physical subsystem while others simply describe a mathematical relationship. It is shown that, once the residual structure has been selected, the model for each scalar parity relation may be obtained directly, by identification from experimental data. *The identified models may be linear or nonlinear, and of any complexity.* Direct identification eliminates the need for algebraic model transformation - which is particularly advantageous in connection with nonlinear models where such transformations may be prohibitively complex.

12.1. MODELING ERRORS IN THE RESIDUALS

The effect of modeling errors is a central issue in model-based fault detection and diagnosis. We have addressed this problem at various places in this book. In Chapter 4, the genesis of model errors was described in the context of least squares identification. In Chapter 5, we presented modeling errors, in the general framework of model-based diagnosis, as multiplicative disturbances. In Chapter 9, perfect decoupling from selected parametric disturbances was discussed while robust techniques, involving approximate decoupling from model errors, were described in Chapter 10. In the present section, we will address the issue from the point of view of the interaction between model identification and the diagnosis of additive faults.

Before proceeding, we need to make some comments about our notation.

1. In Chapter 4, we followed the general convention of the identification literature, in that
   - we used unmarked letter-symbols to denote the *true parameters*, and models (transfer functions) formed of them, such as π and $g(\phi)$;
   - we used the ^ symbol to indicate their *estimates*, such as $\hat{\pi}$ and $\hat{g}(\phi)$.
2. In the rest of the book, in order to keep the notation simple when dealing with model-based diagnosis,
   - we used unmarked letters for the *model*, such as $M(\phi)$;
   - at the few places where we had to refer to the *true system*, we did so by using the ° symbol, such as $M°(\phi)$.

In this chapter, we will follow the second convention.

12.1.1. Residuals and Prediction Error

Here we will show that the moving average residuals introduced in Subsection 6.1.2 are identical with the prediction error used in least squares identification, see Subsection 4.2.3. Consider Eq. (6.13) and write the computational form for the i-th primary residual

$$o_i^*(t) = h_i(\phi)y_i(t) - \mathbf{g}_{i.}(\phi)\mathbf{u}(t) \tag{12.1}$$

This can be rearranged as

$$o_i^*(t) = y_i(t) - [\mathbf{g}_{i.}(\phi)\mathbf{u}(t) - (h_i(\phi) - I)y_i(t)] \tag{12.2}$$

Now according to (4.38), with (4.39) and (4.40), the prediction of the output in a multiple-input single-output system is obtained as

$$\hat{y}_i(t) = \boldsymbol{\varphi}_i'(t)\pi_i = \mathbf{g}_{i.}(\phi)\mathbf{u}(t) - (h_i(\phi) - I)y_i(t) \tag{12.3}$$

Thus the bracketed expression in (12.2) can be recognized as the prediction $\hat{y}_i(t)$

and thus the residual $o_i^*(t)$ as the (negative) prediction error

$$o_i^*(t) = y_i(t) - \hat{y}_i(t) \tag{12.4}$$

(12.4) is an important result for the following reason. The ordinary (prediction error) least squares identification algorithm estimates the system parameters by minimizing the sum of the prediction error squares. In the presence of noise, the parameter estimates are biased, but the bias is exactly what makes the sum of the squares minimal, in the face of the noise and under the given inputs. What (12.4) reveals is that the biased estimates are optimal also from the point of view of the diagnostic residuals. With the biased model, the moving average primary residuals have minimal variance, as long as the signal and noise conditions are the same in the course of diagnostic monitoring as they were during model identification.

The above result does not hold, however, for the ARMA residuals. By (6.14), the i-th ARMA primary residual is

$$o_i(t) = \frac{1}{h_i(\phi)} o_i^*(t) = y_i(t) - \frac{g_{i.}(\phi)}{h_i(\phi)} u(t) \tag{12.5}$$

By the terminology of parameter estimation, this is an "output error." The parameter estimates returned by the prediction error least squares algorithm do not minimize the sum of the output error squares. In fact, there are variants of the least squares algorithm built around the output errors which yield, potentially, the unbiased estimates. These would then minimize the variance of the ARMA residual. However, these algorithms are nonlinear, involving significant numerical complexity and the possibility of local minima.

So far, we have been considering only primary residuals (whether MA or ARMA). When an algebraic transformation is applied to a set of primary residuals, to achieve disturbance decoupling and/or isolation enhancement, whatever optimality properties the primary residuals have are lost. However, if the decoupled and/or enhanced residuals are generated by parity relations obtained by direct identification, as described in Section 12.3 below, such optimality properties may again be present.

12.1.2. The Effects of Modeling Error

The effects of modeling error on the residuals will be reviewed here, in the framework of the generic residual generator structure. We will start by describing the first order effect, caused by the error in the input-to-output transfer function. This will be followed by the second order effect, which is due to discrepancies in the fault-to-output transfer function. The interaction of sensor and actuator faults with modeling errors will be addressed separately; this case

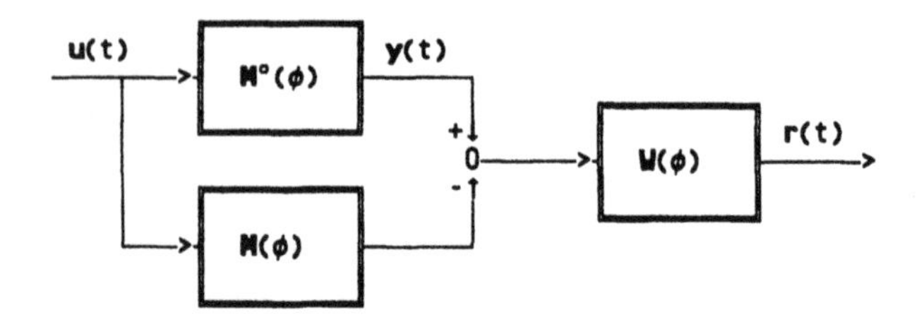

FIGURE 12.1. First order model error effect.

is special because now the two transfer functions are the same.

First order effect. The most obvious consequence of modeling errors is that while the observed inputs $\boldsymbol{u}(t)$ act on the true plant $\boldsymbol{M}^{\circ}(\phi)$, so that

$$y(t) = \boldsymbol{M}^{\circ}(\phi)\, \boldsymbol{u}(t) \tag{12.6}$$

the predicted outputs are computed using the model $\boldsymbol{M}(\phi)$ (see Fig. 12.1). This basic situation was introduced already in Chapter 5. As it was seen there, the computed residual, in the absence of faults and noise, is

$$r(t) = \boldsymbol{W}(\phi)[y(t) - \boldsymbol{M}(\phi)\boldsymbol{u}(t)] = \boldsymbol{W}(\phi)\, \Delta\boldsymbol{M}_D(\phi)\, \boldsymbol{u}(t) \tag{12.7}$$

where

$$\Delta\boldsymbol{M}_D(\phi) = \boldsymbol{M}^{\circ}(\phi) - \boldsymbol{M}(\phi) \tag{12.8}$$

What Eq. (12.7) says is that, in the presence of modeling errors, the residual is in general nonzero, even when there are no faults and noise. (12.7) also reveals that this residual error depends on the inputs $\boldsymbol{u}(t)$ and is thus time-varying. To protect the residual, as much as possible, against this first order effect of modeling errors is the fundamental problem of robustness in residual generation. Several methods aimed at this have been discussed in Chapter 10.

Second order effect. An additional effect of model errors follows from the discrepancy between the true fault-to-output transfer function $\boldsymbol{S}_F^{\circ}(\phi)$ and its model $\boldsymbol{S}_F(\phi)$ (see Fig. 12.2). Let us consider here a single residual $r_i(t)$. While the generator $\boldsymbol{w}_i'(\phi)$ is designed, for a given response specification $\boldsymbol{z}_{Fi}(\phi)$, using the model as

$$\boldsymbol{w}_i'(\phi)\, \boldsymbol{S}_F(\phi) = \boldsymbol{z}_{Fi}'(\phi) \tag{12.9}$$

the faults actually act on the plant output as

$$y(t|\boldsymbol{p}) = \boldsymbol{S}_F^{\circ}(\phi)\, \boldsymbol{p}(t) \tag{12.10}$$

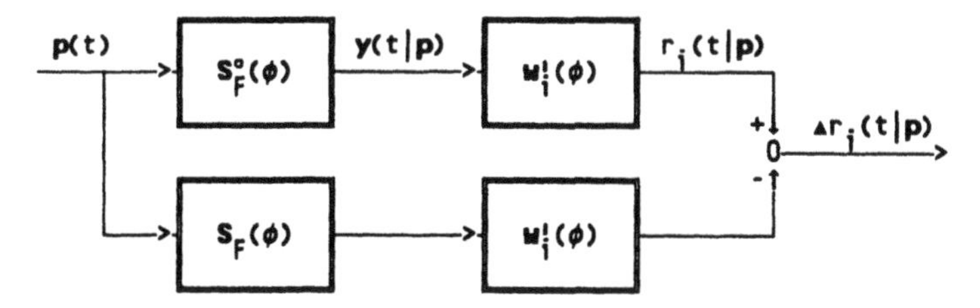

FIGURE 12.2. Second order model error effect.

yielding the fault response in the residual as

$$r_i(t|p) = w_i'(\phi)\, S_F^\circ(\phi)\, p(t) \tag{12.11}$$

Thus there is a discrepancy between the intended and actual fault responses, namely

$$\blacktriangle r_i(t|p) = r_i(t|p) - z_{Fi}'(\phi)p(t) = w_i'(\phi)\, \blacktriangle S_F(\phi)\, p(t) \tag{12.12}$$

where

$$\blacktriangle S_F(\phi) = S_F^\circ(\phi) - S_F(\phi) \tag{12.13}$$

Eq. (12.12) describes an interaction between faults and modeling errors therefore we consider this a second order effect. What (12.12) means is that

- residuals designed to exhibit directional behavior will not have the exact directional response, and
- structured residuals, designed to exhibit zero response to particular faults, will actually have (slightly) nonzero responses.

Sensor and actuator faults. The interaction between sensor and actuator faults and model errors is a special case of the second order effects because

- for output sensor faults, the fault-to-output transfer is the unit matrix which is not subject to modeling error;
- for input sensor and actuator faults, the fault-to-output transfer matrix is identical with (a part of) the input-to-output transfer matrix, which results in an interplay between the first and second order effects.

Output sensor fault (Fig. 12.3). Consider a single output sensor fault $\blacktriangle y_j(t)$. Now

$$p(t) = \blacktriangle y(t) = [0 \ldots 0 \;\; \blacktriangle y_j(t) \;\; 0 \ldots 0]' \tag{12.14}$$

With this, the input-fault-to-output relationship is

$$y(t) = M^\circ(\phi)u(t) + \blacktriangle y(t) \tag{12.15}$$

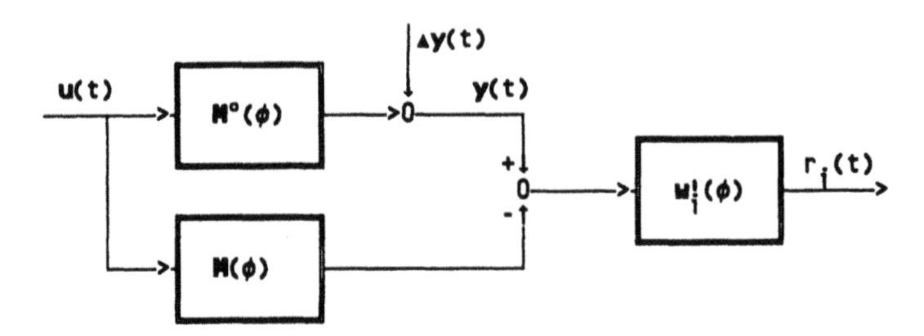

FIGURE 12.3. Output sensor fault.

thus the column of the $S_F(\phi)$ matrix which accompanies $\blacktriangle y_j(t)$ is

$$s_{F\cdot j}(\phi) = [0 \ldots 0 \ 1 \ 0 \ldots 0]' \tag{12.16}$$

This is clearly not subject to modeling error. Thus

$$\blacktriangle r_i(t|\blacktriangle y_j) = 0 \tag{12.17}$$

Input sensor fault (Fig. 12.4). With a single input sensor fault

$$p(t) = \blacktriangle \boldsymbol{u}(t) = [0 \ldots 0 \ \blacktriangle u_j(t) \ 0 \ldots 0]' \tag{12.18}$$

and

$$y(t) = \boldsymbol{M}°(\phi)\boldsymbol{u}°(t) \qquad \boldsymbol{u}(t) = \boldsymbol{u}°(t) + \blacktriangle \boldsymbol{u}(t) \tag{12.19}$$

so that

$$r_i(t) = w_i'(\phi)\{\boldsymbol{M}°(\phi)\boldsymbol{u}°(t) - \boldsymbol{M}(\phi)[\boldsymbol{u}°(t) + \blacktriangle \boldsymbol{u}(t)]\} \tag{12.20}$$

Now observe that $\boldsymbol{u}°(t)$ is invariant with respect to $\blacktriangle \boldsymbol{u}(t)$ so, to separate the second order effect, we need to express the first order effect in terms of $\boldsymbol{u}°$ (rather than $\boldsymbol{u}$). This yields

$$r_i(t) = w_i'(\phi) \ \blacktriangle \boldsymbol{M}_D(\phi) \ \boldsymbol{u}°(t) - w_i'(\phi) \ \boldsymbol{M}(\phi) \ \blacktriangle \boldsymbol{u}(t) \tag{12.21}$$

Finally, observe that

$$\boldsymbol{M}(\phi)\blacktriangle \boldsymbol{u}(t) = \boldsymbol{m}_{\cdot j}(\phi)\blacktriangle u_j(t) \tag{12.22}$$

so that

$$r_i(t|\blacktriangle u_j) = - w_i'(\phi)\boldsymbol{m}_{\cdot j}(\phi)\blacktriangle u_j(t) = z_{ij}(\phi)\blacktriangle u_j(t) \tag{12.23}$$

Here $-\boldsymbol{m}_{\cdot j}(\phi)$ can be recognized as the column of $\boldsymbol{S}_F(\phi)$ which accompanies the fault $\blacktriangle u_j(t)$. Eq. (12.23) reveals that the response is $z_{ij}(\phi)$ as designed,

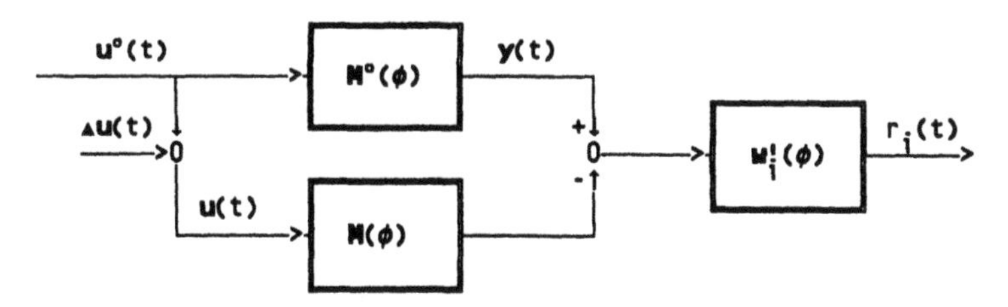

FIGURE 12.4. Input sensor fault.

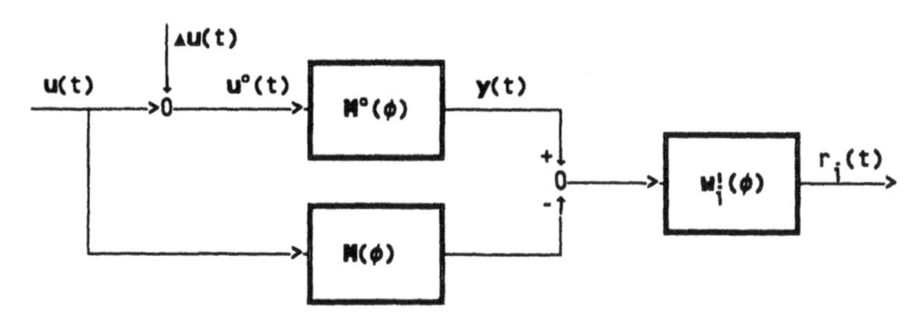

FIGURE 12.5. Input actuator fault.

that is, the second order error effect is zero. One may say that the input sensor fault "does not enter" the physical plant, only the model, thus the design based on the model is exact.

Input actuator fault (Fig. 12.5). With a single input actuator fault, (12.18) holds, but now

$$y(t) = \boldsymbol{M}^{\circ}(\phi)\boldsymbol{u}^{\circ}(t) \qquad \boldsymbol{u}^{\circ}(t) = \boldsymbol{u}(t) + \blacktriangle\boldsymbol{u}(t) \tag{12.24}$$

so that

$$r_i(t) = \boldsymbol{w}_i'(\phi)\{\boldsymbol{M}^{\circ}(\phi)[\boldsymbol{u}(t) + \blacktriangle\boldsymbol{u}(t)] - \boldsymbol{M}(\phi)\boldsymbol{u}(t)\} \tag{12.25}$$

In this case, $\boldsymbol{u}(t)$ is invariant with respect to $\blacktriangle\boldsymbol{u}(t)$, so we need to express the first order effect in terms of $\boldsymbol{u}$ (rather than $\boldsymbol{u}^{\circ}$). This yields

$$r_i(t) = \boldsymbol{w}_i'(\phi)\ \blacktriangle\boldsymbol{M}_D(\phi)\ \boldsymbol{u}(t) + \boldsymbol{w}_i'(\phi)\boldsymbol{M}^{\circ}(\phi)\blacktriangle\boldsymbol{u}(t) \tag{12.26}$$

Finally, with

$$\boldsymbol{M}^{\circ}(\phi)\blacktriangle\boldsymbol{u}(t) = \boldsymbol{m}_{.j}^{\circ}(\phi)\blacktriangle u_j(t) \tag{12.27}$$

the fault response can be written as

$$r_i(t|\blacktriangle u_j) = \boldsymbol{w}_i'(\phi)\boldsymbol{m}_{.j}^{\circ}(\phi)\blacktriangle u_j(t) \neq \boldsymbol{w}_i'(\phi)\boldsymbol{m}_{.j}(\phi)\blacktriangle u_j(t) = z_{ij}(\phi)\blacktriangle u_j(t) \tag{12.28}$$

Now $m_{.j}(\phi)$ is the column of $S_F(\phi)$ which accompanies the fault $\blacktriangle u_j(t)$. Eq. (12.28) reveals that now the second order error effect $\blacktriangle r_i(t|\blacktriangle u_j)$ is not zero. The actuator fault "does enter" the physical plant, thus the design based on the model is, in general, not exact.

12.2. IDENTIFICATION WITH LINEARLY DEPENDENT INPUTS

In practice, inputs to a particular subsystem for which we wish to establish a model are frequently linearly dependent. This may be due to the effect of technological units "upstream", which generate those inputs, or to various control actions. As shown in Chapter 4, this is a typical case of insufficient excitation, which results in the model being unidentifiable. However, for the purpose of fault detection and isolation, we need only *valid consistency relations*, not necessarily the true description of the plant. In such a model, while the individual transfer functions may be incorrect, the overall behavior of the plant is correctly characterized, as long as the relationship among the inputs remains the same as it was during the course of identification. We may say that, in this case, the *gains shift* between the various inputs, while the overall consistency relation remains valid.

Gain-shifting may be utilized to preset some gains arbitrarily, in order to achieve desired fault sensitivity properties. However, it also leads to residual errors whenever the relationship among the inputs changes relative to the reference situation. Unfortunately, such changes include the case of actuator faults. Therefore, gain-shifting should be applied with extreme care if the model is known. If the model needs to be identified from linearly related data then the shifting of gains may be unavoidable. Therefore, whenever possible, the experimental conditions should rather be modified so that data is collected in the absence of linear relations among the inputs.

In this section, we will first introduce the phenomenon of gain shifting and show how it affects the identification procedure. Then we will examine the residual errors which arise from shifted gains under various conditions.

12.2.1. Analytical Gain Shifting

First we will discuss the concept of gain shifting under the assumptions that the true system transfer functions and the linear relationships among the inputs are known. Of course, this is usually unrealistic to assume in practice. So we will then proceed to the more realistic situation when the model is unknown and is to be identified from data containing inputs which are subject to unknown linear relations.

Consider a (sub)system with κ inputs $u_1(t) \ldots u_\kappa(t)$, and assume that the inputs are subject to σ linear relationships, in the form

$$C(\phi)u(t) = 0 \tag{12.29}$$

where $C(\phi)$ is a polynomial matrix containing σ independent rows. Using (12.29), σ of the inputs may be expressed in terms of the others. Assume that we express the last σ inputs:

$$u_j(t) = \sum_{l=1}^{\kappa-\sigma} \gamma_{jl}(\phi)\, u_l(t) \qquad j=\kappa-\sigma+1 \ldots \kappa \tag{12.30}$$

where $\gamma_{jl}(\phi)$ is, in general, a rational function in ϕ. Consider a single output $y_i(t)$. The true input-output relationship is

$$y_i(t) = m^\circ_{i1}(\phi)u_1(t) + \ldots + m^\circ_{i\kappa}(\phi)u_\kappa(t) \tag{12.31}$$

where $m^\circ_{ij}(\phi)$, $j=1\ldots\kappa$, are the true transfer functions. Now substitute the inputs $u_{\kappa-\sigma+1}(t) \ldots u_\kappa(t)$, using (12.30). This yields

$$y_i(t) = [m^\circ_{i1}(\phi) + \sum_{l=\kappa-\sigma+1}^{\kappa} \gamma_{l1}(\phi)m^\circ_{il}(\phi)]\, u_1(t)$$

$$+ \ldots + [m^\circ_{i,\kappa-\sigma}(\phi) + \sum_{l=\kappa-\sigma+1}^{\kappa} \gamma_{l,\kappa-\sigma}(\phi)m^\circ_{il}(\phi)]\, u_{\kappa-\sigma}(t) \tag{12.32}$$

Observe that (12.32) is a valid consistency relation, and remains so even if individual transfer functions $m^\circ_{ij}(\phi)$ are replaced by "models" $m_{ij}(\phi)\neq m^\circ_{ij}(\phi)$, as long as the bracketed coefficients of the set of independent inputs $u_1(t) \ldots u_{\kappa-\sigma}(t)$ remain unchanged. That is, σ of the models, possibly $m_{i,\kappa-\sigma+1}(\phi) \ldots m_{i,\kappa}(\phi)$, may be chosen arbitrarily, and the remaining models then need to be computed from the relationships

$$m_{ij}(\phi) + \sum_{l=\kappa-\sigma+1}^{\kappa} \gamma_{lj}(\phi)m_{il}(\phi) = m^\circ_{ij}(\phi) + \sum_{l=\kappa-\sigma+1}^{\kappa} \gamma_{lj}(\phi)m^\circ_{il}(\phi) \qquad j=1\ldots\kappa-\sigma \tag{12.33}$$

So the gains are effectively shifted among the inputs while the overall validity of the consistency relation is maintained.

In the special case of a single linear relation, (12.29) is

$$c'(\phi)u(t) = c_1(\phi)u_1(t) + \ldots + c_\kappa(\phi)u_\kappa(t) = 0 \tag{12.34}$$

From this, provided that $c_\kappa(\phi)\neq 0$,

$$u_K(t) = -\frac{c_1(\phi)}{c_K(\phi)} u_1(t) - \ldots - \frac{c_{K-1}(\phi)}{c_K(\phi)} u_{K-1}(t) \tag{12.35}$$

Substituting this into (12.31) yields

$$y_i(t) = [m_{i1}^{\circ}(\phi) - \frac{c_1(\phi)}{c_K(\phi)} m_{iK}^{\circ}(\phi)]\, u_1(t)$$

$$+ \ldots + [m_{i,K-1}^{\circ}(\phi) - \frac{c_{K-1}(\phi)}{c_K(\phi)} m_{iK}^{\circ}(\phi)]\, u_{K-1}(t) \tag{12.36}$$

Now one model, suitably $m_{iK}(\phi)$, can be chosen arbitrarily, and the others have to satisfy the consistency conditions

$$m_{ij}(\phi) - \frac{c_j(\phi)}{c_K(\phi)} m_{iK}(\phi) = m_{ij}^{\circ}(\phi) - \frac{c_j(\phi)}{c_K(\phi)} m_{iK}^{\circ}(\phi) \qquad j=1 \ldots K-1 \tag{12.37}$$

Notice that the consistency conditions (12.33) and (12.37) constrain the structure (complexity) of the possible transfer functions. If the $\gamma(\phi)$ coefficients are rational or polynomial then the model transfer functions may need to be more complex (higher order) than the true transfer functions, unless they are chosen to be identical with the latter.

12.2.2. Identification with Gain Assignment

In reality, the true transfer functions are usually not known, so the model is obtained by identification. Any linear relationship among the inputs is seldom known explicitly either, but is of course present in the data used for identification. This may cause the data matrix to become singular and thus the model may not be identified. (Note that, due to the presence of noise, the data matrix is seldom outright singular, but may be poorly conditioned and the parameter estimates become very unreliable.)

If the relationships among the inputs are dynamic (polynomial) then they may or may not cause rank defect in the data matrix, depending on how they relate to the selected dynamic model structure. For the sake of simplicity, we will assume first that the input relations are static; in this case, the rank defect clearly follows. We will return to the more general case afterwards.

Let us write the input-output relationship (12.31) as

$$h_i(\phi) y_i(t) = g_{i1}(\phi) u_1(t) + \ldots + g_{iK}(\phi) u_K(t) \tag{12.38}$$

where we utilized that

$$m_{ij}(\phi) = g_{ij}(\phi) / h_i(\phi) \qquad j=1 \ldots \kappa \tag{12.39}$$

Note that we wrote (12.38) in terms of model transfer functions rather than the "true" transfer functions $m^{\circ}_{ij}(\phi)$. Now recall from Chapter 4, Eq. (4.38), that (12.38) can also be written as

$$y_i(t) = \boldsymbol{\varphi}'_i(t)\, \boldsymbol{\pi}_i \tag{12.40}$$

where

$$\boldsymbol{\varphi}'_i(t) = [u_1(t) \ldots u_1(t-\nu); \ \ldots \ \ldots; u_\kappa(t) \ldots u_\kappa(t-\nu); -y_i(t-1) \ldots y_i(t-\nu)] \tag{12.41}$$

$$\boldsymbol{\pi}_i = [g^0_{i1} \cdots g^{\nu}_{i1}; \ \ldots \ \ldots; g^0_{i\kappa} \cdots g^{\nu}_{i\kappa}; h^1_i \ldots h^{\nu}_i] \tag{12.42}$$

(and where we ignored the noise-term from (4.38)). Recall also from (4.11) that the data matrix is formed as

$$\boldsymbol{\Phi}_i = \begin{bmatrix} \boldsymbol{\varphi}'_i(t) \\ \cdot \\ \cdot \\ \cdot \\ \boldsymbol{\varphi}'_i(t-K+1) \end{bmatrix} \tag{12.43}$$

where K is the length of the dataset. Now assume that the inputs are subject to the linear relationships

$$C\, \boldsymbol{u}(t) = 0 \tag{12.44}$$

where C has σ linearly independent rows. Assume also that $u_j(t)$, $j=1 \ldots \kappa$, are persistently exciting of sufficient order. Then since (12.44) makes σ inputs (and their time-shifted values) linear functions of the remaining inputs (and their time-shifted values), we have the rank reduction

$$Rank\, \boldsymbol{\Phi}_i \leq (\kappa - \sigma)\cdot(\nu+1) + \nu \tag{12.45}$$

Clearly, the matrix $\boldsymbol{\Phi}'_i \boldsymbol{\Phi}_i$ is not invertible.

To make the identification problem solvable, we need to remove $\sigma\cdot(\nu+1)$ elements from $\boldsymbol{\varphi}'_i(t)$, and move them to the "other side" of (12.40), with the corresponding parameters assumed known. Choose the last σ inputs (and their time-shifted values) for removal and define

$$\tilde{\varphi}_i'(t) = [u_1(t) \ldots u_1(t-\nu); \ldots \ldots; u_{\kappa-\sigma}(t) \ldots u_{\kappa-\sigma}(t-\nu); -y_i(t-1) \ldots y_i(t-\nu)] \tag{12.46}$$

$$\tilde{\pi}_i = [g_{i1}^0 \ldots g_{i1}^\nu; \ldots \ldots; g_{i,\kappa-\sigma}^0 \ldots g_{i,\kappa-\sigma}^\nu; h_i^1 \ldots h_i^\nu] \tag{12.47}$$

$$\tilde{y}_i(t) = y_i(t) - g_{i,\kappa-\sigma+1}(\phi)u_{\kappa-\sigma+1}(t) - \ldots - g_{i\kappa}(\phi)u_\kappa(t) \tag{12.48}$$

Now the "known" polynomials $g_{i,\kappa-\sigma+1}(\phi) \ldots g_{i\kappa}(\phi)$ are chosen arbitrarily and the remaining parameters are obtained from the least squares estimate

$$\tilde{\pi}_i = [\tilde{\Phi}_i' \tilde{\Phi}_i]^{-1} \tilde{\Phi}_i' \tilde{Y}_i \tag{12.49}$$

where

$$\tilde{\Phi}_i = \begin{bmatrix} \tilde{\varphi}_i'(t) \\ \cdot \\ \cdot \\ \tilde{\varphi}_i'(t-K+1) \end{bmatrix} \qquad \tilde{Y}_i = \begin{bmatrix} \tilde{y}_i(t) \\ \cdot \\ \cdot \\ \tilde{y}_i(t-K+1) \end{bmatrix} \tag{12.50}$$

The identification procedure described here is the empirical equivalent of analytical gain-shifting discussed in the previous subsection. Just like in the analytical procedure, the number of arbitrarily chosen transfer functions is the same as the number of linear dependencies. The latter are now implicit in the data. The consistency criteria, which were enforced explicitly in the analytical approach, are now implemented automatically by the least squares algorithm.

Now let us consider a polynomial linear relationship among the inputs

$$c'(\phi)u(t) = c_1(\phi)u_1(t) + \ldots + c_j(\phi)u_j(t) + \ldots + c_\kappa(\phi)u_\kappa(t) = 0 \tag{12.51}$$

where

$$c_j(\phi) = \phi^{-dj}[c_j^0 + c_j^1\phi^{-1} + \ldots + c_j^{nj}\phi^{-nj}] \qquad j=1 \ldots \kappa \tag{12.52}$$

Provided that no coefficient in (12.52) is zero, (12.51) includes, for each j, the time-shifted variables $u_j(t-d_j) \ldots u_j(t-d_j-n_j)$. Now assume that a structure has been chosen for the model which leads to the regression vector $\varphi_i'(t)$ as seen in (12.41). Recall that each element of the regression vector corresponds to a column in the data matrix Φ_i. Assume further that each time-sequence $u_j(t), j=1\ldots\kappa$, is persistently exciting of sufficiently high order. Consider the time-shifted version of (12.51), $\phi^k c'(\phi)u(t)$, where k is any integer. Then the linear relationship (12.51) causes as many rank-losses in the data matrix Φ_i as the number of different values of k for which all time-shifted values of all inputs which appear in $\phi^k c'(\phi)u(t)$ are also present in $\varphi_i'(t)$.

Example 12.1.

Consider a system with a single output and two inputs. Seek the model as second order transfer functions with a one-step delay each. Thus the regression vector is

$$\varphi'(t) = [u_1(t-1) \quad u_1(t-2) \quad u_2(t-1) \quad u_2(t-2) \quad -y(t-1) \quad -y(t-2)]$$

If there are no linear relations between the inputs, and the input sequences are persistently exciting of order 3 or more, then $Rank\Phi=6$.

a. Assume first that there is a linear relation between the inputs, in the form

$$c'(\phi)u(t) = c_{10}u_1(t) + c_{20}u_2(t) = 0$$

Though this has no match in the regression vector, the following shifted versions do:

$$\phi^{-1}c'(\phi)u(t) = c_{10}u_1(t-1) + c_{20}u_2(t-1) = 0$$

$$\phi^{-2}c'(\phi)u(t) = c_{10}u_1(t-2) + c_{20}u_2(t-2) = 0$$

Thus there are two linear relations among the six columns, that is, $Rank\Phi=4$.

b. Now assume that the linear relation is

$$c'(\phi)u(t) = c_{10}u_1(t) + c_{11}u_1(t-1) + c_{20}u_2(t) + c_{21}u_2(t-1) = 0$$

Again, this has no match in the regression vector, but the following shifted version does:

$$\phi^{-1}c'(\phi)u(t) = c_{10}u_1(t-1) + c_{11}u_1(t-2) + c_{20}u_2(t-1) + c_{21}u_2(t-2) = 0$$

This is a single relationship among the six columns so $Rank\Phi=5$.

c. Assume finally that the linear relation is

$$c'(\phi)u(t)$$
$$=c_{10}u_1(t)+c_{11}u_1(t-1)+c_{12}u_1(t-2)+c_{20}u_2(t)+c_{21}u_2(t-1)+c_{22}u_2(t-2)= 0$$

Both this, and any of its shifted versions, reach outside $\varphi'(t)$. Thus there are no linear relations among the columns of Φ and no rank reduction. ▒

The above results indicate that a polynomial relationship among the inputs does leave the rank of the data-matrix intact if the $c_j(\phi)$ polynomials are "longer" than the numerator polynomials $g_{ij}(\phi)$ in the selected model. Also, even if there is a rank-loss, it does not necessarily account for the full $g_{ij}(\phi)$ polynomial (as it does in the case of non-polynomial c coefficients). Therefore, the $g_{ij}(\phi)$ polynomials "moved to the other side" in (12.48) may only be partially assumed or the model structure needs to be changed. Note that this is consistent with the properties observed in connection with analytical gain-shifting, where the conditions (12.33) and (12.37) did not allow arbitrary model structures if the c coefficients were polynomial.

12.2.3. Errors Due to Gain Shifting

The models obtained by gain-shifting, whether analytically or by identification, are of course not correct. They provide valid consistency relations, but only as long as the relationship among the inputs remains unchanged relative to the reference situation. Here we will investigate the residual errors arising when this is not the case. We will explore the first-order error effect due to changed input relations. Then we will look into the second order effect resulting from the interaction between gain-shifting and input sensor and actuator faults. We will see that while the former causes no error, the latter does.

Consider a multiple-output system with the usual output vector $y(t)=[y_1(t) \ldots y_\mu(t)]'$. Assume that a single polynomial linear relationship

$$c'(\phi)u(t) = 0 \tag{12.53}$$

exists among the inputs, as described in (12.34). The model has been so built (analytically or by identification) that (12.37) holds for each row of the model matrix $M(\phi)$. Then it holds also for the entire $M(\phi)=[m_{.1}(\phi) \ldots m_{.\kappa}(\phi)]$ matrix, so that it can be re-written in terms of the columns $m_{.j}(\phi)$ as

$$m_{.j}(\phi) - \frac{c_j(\phi)}{c_\kappa(\phi)} m_{.\kappa}(\phi) = m^\circ_{.j}(\phi) - \frac{c_j(\phi)}{c_\kappa(\phi)} m^\circ_{.\kappa}(\phi) \qquad j=1 \ldots \kappa-1 \tag{12.54}$$

where $m_{.\kappa}(\phi)$ is chosen arbitrarily and $c_\kappa(\phi)\neq 0$. Now consider a single residual computed the usual way as

$$r_i(t) = w_i'(\phi)[y(t) - M(\phi)u(t)] = w_i'(\phi)[M^\circ(\phi) - M(\phi)]\, u(t)$$

$$= w_i'(\phi) \sum_{j=1}^{\kappa} [m^\circ_{.j}(\phi) - m_{.j}(\phi)]\, u_j(t) \tag{12.55}$$

Clearly, if (12.53) and (12.54) hold then

$$[M^\circ(\phi) - M(\phi)]\, u(t) = 0 \tag{12.56}$$

for any choice of $m_{\cdot\kappa}(\phi)$.

First order error effect. Assume that while the model $M(\phi)$ has been created based on the linear relationship (12.53) in its original from, the actual relationship among the inputs is

$$c^{\circ\prime}(\phi)\, u(t) = 0 \qquad where \qquad c^{\circ\prime}(\phi) \neq c'(\phi) \tag{12.57}$$

Expressing $u_\kappa(t)$ from (12.35), $c_j(\phi)$, $j=1\ldots\kappa$, replaced with $c_j^\circ(\phi)$, the residual can be written as

$$r_i(t) = w_i'(t) \sum_{j=1}^{\kappa-1} \left[[m_{\cdot j}^\circ(\phi) - m_{\cdot j}(\phi)] - \frac{c_j^\circ(\phi)}{c_\kappa^\circ(\phi)} [m_{\cdot\kappa}^\circ(\phi) - m_{\cdot\kappa}(\phi)] \right] u_j(t) \tag{12.58}$$

Now expressing $m_{\cdot j}^\circ(\phi) - m_{\cdot j}(\phi)$ from (12.54) yields

$$r_i(t) = w_i'(t) \sum_{j=1}^{\kappa-1} \left[\frac{c_j(\phi)}{c_\kappa(\phi)} - \frac{c_j^\circ(\phi)}{c_\kappa^\circ(\phi)} \right] [m_{\cdot\kappa}^\circ(\phi) - m_{\cdot\kappa}(\phi)]\, u_j(t) \tag{12.59}$$

Equation (12.59) clearly shows that there is a residual error, unless one of the following conditions is satisfied:

- the model $m_{\cdot\kappa}(\phi)$ is chosen "correctly," that is, to be equal to the true transfer function $m_{\cdot\kappa}^\circ(\phi)$;
- the actual relationship among the inputs, as expressed by the $c_j^\circ(\phi)/c_\kappa^\circ(\phi)$ coefficient ratios, is identical with the design conditions (assumptions) represented by the $c_j(\phi)/c_\kappa(\phi)$ ratios.

Second order error effects. Consider a single input sensor or actuator fault (c.f. (12.18))

$$\Delta u(t) = [0 \ldots 0 \quad \Delta u_j(t) \quad 0 \ldots 0]' \tag{12.60}$$

Sensor fault (Fig. 12.4). Now

$$y(t) = M^\circ(\phi)\, u^\circ(t) \qquad\qquad u(t) = u^\circ(t) + \Delta u(t) \tag{12.61}$$

(c.f. (12.19)). The generator is designed as

$$- w_i'(\phi)\, m_{.j}(\phi) = z_{ij}(\phi) \tag{12.62}$$

where $z_{ij}(\phi)$ is the specified response. The linear relation (12.53) now applies to the true inputs $u°(\phi)$, as

$$c'(\phi)\, u°(t) = 0 \tag{12.63}$$

implying

$$[M°(\phi) - M(\phi)]\, u°(t) = 0 \tag{12.64}$$

Now the residual in response to the sensor fault $\blacktriangle u_j(t)$ is

$$r_i(t | \blacktriangle u_j) = w_i'(\phi)[M°(\phi)\, u°(t) - M(\phi)\, u(t)]$$

$$= w_i'(\phi)[M°(\phi) - M(\phi)]\, u°(t) - w_i'(\phi) m_{.j}(\phi)\, \blacktriangle u_j(t) \tag{12.65}$$

Here the first term is zero by (12.64) and the second is $z_{ij}(\phi) \blacktriangle u_j(t)$ by (12.62). Thus the response is as specified, that is, no error arises from the interaction between gain-shifting and an input sensor fault.

Actuator fault (Fig. 12.5). Now

$$y(t) = M°(\phi)\, u°(t) \qquad u°(t) = u(t) + \blacktriangle u(t) \tag{12.66}$$

(c.f. (12.24)). The generator is designed as

$$w_i'(\phi)\, m_{.j}(\phi) = z_{ij}(\phi) \tag{12.67}$$

The linear relation now applies to the command inputs $u(t)$, so (12.53) and (12.56) hold as originally stated. The residual in response to the actuator fault $\blacktriangle u_j(t)$ is

$$r_i(t | \blacktriangle u_j) = w_i'(\phi)[M°(\phi)\, u°(t) - M(\phi)\, u(t)]$$

$$= w_i'(\phi)[M°(\phi) - M(\phi)]\, u(t) + w_i'(\phi) m°_{.j}(\phi)\, \blacktriangle u_j(t) \tag{12.68}$$

Here the first term is zero by (12.56) but the second is different from $z_{ij}(\phi) \blacktriangle u_j(t)$, see (12.67), unless $m_{.j}(\phi) = m°_{.j}(\phi)$. Thus the interaction between gain-shifting and an input actuator fault in general does result in a second order error.

12.3. DIRECT IDENTIFICATION OF STRUCTURED PARITY RELATIONS

So far we usually assumed that we start the design of the diagnostic algorithm from the input-fault-disturbance-output relationship

$$y(t) = M(\phi)u(t) + S_F(\phi)p(t) + S_D(\phi)q(t) \tag{12.69}$$

Residuals with the desired fault isolation and disturbance rejection properties were then obtained by applying an algebraic transformation to the above equation. This implied that whatever model identification needed to take place, it concerned the transfer functions appearing in (12.69).

We mentioned, however, in Section 7.4 that in many cases structured parity relations derived by transformation may also be recognized as models of particular units in the physical system. We will utilize and extend this observation here, and will show that parity relations designed to be structured for sensor and actuator faults can be identified directly from experimental data. In this case, there is no need for algebraic transformation. This becomes especially important when structured parity relation design is applied to nonlinear models, since algebraic manipulations with such models may be quite difficult.

Before proceeding, we need to introduce some definitions:

By the *coarse structure* of a model (or parity relation) we mean *which variables* (and/or faults) are present. The course structure of a structured parity relation is defined by the "1"s and "0"s in the respective row of the structure matrix.

By the *fine structure* of a model we mean *how* those variables (faults) are present. In the case of transfer function models, the fine structure includes the orders of the numerators and denominators and the dead-times if any. In connection with a polynomial nonlinear model, the fine structure concerns the polynomial degrees.

As for the model in (12.69), the coarse structure is determined by the physical setup and usually may be found by inspection. The model outputs are the plant outputs, one appearing in each scalar equation. The full set of observed plant inputs is present as model inputs, though some may be missing from each particular row. The full set of faults and disturbances is also shown in the equation but the actual presence of particular faults/disturbances in each scalar equation depends on the physical setup.

If the model needs to be found empirically, then the fine structure may only be partially known or not known at all. In this case, one needs to resort to some of the guided-search-type structure estimation methods outlined in Section 4.6. Usually we do not aim at finding the "true" system structure (which may be extremely complex) but settle with a reasonably good representation of the plant, as measured by some performance index.

12.3.1. Structured Parity Relations for Sensor and Actuator Faults

We will concern ourselves with the situation when the disturbances, if any, are ignored and the faults of interest are associated with sensors and actuators. Then $q(t)=0$ and $S_F(\phi)$ is composed of columns of the unit matrix (in connection with output sensor faults) and of the input-output matrix $M(\phi)$ (in connection with input sensor and actuator faults).

Let us consider a parity relation

$$r_i(t) = w_i'(\phi)[y(t) - M(\phi)u(t)] = w_i'(\phi)S_F(\phi)p(t) \tag{12.70}$$

Choose $w_i'(\phi)$ so that

$$w_i'(\phi)S_{Fi}(\phi) = 0 \tag{12.71}$$

where $S_{Fi}(\phi)$ contains those columns of $S_F(\phi)$ which belong to the faults selected for elimination. Note that (12.71) is a homogeneous condition; assume it is also almost full, that is, we eliminate μ-1 faults (where μ is the number of plant outputs).

Since the columns of $S_F(\phi)$ are those of I and $M(\phi)$, when eliminating a fault from the internal form of the residual, the respective variable is also eliminated from the computational form. So in the generic residual expression

$$r_i(t) = w_{i1}(\phi)y_1(t)+ \dots +w_{i\mu}(\phi)y_\mu(t) + v_{i1}(\phi)u_1(t)+ \dots +v_{i\kappa}(\phi)u_\kappa(t) \tag{12.72}$$

$w_{ij}(\phi)=0$ for those outputs whose sensor fault has been eliminated and $v_{ij}(\phi)=-w_i'(\phi)m_{.j}(\phi)=0$ for those inputs whose sensor or actuator fault has been eliminated. Thus actually

$$r_i(t) = \tilde{w}_i'(\phi)y^{(i)}(t) + \tilde{v}_i'(\phi)u^{(i)}(t) \tag{12.73}$$

where $y^{(i)}(t)$ and $u^{(i)}(t)$ contain the remaining outputs and inputs and $\tilde{w}_i'(\phi)$ and $\tilde{v}_i'(\phi)$ contain the respective elements of $w_i'(\phi)$ and $v_i'(\phi)$.

Now recall from Subsection 6.6.3 that, under almost full homogeneous specification, one element of the $\tilde{w}_i'(\phi)$ vector may be assigned. Select the k-th element and assign it as $w_{ik}(\phi)=1$. Then (12.73) becomes

$$r_i(t) = y_k(t) + \hat{w}_i^{(k)}{}'(\phi)\mathbf{y}^{(ik)}(t) + \hat{v}_i'(\phi)\mathbf{u}^{(i)}(t) \tag{12.74}$$

where $\mathbf{y}^{(ik)}(t)$ is $\mathbf{y}^{(i)}(t)$ without $y_k(t)$ and $\hat{w}_i^{(k)}{}'(\phi)$ is $\hat{w}_i'(\phi)$ without $w_{ik}(\phi)$. This can be re-written as

$$r_i(t) = y_k(t) - [- \hat{w}_i^{(k)}{}'(\phi)\mathbf{y}^{(ik)}(t) - \hat{v}_i'(\phi)\mathbf{u}^{(i)}(t)] \tag{12.75}$$

Eq. (12.75) looks like a primary residual, with the bracketed expression serving as the prediction of $y_k(t)$. Formally, this is a multiple-input single-output subsystem, having $y_k(t)$ as its output and the elements of $\mathbf{y}^{(ik)}(t)$ and $\mathbf{u}^{(i)}(t)$ as its inputs. The "model" of this subsystem consists of the transfer functions in $\hat{w}_i^{(k)}{}'(\phi)$ and $\hat{v}'_i(\phi)$. As it was shown in Subsection 6.6.3, these are uniquely defined, once $w_{ik}(\phi)$ has been assigned. It is this subsystem model, namely

$$y_k(t) = - \hat{w}_i^{(k)}{}'(\phi)\mathbf{y}^{(ik)}(t) - \hat{v}_i'(\phi)\mathbf{u}^{(i)}(t) \tag{12.76}$$

which we want to identify from experimental data.

To gain a slightly different perspective, recall that the algebraic generation of structured parity relations can also be viewed as a sequential elimination procedure. Initially, we have $\kappa+\mu$ variables (κ inputs and μ outputs) and μ input-output equations. We may use the equations sequentially, to express one of the variables from one of the equations, and substitute it into the remaining equations. Normally, after μ-1 steps, we are left with $\kappa+1$ variables and a single equation linking them. This equation constitutes the transformed subsystem and this is what we seek by direct identification.

Note that (12.76) may actually represent a physically meaningful subsystem of the monitored plant, in which case direct identification of the subsystem parameters occurs naturally. At other instances, (12.76) simply describes a mathematical relationship among system variables, such as certain variables expressed from some model equations and substituted into others. But even if this is the case, it is a meaningful relationship which is fully represented in the experimental data (provided the latter satisfies the usual excitation requirements).

Example 12.2.

Consider a three-input two-output system shown in Fig. 12.6.a. The overall system representation is given in Fig. 12.6.b. The structure of the primary parity relations is

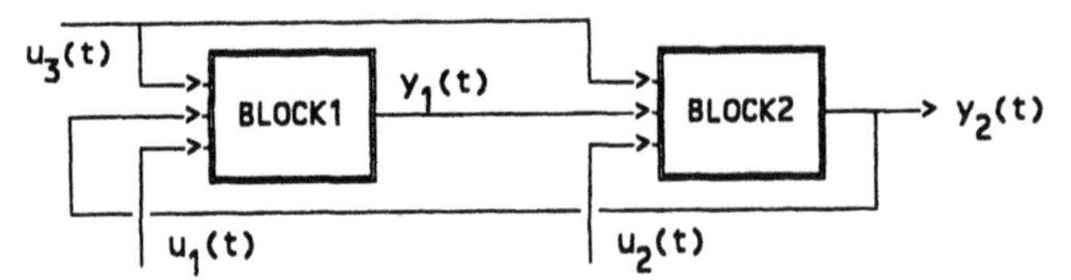

FIGURE 12.6.a. Three-input two-output system.

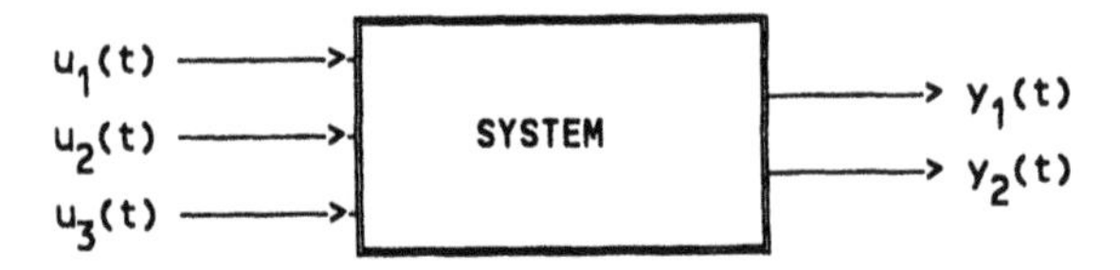

FIGURE 12.6.b. Overall representation of the system of Fig. 12.6.a.

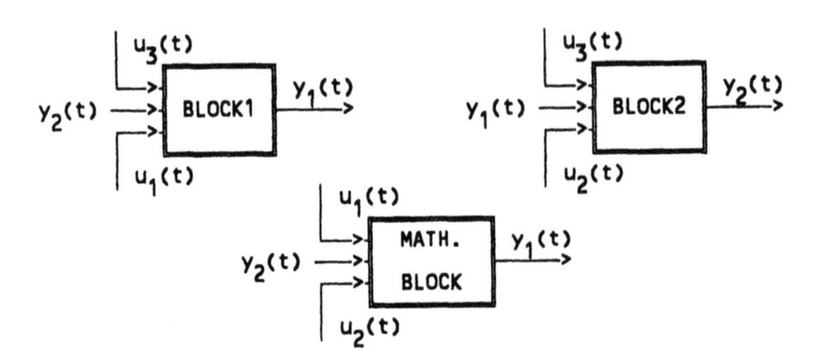

FIGURE 12.6.c. Parity relations decoupled from input faults.

	$\blacktriangle u_1$	$\blacktriangle u_2$	$\blacktriangle u_3$	$\blacktriangle y_1$	$\blacktriangle y_2$
r_1	1	1	1	0	1
r_2	1	1	1	1	0

Given that there are two outputs, structured parity relations decoupled from one fault each can be generated. Actually, the primary equations are already part of this set, in that they are decoupled from an output sensor fault each. Additional structures, decoupled from input sensor or actuator faults are:

	$\blacktriangle u_1$	$\blacktriangle u_2$	$\blacktriangle u_3$	$\blacktriangle y_1$	$\blacktriangle y_2$
r_3	0	1	1	1	1
r_4	1	0	1	1	1
r_5	1	1	0	1	1

Clearly, r_3 is the structure of the physical subsystem Block2, while r_4 is that of Block1 (Fig. 12.6.c). However, there is no physical subsystem which corresponds to r_5; this is just $u_3(t)$ expressed from one of the primary equations (or from one of the block equations) and substituted into the other.

12.3.2. Coarse Structure Selection for Identification

Before the actual identification of the transformed subsystems may take place, one needs to select the coarse structure of the models. To assign the variables appearing in the various transformed subsystems, the structure matrix for the set of parity relations has to be designed. This follows the rules described in Chapter 7. Each row of the structure matrix will yield a transformed subsystem with a single (subsystem) output and multiple inputs.

If the transformation is implemented algebraically, and the $S_F(\phi)$ matrix is known, then structure selection is subject to the following conditions:

Condition 1. $Rank S_{Fi}(\phi) = \mu - 1$ for each row i, where $S_{Fi}(\phi)$ contains 1_j (the j-th column of the unit matrix) for each output fault $\blacktriangle y_j(t)$ and $m_{.j}(\phi)$ for each input fault $\blacktriangle u_j(t)$, which are eliminated from the row. In most cases, but not always, this implies that

Condition 1a. Each row contains exactly $\kappa + 1$ variables, where κ is the number of plant inputs.

Condition 2. $Rank[S_{Fi}(\phi) \quad m_{.l}(\phi)] = Rank S_{Fi}(\phi) + 1$, for each input fault $\blacktriangle u_l(t)$ included in the row.

When the subsystems represented by the selected row-structures are to be identified directly, then Condition 1a may be observed in the structure selection but $S_{Fi}(\phi)$ is of course not known beforehand. Any violation of the rank conditions affects the identification procedure, nevertheless. Thus the presence of any problem with the rank conditions needs to be deduced from symptoms observed during identification. Some insight into the implications of the rank conditions may be gained by examining how they act in the algebraic transformation.

- If Condition 1 is violated so that $Rank S_{Fi}(\phi) = \mu$ (too few variables in the target row-structure) then the algebraic transformation cannot be executed. Direct identification can be performed, but the variables ignored in the structure act as disturbances and may result in significant model errors.
- If Condition 1 is violated so that $Rank S_{Fi}(\phi) < \mu - 1$ (too many variables in the target row-structure) then the algebraic procedure yields multiple relationships among the selected variables. If one of these is picked as the "model" then the remaining relationships act as constraints among the

subsystem inputs. Direct identification in this case is clearly performed with linearly dependent subsystem inputs, requiring the assignment of some of the parameters.

- If Condition 2 is violated then, in the course of the algebraic procedure, some of the variables not specified for elimination will disappear together with ones selected for elimination. In direct identification, those inputs will appear with relatively small (in theory zero) estimated gains, resulting in insufficient sensitivity to the respective faults.

It is easy to assure that the number of variables in the target row-structures is chosen correctly. Difficulties experienced then in the direct identification (poorly conditioned data matrix or small sensitivities to some faults) indicate the presence of rank defects in the $\boldsymbol{M}(\phi)$ matrix. If such problems are detected then the structure matrix needs to be redesigned accordingly and a new identification attempted. This procedure may take several iterations.

Of course, noise will make the ideally exact estimates biased. However, as shown in Subsection 12.1.1, the prediction error in identification is identical with the parity relation residual if a MA residual generator is implemented from the directly identified model. Thus, in this case, the bias will minimize the residual variance. There will be additional modeling errors if the selected model-class or fine structure do not match the true relationship. On the other hand, noise and model-class mismatch may obscure existing linear relations among the variables and make the identification problem seemingly solvable without parameter assignment. However, the estimates obtained under such circumstances may be very inaccurate or completely meaningless.

Example 12.3.

Consider the three-input two-output system:

$$y_1(t) = m_{11}(\phi)u_1(t) + m_{12}(\phi)u_2(t) + m_{13}(\phi)u_3(t) \tag{1}$$

$$y_2(t) = m_{21}(\phi)u_1(t) + m_{22}(\phi)u_2(t) + m_{23}(\phi)u_3(t) \tag{2}$$

a. Derive a transformed equation so that $u_1(t)$ is eliminated. In general, this does not violate any of the conditions. Expressing $u_1(t)$ from (1) and substituting it into (2) yields

$$y_2 = - w_{21}y_1 - v_{22}u_2 - v_{23}u_3$$

$$= \frac{m_{21}}{m_{11}} y_1 + [m_{22} - \frac{m_{21}m_{12}}{m_{11}}]u_2 + [m_{23} - \frac{m_{21}m_{13}}{m_{11}}]u_3 \tag{3}$$

b. Seek the transformed relationship as

$$y_2 = - w_{21} y_1 - v_{21} u_1 - v_{22} u_2 - v_{23} u_3 \tag{4}$$

Now Condition 1a is violated. Since (1) expresses y_1 in terms of u_1, u_2 and u_3, the inputs of the transformed relationship are linearly dependent.

c. Consider (3) and assume that $m_{.3} = cm_{.1}$ (Condition 2 is violated). Then

$$m_{23} - \frac{m_{21} m_{13}}{m_{11}} = cm_{21} - \frac{m_{21} cm_{11}}{m_{11}} = 0 \tag{5}$$

d. Now consider a system in which there is a third input-output relationship, in addition to (1) and (2),

$$y_3(t) = m_{31}(\phi) u_1(t) + m_{32}(\phi) u_2(t) + m_{33}(\phi) u_3(t) \tag{6}$$

Obtain a transformed relationship which does not contain $u_1(t)$ and $u_2(t)$. These can be expressed, for example, from (1) and (2) as

$$\begin{bmatrix} u_1 \\ u_2 \end{bmatrix} = \begin{bmatrix} m_{11} & m_{12} \\ m_{21} & m_{22} \end{bmatrix}^{-1} \begin{bmatrix} y_1 \\ y_2 \end{bmatrix} - \begin{bmatrix} m_{11} & m_{12} \\ m_{21} & m_{22} \end{bmatrix}^{-1} \begin{bmatrix} m_{13} \\ m_{23} \end{bmatrix} u_3 \tag{7}$$

and then substituted into (6). Now assume that $m_{.2} = cm_{.1}$ (Condition 1 violated); then the inverse matrix in (7) does not exist. Expressing only $u_1(t)$ e.g., from (1) would remove this problem but then $u_2(t)$ would also disappear from (2) and (6), making the remaining inputs in (6) linearly dependent. ⁝⁝⁝

12.3.3. Fine Structure, Causality and Stability

Here we will address the issues of selecting the fine structures of the directly identified models, and the handling of potential problems concerning their causality and stability.

Fine structure. Let us re-write (12.76) as

$$y_k(t) = \sum_l \frac{\beta_{kl}^+(\phi)}{\alpha_k^+(\phi)} y_l(t) + \sum_j \frac{\gamma_{kj}^+(\phi)}{\alpha_k^+(\phi)} u_j(t) \tag{12.77}$$

The fine structure of the transformed subsystem concerns the degrees of the polynomials appearing in this equation. According to the results of Subsection 6.6.3, Equation (6.175), these degrees are subject to the following bounds:

$$Deg\alpha_k^+(\phi) \leq \nu - \rho_I$$

$$Deg\beta^{+}_{kl}(\phi) \le \nu - \rho_I \qquad \textit{for all } l$$

$$Deg\gamma^{+}_{kj}(\phi) \le \nu - \rho_I \qquad \textit{for all } j \tag{12.78}$$

Here ν is the true (McMillan) order of the plant and ρ_I is the number of strictly input faults (input sensor and actuator faults) eliminated from the equation, subject also to Condition 1 above. The value of ρ_I is obvious from the structure (apart from the validity of Condition 1). If ν is also known, the polynomial degrees may be set for the identification. However, ν may not be known, especially in the framework of direct identification. Then the polynomial degrees need to be determined empirically, resorting to the order estimation techniques outlined in Section 4.6.

As it was seen in Subsection 6.6.3, Equations (6.172) and (6.174), an arbitrary choice of $w_{ik}(\phi)$ does not guarantee that the elements of $\hat{w}_i^{(k)}{}'(\phi)$ and $\hat{v}_i{}'(\phi)$ will be causal and stable. In fact, in the design algorithm described there, $w_{ik}(\phi)$ was chosen exactly in such a way that it ascertained the causality and stability of the residual generator. To do so, however, would require the full knowledge of the model, which of course is not available when the transformed model is sought by direct identification.

Causality. The problem of causality can be solved by the proper choice of the output for which $w_{ik}(\phi)=1$ is assigned. If the transformed model is that of a single-output physical subsystem then the output of this subsystem is the natural choice for $y_k(t)$, which of course assures causality. However, even if such a physical subsystem cannot be found, there is always at least one choice of y_k which results in a causal model. Recall (12.77); clearly, this subsystem is non-causal if any of the numerators $\beta^{+}_{kl}(\phi)$ and $\gamma^{+}_{kj}(\phi)$ has higher polynomial degree than the common denominator $\alpha^{+}_{k}(\phi)$.

Consider first the $\beta^{+}_{kl}(\phi)/\alpha^{+}_{k}(\phi)$ transfer functions. If any of these, with the arbitrarily selected $y_k(t)$ as output, is non-causal, then one needs to pick the output $y_l(t)$ whose accompanying polynomial $\beta^{+}_{kl}(\phi)$ is the highest degree. Making this the new output $y_k(\phi)$ of the transformed subsystem and rearranging (12.77) accordingly, the new $\beta^{+}_{kl}(\phi)/\alpha^{+}_{k}(\phi)$ transfer functions will be causal.

Consider now the $\gamma^{+}_{kj}(\phi)/\alpha^{+}_{k}(\phi)$ transfer functions. These are obtained as

$$\frac{\gamma^{+}_{kj}(\phi)}{\alpha^{+}_{k}(\phi)} = -\, v^{+}_{j}(\phi) = w_i^{+}{}'(\phi) m^{+}_{\cdot j}(\phi) = -\sum_{l} \frac{\beta^{+}_{kl}(\phi)}{\alpha^{+}_{k}(\phi)} \frac{g^{+}_{lj}(\phi)}{h^{+}(\phi)} \tag{12.79}$$

Since the transfer functions $g^{+}_{ij}(\phi)/h^{+}(\phi)$, which describe the natural input-output relations in the physical plant, are causal, $\gamma^{+}_{kj}(\phi)/\alpha^{+}_{k}(\phi)$ will be causal if all the $\beta^+_{kl}(\phi)/\alpha^+_k(\phi)$ functions are causal (or zero). That is, finding the transformed subsystem output $y_k(t)$ so that all the y_l to y_k transfers are causal guarantees that all the u_j to y_k transfers are causal as well.

To actually find the proper output for the transformed subsystem, one needs to start with an arbitrary output and identify the subsystem allowing non-causal transfer functions. While the selected output y_k is present in the regression vector with the usual shifted values $y_k(t\text{-}1) \ldots y_k(t\text{-}\nu+\rho_l)$, the inputs u_j and the other outputs y_l are present as $u_j(t+d) \ldots u_j(t) \ldots u_j(t\text{-}\nu+\rho_l)$ and $y_l(t+d) \ldots y_l(t) \ldots y_l(t\text{-}\nu+\rho_l)$, where d is the largest foreseen lead-time. The corresponding transfer function coefficients are then identified and the output with the greatest lead-time marked. This will be the selected output of the transformed subsystem. The transfer functions may finally be obtained from the results of the first attempt algebraically, or by re-identification in the new structure.

Stability. There is no guarantee that a stable transformed subsystem can be found with the assignment $w_{ik}(\phi)=1$. The transformed subsystem may be stable with some choices of y_k and unstable with others, and we do not know of any systematic procedure to find the right output if this is the case. One may, of course, try to identify the transformed subsystem with various output choices and see if a stable variant can be found.

Even if there is no stable representation of the transformed subsystem, the residual generator is easy to make stable. One needs to resort to the logic of Subsection 6.6.3 where, in Algorithm III, Eq. (6.173) and on, $w_{ik}(\phi)$ was so chosen that it canceled the undesired poles. Recall that, while all poles were canceled there to obtain a polynomial generator, it suffices to cancel the unstable poles if only the stability of the generator is the objective. In the framework of direct identification, this corresponds to identifying the transformed subsystem with $w_{ik}(\phi)=1$, and decomposing the denominator of the identified transfer functions into stable and unstable factors as

$$\alpha_k(\phi) = \alpha_{Sk}(\phi)\,\alpha_{Uk}(\phi) \tag{12.80}$$

Then a stable residual generator is obtained as

$$r_i(t) = \alpha_{Uk}(\phi)y_k(t) - \sum_l \frac{\beta_{kl}(\phi)}{\alpha_{Sk}(t)}\, y_l(t) - \sum_j \frac{\gamma_{kj}(\phi)}{\alpha_{Sk}(\phi)}\, u_j(t) \tag{12.81}$$

This corresponds to changing from $w_{ik}(\phi)=1$ to $w_{ik}(\phi)=\alpha_{Uk}(\phi)$. Notice that there is no numerical pole-zero cancellation in (12.81).

Example 12.4.

Consider a stable two-input two-output system

$$y_1(t) = \frac{g_{11}(\phi)}{h_1(\phi)} u_1(t) + \frac{g_{12}(\phi)}{h_1(\phi)} u_2(t)$$

$$y_2(t) = \frac{g_{21}(\phi)}{h_2(\phi)} u_1(t) + \frac{g_{22}(\phi)}{h_2(\phi)} u_2(t)$$

Design a residual

$$r_i(t) = w_{i1}(t)[y_1(t) - \frac{g_{11}(\phi)}{h_1(\phi)} u_1(t) - \frac{g_{12}(\phi)}{h_1(\phi)} u_2(t)]$$

$$+ w_{i2}(t)[y_2(t) - \frac{g_{21}(\phi)}{h_2(\phi)} u_1(t) - \frac{g_{22}(\phi)}{h_2(\phi)} u_2(t)]$$

so that it is insensitive to sensor/actuator faults on $u_1(t)$, that is

$$w_{i1}(\phi) \frac{g_{11}(\phi)}{h_1(\phi)} + w_{i2}(\phi) \frac{g_{21}(\phi)}{h_2(\phi)} = 0$$

a. Choose $w_{i1}(\phi)=1$. Then $w_{i2} = -g_{11}h_2/g_{21}h_1$, so that

$$r_i(t) = y_1(t) - \frac{g_{11}(\phi)h_2(\phi)}{g_{21}(\phi)h_1(\phi)} y_2(t) - \frac{g_{12}(\phi)g_{21}(\phi) - g_{11}(\phi)g_{22}(\phi)}{g_{21}(\phi)h_1(\phi)} u_2(t)$$

Clearly, the transformed subsystem is
- non-causal if $g_{21}(\phi)$ has more time-lag than $g_{11}(\phi)$;
- unstable if $g_{21}(\phi)$ has zeroes on or outside the unit circle.

b. Now choose $w_{i2}(\phi)=1$. Then $w_{i1} = -g_{21}h_1/g_{11}h_2$ and

$$r_i(t) = y_2(t) - \frac{g_{21}(\phi)h_1(\phi)}{g_{11}(\phi)h_2(\phi)} y_1(t) - \frac{g_{12}(\phi)g_{21}(\phi) - g_{11}(\phi)g_{22}(\phi)}{g_{11}(\phi)h_2(\phi)} u_2(t)$$

Now the transformed subsystem is
- non-causal if $g_{11}(\phi)$ has more time-lag than $g_{21}(\phi)$;
- unstable if $g_{11}(\phi)$ has zeroes on or outside the unit circle.

Comparing the two cases, one can conclude that
- at least one of the two choices yields a causal subsystem;
- either or both of the choices may lead to an unstable subsystem (in which

case one needs to cross-multiply with the unstable pole-factors to obtain a stable residual generator).

12.4. NOTES AND REFERENCES

The subjects of this chapter have great practical significance yet they have attracted very little attention in the literature.

The effect of modeling errors on the residuals was examined in (Gertler et al, 1990). The relationship between prediction error (identification) and parity relation residuals was pointed out in (Gertler, 1995b). Direct identification of structured parity relations was described in (Gertler et al, 1994). The idea, which probably originated from Qiang Luo, was used extensively in a car engine diagnosis project undertaken by the author and coworkers (see Gertler and Costin, 1994; Gertler et al, 1995). The former paper (Gertler et al, 1994) also includes an analysis of the effect of linearly dependent data in the identification of parity relations, a problem also encountered in the car engine project. The structure selection, causality and stability considerations are given here for the first time.

13 Diagnosing Multiplicative Faults by Parameter Estimation

13.0. INTRODUCTION

As we pointed out earlier, many plant faults are best characterized as changes in some plant parameters. Also, model errors resulting from shifting operating points may be described in terms of changes in model parameters. In Chapter 9, such parameter discrepancies were handled in the framework of parity relations. In this chapter, an alternative approach will be discussed which relies on the estimation of the concerned parameters using the usual systems identification methods. The detection and isolation of parametric discrepancies by parameter estimation involves the identification of a reference model, in a situation when it is known (assumed) that no discrepancies are present, followed by repeated re-identification on-line. The comparison of the on-line estimates to the reference model is to reveal any discrepancy in the parameters.

First we will review the basic least squares algorithm from Chapter 4. We will describe multiple-input multiple-output systems and handle multiple-input single-output systems as a special case.

The next section will be devoted to the estimation of the parameter discrepancy. We will show how this can be done using parity relation residuals. Further, we will discuss the estimation transients following a sudden change in one of the parameters, and will introduce the sliding window identification algorithm to reduce their effect. Bias and uncertainty in the framework of change estimation will also be explored.

The following section deals with the estimation of changes in the underlying (physical) parameters of the plant. We will first review a two-step procedure in which the model parameters are estimated first. Then we will show how the underlying parameter changes can be identified in a single step. We will also comment on the excitation requirements of the two procedures. Finally, we will show that the parity relation approach (under certain conditions) is a limiting case of parameter change estimation, which uses the minimal dataset under which estimates can still be obtained. While the parameter estimation approach may be rather demanding in terms of on-line computations, parity relations are less expensive. On the other hand, the latter are much more sensitive to noise, resulting in larger thresholds and thus smaller fault sensitivity. Also, parity relations have somewhat more stringent (shorter term) excitation conditions than conventional parameter estimation.

The physical plant usually operates in continuous time. While the physical parameters may serve directly as the model parameters in the continuous-time model, in the discrete-time model they often appear in a complicated way. As an alternative to the fully discrete-time approach, one may attempt to identify the physical parameters directly from the continuous-time model. An algorithm doing this, by the use of a sliding integral operator, is described in the fourth section of the chapter.

13.1. THE BASIC LEAST SQUARES ALGORITHM

The following equations are recalled from Chapter 4, Subsections 4.2.3 and 4.2.4, slightly revised and with the noise-term ignored. We will denote the transfer function matrix of the true system (assuming such transfer function exists) as $\boldsymbol{M}^{\circ}(\phi)$ and its associated parameter vector as $\boldsymbol{\pi}^{\circ}$. Thus the input-output relation of a multiple-input multiple-output system is

$$y(t) = \boldsymbol{M}^{\circ}(\phi)\, \boldsymbol{u}(t) \tag{13.1}$$

Eq. (13.1) can be written in more detail as

$$y_i(t) = \sum_{j=1}^{K} \frac{g_{ij}^{\circ}(\phi)}{h_i^{\circ}(\phi)} u_j(t) = \sum_{j=1}^{K} \frac{\bar{g}_{ij}^{\circ}(\phi)}{\bar{h}^{\circ}(\phi)} u_j(t) \qquad i=1\ldots\mu \tag{13.2}$$

where $\bar{h}^{\circ}(\phi)$ is the least common multiple of $h_i^{\circ}(\phi)$, $i=1\ldots\mu$.

Single output case. Consider first a single output, desribed by the middle section of (13.2) above. Define the regression vector

$$\boldsymbol{\varphi}_i'(t) = [u_1(t) \ldots u_1(t-\nu); \ldots \ldots ; u_K(t) \ldots u_K(t-\nu); -y_i(t-1) \ldots -y_i(t-\nu)] \tag{13.3}$$

and the parameter vector

$$\pi_i^\circ = [g_{i1}^0 \ldots g_{i1}^\nu; \ldots \ldots ; g_{i\kappa}^0 \ldots g_{i\kappa}^\nu;\ h_i^1 \ldots h_i^\nu]^{\circ\prime} \tag{13.4}$$

Then the output $y_i(t)$ can be expressed as

$$y_i(t) = \varphi_i'(t)\, \pi_i^\circ \tag{13.5}$$

This leads to the least squares estimate of the parameters

$$\hat{\pi}_i = [\sum_{k=0}^{K-1} \varphi_i(t-k)\, \varphi_i'(t-k)]^{-1} [\sum_{k=0}^{K-1} \varphi_i(t-k)\, y_i(t-k)] = [\Phi_i' \Phi_i]^{-1} \Phi_i' Y_i \tag{13.6}$$

where

$$\Phi_i = \begin{bmatrix} \varphi_i'(t) \\ \cdot \\ \cdot \\ \cdot \\ \varphi_i'(t-K+1) \end{bmatrix} \qquad Y_i = \begin{bmatrix} y_i(t) \\ \cdot \\ \cdot \\ \cdot \\ y_i(t-K+1) \end{bmatrix} \tag{13.7}$$

Multiple output system. Now consider the system with outputs $y_i(t)$, $i=1\ldots\mu$. Introduce the regression vectors

$$\psi_i'(t) = [\underbrace{0' \ldots 0'}_{1} \ldots \ldots \underbrace{0' \ldots 0'}_{i-1}\ \underbrace{u_1(t) \ldots u_1(t-\nu) \ldots \ldots u_\kappa(t) \ldots u_\kappa(t-\nu)}_{i}$$

$$\underbrace{0' \ldots 0'}_{i+1} \ldots \ldots \underbrace{0' \ldots 0'}_{\mu}\ -y_i(t-1) \ldots y_i(t-\nu)] \qquad i=1\ldots\mu \tag{13.8}$$

and the parameter vector

$$\pi^\circ = [[\tilde{g}_{11}]' \ldots [\tilde{g}_{1\kappa}]' \ldots \ldots \ldots [\tilde{g}_{\mu 1}]' \ldots [\tilde{g}_{\mu\kappa}]'; [\tilde{h}]']^{\circ\prime} \tag{13.9}$$

where

$$[\tilde{g}_{ij}]' = [\tilde{g}_{ij}^0, \tilde{g}_{ij}^1 \ldots \tilde{g}_{ij}^\nu] \qquad i=1\ldots\mu,\ \ j=1\ldots\kappa \tag{13.10}$$

$$[\tilde{h}]' = [\tilde{h}^1 \ldots \tilde{h}^\nu] \tag{13.11}$$

With these, (13.1) may also be written as

$$y(t) = \Psi'(t)\, \pi^\circ \tag{13.12}$$

where

$$\boldsymbol{\psi}'(t) = \begin{bmatrix} \boldsymbol{\psi}_1'(t) \\ \cdot \\ \cdot \\ \cdot \\ \boldsymbol{\psi}_\mu'(t) \end{bmatrix} \tag{13.13}$$

This formulation leads to the least squares estimation algorithm

$$\hat{\boldsymbol{\pi}} = [\sum_{k=0}^{K-1} \boldsymbol{\psi}(t-k)\, \boldsymbol{\psi}'(t-k)]^{-1} [\sum_{k=0}^{K-1} \boldsymbol{\psi}(t-k)\, y(t-k)] = [\boldsymbol{\Phi}'\, \boldsymbol{\Phi}]^{-1} \boldsymbol{\Phi}'\, Y \tag{13.14}$$

where

$$\boldsymbol{\Phi} = \begin{bmatrix} \boldsymbol{\psi}'(t) \\ \cdot \\ \cdot \\ \cdot \\ \boldsymbol{\psi}'(t-K+1) \end{bmatrix} \qquad Y = \begin{bmatrix} y(t) \\ \cdot \\ \cdot \\ \cdot \\ y(t-K+1) \end{bmatrix} \tag{13.15}$$

13.2. ESTIMATING THE PARAMETER CHANGE

We will first concern ourselves with the situation when the nominal model $M(\phi)$, and its associated parameter vector $\boldsymbol{\pi}$, represent the initial system correctly. When the system changes, due to a parametric fault or a change in the operating point, then the new "true" system $M^\circ(\phi)$ is described as

$$M^\circ(\phi) = M(\phi) + \Delta M(\phi) \tag{13.16}$$

with the associated parameters

$$\boldsymbol{\pi}^\circ = \boldsymbol{\pi} + \Delta\boldsymbol{\pi} \tag{13.17}$$

We wish to determine the parameter change $\Delta\boldsymbol{\pi}$. Of course, this can be done by collecting new data after the change and performing a new identification which would yield an estimate of the new $\boldsymbol{\pi}^\circ$. Here we will show an alternative way which utilizes the parity equation residuals obtained with the nominal model. Also, we will explore the transient behavior of the estimation algorithm following a parameter jump. Finally, we will consider how parameter bias and uncertainty, caused by the presence of noise, affect the change estimation.

13.2.1. Change Estimation from Parity Equation Residuals

Consider first a single moving average primary residual (see (6.13)):

$$o_i^*(t) = h_i(\phi)y_i(t) - \sum_{j=1}^{K} g_{ij}(\phi)u_j(t) \tag{13.18}$$

Observe that while the output arises from the true system $M^\circ(\phi)$, the residual is computed with the parameters of the nominal model $M(\phi)$. With the regression vector (13.3) and the model parameter vector $\boldsymbol{\pi}$, formulated according to (13.4), this can be written as

$$o_i^*(t) = y_i(t) - \boldsymbol{\varphi}_i'(t)\boldsymbol{\pi}_i \tag{13.19}$$

Now substitute $y_i(t)$ from (13.5) and take (13.17) into account:

$$o_i^*(t) = \boldsymbol{\varphi}_i'(t)\boldsymbol{\pi}_i^\circ - \boldsymbol{\varphi}_i'(t)\boldsymbol{\pi}_i = \boldsymbol{\varphi}_i'(t)\blacktriangle\boldsymbol{\pi}_i \tag{13.20}$$

Similarly, the primary residual vector

$$\mathbf{o}^*(t) = \tilde{h}(\phi)\mathbf{y}(t) - \sum_{j=1}^{K} \tilde{\mathbf{g}}_{\cdot j}(\phi)u_j(t) \tag{13.21}$$

may be written, with (13.8), (13.9) and (13.13), as

$$\mathbf{o}^*(t) = \mathbf{y}(t) - \boldsymbol{\Psi}'(t)\,\boldsymbol{\pi} = \boldsymbol{\Psi}'(t)\blacktriangle\boldsymbol{\pi} \tag{13.22}$$

Now observe that (13.20) is the same relationship between $\blacktriangle\boldsymbol{\pi}_i$ and $o_i^*(t)$ as (13.5) is between $\boldsymbol{\pi}_i^\circ$ and $y_i(t)$. Thus replacing $y_i(t\text{-}k)$, $k=0\ldots K\text{-}1$, with $o_i^*(t)$ in the least squares algorithm (13.6), it will return the estimate of $\blacktriangle\boldsymbol{\pi}_i$. Similarly, because (13.22) is formally identical with (13.12), replacing $\mathbf{y}(t\text{-}k)$, $k=0\ldots K\text{-}1$, with $\mathbf{o}^*(t)$ in (13.14), the least squares algorithm will yield the estimate of $\blacktriangle\boldsymbol{\pi}$. That is

$$\blacktriangle\hat{\boldsymbol{\pi}}_i = (\boldsymbol{\Phi}_i'\boldsymbol{\Phi}_i)^{-1}\boldsymbol{\Phi}_i'\,\mathbf{O}_i^* \tag{13.23}$$

$$\blacktriangle\hat{\boldsymbol{\pi}} = (\boldsymbol{\Phi}'\,\boldsymbol{\Phi})^{-1}\boldsymbol{\Phi}'\,\mathbf{O}^* \tag{13.24}$$

where

$$O_i^* = \begin{bmatrix} o_i^*(t) \\ \cdot \\ \cdot \\ \cdot \\ o_i^*(t-K+1) \end{bmatrix} \qquad O^* = \begin{bmatrix} o^*(t) \\ \cdot \\ \cdot \\ \cdot \\ o^*(t-K+1) \end{bmatrix} \tag{13.25}$$

The existence conditions for (13.23) and (13.24) are

$$Rank\, \Phi_i = N_i \qquad Rank\, \Phi = N \tag{13.26}$$

where N_i and N are the number of parameters in π_i and π.

Equations (13.23) and (13.24) may also be derived by expressing $y_i(t)$ and $y(t)$ from (13.19) and (13.22) and substituting them into the basic least squares algorithms. We will show this for the multiple output case. From (13.22), with (13.25),

$$Y = \Phi\pi + O^* \tag{13.27}$$

Substituting this into (13.14) yields

$$\hat{\pi} = [\Phi'\Phi]^{-1}\Phi'\Phi\pi + [\Phi'\Phi]^{-1}\Phi' O^* = \pi + \Delta\hat{\pi} \tag{13.28}$$

Equations (13.23) and (13.24) constitute an important relationship between the primary parity equation residuals and the parameter changes. However, (13.28) reveals that the estimates $\Delta\hat{\pi}_i$ and $\Delta\hat{\pi}$ they provide are algebraically identical with those computed as $\hat{\pi}_i - \pi_i$ or $\hat{\pi} - \pi$. Thus (13.23) and (13.24) may have limited practical advantage over the re-identification of the full parameters and no accuracy improvement may be expected from one approach or the other. This implies that if the nominal model contains errors, due to noise or to mismatch in the model structure, those errors cannot be estimated from the residuals. More about this will be said in Subsection 13.2.3.

13.2.2. Parameter Transients and the Sliding Window Algorithm

Now consider the situation when one (or more) of the parameters changes suddenly. Such a change triggers a transient in the identification algorithm. The estimate of the changed parameter does not follow the change immediately and, in some cases, not even smoothly. Besides, as we will show here, the sudden change of a single parameter causes transient errors in the estimate of the other parameters as well. This, of course, may be a major problem in the isolation of the change. To alleviate this problem, we will propose a sliding window variant of the least squares algorithm.

We will explore the transient phenomenon in connection with a single-output system; the results carry over to multiple-output systems as well. Assume that the parameter vector is π_i up to a certain point in time, t-τ, and it becomes $\pi_i + \blacktriangle\pi_i$ immediately after that sample. Assume further that the change involves only a single parameter, that is,

$$\blacktriangle\pi_i = [0 \dots 0 \quad \blacktriangle\pi_{il} \quad 0 \dots 0]' \tag{13.29}$$

Then the output is

$$y_i(t\text{-}k) = \varphi_i'(t\text{-}k)\pi_i \qquad k \geq \tau$$

$$y_i(t\text{-}k) = \varphi_i'(t\text{-}k)(\pi_i + \blacktriangle\pi_i) \qquad k < \tau \tag{13.30}$$

Seeking the parameter estimates from (13.6) and substituting the outputs from (13.30), we obtain

$$\hat{\pi}_i = \tag{13.31}$$

$$[\sum_{k=0}^{K-1} \varphi_i(t\text{-}k)\varphi_i'(t\text{-}k)]^{-1} [\sum_{k=0}^{min(K,\tau)-1} \varphi_i(t\text{-}k)\varphi_i'(t\text{-}k)(\pi_i + \blacktriangle\pi_i) + \sum_{k=min(K,\tau)}^{K-1} \varphi_i(t\text{-}k)\varphi_i'(t\text{-}k)\pi_i]$$

From this,

$$\blacktriangle\hat{\pi}_i = \hat{\pi}_i - \pi_i = [\sum_{k=0}^{K-1} \varphi_i(t\text{-}k)\varphi_i'(t\text{-}k)]^{-1} [\sum_{k=0}^{min(K,\tau)-1} \varphi_i(t\text{-}k)\varphi_i'(t\text{-}k)] \blacktriangle\pi_i \tag{13.32}$$

One may also write (13.32) as

$$\blacktriangle\hat{\pi}_i = (\Phi_i' \Phi_i)^{-1} \Phi_{i\tau}' \Phi_{i\tau} \blacktriangle\pi_i \tag{13.33}$$

where

$$\Phi_{i\tau} = \begin{bmatrix} \varphi_i'(t) \\ \cdot \\ \cdot \\ \cdot \\ \varphi_i'(t - min(K,\tau)+1) \end{bmatrix} \tag{13.34}$$

Clearly, if $\tau \geq K$, that is, when all the data used in the identification comes from samples after the change, then $\Phi_{i\tau} = \Phi_i$ and $\blacktriangle\hat{\pi}_i = \blacktriangle\pi_i$. However, if $\tau < K$ then the matrix $(\Phi_i' \Phi_i)^{-1} \Phi_{i\tau}' \Phi_{i\tau}$ in (13.33) is not the unit matrix and it is not even diagonal. Therefore, the effect of a single nonzero element in $\blacktriangle\pi_i$ spreads to other elements of $\blacktriangle\hat{\pi}_i$.

The sliding window algorithm. We have not said anything so far about the value of K. (13.6) and (13.14) are basically batch algorithms, which are used off-line. Then the value of K is large, in the order of magnitude of hundreds or thousands. For on-line estimation, generally the recursive algorithm is used, usually with forgetting factor (see Section 4.3). The recursive algorithm has infinite memory ($K=\infty$), therefore the transient is of infinite duration, though it tapers off exponentially due to the forgetting factor.

Because of the above difficulty, it is more advantageous to use a semi-batch algorithm for on-line parameter estimation, if fault isolation is an objective. This algorithm is basically the batch version, with a relatively short sliding data-window. Thus we may re-write (13.6) and (13.14) as

$$\hat{\pi}_i(t,K) = [\Phi_i'(t,K)\, \Phi_i(t,K)]^{-1}\, \Phi_i'(t,K)\, Y_i(t,K) \tag{13.35}$$

$$\hat{\pi}(t,K) = [\Phi'(t,K)\, \Phi(t,K)]^{-1}\, \Phi'(t,K)\, Y(t,K) \tag{13.36}$$

where the arguments t and K represent that the estimate is taken at time t from a data window of K length. Of course, (13.23) and (13.24) may be re-written in a similar way.

Computationally, the sliding window algorithm can be made at least semi-recursive. Consider the multiple output case and recall (4.58) and (4.59):

$$Q(t,K) = \sum_{k=0}^{K-1} \Psi(t-k)\, \Psi'(t-k) \qquad p(t,K) = \sum_{k=0}^{K-1} \Psi(t-k)\, y(t-k) \tag{13.37}$$

Then $\hat{\pi}(t,K) = Q^{-1}(t,K)p(t,K)$ and

$$Q(t,K) = Q(t-1,K) + \Psi(t)\, \Psi'(t) - \Psi(t-K)\, \Psi'(t-K) \tag{13.38}$$

$$p(t,K) = p(t-1,K) + \Psi(t)\, y(t) - \Psi(t-K)\, y(t-K) \tag{13.39}$$

Thus the updating of $Q(t,K)$ and $p(t,K)$ is recursive but it is necessary to store K previous values of the matrix $\Psi(t-k)\Psi'(t-k)$ and of the vector $\Psi(t-k)y(t-k)$. Note that $Q(t,K)$ may also be inverted recursively, by applying two rank-one updates at every sample.

With a sliding data-window, one knows that the transient following a parameter jump lasts exactly $K-1$ samples. Therefore, any isolation decision has to be delayed by $K-1$ samples following the detection of a change. This, of course, introduces a delay, but then one may be certain that there is no spreading of the parameter jump to the estimates of other parameters.

While a shorter window is clearly advantageous from the point of view of isolation delay, this is balanced off by higher noise sensitivity as the window length decreases. The variance of the parameter estimates will be treated in the next subsection, see (13.43). Though this formula is approximate, especially for small K's, it clearly reflects the obvious fact that the variance of the estimates increases with decreasing window length.

Another aspect of trade-off is the persistent excitation requirement. The input sequence needs to be persistently exciting of sufficient order so that the parameters (parameter changes) can be identified (see Section 4.4). If the input is strongly persistently exciting then this is guaranteed for any window length greater than or equal to the number of parameters. However, if persistent excitation is present only in the weak sense then it is more likely that the data matrix over some windows will exhibit rank defect as the window length decreases.

Example 13.1.

Consider the simple system

$$y(t) = \frac{g}{1+h\phi^{-1}} u(t)$$

driven by $u(t)=\sin t,\ t\geq 0$. Initially, $g=0.2$ and $h=-0.8$. Following the 3rd sample, h is changed to $h=-0.6$. The input and output sequences are

t	$u(t)$	$-y(t-1)$	$y(t)$
1	0.8415	0	0.1683
2	0.9093	-0.1683	0.3165
3	0.1411	-0.3165	0.2814
4	-0.7668	-0.2814	0.0175
5	-0.9589	-0.0175	-0.1813
6	-0.2794	0.1813	-0.1647

The first two columns provide the $\Phi(t,K)$ matrix while the third column the $Y(t,K)$ vector, as the window slides over the data. Choose a window length of $K=3$. Then the parameter estimates are obtained from (13.35) as

$$\hat{\pi}(3,3) = [0.2000 \quad -0.8000]$$

$$\hat{\pi}(4,3) = [0.2292 \quad -0.7259]$$

$$\hat{\pi}(5,3) = [0.4516 \quad -1.4604]$$

$$\hat{\pi}(6,3) = [0.2000 \quad -0.6000]$$

The transient is quite obvious.

13.2.3. Bias and Uncertainty in Change Estimation

If the observations are noisy then the parameter estimates are, in general, subject to bias (consistency error) and uncertainty. Here we will explore how these affect the estimation of parameter changes. A single-output subsystem will be considered; the results generalize to full multiple-output systems.

In general, the parameter estimates in a single-output subsystem can be described as

$$\hat{\pi}_i = \pi_i^{\circ} + \Delta\pi_i^{\circ} + \delta\pi_i \tag{13.40}$$

where π_i° is the true parameter vector, $\Delta\pi_i^{\circ}$ is the estimation bias (the expectation of the estimates), which depends on the true parameters, and $\delta\pi_i$ is the uncertainty (characterized by the variance of the estimates). For more on this the reader should refer back to Section 4.5.

Initial estimate. Assume that the true fault-free plant was described by the parameters $\pi_i^{\circ\circ}$ and the nominal model π_i was obtained by identification, from a large and noisy dataset. Thus, in general, π_i is subject to bias $\Delta\pi_i^{\circ\circ}$ but its uncertainty may be assumed zero, that is,

$$\pi_i = \pi_i^{\circ\circ} + \Delta\pi_i^{\circ\circ} \tag{13.41}$$

On-line estimates are then obtained from shorter noisy datasets and are subject to both bias and uncertainty. However, if there is no parameter change, and the signal-to-noise ratio is the same as it was when the reference model was identified then the bias does not change relative to the reference situation, that is,

$$\blacktriangle\hat{\pi}_i = \hat{\pi}_i - \pi_i = [\pi_i^{\circ\circ} + \Delta\pi_i^{\circ\circ} + \delta\pi_i] - [\pi_i^{\circ\circ} + \Delta\pi_i^{\circ\circ}] = \delta\pi_i \tag{13.42}$$

That is, as long as there is no parameter change, the change estimates are not affected by bias, but they are subject to the full parameter uncertainty. Approximate formulas for the size of the uncertainty were discussed in Subsection 4.5.2. By equation (4.137), the approximate variance of the uncertainty is:

$$E\{\delta\pi_i\,\delta\pi_i'\} = Var\{\hat{\pi}_i(t,K)\} \approx \frac{\sigma_{\epsilon i}^2}{K}\left[\frac{1}{K}\sum_{k=0}^{K-1}\varphi_i(t-k)\varphi_i'(t-k)\right]^{-1} \tag{13.43}$$

where $\sigma^2_{\epsilon i}$ is the variance of the noise-induced equation error.

Now assume that there has been a parameter change $\blacktriangle\pi_i$ so that the new true system is $\pi_i^o = \pi_i^{oo} + \blacktriangle\pi_i$. Then the change estimate becomes

$$\blacktriangle\hat{\pi}_i = \hat{\pi}_i - \pi_i = [\pi_i^{oo} + \blacktriangle\pi_i + \Delta(\pi_i^{oo} + \blacktriangle\pi_i) + \delta\pi_i] - [\pi_i^{oo} + \Delta\pi_i^{oo}]$$

$$= \blacktriangle\pi_i + \Delta(\pi_i^{oo} + \blacktriangle\pi_i) - \Delta\pi_i^{oo} + \delta\pi_i \tag{13.44}$$

The consistency error was analyzed in Subsection 4.5.1. Applying Eq. (4.128) to the change estimation problem one can see that

$$\Delta(\pi_i^{oo} + \blacktriangle\pi_i) \approx \Delta\pi_i^{oo} - (Q^*_i)^{-1} \Delta Q^*_i \blacktriangle\pi_i \tag{13.45}$$

where Q^*_i is the data matrix defined in (4.122) and (4.123) and where we ignored the fact that Q^*_i also changes somewhat as we move from π_i^{oo} to $\pi_i^{oo} + \blacktriangle\pi_i$. With this,

$$\boxed{\blacktriangle\hat{\pi}_i = \blacktriangle\pi_i - (Q^*_i)^{-1} \Delta Q^*_i \blacktriangle\pi_i + \delta\pi_i} \tag{13.46}$$

That is, the estimate of the change is subject to bias and uncertainty. Further:

a. the change estimate inherits the uncertainty of the full parameter estimate;

b. the size of the bias is proportional to the size of the change;

c. since the matrix $(Q^*_i)^{-1} \Delta Q^*_i$ is not diagonal, a change in a single parameter causes bias (that is, a nonzero change estimate) in all the other parameters.

Of the above properties, a, affects the sensitivity of both detection and isolation while c, is an additional effect to take into account with isolation. Of course, if the identification method is unbiased, so is the change estimate.

13.3. ESTIMATING CHANGES IN THE UNDERLYING PARAMETERS

Faults are usually associated with physical parameters of the plant which lie behind the parameters of a transfer function or similar model. Therefore one may be interested in the changes of these underlying parameters.

The relationship between the underlying parameters θ and the model parameters π is usually nonlinear:

$$\pi = f(\theta) \tag{13.47}$$

The relationship between the changes can be approximated as

$$\Delta \boldsymbol{\pi} \approx \boldsymbol{R} \, \Delta \boldsymbol{\theta} \tag{13.48}$$

where

$$\boldsymbol{R} = \left[\frac{\partial \boldsymbol{\pi}}{\partial \boldsymbol{\theta}} \right]_{\Delta \boldsymbol{\theta}=0} = \begin{bmatrix} \frac{\partial \pi_1}{\partial \theta_1} & \cdots & \frac{\partial \pi_1}{\partial \theta_L} \\ \vdots & & \\ \frac{\partial \pi_N}{\partial \theta_1} & \cdots & \frac{\partial \pi_N}{\partial \theta_L} \end{bmatrix}_{\Delta \boldsymbol{\theta}=0} \tag{13.49}$$

Here we implied that $\boldsymbol{\pi}$ has N elements while $\boldsymbol{\theta}$ has L elements. Observe also that the derivative is taken at the *nominal* value of the parameters.

Below we will discuss two procedures to estimate the changes in the underlying parameters. In the first procedure, the model parameter changes are first estimated and then the underlying parameter changes are computed from these in a separate step. The other procedure involves a modification of the least squares algorithm so that it estimates the underlying parameter changes directly from the observed data.

13.3.1. Indirect Estimation of Underlying Parameter Changes

Assume that $\Delta \boldsymbol{\pi}$ has been estimated. If $L=N$ then $\boldsymbol{R}$ is square. If further $f(\boldsymbol{\theta})$ is a linear relationship then $\Delta \hat{\boldsymbol{\theta}}$ is obtained by inverting (13.48):

$$\Delta \hat{\boldsymbol{\theta}} = \boldsymbol{R}^{-1} \Delta \hat{\boldsymbol{\pi}} \tag{13.50}$$

(13.50) may be used, as an approximation, even if $f(\boldsymbol{\theta})$ is nonlinear, provided $\Delta \hat{\boldsymbol{\pi}}$ is relatively small. If greater accuracy is needed, then $\Delta \hat{\boldsymbol{\theta}}$ has to be computed numerically, in an iterative algorithm. By the popular Newton-Raphson technique, the $k+1$-th iteration $\Delta \hat{\boldsymbol{\theta}}^{k+1}$ is obtained as

$$\Delta \hat{\boldsymbol{\theta}}^{k+1} = \Delta \hat{\boldsymbol{\theta}}^{k} + \boldsymbol{R}^{-1}(\Delta \hat{\boldsymbol{\theta}}^{k}) \, [\Delta \hat{\boldsymbol{\pi}} - f(\Delta \hat{\boldsymbol{\theta}}^{k})] \tag{13.51}$$

with $\Delta \hat{\boldsymbol{\theta}}^{0}=\boldsymbol{0}$. Observe that the partial derivative matrix now is computed in each step at $\Delta \hat{\boldsymbol{\theta}}^{k}$, instead of the nominal point. The iteration stops when $|\Delta \hat{\boldsymbol{\theta}}^{k+1} - \Delta \hat{\boldsymbol{\theta}}^{k}| < \epsilon$, where ϵ is a preset threshold. For the algorithm to converge, $f(\boldsymbol{\theta})$ has to be monotonous between $\Delta \hat{\boldsymbol{\theta}}^{0} = \boldsymbol{0}$ and the solution $\Delta \hat{\boldsymbol{\theta}}$; if the latter is relatively small then there is a good chance this condition is satisfied.

If $L<N$ then a least squares fit may be computed as

$$\Delta\hat{\boldsymbol{\theta}} = (\boldsymbol{R}'\boldsymbol{R})^{-1}\boldsymbol{R}'\Delta\hat{\boldsymbol{\pi}} \tag{13.52}$$

For the solution to exist, $\boldsymbol{R}$ must have full column rank. Note that (13.52) implies a linear relationship. If the nonlinearity also needs to be taken into account then (13.52) may be combined with (13.51), so that the inverse matrix $\boldsymbol{R}^{-1}$ in (13.51) is replaced with the pseudo-inverse $(\boldsymbol{R}'\boldsymbol{R})^{-1}\boldsymbol{R}'$. The latter then has to be re-computed at every step of the iteration.

13.3.2. Direct Estimation of Underlying Parameter Changes

Consider the relationship (13.20) and substitute

$$\Delta\boldsymbol{\pi}_i = \boldsymbol{R}_i\Delta\boldsymbol{\theta}_i \quad \textit{where} \quad \boldsymbol{R}_i = [\partial\,\boldsymbol{\pi}_i/\partial\,\boldsymbol{\theta}_i]_{\Delta\pi i=0} \tag{13.53}$$

and $\boldsymbol{\theta}_i$ is the set of underlying parameters associated with the i-th single-output subsystem. This yields

$$o_i^*(t) = \boldsymbol{\varphi}_i'(t)\,\boldsymbol{R}_i\Delta\boldsymbol{\theta}_i \tag{13.54}$$

Comparing this to (13.20) reveals that if we replace $\boldsymbol{\varphi}_i'(t\text{-}k)$, $k=0...K\text{-}1$, with $\boldsymbol{\varphi}_i'(t\text{-}k)\boldsymbol{R}_i$ (that is, $\boldsymbol{\Phi}_i$ with $\boldsymbol{\Phi}_i\boldsymbol{R}_i$) in (13.23) then the least squares algorithm provides a direct estimate of $\Delta\boldsymbol{\theta}_i$. Similarly, substituting $\Delta\boldsymbol{\pi}=\boldsymbol{R}\Delta\boldsymbol{\theta}$ into (13.22) yields

$$o^*(t) = \boldsymbol{\Psi}'(t)\,\boldsymbol{R}\Delta\boldsymbol{\theta} \tag{13.55}$$

Thus replacing $\boldsymbol{\Phi}$ with $\boldsymbol{\Phi}\boldsymbol{R}$ in (13.24) provides a direct estimate of $\Delta\boldsymbol{\theta}$. That is:

$$\Delta\hat{\boldsymbol{\theta}}_i = (\boldsymbol{R}_i'\boldsymbol{\Phi}_i'\boldsymbol{\Phi}_i\boldsymbol{R}_i)^{-1}\boldsymbol{R}_i'\boldsymbol{\Phi}_i'\boldsymbol{O}_i^* \tag{13.56}$$

$$\Delta\hat{\boldsymbol{\theta}} = (\boldsymbol{R}'\boldsymbol{\Phi}'\boldsymbol{\Phi}\boldsymbol{R})^{-1}\boldsymbol{R}'\boldsymbol{\Phi}'\boldsymbol{O}^* \tag{13.57}$$

Notice that while (13.56) estimates the changes of the underlying parameter set from the observations of a single output, (13.57) does this from a set of μ outputs.

The existence of the solutions in (13.56) and (13.57) requires that

$$Rank\,(\Phi_i R_i) = L_i \qquad\qquad Rank\,(\Phi R) = L \tag{13.58}$$

This implies the condition

$$Rank\, R_i = L_i \text{ and } Rank\, \Phi_i \geq L_i \quad Rank\, R = L \text{ and } Rank\, \Phi \geq L \tag{13.59}$$

The satisfaction of (13.59) does not guarantee that (13.58) is also satisfied. Still, (13.58) poses milder persistent excitation conditions than (13.26), which applies if $\Delta\pi$ is estimated first.

13.3.3. An Important Link to Parity Relations

Here we will show that parity relations designed for diagonal (unit) response to underlying parameter faults are special cases of the sliding window algorithm.

Consider first the multiple output case and substitute $o^*(t)$ from (13.21) into (13.55)

$$o^*(t) = \tilde{h}(\phi)y(t) - \tilde{G}(\phi)u(t) = \Psi'(t)\, R\, \Delta\theta \tag{13.60}$$

This is a set of moving average primary residuals, the internal part shown as a function of the underlying parameter discrepancies. While it is expressed in terms of the regression matrix Ψ and the parameter-to-parameter derivatives R, it is algebraically identical with equation (9.29) in Chapter 9 where the internal form was expressed in terms of the variable vectors $u(t)$ and $y(t)$ and the derivatives of the transfer functions with respect to the parameters, $\partial\tilde{G}(\phi,\theta)/\partial\theta_l$ and $\partial\tilde{h}(\phi,\theta)/\partial\theta_l$. Assume that $\mu=L$, so that the matrix $\Psi'(t)R$ is square. Design a transformed residual vector $r(t)=W(t)o^*(t)$ with unit diagonal response to the underlying parameter discrepancies. This implies

$$W(t)\, \Psi'(t)\, R = I \tag{13.61}$$

from which

$$r(t) = [\Psi'(t)\, R]^{-1}\, o^*(t) \tag{13.62}$$

Now apply the least squares algorithm (13.57) to the $\mu=L$ case, with a sliding window setting $K=1$. Then $\Phi(t,1) = \Psi'(t)$ and $O^*(t,1)=o^*(t)$. Further, $\Phi(t,1)R$ is square so that

$$[R'\, \Phi'(t,1)\, \Phi(t,1)\, R]^{-1} R'\, \Phi'(t,1) = [\Phi(t,1)\, R]^{-1} \tag{13.63}$$

Thus

$$\Delta\hat{\theta}(t,1) = [\Psi'(t)\, R]^{-1}\, o^*(t) \tag{13.64}$$

Clearly,

$$\boxed{r(t) = \Delta\hat{\theta}(t,1)} \tag{13.65}$$

that is, *the set of parity relation residuals, designed for diagonal response to underlying parameter faults, is identical with the minimum data-length sliding window least squares estimate of the same.*

Similar results can be obtained from a sequence of L samples of a single residual $o_i^*(t)$. Consider (13.54) and substitute $o_i^*(t)$ from (13.18):

$$o_i^*(t) = h_i(\phi)y_i(t) - g_{i.}(\phi)u(t) = \varphi_i'(t)\, R_i\, \Delta\theta_i \tag{13.66}$$

Now take a series of $K=L$ such residuals

$$O_i^*(t) = \begin{bmatrix} o_i^*(t) \\ \cdot \\ \cdot \\ \cdot \\ o_i^*(t-L+1) \end{bmatrix} = \begin{bmatrix} \varphi'(t) \\ \cdot \\ \cdot \\ \cdot \\ \varphi'(t-L+1) \end{bmatrix} R_i\, \Delta\theta_i = \Phi_i(t,L)\, R_i\, \Delta\theta_i \tag{13.67}$$

Observe that $\Phi_i(t,L)R_i$ is square. Design a set of transformed residuals $r(t)=W(t)O_i^*(t)$, so that their response to $\Delta\theta_i$ is diagonal. Then $W(t)\Phi_i(t,L)R_i=I$, yielding

$$r(t) = [\Phi_i(t,L)\, R_i]^{-1}\, O_i^*(t) \tag{13.68}$$

Now apply (13.56) with a sliding window of $K=L$ samples. Observe that this is the minimum data-length from which estimates can be computed. Just like in (13.63), the pseudo inverse of $\Phi_i(t,L)R_i$ becomes the ordinary inverse and the change estimate is obtained as

$$\Delta\hat{\theta}_i(t,L) = [\Phi_i(t,L)\, R_i]^{-1}\, O_i^*(t) \tag{13.69}$$

Thus

$$r(t) = \blacktriangle \hat{\boldsymbol{\theta}}_i(t,L) \tag{13.70}$$

So again, the residual set designed for diagonal response to the underlying parameter changes is identical with the minimum dataset least squares estimate.

We have shown two cases above: (i) a single sample of residuals obtained from several outputs ($\mu=L$, $K=1$), and (ii) a set of samples of a residual obtained from a single output ($\mu=1$, $K=L$). It is, of course, possible to create residuals with $\mu \neq 1$ and $K \neq 1$, as long as $\mu \cdot K \geq L$.

With minimum data-length, the "estimates" of course are extremely noise sensitive. The persistent excitation conditions are the same as in (13.58)-(13.59), but now they have to be satisfied within a minimum-length sample, that is, in the strong sense.

Example 13.2.

Simulation studies have been performed (DiPierro, 1997) on an eight-order linear model of an automobile engine with two inputs and four outputs (Kamei et al, 1987; Gertler and Kunwer, 1995). Four of the system poles were considered as underlying parameters and all four outputs were used to generate residuals. The behavior of the single change estimate $\blacktriangle \hat{\theta}_1$ was studied while a 10% change was applied to each of the four parameters, one at a time. Fig. 13.1 shows the estimates obtained without noise, with sliding windows of $K=1$, $K=5$ and $K=20$, and with exponential forgetting, $\lambda=0.95$. The transient interaction is clearly visible, as well as the advantage of a finite window over the infinite exponential algorithm. In Fig. 13.2, the same experiments were repeated with 1% random noise added to the outputs. The effect of the noise is quite obvious. Clearly, a compromise is needed which, in this example, is somewhere between $K=5$ and $K=20$.

Note that the two inputs were excited with identical signals, the combination of five cosine functions with relative prime frequencies. Clearly, with $u_1(t)=u_2(t)$, we have a gain-shifting situation, so the parameters of the model numerator could not be uniquely identified. This, however, does not interfere with the identification of the (changes of the) four underlying parameters.

13.4. CONTINUOUS-TIME MODEL IDENTIFICATION

The relationship between the physical parameters of the continuous-time plant and the parameters of its discrete-time model are usually complex and nonlinear. We have shown in the preceding sections how to approach this problem in the framework of discrete-time models. These techniques have been straight-

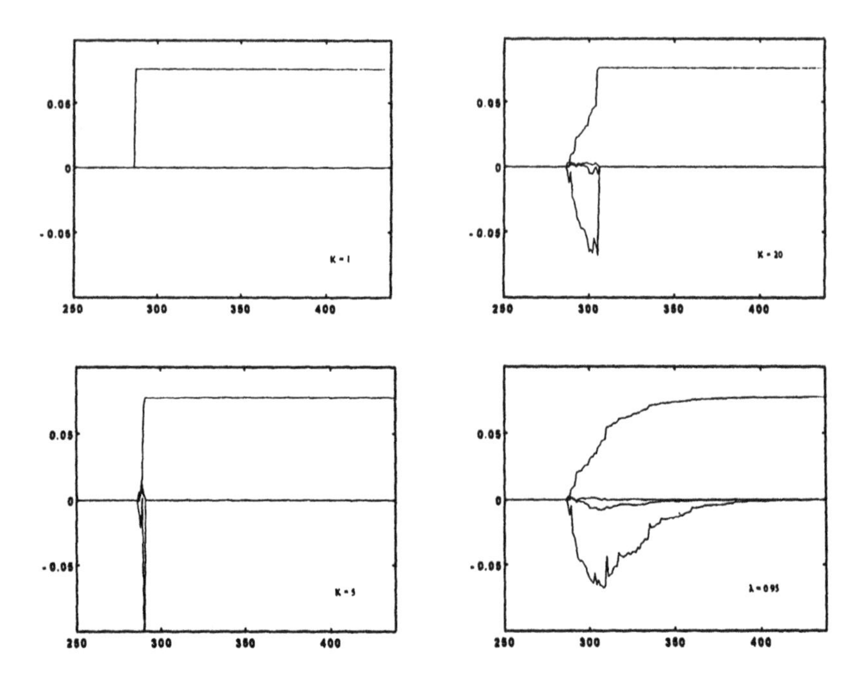

FIGURE 13.1. Estimates of underlying parameter changes: no noise.

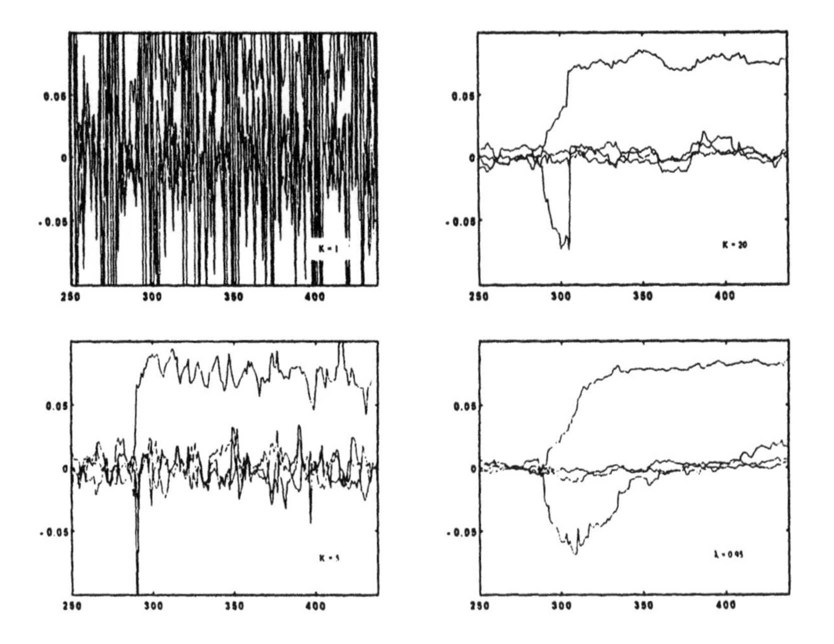

FIGURE 13.2. Estimates of underlying parameter changes: 1% output noise.

forward extensions of the standard discrete-time identification procedure but, because of the nonlinear relationship, they usually involved approximation. An alternative approach is to find the parameters of the continuous-time model by direct identification. In many cases, those parameters are the physical parameters proper, or some simple combinations thereof. The identification procedure still relies on sampled data, but it is more complex than model parameter identification in the discrete-time case.

Continuous-time systems are described by differential equations. With the usual assumptions that the system is linear and time-invariant, these differential equations are also linear with constant parameters, and directly correspond to continuous-time (Laplace transformed) transfer functions (see Subsection 2.4.1). For a single-input single-output linear time-invariant system, the differential equation is written as

$$y(\tau) + a^1 \frac{dy(\tau)}{d\tau} + \ldots + a^n \frac{d^n y(\tau)}{d\tau^n} = b^o u(\tau) + b^1 \frac{du(\tau)}{d\tau} + \ldots + b^n \frac{d^n u(\tau)}{d\tau^n} \qquad (13.71)$$

The data is available in sampled form, therefore the derivatives in (13.71) need to be approximated from discrete samples. The main difficulty with this approach is that it is very noise sensitive, since the higher derivatives amplify the high-frequency noise. Therefore an integral equation equivalent of (13.71) is usually used for identification. While this resolves the noise problem, it in turn requires the knowledge (or estimation) of initial conditions. The following procedure, which is due to Sagara and Zhao (1990), eliminates the initial condition problem as well, by using finite integration over a sliding window. We will present the algorithm here in a slightly simplified form.

The sliding integral operator. Define the sliding integral of a continuous-time variable $x(\tau)$ as

$$I_1[x, t] = \int_{\tau=t\Delta-m\Delta}^{t\Delta} x(\tau)\, d\tau \qquad (13.72)$$

where Δ is the sampling interval and m is the window length in sample counts. With rectangular approximation, this can be written as

$$I_1[x, t] = \Delta \sum_{k=1}^{m} x(t\Delta - k\Delta) = \Delta\, Q(\phi)\, x(t) \qquad (13.73)$$

where

$$Q(\phi) = (\phi^{-1} + \phi^{-2} + \ldots + \phi^{-m}) \qquad (13.74)$$

With this, the second sliding integral is

$$I_2[x, t] = I_1[I_1(x, t'), t] = \Delta^2 Q^2(\phi)\, x(t) \tag{13.75}$$

Here t' is a sliding discrete time variable varying over the range $1...m$. Similarly, the j-th sliding integral is

$$I_j[x, t] = \Delta^j Q^j(\phi)\, x(t) \tag{13.76}$$

The sliding integrals of the derivatives. Now apply the sliding integral to the differential equation (13.71) n times:

$$\begin{aligned} I_n[y, t] + a^1 I_n[y^{(1)}, t] + \ldots + a^n I_n[y^{(n)}, t] \\ = b^0 I_n[u, t] + b^1 I_n[u^{(1)}, t] + \ldots + b^n I_n[u^{(n)}, t] \end{aligned} \tag{13.77}$$

where $y^{(j)}$ and $u^{(j)}$ are the j-th derivative of $y(\tau)$ and $u(\tau)$. To compute the sliding integrals of these derivatives, consider the following relationship:

$$I_1[x^{(j)}, t] = x^{(j-1)}(t\Delta) - x^{(j-1)}(t\Delta - m\Delta) \tag{13.78}$$

where we utilized the exact definition (13.72) of the sliding integral and the fact that the integral of $x^{(j)}(\tau)$ is $x^{(j-1)}(\tau)$. This equation can also be written as

$$I_1[x^{(j)}, t] = (1 - \phi^{-m})\, x^{(j-1)}(t) \tag{13.79}$$

With this, assuming $j \leq 2$,

$$I_2[x^{(j)}, t] = I_1[I_1(x^{(j)}, t'), t] = (1 - \phi^{-m})^2 x^{(j-2)}(t) \tag{13.80}$$

Further, utilizing that $x^{(j-j)}(\tau) = x(\tau)$,

$$I_j[x^{(j)}, t] = (1 - \phi^{-m})^j x(t) \tag{13.81}$$

Finally, by combining (13.81) with (13.76),

$$I_n[x^{(j)}, t] = I_{n-j}[I_j(x^{(j)}, t'), t] = \rho_j(\phi)\, x(t) \tag{13.82}$$

where

$$\rho_j(\phi) = \Delta^{n-j} Q^{n-j}(1 - \phi^{-m})^j \qquad j = 0...n \tag{13.83}$$

Observe that

$$Deg\ \rho_j(\phi) = n \cdot m \qquad for\ j=0 \ldots n \tag{13.84}$$

Identification from sliding integrals. Let us rewrite (13.77) with (13.82):

$$\rho_o(\phi)y(t) + a^1 \rho_1(\phi)y(t) + \ldots + a^n \rho_n(\phi)y(t) \tag{13.85}$$
$$= b^o \rho_o(\phi)u(t) + b^1 \rho_1(\phi)u(t) + \ldots + b^n \rho_n(\phi)u(t)$$

This can also be written as

$$v(t) = \boldsymbol{\varphi}'(t)\,\boldsymbol{\theta} \tag{13.86}$$

where

$$v(t) = \rho_o(\phi)y(t) \tag{13.87a}$$

$$\boldsymbol{\varphi}'(t) = [\rho_o(\phi)u(t) \ldots \rho_n(\phi)u(t); -\rho_1(\phi)y(t) \ldots -\rho_n(\phi)y(t)] \tag{13.87b}$$

$$\boldsymbol{\theta} = [b^o \ldots b^n ; a^1 \ldots a^n]' \tag{13.87c}$$

This leads to the least squares algorithm

$$\hat{\boldsymbol{\theta}} = (\boldsymbol{\Phi}'\boldsymbol{\Phi})^{-1}\boldsymbol{\Phi}'\,V \tag{13.88}$$

where

$$\boldsymbol{\Phi} = \begin{bmatrix} \boldsymbol{\varphi}'(t) \\ \cdot \\ \cdot \\ \cdot \\ \boldsymbol{\varphi}'(t-K+1) \end{bmatrix} \qquad V = \begin{bmatrix} v(t) \\ \cdot \\ \cdot \\ \cdot \\ v(t-K+1) \end{bmatrix} \tag{13.89}$$

The size of the $\boldsymbol{\Phi}'\boldsymbol{\Phi}$ matrix is $(2n+1)\cdot(2n+1)$. An analysis of the columns of the $\boldsymbol{\Phi}$ matrix shows that the input signal needs to be persistently exciting of order $2n+1$. However, computationally the algorithm is significantly more complex than the standard discrete-time algorithms. This is because each element of the $\boldsymbol{\Phi}$ and V matrices is the weighted sum of $n \cdot m$ observations. Clearly there is a trade-off between computational complexity and noise rejection, but for the latter to be acceptable one may need to choose the window length as large as 20. This may create a problematic burden in on-line applications.

13.5. NOTES AND REFERENCES

Early work on the application of parameter estimation methods to fault detection and isolation, much of it done by himself and coworkers, was surveyed by Rolf Isermann (1984). Additional early contributions are due, among others, to A. Rault and colleagues (Rault et al, 1984). The Isermann school continued to play a leading role in this area, marked by a long succession of papers and practical applications, see for example (Isermann and Freyermuth, 1991; Isermann, 1993). Of the multitude of other contributors, we mention the interesting work of B. Ninness and coworkers (Ninness et al, 1991) and K. Kumamaru and colleagues (Kumamaru et al, 1985, 1996, 1997).

The idea of estimating parameter changes from parity relation residuals was first proposed in 1985, at a little-known conference (Gertler, 1985), and then forgotten. It resurfaced ten years later, simultaneously in (Gertler, 1995b) and (Höfling and Isermann, 1995). The same estimation formula was shown also in (Delmaire et al, 1994), as a link between parity relations and parameter estimation. The full depth of the relationship was exposed in (Gertler, 1995b), see also (Gertler, 1997). Zhang and colleagues (1994) used parity relation residuals to detect small parameter changes, without explicit change estimation. T. Höfling explored the combination of parity relations and parameter estimation in fault detection and diagnosis in several papers; for a summary see (Höfling, 1996). Relations between diagnostic observers and parameter estimation were investigated in (Magni, 1995; Alcorta Garcia and Frank, 1996).

The indirect procedure to estimate underlying parameter changes was described in (Schumann et al, 1992). Direct estimation of (small) changes in the underlying parameters was introduced in (Gertler, 1995b), see also (Gertler, 1997). Estimation transients and the sliding window algorithm were investigated by Gus DiPierro (1997) and reported in (Gertler and DiPierro, 1997). The considerations on bias and uncertainty in change estimation appear here for the first time.

Parameter estimation from the continuous-time model was addressed in (Schumann et al, 1992), using earlier results by Young (1981). The sliding integral algorithm described here is due to Sagara and Zhao (1990). An extension to fault detection and isolation was reported in (Li and Gertler, 1997).

References

Alcorta Garcia, E. and Frank, P.M. (1996). On the relationship between observer based and parameter identification based approaches to fault detection. *Proc. IFAC 13th World Congress,* (San Francisco, CA), Vol. N, pp. 25-30.

Almásy, G.A. and Sztanó, T. (1975). Checking and correction of measurements on the basis of linear system model. *Problems of Control and Information Theory, Vol. 4,* pp. 57-69.

Anderson, B.D.O. and Moore, J.B. (1979). *Optimal Filtering.* Prentice Hall.

Åström, K.J. and Bohlin, T. (1965). Numerical identification of linear dynamic systems from normal operating reports. *IFAC Symposium on Self-Adaptive Control Systems* (Teddington, England).

Åström, K.J. and Eykhoff, P. (1971). System identification - A survey. *Automatica, Vol. 7,* pp. 123-162.

Åström, K.J. and Wittenmark, B. (1989). *Adaptive Control.* Addison-Wesley.

Åström, K.J. and Wittenmark, B. (1997). *Computer Controlled Systems - Theory and Design.* Third Edition. Prentice Hall.

Basseville, M. and Benveniste, A. (1983). Design and comparative study of some sequential jump detection algorithms for digital signals. *IEEE Trans. on Acoustics, Speech, Signal Proc., ASSP-31,* pp. 521-535.

Basseville, M. and Benveniste, A. (Eds.) (1986). *Detection of Abrupt Changes in Signals and Dynamical Systems, LNCIS 77,* Springer Verlag.

Basseville, M. (1986). Two examples of the application of the GLR method in signal processing. In: *Detection of Abrupt Changes in Signals and Dynamical Systems, LNCIS 77,* (Basseville, M. and Benveniste, A., Eds.; Springer Verlag) pp. 50-74.

Basseville, M., Benveniste, A., Moustakides, G. and Rouge, A. (1987). Detection and diagnosis of changes in the eigenstructure of nonstationary multivariable systems. *Automatica, Vol. 23,* pp. 479-489.

Basseville, M. (1988). Detecting changes in signals and systems - A survey. *Automatica, Vol. 24,* pp. 309-326.

Basseville, M. and Nikiforov, I.V. (1993). *Detection of Abrupt Changes - Theory and Application.* Prentice Hall, Information and System Sciences Series.

Beard, R.V. (1971). Failure accommodation in linear systems through self-reorganization. *Rept. MTV-71-1,* Man Vehicle Laboratory, MIT, Cambridge, MA.

Ben-Haim, Y. (1980). An algorithm for failure location in a complex network. *Nuclear Science and Engineering, Vol. 75,* pp. 191-199.

Ben-Haim, Y. (1983). Malfunction isolation in linear stochastic systems: Application to nuclear power plants. *Nuclear Science and Engineering, Vol. 85,* pp. 156-166.

Benveniste, A., Basseville, M. and Moustakides, G. (1987). The asymptotic local approach to change detection and model validation. *IEEE Trans. on Automatic Control, Vol. AC-32,* pp. 583-592.

Bokor, J. and Keviczky, L. (1989). Detection filter design for eigenvalue changes in dynamic systems. *Proc. of 28th IEEE Conference on Decison and Control,* (Tampa, FL), pp. 1807-1808.

Brogan, W.L. (1985). *Modern Control Theory.* Second Edition. Prentice Hall.

Chen, C.T. (1984). *Linear System Theory and Design.* Holt, Rinehart and Winston, Inc.

Chen, J., Patton, R.J. and Zhang, H.Y. (1996). Design of unknown input observers and robust fault detection filters. *Int. Journal of Control, Vol. 63,* pp. 85-105.

Chien, T.T. and Adams, M.B. (1976). A sequential failure detection approach and its application. *IEEE Trans. on Automatic Control, Vol. AC-21,* pp. 750-757.

Chow, E.Y. (1980). Failure detection system design methodology. D.Sc. Thesis, MIT, Cambridge, MA.

Chow, E.Y. and A.S. Willsky (1984). Analytical redundancy and the design of robust failure detection systems. *IEEE Trans. on Automatic Control, Vol. AC-29*, pp. 603-614.

Clark, R.N., Fosth, D.C. and Walton, W.M. (1975). Detecting instrument malfunctions in control systems. *IEEE Trans. on Aerospace and Electronic Systems, Vol. AES-11*, pp. 465-473.

Clark, R.N. (1979). The dedicated observer approach to instrument fault detection. *Proc. 15th IEEE Conference on Decision and Control*, (Fort Lauderdale, FL), pp. 237-241.

Clark, R.N. and Campbell, B. (1982). Instrument fault detection in a pressurized water reactor pressurizer. *Nuclear Technology, Vol. 56*, pp. 23-32.

Clark, R.N. (1989). State estimation schemes for instrument fault detection. In: *Fault Diagnosis in Dynamic Systems* (R.J. Patton, P.M. Frank and R.N. Clark, Eds.; Prentice Hall) pp. 21-45.

Clarke, D.W. (1967). Generalized-least-squares estimation of the parameters of a dynamic model. *First IFAC Symposium on Identification in Automatic Control Systems*, (Prague, Czechoslovakia), paper 3.17.

Crowe, C.M. (1988). Recursive identification of gross errors in linear data reconciliation. *AIChE Journal, Vol. 34*, pp. 541-550.

Deckert, J.C., Desai, M.N., Deyst, J.J. and Willsky, A.S. (1977). F-8 DFBW sensor failure detection using analytical redundancy. *IEEE Trans. on Automatic Control, Vol. AC-22*, pp. 795-803.

Delmaire, G., Cassar, J.P. and Staroswiecki, M. (1994). Comparison of identification and parity space approaches for failure detection in single-input single-output systems. *Proc. 33rd IEEE Conference on Decision and Control*, (Orlando, Florida), pp. 2279-2285.

Desai, M. and Ray, A. (1984). A fault detection and isolation methodology - Theory and application. *Proc. 1984 American Control Conference*, (San Diego, CA), pp. 262-270.

Ding, X. and Frank, P.M. (1989). Fault detection via optimally robust detection filters. *Proc. of the 28th IEEE Conference on Decision and Control*, (Tampa, FL), pp. 1767-1772.

Ding, X. and Frank, P. (1990). Fault detection via factorization approach. *Syst. Cont. Lett., Vol. 14*, pp. 431-436.

Ding, X. and Frank, P.M. (1991). Frequency domain approach and threshold selector for robust model-based fault detection and isolation. *IFAC Safeprocess Conference,* (Baden-Baden, Germany), pp. 307-312.

DiPierro, G. (1997). An on-line method of detecting parametric faults. M.S. Thesis, George Mason University, Fairfax, VA.

Downs, J.J. and Vogel, E.F. (1993). A plant-wide industrial process control problem. *Computers and Chemical Engineering, Vol. 17,* pp. 245-255.

Eckart, C. and Young, G. (1936). The approximation of one matrix by another of lower rank. *Psychometrica, Vol. 1,* pp. 211-218.

Edelmayer, A., Bokor, J., Szigeti, F. and Keviczky, L. (1997). Robust detection filter design in the presence of time-varying system perturbations. *Automatica, Vol. 33,* pp. 471-476.

Emami-Naeini, A., Akhter, M.M. and Rock, S.M. (1988). Effect of model uncertainty on failure detection: The threshold selector. *IEEE Trans. on Automatic Control, Vol. AC-33,* pp. 1106-1115.

Eykhoff, P. (1974). *System Identification.* John Wiley, London.

Fang, X. (1993). An intelligent and robust scheme for failure detection and diagnosis in dynamic systems. Ph.D. Dissertation, George Mason University, Fairfax, VA.

Feller, W. (1950). *An Introduction to Probability Theory and Its Applications.* Second edition. John Wiley, New York.

Fisher, R.A. (1950). *Contributions to Mathematical Statistics.* Collected works. John Wiley, New York.

Francis, B.A. (1987). *A Course in H_∞ Control Theory, LNCIS 88*, Springer Verlag.

Frank, P.M. and Keller, L. (1980). Sensitivity discriminating observer design for instrument failure detection. *IEEE Trans. on Aerospace and Electronic Systems, Vol. AES-16,* pp. 460-467.

Frank, P.M. and Wünnenberg, J. (1989). Robust fault diagnosis using unknown input observer schemes. In *Fault Diagnosis in Dynamic Systems* (R.J. Patton, P.M. Frank and R. Clark, Eds.; Prentice Hall), pp. 47-98.

Frank, P.M. (1989). Evaluation of analytical redundancy for fault diagnosis in dynamic systems. *Preprints of IFAC Symposium on Advanced Information Processing in Automatic Control,* (Nancy, France), pp. I.7-I.19.

Frank, P. (1990). Fault diagnosis in dynamic systems using analytical and knowledge-based redundancy. *Automatica, Vol. 26*, pp. 459-474.

Frank, P. (1991). Enhancement of robustness in observer based fault detection. *Proc. IFAC Safeprocess Conference*, (Baden-Baden, Germany), pp. 99-111.

Frank, P. (1992). Robust model-based fault detection in dynamic systems. *Preprints of IFAC Symposium on On-Line Fault Detection in the Chemical Process Industries*, (Newark, DE), pp. 1-13.

Frank, P. (1993). Advances in observer-based fault diagnosis. *Proc. Tooldiag Conference*, (Toulouse, France), pp. 817-836.

Friedland, B. (1979). Maximum likelihood estimation of a process with random transitions (failures). *IEEE Trans. on Automatic Control, Vol. AC-24*, pp. 932-937.

Friedland, B. (1981). Multidimensional maximum likelihood failure detection and estimation. *IEEE Trans. on Automatic Control, Vol. AC-26*, pp. 567-570.

Friedland, B. (1982). Maximum likelihood failure detection of aircraft flight control sensors. *Journal of Guidance, Control and Dynamics, Vol. 5*, pp. 498-503.

Gantmakher, F.R. (1959). *Theory of Matrices*. Chelsea Publishing Co., Inc., New York.

Ge, W. and Fang, C.Z. (1988). Detection of faulty components via robust observation. *Int. Journal of Control, Vol. 47*, pp. 581-599.

Gertler, J. and Almásy, G.A. (1973). Balance calculations through dynamic system modeling. *Automatica, Vol. 9*, pp. 79-85.

Gertler, J. and Singer, D. (1985). Augmented models for statistical fault isolation in complex dynamic systems. *Proc. American Control Conference*, (Boston, MA.), Vol. 1, pp. 317-322.

Gertler, J., Singer, D. and Sundar, A. (1985). A robustified linear fault isolation technique for complex dynamic systems. *Preprints of IFAC/IFIP Conference on Digital Computer Applications to Process Control*, (Vienna, Austria), pp. 493-498.

Gertler, J. (1985). Fault detection and isolation in complex technical systems - A new model error approach. *Proc. Conference on Information Sciences and Systems, Johns Hopkins University*, (Baltimore, MD), pp. 68-73.

Gertler, J. (1988). A survey of model based failure detection and isolation in complex plants. *IEEE Control Systems Magazine, Vol. 8*, No. 6, pp. 3-11.

Gertler, J. and Luo, Q. (1989). Robust isolable models for failure diagnosis. *AIChE Journal, Vol. 35*, pp. 1856-1868.

Gertler, J. and Singer, D. (1990). A new structural framework for parity equation based failure detection and isolation. *Automatica, Vol. 26*, pp. 381-388.

Gertler, J., Fang, X. and Luo, Q. (1990). Detection and diagnosis of plant failures - The orthogonal parity equation approach. In *Advances in Control and Dynamic Systems*, (C.T. Leondes, Ed.; Academic Press), *Vol. 37*, pp. 159-216.

Gertler, J. (1991). Analytical redundancy methods in fault detection and isolation - Survey and synthesis. *Preprints of IFAC Safeprocess Conference*, (Baden-Baden, Germany), Vol. 1, pp. 9-22.

Gertler, J., Costin, M., Luo, Q., Fang, X., Hira, R. and Kowalczuk, Z. (1991). On-board fault detection and isolation for automotive engines using orthogonal parity equations. *Preprints of IFAC Safeprocess Conference*, (Baden-Baden, Germany), Vol. 2, pp. 241-246.

Gertler, J. (1992). Structured residuals for fault isolation, disturbance decoupling and modeling error robustness. *Preprints of IFAC Symposium on On-Line Fault Detection and Supervision in the Chemical Process Industries*, (Newark, DE), pp. 111-119.

Gertler, J. and Anderson, K.C. (1992). An evidential reasoning extension to quantitative model-based failure diagnosis. *IEEE Trans. on Systems, Man and Cybernetics, Vol. 22*, pp. 275-289.

Gertler, J. (1993). Residual generation in model based fault diagnosis. *Control - Theory and Advanced Technology, Vol. 9*, pp. 259-285.

Gertler, J. and Kunwer, M. (1993). Optimal residual decoupling for structured diagnosis and disturbance insensitivity. *Proc. TOOLDIAG Symposium*, (Toulouse, France), pp. 436-452.

Gertler, J. and Monajemy, R. (1993). Generating fixed direction residuals with dynamic parity equations. *Proc. IFAC 12th World Congress*, (Sydney, Australia), Vol. 5, pp. 507-512.

Gertler, J., Costin, M., Fang, X., Hira, R., Kowalczuk, Z. and Luo, Q. (1993). Model-based on-board fault detection and diagnosis for automotive engines. *Control Engineering Practice, Vol. 1*, pp. 3-17.

Gertler, J. (1994). Modeling errors as unknown inputs. *Preprints of 2nd IFAC Safeprocess Symp.*, (Helsinki, Finland), pp. 266-271.

Gertler, J. and Costin, M. (1994). Model-based diagnosis of automotive engines - Case study on a physical vehicle. *Preprints of 2nd IFAC Safeprocess Symp.*, (Helsinki, Finland), pp. 421-430.

Gertler, J., Fang, X., Luo, Q. and Costin, M. (1994). Direct identification of structured parity equations. *Preprints of IFAC 10th SYSID Symp.*, (Copenhagen, Denmark), Vol. 3, pp. 95-100.

Gertler, J. and Monajemy, R. (1995). Generating fixed direction residuals with dynamic parity equations. *Automatica, Vol. 31*, pp. 627-635.

Gertler, J. and Kunwer, M. (1995). Optimal residual decoupling for structured diagnosis and disturbance insensitivity. *Int. Journal of Control, Vol. 61*, pp. 395-421.

Gertler, J., Costin, M., Fang, X., Kowalczuk, Z., Kunwer, M. and Monajemy, R. (1995). Model-based diagnosis for automotive engines - Algorithm development and testing on a production vehicle. *IEEE Trans. on Control System Technology, Vol. 3*, pp. 61-69.

Gertler, J. (1995a). Towards a theory of dynamic consistency relations. *Preprints of IFAC Workshop on On-Line Fault Detection and Supervision in the Chemical Process Industries*, (Newcastle, England), pp. 143-156.

Gertler, J. (1995b). Diagnosing parametric faults - From identification to parity relations. *Proc. American Control Conference*, (Seattle, WA), pp. 1615-1620.

Gertler, J. and Yin, K. (1996). Statistical decision making for dynamic parity relations. *Proc. IFAC 13th World Congress*, (San Francisco, CA), Vol. N, pp. 13-18.

Gertler, J. (1997). Fault detection and diagnosis using parity relations. *Control Engineering Practice, Vol. 5*, pp. 653-670.

Gertler, J. and DiPierro, G. (1997). On the link between parity relations and parameter estimation. *Preprints of IFAC 3rd Safeprocess Symp.*, (Hull, England), pp. 468-473.

Gilbert, E.G. (1963). Controllability and observability in multivariable control systems. *SIAM J. Control, Vol. 1*, pp. 128-151.

Goodwin, G.C. and Sin, K.S. (1984). *Adaptive Filtering, Prediction and Control*. Prentice Hall.

Halme, A. and Selkainaho, J. (1984). Instrument fault detection using an adaptive filtering method. *Proc. IFAC 9th World Congress*, (Budapest, Hungary), pp. 1765-1770.

Himmelblau, D.M. (1978). *Fault Detection and Diagnosis in Chemical and Petrochemical Processes*. Chemical Engineering Monograph 8, Elsevier.

Höfling, T. (1993). Detection of parameter variations by continuous-time parity equations. *Proc. IFAC 12th World Congress*, (Sydney, Australia), Vol. 5, pp. 513-518.

Höfling, T. and Pfeufer, T. (1994). Detection of additive and multiplicative faults - Parity space vs. parameter estimation. *Preprints of 2nd IFAC Safeprocess Symposium,* (Helsinki, Finland), Vol. 2, pp. 539-544.

Höfling, T. and Isermann, R. (1995). Parameter estimation triggered by continuous-time parity equations. *Proc. American Control Conference,* (Seattle, WA), pp. 1145-1146.

Höfling, T. (1996). Methoden zur Fehlererkennung mit Parameterschätzung und Paritätsgleichungen. Dr.Ing. Dissertation, Technische Hochschule Darmstadt, Germany.

Horak, D.T. and Goblirsch, D.M. (1986). Outputs in systems with bounded uncertainties: Application to failure detection. *Proc. 1986 American Control Conference*, (Seattle, WA), pp. 301-308.

Horak, D.T. (1988). Failure detection in dynamic systems with modeling errors. *Journal of Guidance, Control and Dynamics, Vol. 11,* pp. 508-516.

Horn, R.A. and Johnson, C.A. (1985). *Matrix Analysis.*Cambridge University Press, Cambridge, England.

Hou, M. and Müller, P.C. (1992). Design of observers for linear systems with unknown inputs. *IEEE Trans. on Automatic Control, Vol. AC-37,* pp. 871-875.

Isermann, R. (1984). Process fault detection based on modelling and estimation methods. *Automatica, Vol. 20,* pp. 387-404.

Isermann, R. (1985). Process failure diagnosis based on modeling and identification methods. *Preprints of IFAC-IFIP Conference on Digital Computer Applications to Process Control,* (Vienna, Austria), pp. 49-58.

Isermann, R. and Freyermuth, B. (1991). Process fault diagnosis based on process model knowledge. Parts I and II. *ASME Journal of Dynamic Systems, Measurement and Control, Vol. 113*, pp. 620-626 and 627-633.

Isermann, R. (1993). Fault diagnosis of machines via parameter estimation and knowledge processing. *Automatica, Vol. 29,* pp. 815-835.

Jones, H.L. (1973). Failure detection in linear systems. Ph. D. Thesis, Dept. of Aero and Astro, MIT, Cambridge, MA.

Kailath, T. (1980). *Linear Systems*. Prentice Hall.

Kalman, R.E. (1960). On the general theory of control systems. *Proc. First IFAC World Congress,* (Moscow, USSR), pp. 481-493.

Kalman, R.E. (1963). Mathematical description of linear dynamical systems. *SIAM J. of Control, Vol. 1,* pp. 152-192.

Kalman, R.E. (1965). Irreducible realizations and the degree of a rational matrix. *SIAM J. of Applied Mathematics, Vol. 13,* pp. 520-544.

Kamei, E., Namba, H., Osaki, K. and Ohba, M. (1987). Application of reduced order model to automotive engine control systems. *Proc. American Control Conference,* (Minneapolis, MN), pp. 1815-1820.

Kendall, M.G. and Stuart, A. (1969). *The Advanced Theory of Statistics - Volume 1.* Third Edition. Hafner Publishing Company, New York.

Kinnaert, M., Hanus, R. and Arte, Ph. (1995). Fault detection and isolation for unstable linear systems. *IEEE Trans. on Automatic Control, Vol. 40,* pp. 740-742.

Kramer, M.A. (1987). Malfunction diagnosis using quantitative models with non-boolean reasoning in expert systems. *AIChE Journal, Vol. 33,* pp. 130-140.

Kumamaru, K., Sagara, S., Yanagida, H. and Söderström, T. (1985). Fault detection of dynamical systems based on a recognition approach to model discrimination. *Proc. of 7th IFAC-IFORS Symposium on Identification and Process Parameter Estimation,* (York, England), pp. 1625-1630.

Kumamaru, K., Hu, J., Inoue, K. and Söderström, T. (1996). Robust fault detection using index of Kullback discrimination information. *Proc. IFAC 13th World Congress* (San Francisco, CA), Vol. N. pp. 205-210.

Kumamaru, K., Hu, J., Inoue, K. and Söderström, T. (1997). Fault detection of nonlinear systems by using hybrid quasi-ARMAX models. *Preprints of 3rd IFAC Safeprocess Symp.* (Hull, England), pp. 1126-1131.

Kunwer, M.M. (1992). Modeling error robustness in failure detection algorithms using imperfect decoupling. M.S. Thesis, George Mason University, Fairfax, VA.

Leininger, G.G. (1981). Model degradation effects on sensor failure detection. *Proc. American Control Conference,* (Blacksburg, VA), paper FP-3a.

Li, W. and Gertler, J. (1997). Detection and isolation of slight parametric faults in continuous-time systems. *Preprints of 11th IFAC SYSID Symposium,* (Kitakyushu, Japan), pp. 1161-1166.

Ljung, L. and Söderström, T. (1983). *Theory and Practice of Recursive Identification.* MIT Press, Cambridge, MA.

Ljung, L. (1987) *Systems Identification - Theory for the User.* Prentice Hall.

Lou, X.C., Willsky, A.S. and Verghese, G.C. (1986). Optimally robust redundancy relations for failure detection in uncertain systems. *Automatica, 22,* pp. 333-344.

Luo, Q. (1990). Parity equation approach to failure detection and isolation in dynamic systems. Master's Thesis, George Mason University, Fairfax, VA.

MacFarlane, A.G.J. and Karcanias, N. (1976). Poles and zeroes of linear multivariable systems: a survey of the algebraic, geometric and complex-variable theory. *Int. Journal of Control, Vol. 24,* pp. 33-74.

Magni, J.F. and Mouyon, P. (1994). On residual generation by observer and parity space approaches. *IEEE Trans. on Automatic Control, Vol. AC-39,* pp. 441-447.

Magni, J.F. (1995). On continuous-time parameter identification using observers. *IEEE Trans. on Automatic Control, Vol. AC-40,* pp. 1789-1792.

Mah, R.S., Stanley, G.M. and Downing, D.M. (1976). Reconciliation and rectification of process flow and inventory data. *Ind. Eng. Chemistry, Process Design, Vol. 15,* pp. 175-183.

Massoumnia, M.A. (1986). A geometric approach to the synthesis of failure detection filters. *IEEE Trans. on Automatic Control, Vol. AC-31,* pp. 839-846.

Massoumnia, M.A. and W.E. Vander Velde (1988). Generating parity relations for detecting and identifying control system component failures. *Journal of Guidance, Control and Dynamics, Vol. 11,* No. 1, pp. 60-65.

Massoumnia, M.A., Verghese, G.C. and Willsky, A.S. (1989). Failure detection and identification. *IEEE Trans. on Automatic Control, Vol. 34,* pp. 316-321.

McAvoy, T. and Ye, N. (1994). Base control for the Tennessee Eastman problem. *Computers and Chemical Engineering, Vol. 18,* pp. 383-413.

McMillan, B. (1952). Introduction to formal realization theory. *Bell Syst. Tech. Journal, Vol. 31,* pp. 217-279, 541-600.

Mehra, R.K. and Peschon, J. (1971). An innovations approach to fault detection in dynamic systems. *Automatica, Vol. 7,* pp. 637-640.

Mironovskii, L.A. (1979). Functional diagnosis of linear dynamic systems. *Avtomatika i Telemekhanika, Vol. 40,* pp. 1198-1205.

Mironovskii, L.A. (1980). Functional diagnosis of dynamic systems - A survey. *Avtomatika i Telemekhanika, Vol. 41,* pp. 1122-1143.

Monajemy, R. (1993). Residual generation in model-based fault diagnosis - The fault system matrix. MS Thesis, George Mason University, ECE Dept.

Neyman, J. and Pearson, E.S. (1967). *Collected Joint Statistical Papers.* Cambridge University Press, Cambridge, England.

Nikoukhah, R. (1994). Innovations generation in the presence of unknown inputs: application to robust failure detection. *Automatica, Vol. 30,* pp. 1851-1867.

Ninness, B., Goodwin, G.C., Kwon, O.K. and Carlsson, B. (1991). Robust fault detection based on low order models. *Preprints of Safeprocess Conference,* (Baden-Baden, Germany), pp. 199-204.

Park, J. and Rizzoni, G. (1994). A new interpretation of the fault detection filter. Part 1: Closed-form algorithm. *Int. Journal of Control, Vol. 60,* pp. 767-787.

Park, J., Halevi, Y. and Rizzoni, G. (1994a). A new interpretation of the fault-detection filter. Part 2: The optimal detection filter. *Int. Journal of Control, Vol. 60,* pp. 1339-1351.

Park, J., Rizzoni, G. and Ribbens, W.B. (1994b). On the representation of sensor faults in fault detection filters. *Automatica, Vol. 30,* pp. 1793-1796.

Patton, R.J., Frank, P.M. and Clark, R.N. (Eds.) (1989). *Fault Diagnosis in dynamic Systems - Theory and Application.* Prentice Hall International.

Patton, R.J. and Kangethe, S.M. (1989). Robust fault diagnosis using eigenstructure assignment. In *Fault Diagnosis in Dynamic Systems* (R.J. Patton, P.M. Frank and R. Clark, Eds.; Prentice Hall), pp. 99-154.

Patton, R.J. and Chen, J. (1991). A review of parity space approaches to fault diagnosis. *Preprints of IFAC Safeprocess Conference,* (Baden-Baden, Germany), pp. 65-81.

Patton, R.J. and Chen, J. (1993). Optimal unknown input distribution matrix selection for robust fault diagnosis. *Automatica, 29,* pp. 837-841.

Patton, R.J. (1994). Robust model-based fault diagnosis: the state of the art. *Preprints of 2nd IFAC Safeprocess Symp.*, (Helsinki, Finland), pp. 1-24.

Patton, R.J. (1995). Robustness in model-based fault diagnosis - The 1995 situation. *Preprints of IFAC Workshop on On-Line Fault Detection in the Chemical Process Industries,* (Newcastle, England), pp. 55-77.

Pau, L. (1981). *Failure Diagnosis and Performance Monitoring.* Marcel Dekker.

Peng, Y., Youssouf, A., Arte, P. and Kinnaert, M. (1997). A complete procedure for residual generation and evaluation with application to heat exchanger. *IEEE Trans. on Control Systems Technology, Vol. 5,* pp. 542-555.

Potter, J.E. and Suman, M.C. (1977). Thresholdless redundancy management with arrays of skewed instruments. *Integrity in Electronic Flight Control Systems, AGARDograph-224,* pp. 15-1 to 15-25.

Pouliezos, A.D. and Stavrakakis, G.S. (1994). *Real Time Fault Monitoring of Industrial Processes*. Kluwer Academic Publishers, Dordrecht/Boston/London.

Prékopa, A. (1962). *Probability Theory*. (In Hungarian.) Engineering Publishing House, Budapest, Hungary.

Qiu, Z. and Gertler, J. (1993). Robust FDI systems and H_∞ optimization - Disturbances and tall fault case. *Proc. 32nd IEEE Conference on Decision and Control,* (San Antonio, TX), pp. 1710-1715.

Qiu, Z. (1994). Robust FDI systems and H_∞ optimization. Ph.D. Dissertation, George Mason University, Fairfax, VA.

Qiu, Z. and Gertler, J. (1994). Robust FDI systems and H_∞ optimization - Disturbances and case study. *Preprints of 2nd IFAC Safeprocess Symp.*, (Helsinki, Finland), pp. 260-265.

Rault, A., Jaume, D. and Verge, M. (1984). Industrial process fault detection and localization. *Proc. IFAC 9th World Congress* (Budapest, Hungary), Vol.4, pp. 1789-1794.

Ray, A. and Desai, M. (1986). A redundancy management procedure for fault detection and isolation. *Transactions of the ASME, Vol. 108,*, pp. 248-254.

Ray, A. and Luck, R. (1991). An introduction to sensor signal validation in redundant measurement systems. *IEEE Control Systems Magazine, Vol. 11,* No. 2, pp. 43-49.

Romagnoli, J.A. and Stephanopoulos, G. (1981). Rectification of process measurement data in the presence of gross errors. *Chemical Engineering Science, Vol. 36,* pp. 1849-1863.

Rosenbrock, H.H. (1970). *State-Space and Multivariable Theory*, John Wiley and Sons.

Sagara, S. and Zhao, Z.Y. (1990). Numerical integration approach to on-line identification of continuous-time systems. *Automatica, Vol. 26,* pp.63-74.

Sawaragi, Y. and Sagara, S. (Eds.) (1997). *Preprints of the 11th IFAC Symposium on System Identification* (Kitakyushu, Japan).

Scharf, L.L. (1991). *Statistical Signal Processing - Detection, Estimation and Time-Series Analysis*. Addison Wesley.

Schrader, C.B. and Sain, M.K. (1989). Research on system zeros: A survey. *Int. Journal of Control, Vol. 50,* pp. 1407-1433.

Schumann, A., Isermann, R. and Freyermuth, B. (1992). Determination of physical parameters for dynamical processes. *Control - Theory and Advanced Technology, Vol. 9*, pp. 5-26.

Shutty, J. (1985). A multilevel approach to fault detection. MS. Thesis, Case Western Reserve University, Cleveland, OH.

Söderström, T. and Stoica, P. (1987). *System Identification.* Prentice Hall International.

Stanley, G.M. and Mah, R.S. (1977). Estimation of flows and temperatures in process networks. *AIChE Journal, Vol. 23,* pp. 642-650.

Stanley, G.M. and Mah, R.S. (1981). Observability and redundancy in process data estimation. *Chemical Engineering Science, Vol. 36,* pp. 259-272.

Stark, H. and Woods, J.W. (1994). *Probability, Random Processes and Estimation Theory for Engineers.* Second Edition. Prentice Hall.

Staroswiecki, M., Cassar, J.P. and Cocquempot, V. (1993). Generation of optimal structured residuals in the parity space. *Proc. IFAC 12th World Congress* (Sydney, Australia), Vol. 5, pp. 535-542.

Stefanidis, P., Paplinski, A.P. and Gibbard, M.J. (1992). *Numerical Operations with Polynomial Matrices, LNCIS No. 171,* Springer Verlag.

Strassel, A. (1996). Statistical testing of residuals arising from dynamic parity relations. M.S. Project, George Mason University, Fairfax, VA.

Strejc, V. (Ed.) (1967). *Preprints of the First IFAC Symposium on Identification in Automatic Control Systems* (Prague, Czechoslovakia).

Stuart,A. and Ord, J.K. (1991). *Kendall's Advanced Theory of Statistics, Volume 2.* Oxford University Press, New York.

Sundar, A. (1985). Process fault detection using the augmented system model approach. Master's Thesis, Case Western Reserve University, Cleveland, OH.

Tylee, J.L. (1983). On-line failure detection in nuclear power plant instrumentation. *IEEE Trans. on Automatic Control, Vol. AC-28,* pp. 406-415.

Vaclavek, V. (1974). Gross systematic errors or biases in the balance calculations. *Papers of the Prague Institute of Technology,* Prague, Czechoslovakia.

Van Dooren, P.M. (1981). The generalized eigenstructure problem in linear system theory. *IEEE Trans. on Automatic Control, Vol. AC-26,* pp. 111-129.

Vidyasagar, M. (1985). *Control System Synthesis - A Factorization Approach.* MIT Press, Cambridge, MA.

Viswanadham, N. and Srichander, R. (1987). Fault detection using unknown input observers. *Control - Theory and Advanced Technology, Vol. 3,* pp. 91-101.

Viswanadham, N., Taylor, J.H. and Luce, E.C. (1987a). A frequency domain approach to failure detection and isolation with application to GE-21 turbine engine control system. *Control - Theory and Advanced Technology, Vol. 3,* pp. 45-72.

Viswanadham, N., Sarma, V.V.S and Singh, M.G. (1987b). *Reliability of Computer Control Systems.* Systems and Control Series, No.8, North Holland.

Viswanadham, N. and Minto, K.D. (1988). Robust observer design with application to fault detection. *Proc. American Control Conference,* (Atlanta, GA), pp. 1393-1399.

Wald, A. (1955). *Selected Papers in Statistics and Probability.* McGraw-Hill, New York.

Watanabe, U. and Himmelblau, D.M. (1982). Instrument fault detection in systems with uncertainties. *Int. Journal of Systems Science, Vol. 13,* pp. 137-158.

Watanabe, U. and Himmelblau, D.M. (1983). Fault diagnosis in nonlinear chemical processes. *AIChE Journal, Vol. 29,* pp. 243-249, 250-261.

White, J. E. and Speyer, J.L. (1987). Detection filter design: Spectral theory and algorithms. *IEEE Trans. on Automatic Control, Vol. AC-32,* pp. 593-603.

Willsky, A.J. (1976). A survey of design methods for failure detection in dynamic systems. *Automatica, Vol. 12,* pp. 601-611.

Willsky, A.S. and Jones, H.L. (1976). A generalized likelihood ratio approach to detection and estimation of jumps in linear systems. *IEEE Trans. on Automatic Control, AC-21,* pp. 108-112.

Willsky, A.S. (1986). Detection of abrupt changes in dynamic systems. In *Detection of Abrupt Changes in Signals and Dynamic Systems* (M. Basseville and A. Benveniste, Eds.; Springer LNCIS 77), pp. 27-49.

Young, P.C. (1981). Parameter estimation for continuous-time models - A survey. *Automatica, Vol.17,* pp.23-39.

Zhang, Q., Basseville, M. and Benveniste, A. (1994). Early warning of slight changes in systems. *Automatica, Vol. 30,* pp. 95-113.

Subject Index

9 780824 794279

For Product Safety Concerns and Information please contact our EU representative GPSR@taylorandfrancis.com
Taylor & Francis Verlag GmbH, Kaufingerstraße 24, 80331 München, Germany

www.ingramcontent.com/pod-product-compliance
Ingram Content Group UK Ltd.
Pitfield, Milton Keynes, MK11 3LW, UK
UKHW021605240425
457818UK00018B/392